P9-CMX-615

Human Diseases from Wildlife

Human Diseases from Wildlife

Michael R. Conover
Director, Wildlife Resources Department
Jack H. Berryman Institute
Utah State University, Logan, USA

Rosanna M. Vail
Kansas State University
Manhattan, USA

CRC Press
Taylor & Francis Group
Boca Raton London New York

CRC Press is an imprint of the
Taylor & Francis Group, an **informa** business

CRC Press
Taylor & Francis Group
6000 Broken Sound Parkway NW, Suite 300
Boca Raton, FL 33487-2742

Printed on acid-free paper
Version Date: 20140728

International Standard Book Number-13: 978-1-4665-6214-1 (Hardback)

Library of Congress Cataloging-in-Publication Data

Conover, Michael R.
Human diseases from wildlife / Michael R. Conover, Rosanna M. Vail.
pages cm
"A CRC title, part of the Taylor & Francis imprint, a member of the Taylor & Francis Group, the academic division of T&F Informa plc."
Includes bibliographical references and index.
ISBN 978-1-4665-6214-1 (hardback)
1. Zoonoses--Handbooks, manuals, etc. I. Vail, Rosanna M. II. Title.

RC113.5.C67 2015
616.95'9--dc23 2014028239

Visit the Taylor & Francis Web site at
http://www.taylorandfrancis.com

and the CRC Press Web site at
http://www.crcpress.com

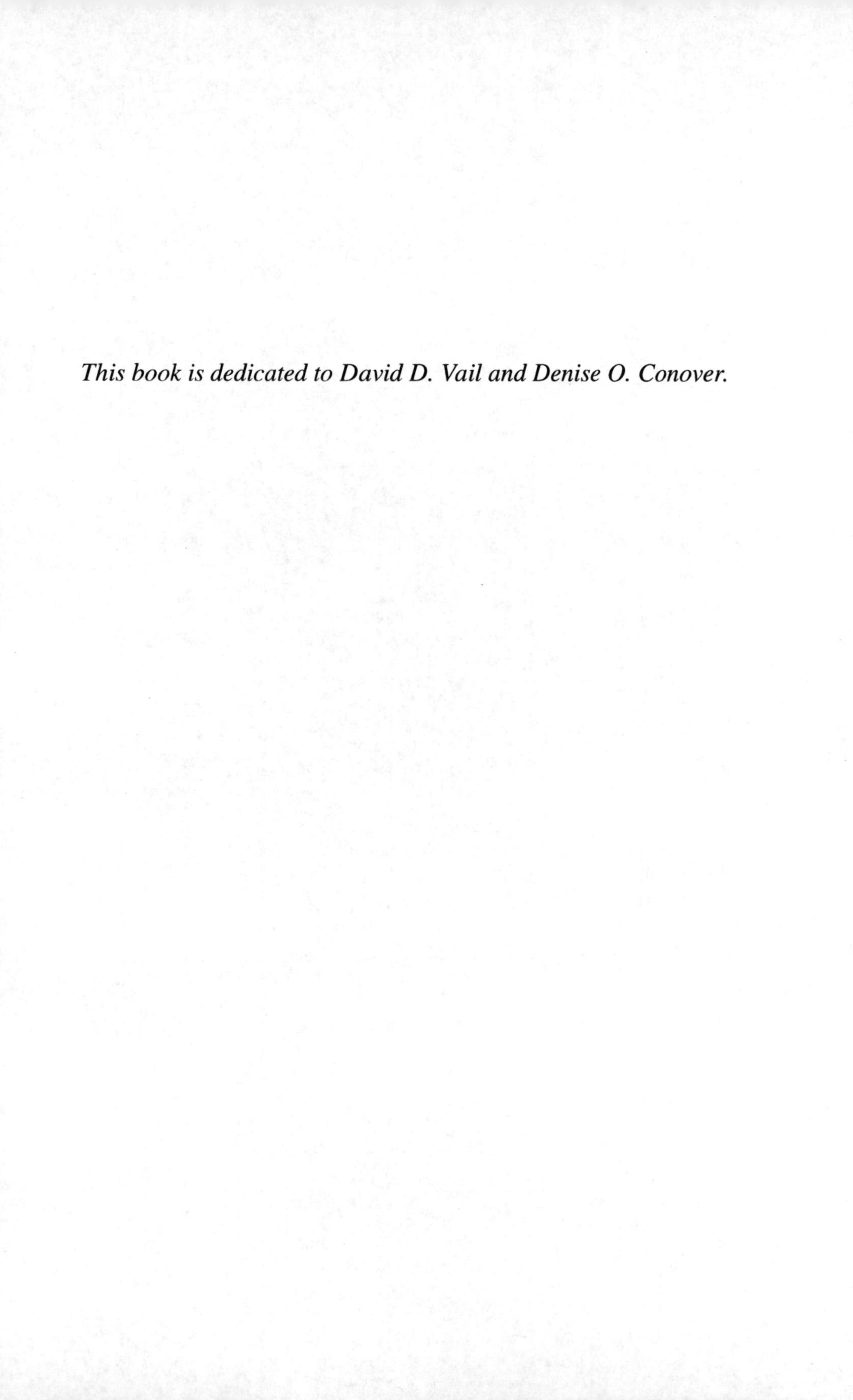

This book is dedicated to David D. Vail and Denise O. Conover.

Contents

SECTION V *Fungal Diseases*

SECTION VI *Prions*

SECTION VII Parasites

Preface

We have titled this book *Human Diseases from Wildlife*, but an equally appropriate title would be *Wildlife Diseases from Humans*. Diseases that are caused by pathogens with the ability to infect both humans and animals are known as zoonotic diseases, which literally means "disease from animals." In this book, we have concentrated on zoonotic diseases in which wild animals play an important role as a reservoir and/or a vector for the pathogen. For some of these diseases, livestock or companion animals (i.e., pets) are often involved and may serve as a bridge that allows a pathogen of wildlife to infect people. That is, livestock may become infected from a wildlife source, and humans become ill from the infected livestock.

We have written this book because too many people are uninformed about zoonotic diseases. This lack of information causes some people to have heightened fear of zoonotic diseases, preventing them from enjoying wildlife or spending time outdoors. Other people needlessly expose themselves to zoonotic diseases by neglecting simple precautions.

The diseases that people can contract from wild animals are fascinating, and we have included interesting information to make the book more enjoyable to read. Concomitantly, each disease is covered in an individual chapter, enabling readers to easily find information about a particular disease. All chapters are organized in the same manner so readers can quickly access the information they seek. We have placed references at the end of paragraphs for the readers' convenience.

To the extent possible, we have avoided medical terms. When we had to use a medical term because there was no general term for it, we have defined it when first used in the text. These terms are also defined in Appendix A so readers can quickly look up their meanings. We have used the common name for species in the text because most readers are more familiar with them than the scientific names. Appendix B contains the scientific name for each species mentioned in the text. Appendix C shows photos of North American ticks, lice, and fleas that can transmit zoonotic diseases to humans. Information on each disease is also provided in Appendix D. We have provided an index so that readers can rapidly find the information they seek.

Michael R. Conover
College of Natural Resources
Utah State University
Logan, Utah

Rosanna M. Vail
Global Campus
Kansas State University
Manhattan, Kansas

Acknowledgments

We thank the many people who were kind enough to provide comments on earlier drafts of the chapters or helped edit them: Caroline Cameron, Joe Caudell, Mary Conner, Denise Conover, Thomas Deliberto, Maureen Frank, Francis Gulland, Marilyn Haskell, Laura Howell, Morgan Hunter, Skip Jack, Patrick Jackson, Michelle Jay-Russell, Jeffrey Langholz, Ricky Langley, George Linz, Kristen Page, Didier Raoult, William Reisen, Anthony Roberts, Brant Schumaker, Thomas Schwan, Kirby Stafford, and James Walton. Anthony Roberts wrote the first draft of Chapter 27 on the prion diseases, and Denise Conover wrote the rough draft of other parts. We also thank the numerous photographers and artists who allowed us to use their work; each is credited in the respective figure caption.

1 Introduction

1.1 INTRODUCTION

Diseases can be divided into three groups: those that only infect humans and not animals, those that only infect animals and not humans, and those that infect both humans and animals. The scientific term for this last group of diseases is zoonotic diseases, and they are the subject of this book. Zoonotic diseases are much more common than most people realize; approximately 58% of all human diseases are zoonotic diseases. This book will concentrate on zoonotic diseases that occur in North America and those involving wildlife species.

1.2 LINKAGES BETWEEN HUMANS AND WILDLIFE

There are many elements that link humans to the natural world and to wildlife species in particular. Humans and wildlife share the same basic needs and challenges: securing food and shelter, avoiding predators and rivals, seeking safety, and making sure that their children survive and prosper. Humans and wildlife share the same world, environment, and habitat. Habitat can be defined as that part of an environment where a species lives and which provides the animal with the food, shelter, and protection needed to survive and reproduce. Historically, people viewed human habitat—the rural, suburban, and urban areas where humans live and work—to be separate from wildlife habitat. Wildlife habitats were those areas that humans had set aside for them. This might be as small as a patch of cover on a farm or as large as a national park or national forest. This dichotomy of human habitat (i.e., human-dominated areas) and wildlife habitat (i.e., wildlands) was always an illusion because both humans and wild animals are mobile. Instead, human development is constantly expanding into many new areas, and people are hiking and camping in the most remote parts of North America. Likewise, wild animals are moving into towns and cities. Today, even our biggest cities have large and growing populations of wildlife, including Canada geese, deer, raccoons, and coyotes. The end result of this two-way movement of people and wildlife is a closer association of humans and wild animals. On balance, this close association is a positive development because humans enjoy seeing wildlife in their neighborhoods, and wild animals have more areas where they can thrive. But sometimes, this close association has negative consequences for either humans or wildlife, or for both.

1.3 ZOONOTIC DISEASES

One of the most unfortunate consequences of this close association is the opportunity for humans and wildlife to share pathogens. Each year, zoonotic diseases sicken hundreds of millions of people across the world and tens of thousands in the United States and Canada. Too often, these diseases prove deadly. Zoonotic diseases also are a major impediment to the lofty goal of ending world hunger. Millions of poultry and livestock are killed annually by zoonotic diseases. These losses are felt most by the poor who are forced to forego animal protein in their diets.

Most (72%) zoonotic diseases are caused by pathogens of wildlife origins (Jones et al. 2008). These diseases are a great concern because it typically is not possible to eradicate a disease if a wildlife species serves as a reservoir host. For example, two of mankind's deadliest pestilence—plague, which was called the Black Death during the Middle Ages, and typhoid, which killed millions during World Wars I and II—have been eliminated in most parts of the world due to the advances of modern medicine. Today, they are limited to the most impoverished places, such as war-torn areas or refugee camps. There is, unfortunately, one exception to this pattern—they also occur in the United States. They occur in the United States not due to shortcomings in the country's public health agencies but rather because these diseases have spread to North American wildlife (typhus circulates in flying squirrels and plague in rodents and other wild mammalian species).

1.4 WILL THE NEXT PANDEMIC BE CAUSED BY A ZOONOTIC DISEASE?

There was great confidence during the second half of the twentieth century that the widespread use of antibiotics and vaccinations would allow humans to eradicate infectious diseases. That confidence, unfortunately, was misplaced. Rather than decreasing, there has been a resurgence of infectious diseases since 1950. This increase can be attributed to (1) a lapse of funding for public health programs, (2) constant genetic changes in pathogens that allow them to become immune to antibodies and more infective, and (3) spread of HIV virus that impairs immune systems, making its victims susceptible to other diseases, such as cryptococcosis.

Since 1950, there have been hundreds of events involving emerging infectious diseases (EIDs) that threaten human health on a global scale. These EID events include newly evolved strains of pathogens, such as drug-resistant strains, pathogens that recently developed the ability to infect humans and cause disease, and historical pathogens of humans that recently became more infective, prevalent, or deadly (Jones et al. 2008). Most of the EID events involved zoonotic pathogens. Surprisingly, these EID events were not concentrated in the tropics or in developing countries. Instead, they were concentrated in temperate regions and developed countries. The EID hot spots were the United States, western Europe, Japan, and southeastern Australia. The apparent reason for this global pattern is that the frequency of zoonotic EID events increases in areas with high human and wildlife densities (Jones et al. 2008). The World Health Organization (WHO) maintains the Global Outbreak Alert and Response Network to monitor disease outbreaks of international importance. During 2012 and 2013, the network reported outbreaks of Lassa fever, Marburg hemorrhagic fever, Rift Valley fever, dengue fever, ebola, cholera, polio, yellow fever (YF), influenza A (H7N9), avian influenza, and Middle East respiratory syndrome coronavirus (WHO 2013).

In the past, mankind has been plagued by deadly pandemics, which killed a substantial proportion of the human population. One of the most severe was the Spanish influenza a century ago that killed between 50 and 100 million people across the world. Public health agencies and epidemiologists have warned that the question of a deadly pandemic occurring is not if but when such a pandemic will begin. Pandemics are more likely to occur when either of the two events happens: (1) a benign pathogen that infects people today without causing illness mutates into a deadly pathogen and (2) a pathogen that infects another species of animal mutates so that it can infect humans. Many epidemiologists believe the second event is more likely to occur and to be more deadly because humans will not have developed any immunity against a pathogen that had not caused a human disease previously. This type of event started the Spanish influenza pandemic—its causative agent was an avian influenza virus that developed the ability to infect humans.

1.5 WILL THE NEXT ANIMAL EXTINCTION BE CAUSED BY A ZOONOTIC DISEASE?

Humans are not the only species that are threatened by zoonotic diseases—wild animals are also victims. Zoonotic diseases are a major source of mortality in many avian and mammalian species. Sometimes, these diseases attract popular attention when they result in massive die-offs, killing thousands of animals at one time; more insidious is the daily mortality of animals, which fails to attract public attention. Too often, humans only realize the loss when enough animals have died that the species is threatened with extinction. For instance, the black-footed ferret in North America was almost driven to extinction, in part, due to plague killing ferrets directly and eliminating their food (prairie dogs). The eradication of the Allegheny woodrat from some parts of North America, owing to raccoon roundworms, offers yet another example (Cully et al. 1997, Daszak et al. 2000).

A disease is more likely to eradicate a wildlife population or drive a species to extinction when one or more of the following conditions exist (Page 2013):

1. The animal population is already vulnerable due to its small size or a limited range. This is the biological equivalent of having all of one's eggs in one basket. If animal numbers are initially low or if all of the animals are concentrated into a single area, a localized outbreak of a disease can cause extinction. In contrast, large populations that are scattered over an entire continent are robust enough to absorb the additional mortalities caused by a disease outbreak without threatening the survival of the entire species.
2. The animal population is already suffering from mortality rates higher than normal due to habitat degradation or to the arrival of exotic species that prey upon it or compete with it for food or shelter.
3. The animal population has the misfortune of occupying habitat that is coveted by humans, who want to alter its habitat to meet human needs.

1.6 "ONE HEALTH" APPROACH TO CONTROLLING ZOONOTIC DISEASES

The pathogens responsible for zoonotic diseases infect wild animals, domestic animals, and humans. Because of this, a multitude of scientific disciplines are germane to the topic of zoonotic diseases. To name a few, medical science is the study of human disease, veterinary science is the examination of animal diseases, bacteriology and virology are the studies of pathogens, epidemiology is the scrutinization of disease patterns, entomology is the study of insect vectors, biology is the study of host populations, psychology is the examination of human behavior, and sociology is a study that focuses on human society. This diversity of disciplines can hinder the ability to develop a coordinated response to the threat posed by a zoonotic disease. The "One Health" approach was developed to address this issue. It recognizes that improving the health of wildlife, livestock, and companion animals will enhance human health by reducing human exposure to pathogens (Palmer et al. 2012). This book contains numerous examples of how the risk of humans becoming ill can only be reduced by eliminating or controlling the pathogen in wildlife population.

LITERATURE CITED

Cully Jr., J. F., A. M. Barnes, T. J. Quan, and G. Maupin. 1997. Dynamics of plague in a Gunnison's prairie dog colony complex from New Mexico. *Journal of Wildlife Diseases* 33:706–719.

Daszak, P., A. A. Cunningham, and A. D. Hyatt. 2000. Emerging infectious diseases of wildlife—Threats to biodiversity and human health. *Science* 287:443–449.

Jones, K. E., N. G. Patel, M. A. Levy, A. Storeygard et al. 2008. Global trends in emerging infectious diseases. *Nature* 451:990–993.

Page, L. K. 2013. Parasites and the conservation of small populations: The case of *Baylisascaris procyonis*. *International Journal for Parasitology: Parasites and Wildlife* 2:203–210.

Palmer, S. 2012. The global challenge of zoonoses control. In: S. R. Palmer, L. Soulsby, P. R. Torgerson, and D. W. G. Brown, editors. *Oxford Textbook of Zoonoses*, 2nd edition. Oxford University Press, Oxford, U.K., pp. 3–11.

WHO. 2013. Global Alert and Response (GAR). http://www.who.int/csr/outbreaknetwork/en (accessed December 1, 2013).

Section I

Bacterial Diseases

2 Plague

For there was a deathly panic throughout the whole city [Ekron]. The hand of God was very heavy there; the men who did not die were stricken with tumors, and the cry of the city went up to heaven.

Samuel 5:11-12, Revised Standard Version of the Bible

"Ring around the rosy, pocketful of posy, ashes, ashes, we all fall down." A children's nursery rhyme believed to have originated during the Black Death epidemic and which helped children deal with the consequences of the dreaded disease.

Conover (2002)

2.1 INTRODUCTION AND HISTORY

Plague is caused by the bacterium *Yersinia pestis* (Figure 2.1). The *Yersinia* genus contains over 13 species, most of which do not cause human illness or do so rarely. Unfortunately, there are three species that cause human disease: *Y. enterocolitica* and *Y. pseudotuberculosis* cause primarily foodborne diseases and are discussed in Chapter 10. The other human pathogen is *Y. pestis*, which causes plague and is the subject of this chapter. These bacteria are Gram negative, which means that the bacteria do not exhibit a purple color when Gram stain is applied; this trait is one of the methods used to distinguish among different species of bacteria. *Y. pestis* is capable of movement when grown at 77°F (25°C), but not at human body temperature (98.6°F or 35°C) (Prentice 2011).

No disease has been more terrifying or changed human history more than plague. Some scholars believe that the first recorded plague epidemic (i.e., a disease outbreak among humans) was the biblical passage cited at the beginning of this chapter. They note that bubonic plague produces large buboes or tumors, which result when lymph nodes swell dramatically, and that the people in Ekron developed tumors. These scholars also note that the epidemic in Ekron coincided with rodents ravishing the land (Samuel 6:5) and that rodents serve as an important reservoir for plague.

Other scholars believe that the first recorded plague epidemic occurred during the First Peloponnesian War between Sparta and its allies fighting against Athens and its allies. The war was a stalemate; Athens could not defeat the Spartan army, and Sparta could not defeat the Athenian navy. During 430 BC, Sparta was besieging the walled city of Athens but could not prevent ships from bringing supplies to Athens. However, a plague epidemic occurred in Athens during that same year, and the disease spread rapidly in the city packed with sailors, troops, and refugees. Before the epidemic ended, about half of the Athenians, including their leader Pericles, lost their lives to the disease. The fear of plague was so great that Athens' allies refused to send more warriors to the besieged city. Ultimately, Athens had to surrender to

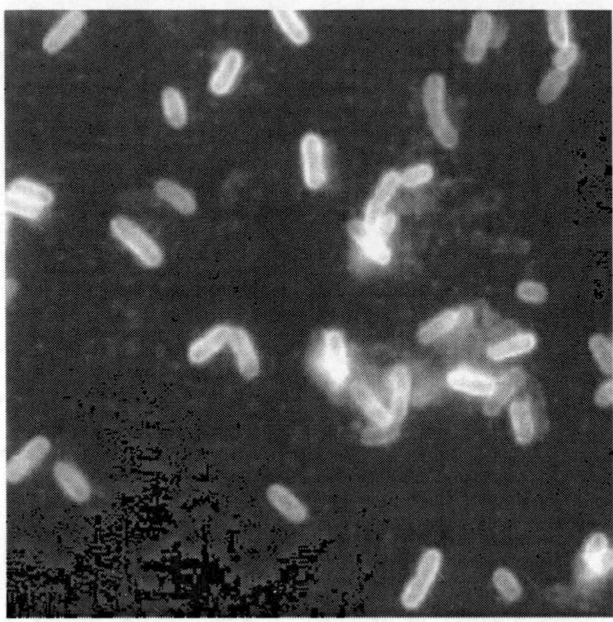

FIGURE 2.1 *Y. pestis* magnified 200 times and highlighted with a direct fluorescent antibody stain. (Courtesy of the CDC.)

Sparta. Some medical historians have questioned whether the disease that devastated Athens was plague or another infectious disease, perhaps typhoid. However, historical descriptions of the disease symptoms are similar to those produced by plague.

The first pandemic that scholars can clearly identify as bubonic plague started during the fifth century AD in China and spread from there along trade routes to Africa. From Africa, grain ships carried it to the crowded city of Constantinople where there were large granaries and, not surprisingly, large populations of rodents. As a consequence, plague devastated Constantinople, killing as many as 10,000 people every day. By the time the epidemic ended, more than one-third of the city's inhabitants had succumbed to plague. From Constantinople, the epidemic spread throughout eastern Europe and the Mediterranean and may have killed a quarter of the human population living in these areas. This pandemic was called the Justinian plague, named after Justinian I, who had the misfortune of being the emperor of Constantinople and the Byzantine Empire during the epidemic. A second major wave of this plague started in 588 AD and spread throughout Europe. It is believed that half of the European population died during this pandemic.

Yet another plague pandemic started in China and India during the 1300 AD and quickly spread along the trade routes to Europe and Africa. Its presence in Europe was first detected in Genoa, Italy, which is not surprising because Genoa was a major trade center at the time. From there, it spread throughout Europe. Europeans called this horrifying epidemic the Black Death. It lasted more than 400 years during which plague outbreaks occurred in various parts of Asia and Europe (Figure 2.2).

FIGURE 2.2 "Plague at North Poussin" by J. Flitter. (Courtesy of the U.S. National Library of Medicine.)

One such outbreak swept through London during 1664, killing 100,000 people. By the time this plague pandemic ended, China may have lost 50% of its population, Europe 33%, and Africa 17%.

The third plague pandemic, called the modern plague epidemic, started during the mid-1800s in Asia. It was a major killer; in India, 13 million people died from the disease. The modern plague reached North America during 1897. In San Francisco, California, 126 people contracted plague, and 122 of them died. In Los Angeles, 41 people contracted plague, and 36 died. The modern plague epidemic never died out; instead, it continues today. During 1994, an outbreak of plague occurred in Surat, India, where it killed over 50 people (Sidebar 2.1). From 1980 through 1994, an average of 1250 people annually contracted plague worldwide, and 123 of them died annually from the disease (Table 2.1). This infection rate is conservative because it is based on reported cases to the WHO, and many cases are not reported. During the same period, an average of 15 people contracted plague annually in the United States, and two people annually died (Figure 2.4). During 2011, there were only three reported cases of plague in the United States. According to the U.S. Centers for Disease Control (CDC), plague currently is endemic (i.e., where a pathogen is able to maintain itself) in California and the southwestern states of Colorado, New Mexico, and Arizona (Figure 2.5). In the United States, 89% of all human plague cases occurred in these states (CDC 1996, 2013).

SIDEBAR 2.1 CONSEQUENCES OF THE 1994 PLAGUE EPIDEMIC IN SURAT, INDIA (BURNS 1994, HAZARIKA 1995)

During September 1994, an outbreak of pneumonic plague occurred in Surat, India (Figure 2.3). Before it ended, plague had infected over 1000 people, killed 52, and challenged the ability of world health authorities to contain the disease. Once news of the epidemic got out, more than 300,000 people fled Surat for other parts of India, fearing both the disease and being quarantined. More than 100 plague patients fled Surat's main hospital, believing that they would be better off on their own than in a hospital where patients were dying of the disease. This massive exodus challenged India health authorities as plague cases were soon being reported across a large part of India. Public health authorities were concerned that modern transportation systems could quickly spread the disease throughout India and the world. In response, Indian military forces established checkpoints and roadblocks to identify, quarantine, and treat people fleeing Surat. Several countries, including the United States, restricted air travel to plague-infected parts of India. Local supplies of tetracycline were soon exhausted, but additional supplies arrived from around the world. Soon, health officials had given antibiotics to almost everyone who was in or from the infected area. Because of rapid intervention, the epidemic was quickly controlled. Despite this, the epidemic took a major toll on India's economy. Worldwide demand for India's products declined; losses to India's economy exceeded $600 million.

2.2 SYMPTOMS IN HUMANS

There are three different types of plague (i.e., bubonic plague, septicemic plague, and pneumonic plague), which vary in the route of infection, which organs are attacked first, and in their symptoms. One type of plague can develop into another type as the infection spreads through the body.

Bubonic plague occurs when people are infected by the bite of an infected flea, and *Y. pestis* first invades the lymphatic system. In trying to fight off the infection, the body's lymph nodes swell, sometimes to the size of a small apple, and become very painful. These swollen lymph nodes are called buboes; they are a symptom of bubonic plague and give bubonic plague its name. There are more than 500 lymph nodes in human bodies, and clusters of them occur in the neck, under the arms, chest, groin, and knees. Often, it is the lymph nodes close to the infection site that swell first. It is worth noting that many different diseases cause lymph nodes to swell; hence, swollen lymph nodes do not mean that a person has plague. Bubonic plague also causes a high temperature of up to 106°F (41°C). Bubonic plague kills 50%–60% of its victims when medical help is unavailable (CDC 1996).

Septicemic plague occurs when *Y. pestis* first invades the bloodstream, usually through the bite of an infected flea, and continues to multiply in the bloodstream. Disease symptoms occur within 2–6 days after the initial infection and include fever

FIGURE 2.3 Map of India showing the location of the city of Surat.

and prominent gastrointestinal symptoms including nausea, vomiting, diarrhea, and abdominal pain. Internal hemorrhaging occurs as the bacteria destroy the linings of capillaries. The disease may progress to coma, septic shock, or meningitis (i.e., inflammation of the protective membrane lining the brain and spinal cord). If left untreated or not diagnosed in time, septicemic plague kills almost all of its victims. Since 1947, 10% of plague victims in the United States have had septicemic plague; half of these victims have died (CDC 1996).

Pneumonic plague occurs when *Y. pestis* is inhaled and, therefore, first infects the lungs. Symptoms of pneumonic plague include high fever, chest pains, shortness of breath, coughing of blood, and severe pneumonia (i.e., inflammation of the lungs). It is the least common form of plague, but the deadliest due to the speed with which this disease progresses. Within 1–3 days of the initial infection, many victims experience flu-like symptoms (fever, coughing, headache, and weakness). On the next day, chest pain, respiratory distress, shortness of breath, and the coughing of blood may occur as *Y. pestis* continues to infect the lungs. The coughing of blood and shortness of breath are the first indications that the illness may be pneumonic plague. Victims who do not seek medical attention within 18 hours after the onset of these respiratory symptoms are likely to die (Poland and Barnes 1994). Death from suffocation can occur within 4 days of the initial infection as the lungs are filled with blood and other fluids. Among untreated humans, mortality rates for pneumonic plague approach 100% (Sidebar 2.2).

TABLE 2.1
Annual Number of People Who Contracted Plague from 1980 through 1994 Based on Cases Reported to the WHO

Continent	Country	No. Annual Cases	No. Annual Deaths
Americas	United States	15	2
	Peru	115	7
	Brazil	47	1
	Others	19	2
	Total	**196**	**12**
Asia	China	17	5
	India	58	4
	Myanmar	77	1
	Vietnam	220	10
	Others	5	2
	Total	**377**	**22**
Africa	Madagascar	93	20
	Tanzania	331	28
	Zaire	149	35
	Others	104	5
	Total	**677**	**88**
Worldwide	**Grand total**	**1250**	**122**

Source: WHO, *Epidemiology Rec.*, 22, 165, 1994.

2.3 *YERSINIA PESTIS* INFECTIONS IN ANIMALS

Plague epidemiology (i.e., study of how diseases or pathogens are maintained and spread) involves both reservoir animal hosts and amplifying animal hosts. Reservoir hosts are those species in which *Y. pestis* is able to maintain itself by infecting them. Reservoir host species are relatively resistant to plague so that many individual animals do not die from plague, and therefore, *Y. pestis* has more time to be passed on to other individuals of the same species. In the western United States, voles and mice, such as the North American deer mouse, are reservoir hosts for *Y. pestis*. Other animal species with less resistance to plague are amplifying hosts to plague. Among amplifying hosts, plague spreads rapidly and then dies back when so many individuals have either died or developed antibodies against *Y. pestis* that the pathogen can no longer maintain itself in the species. Amplifying hosts of plague in the western United States include woodrats, chipmunks, ground squirrels, prairie dogs, and marmots (Figure 2.6). Most people who contract plague become infected during outbreaks of plague in these amplifying host species because it is then that numbers of infected mammals and fleas increase many times over and fleas that were on dead amplifying hosts have to search for other species

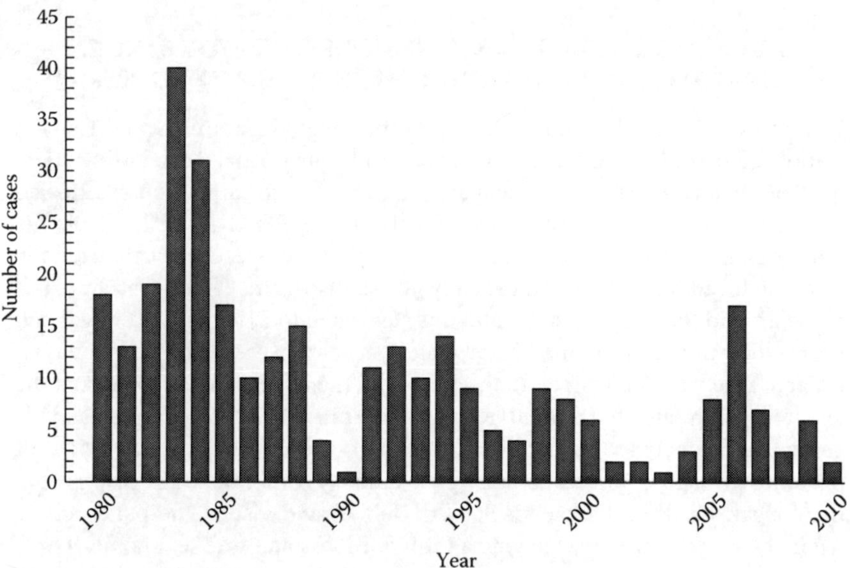

FIGURE 2.4 Number of people who contracted plague annually in the United States from 1980 to 2010. (Data courtesy of the CDC.)

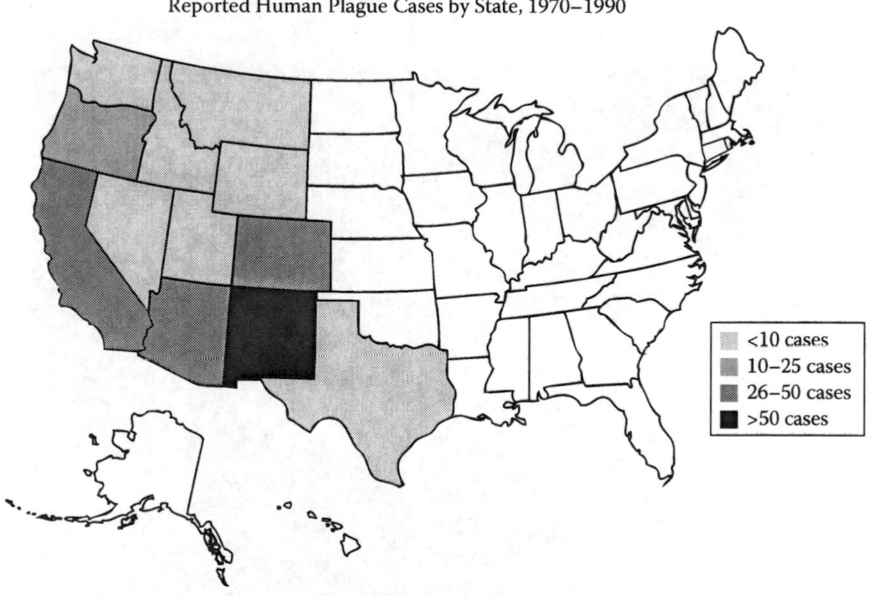

FIGURE 2.5 Number of people who contracted plague in each state during a 20-year period. (Data from the Internet Center for Wildlife Damage Management; http://icwdm.org/handbook/damage/WildlifeDiseases.asp, accessed April 1, 2011.)

SIDEBAR 2.2 PLAGUE KILLS BIOLOGIST WORKING IN GRAND CANYON NATIONAL PARK (STERNBERG 2008)

Eric was a 37-year-old wildlife biologist studying cougars in Grand Canyon National Park during 2008. His job was to trap cougars and place radio collars on them before releasing the cougars unharmed. The radio signals allowed Eric to follow the movements of individual cougars. On October 25, 2008, the radio signal for one female cougar indicated that she was dead. Eric used the signal to locate the carcass on the canyon's southern rim. There was a pool of blood around the cougar's nose, and this suggested to Eric that the cougar had been killed in a fight with a bigger male, but there were no obvious signs of trauma. Eric was determined to discover why the cougar died. He carried the cougar to his home on the South Rim Village, placed it on plastic sheets in his garage, and conducted an autopsy. Unknown to Eric, the cougar had died of pneumonic plague. Sometime during the autopsy, Eric inhaled *Y. pestis* from the cougar. On Friday October 30, Eric felt ill and visited the park's clinic where he was diagnosed as having a flu-like illness and was sent home. Three days later when Eric did not show up for work, his coworkers investigated and found him dead in his house. Physicians learned later that both he and the cougar had succumbed to plague.

FIGURE 2.6 Chipmunks are an amplifying host of plague. (Courtesy of C. R. Madsen.)

FIGURE 2.7 Coyotes can spread *Y. pestis* from one isolated population of small mammals to another. (Courtesy of S. Guymon.)

to victimize. Public health officials often can ascertain when people are especially at risk of contracting plague by monitoring plague in those species that serve as amplifying hosts (Poland and Barnes 1994).

Most of the small mammals that serve as reservoir hosts to plague are sedentary during most of their lives. Several predators of small mammals, including hawks, eagles, coyotes, foxes, bobcats, and cougars, have large home ranges and can move *Y. pestis* across large distances (Figure 2.7). Coyotes, foxes, raccoons, skunks, and ferrets have some resistance to *Y. pestis* and often do not die from it, but bobcats and cougars are highly susceptible (Sidebar 2.3).

The risk of people contracting plague in the United States varies with precipitation. Wet winters and springs in New Mexico and the western United States are followed by an increase in the number of human cases of plague in the local areas. This is not surprising because plant growth and seed production increase during wet years, and more food means more small mammals. Furthermore, humid weather increases flea survival (Enscore et al. 2002, Ari et al. 2008, Brown et al. 2010).

2.4 HOW HUMANS CONTRACT PLAGUE

In the United States, 78% of plague victims contracted plague from bites by infected fleas, 20% from direct contact with an infected animal, and 2% from inhaling the bacteria. Most people in the United States contract plague during the summer, which is not surprising because this is the season when people are more likely to be bitten by fleas. Fleas are small (0.1 in. in length or 0.25 cm) biting insects that feed by

SIDEBAR 2.3 PLAGUE ALMOST CAUSES THE EXTINCTION OF A MAMMAL SPECIES IN THE UNITED STATES (BLACK-FOOTED FERRET RECOVERY PROGRAM 2009)

The black-footed ferret was considered extinct until September 1981 when a small colony of 130 was located near Meeteetse, Wyoming (Figure 2.8). Black-footed ferrets live in prairie dog towns, and the prairie dogs fulfill all of their needs; black-footed ferrets feed upon them and use their burrows for cover and to raise their own young. Prairie dogs are amplifying hosts of plague. When plague sweeps through a prairie dog town, most or all of the prairie dogs are killed. Even if ferrets are not killed by plague, their food base has been destroyed, and they face starvation. Tragically, plague combined with canine distemper killed nearly all of the remaining black-footed ferrets. Between 1985 and 1987, the last 18 ferrets alive in the world were captured by the U.S. Fish and Wildlife Service (USFWS), and a captive breeding program was initiated. This program was successful in increasing the numbers of black-footed ferrets, and some have been released back into the wild. Some of these releases failed when plague outbreaks hit the prairie dog colonies where the ferrets were released, eliminating the prairie dog town and any black-footed ferrets within it.

FIGURE 2.8 Photo of a black-footed ferret. (Courtesy of the USFWS.)

sucking blood from their hosts (see Appendix D for photos of fleas); many wildlife animals have them. Fleas transmit *Y. pestis* from one animal to another when they drink blood from an infected animal and subsequently move to another animal and feed on it. Several flea species serve to spread the pathogen between infected animals and humans. These flea species usually do not bite people, but if plague kills the animal host upon which they are feeding, the hungry fleas are forced to abandon the carcass and seek another mammal on which to feed. Hungry fleas are not too particular about what species to bite, and sometimes they bite a person.

Pets, especially cats, can contract plague by eating small mammals that are infected with *Y. pestis*. Once a pet is infected, the people living in the household have an increased risk of contracting plague. People can also become infected by eating or handling the flesh of an infected animal. This is not a common way for people to contract plague in the United States where most people do not consume small wild mammals, but it is more common in other parts of the world. In the western United States, people who hunt rabbits and squirrels are at risk of contracting plague from cleaning or eating meat from infected animals.

In the United States, 2% of plague victims contracted the disease by inhaling *Y. pestis* from infected animals. Usually, a person has to be within a few yards of the infected animal for this to occur. This might occur if someone was holding an infected pet or inspecting a dead animal. Pneumonic plague is contagious, meaning that it can be spread from one person to another. This happens when coughing by the victim releases bacteria into the air where they can be inhaled by others. Because of this, pneumonic plague can spread rapidly in the human population unless steps are taken to prevent it from doing so. However, person-to-person transmission of plague is very rare; the last such case in the United States occurred during 1925 (CDC 2006).

2.5 MEDICAL TREATMENT

Plague is deadly because *Y. pestis* can quickly overwhelm the body's defenses and cause death before antibiotics can begin to take effect. In the United States, almost all plague fatalities in humans result from victims not seeking medical treatment soon enough or because medical physicians were unable to make the correct diagnosis in time. Plague is a notifiable disease, which means that all cases must be reported to public health authorities (CDC 1996).

Plague is difficult to diagnose because the first symptoms often mimic those caused by flu. The CDC (2006) recommends that medical doctors consider plague to be a possibility when patients (1) have an unexplained fever, inflammation across a large part of the body, pneumonia, or a classic bubo and (2) live in or have traveled to an area where plague is endemic. Whenever plague is suspected, the CDC (2006) recommends that the patient be placed on an appropriate antibiotic treatment immediately without waiting for clinical confirmation of plague. Plague is responsive to several antibiotics, including streptomycin and tetracycline. Special precautions may be used to isolate plague victims, especially those with pneumonic plague, to prevent the disease from spreading to others.

2.6 WHAT PEOPLE CAN DO TO REDUCE THEIR RISK OF CONTRACTING PLAGUE

Most people contract plague near their homes; rats and mice can bring fleas infected with *Y. pestis* into buildings where the fleas are more likely to bite people. This risk can be reduced by eliminating sources of food and shelter for rodents in homes and other buildings. Both rats and mice forage on seeds (i.e., they are granivores). Inside homes, their diet includes grains, cereals, bread, cookies, and crackers. These items should be stored in metal containers with tight lids so that rodents cannot access food. Evidence of mice and rats in a building include the sight of them scurrying about, physical evidence left by their feeding activity (e.g., food containers with holes gnawed in them), and their feces, which are most often found on the floors of basements, attics, and closets or on the shelves of drawers, cupboards, and food pantries. Homeowners may want to kill mice and rats living inside homes and buildings to reduce the danger of people contracting plague and other zoonotic diseases. The best way to do so is to use snap traps that are designed to catch rodents. Poisoned bait should not be used inside homes because some of the mice will likely die in inaccessible locations, and the odor of decaying bodies may permeate the house. Dead rodents will quickly be replaced by others if steps are not taken to keep food in containers that are inaccessible to rodents (Conover 2002).

Populations of rodents and other small mammals in yards, barns, and playgrounds can be reduced by removing brush, piles of junk, or lumber lying on the ground, which small mammals use for cover. Potential food for rodents, opossums, skunks, and raccoons include garbage and pet food (Figure 2.9). Any uneaten dog food and cat food should be disposed of immediately after the pet has finished eating so that uneaten food will not attract opossums, raccoons, and skunks (Conover 2002).

Voles are short-tailed rodents the size of mice that eat grass and can serve as a reservoir host for *Y. pestis* (Figure 2.10). Evidence of their presence includes runways that they make through lawns or along the ground. Voles keep these runways clear of vegetation. These runways are 1–2 in. (3–5 cm) in diameter and end in burrows of similar size that go into the ground. Voles do not invade homes because they only eat grass and other plants. Baits treated with a rodenticide (i.e., a chemical that kills rodents) can be spread on lawns to kill them, but any population reduction will last no more than a month or two. Voles are prolific breeders, and their populations will quickly return to pretreatment levels. Because voles produce so many offspring, dead voles are commonly observed by homeowners, but dead voles are not evidence of a plague amplifying (Conover 2002).

Both domestic cats and dogs can contract plague by handling or eating a plague-infected animal or being bitten by an infected flea (Figure 2.11). Pets should be prevented from killing or eating wild animals. Plague usually is not fatal in dogs but kills 40% of cats (Poland and Barnes 1994). Any cat that is coughing or seriously ill should be handled with care, especially in areas where plague is endemic. From 1959 to 1984, 24 people in the United States have contracted plague from dogs and cats; this number includes four veterinarians and one veterinarian assistant (Sidebar 2.4).

FIGURE 2.9 All food and garbage should be stored in containers that animals cannot open. (Courtesy of the Vertebrate Pest Conference.)

FIGURE 2.10 Voles are small short-tailed mammals, which often occur in yards because they eat fresh grass but rarely occur indoors. (Courtesy of the Vertebrate Pest Conference.)

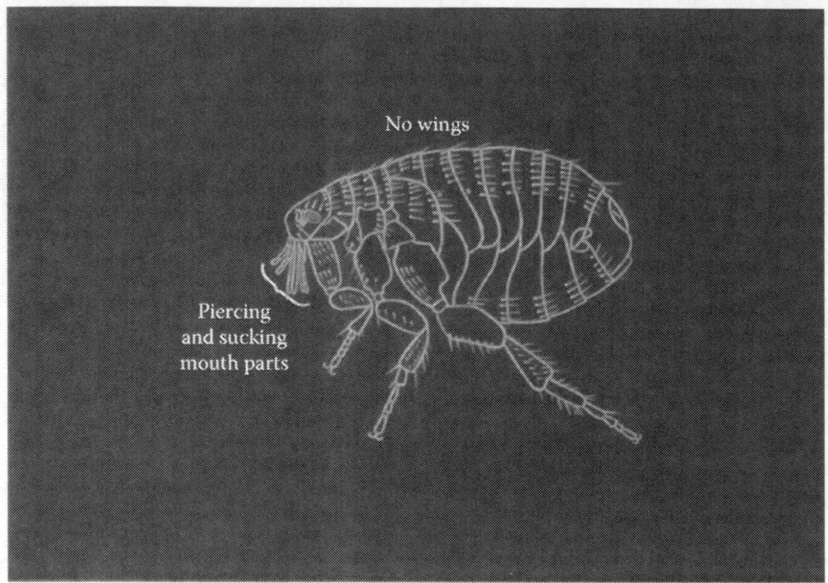

FIGURE 2.11 Fleas can be recognized by their lack of wings and their small mouth parts. (Courtesy of the CDC.)

The risk of contracting plague can be reduced by controlling fleas. Pets living in the western United States should wear flea collars or be washed with an insecticide weekly to reduce the number of fleas being brought into the house. During plague outbreaks, insecticides can be used to reduce flea populations on voles, ground squirrels, chipmunks, or other small mammals that are located in yards and near houses. Insecticides can be delivered either by dusting burrows with them or distributing insecticide-impregnated cotton balls. Small mammals seek nesting material, and many will take the cotton balls back to their burrows where the insecticide will keep any mammals using the burrow free of fleas (Conover 2002).

People living near prairie dog colonies or ground squirrels should especially be on the alert for plague. These mammals die for many reasons, and plague is one of them. Hence, extra caution is warranted when numbers of ground squirrels, prairie dogs, or rodents die within a short period of time. Public health officials should also be alerted when this happens so that they can determine if the animals died of plague.

Hikers, campers, and others who participate in outdoor activities in areas where plague occurs should wear long pants and a long-sleeved shirt and apply N,N-diethyl-m-toluamide (DEET) or another insect repellent to their legs, arms, and the bottom of their pants. People should avoid handling dead animals or those that appear sick and keep their pets away from them. If people must handle sick or dead animals, they should use plastic gloves and discard the gloves immediately after use.

Wildlife biologists, fur trappers, hunters, and other people who work with wild mammals are at risk for contracting plague (Figure 2.12). They should alert their personal physician and family members that they have a heightened risk of contracting

SIDEBAR 2.4 CALIFORNIA VETERINARIAN CONTRACTS BUBONIC PLAGUE (CDC 1984)

On March 30, 1984, a veterinarian with a small-animal practice in Claremont, California, felt fatigued and had a fever; he started coughing the next day. Two days later, the veterinarian visited a medical doctor because his left armpit and forearm were sore. The doctor noted that the patient's lymph nodes were swollen but found no puncture wounds or bites. The doctor prescribed an antibiotic (500 mg of cefadroxil to be taken twice daily). The patient returned to the medical doctor the next morning with a painful swelling of his left arm due to the excessive accumulation of fluid under the skin (i.e., edema); he was hospitalized and given the antibiotic cephalosporin intravenously (i.e., administered through the veins). On April 4, the patient complained of chest pain, shortness of breath, and coughing. An x-ray indicated that both lungs were infected, and the doctors suspected plague. This was confirmed the next day when *Y. pestis* was isolated from tissue samples. The doctors added chloramphenicol to the antibiotic regimen. By April 9, the patient had improved to stable condition and went on to make a full recovery from the disease.

The veterinarian had not traveled outside the local area and had no contact with wildlife during the period prior to the disease. The veterinarian had seen a cat with symptoms consistent with pneumonic plague (difficulty breathing) that had died. Health authorities believed that the veterinarian contracted plague from this cat. Further investigation showed that several die-offs of rodents had been noted in the area where the cat had lived. Blood tests by public health officials showed that several dogs, cats, and coyotes in the area had been exposed to *Y. pestis*.

a zoonotic disease, such as plague, and ask them to alert attending physicians in case of a medical emergency. People living in the western United States should wear plastic gloves and a mask capable of filtering out bacteria when handling wild mammals, and they should shower carefully afterward and wash clothes in hot water. Hunters should take caution when cleaning wild mammals, especially rabbits or hares. During 2006, a woman in California contracted plague after handling raw meat from a rabbit. In Asia, many plague patients are fur trappers who contracted the disease when skinning or eating fur-bearing mammals. All meat from wild game should be thoroughly cooked (CDC 2006).

A vaccine that immunizes people against plague has been developed by Miles Laboratories and distributed by Greer Laboratories. This vaccine is often given to people who are at risk of contracting the disease. A similar vaccine was given to U.S. troops, serving in Southeast Asia during the Vietnam War. The vaccine appeared to be effective because the incidence of plague in U.S. military personnel was much lower than in Vietnam's civilian population. The CDC recommends that people who are at a heightened risk of contracting plague be vaccinated. People at risk include those who have regular contact with free-ranging mammals in areas

FIGURE 2.12 Hunters, fur trappers, and wildlife biologists in the western United States have a heightened risk of contracting plague. (Courtesy of Jack Spencer, Jr.)

where plague is endemic. These people include mammalogists, wildlife biologists, and trappers or others who handle free-ranging mammals on a regular basis. Vaccination is not necessary for people living in the southwestern United States or for travelers to countries where plague is endemic unless they will be handling wild animals or will be engaging in other activities that put them at a higher risk of contracting plague. Vaccinated people should still follow preventative steps to reduce their exposure to *Y. pestis* and seek medical attention if they develop the symptoms of plague because some vaccinated people do not develop resistance to plague (CDC 1996).

2.7 ERADICATING PLAGUE FROM A COUNTRY

Plague is no longer a scourge that kills millions of people in a single pandemic. Yet it continues to occur in many countries, including the United States. It will be difficult to eradicate plague in any country where wild animals serve as reservoir hosts for *Y. pestis*.

LITERATURE CITED

Ari, T. B., A. Gershunov, K. L. Gage, T. Snäll et al. 2008. Human plague in the USA: The importance of regional and local climate. *Biological Letters* 4:737–740.

Black-Footed Ferret Recovery Program. 2009. Black-footed ferret recovery program. http://www.blackfootedferret.org (accessed April 10, 2011).

Brown, H. E., P. Ettestad, P. J. Reynolds, T. L. Brown et al. 2010. Climatic predictors of the intra- and inter-annual distributions of plague cases in New Mexico based on 29 years of animal-based surveillance data. *American Journal of Tropical Medicine and Hygiene* 82:95–102.

Burns, J. F. 1994. With old skills and new, India battles the plague. *New York Times* (September 29, 1994).

CDC. 1984. Plague pneumonia—California. *Morbidity and Mortality Weekly Report* 33:481–483.

CDC. 1996. Prevention of plague: Recommendations of the advisory committee on immunization practices (ACIP). *Morbidity and Mortality Weekly Report* 45(RR-14):1–15.

CDC. 2006. Plague—Four states, 2006. *Morbidity and Mortality Weekly Report* 55:940–943.

CDC. 2013. Summary of notifiable diseases—United States, 2011. *Morbidity and Mortality Weekly Report* 60(53):1–117.

Conover, M. R. 2002. *Resolving Human–Wildlife Conflicts: The Science of Wildlife Damage Management*. Lewis Brothers, Boca Raton, FL.

Enscore, R. E., B. J. Biggerstaff, T. L. Brown, R. E. Fulgham et al. 2002. Modeling relationships between climate and the frequency of human plague cases in the southwestern United States, 1960–1997. *American Journal of Tropical Medicine and Hygiene* 66:186–196.

Hazarika, S. 1995. Plague's origins a mystery. *New York Times* (March 14, 1995).

Poland, J. D. and A. Barnes. 1994. Plague. In: J. H. Steele, editor. *CRC Handbook Series in Zoonoses*, 2nd edition. CRC Press, Boca Raton, FL, pp. 93–112.

Prentice, M. B. 2011. Yersiniosis and plague. In: S. R. Parmer, L. Soulsby, P. R. Torgerson, and D. W. G. Brown, editors. *Oxford Textbook of Zoonoses*, 2nd edition. Oxford University Press, Oxford, U.K., pp. 232–246.

Sternberg, S. 2008. Plague emerges in Grand Canyon, kills biologist. *USA Today* (October 21, 2008).

WHO. 1994. Human plague in 1994. *Epidemiology Record* 22:165–172.

3 Brucellosis

A tremendous volcanic eruption destroyed all life around Mount Vesuvius during the night of August 24, 79 AD and buried the Roman cities of Pompeii and Herculaneum. An examination of 250 preserved bodies in Herculaneum revealed that 17% had bone lesions typical of brucellosis.

Paraphrased from Capasso (2002)

Blood, guts, and knife cuts: a combination that exposes hunters of feral hogs to the risk of contracting brucellosis.

Paraphrased from Massey et al. (2011)

3.1 INTRODUCTION AND HISTORY

Brucellosis, an infectious disease of many mammalian species including humans, is caused by bacteria of the genus *Brucella*. These bacteria are Gram negative, small (0.5–1.5 μm in length), and rod shaped. They exist alone or in small clusters and are not capable of movement (Figure 3.1). There are several species of *Brucella*. *B. abortus* causes mainly a disease of cattle, but in North America, it also infects elk and bison. *B. melitensis* is found in sheep and goats; *B. suis* is found in domestic and feral hogs; *B. ovis* infects sheep; *B. canis* occurs in dogs; *B. neotomae* infects woodrats; *B. pinnipediae* is found in seals, sea lions, and walruses; and *B. cetaceae* infects whales and porpoises (Table 3.1).

The different species of *Brucella* are host specific, with each species maintaining itself in different mammal species that serve as reservoir hosts for the pathogen. The different species of *Brucella* can infect other mammalian species, but the bacteria do not persist in these accidental host species. Four *Brucella* species can infect humans: they are *B. melitensis*, *B. suis*, *B. abortus*, and *B. canis* with the species listed in descending order of importance as a human disease. Humans are an accidental host (i.e., the pathogen cannot maintain itself by infecting accidental hosts) for all of these *Brucella* species meaning that *Brucella* populations are unable to maintain themselves by just infecting humans.

During the 1800s, the human disease caused by *Brucella* was called Crimean War fever, Rock of Gibraltar fever, Malta fever, and Mediterranean fever. These names resulted because the disease first became a concern to medical doctors during the Crimean War when large numbers of British soldiers were stricken with the disease while stationed in the Mediterranean region (where the Rock of Gibraltar and Malta are located). It was also called undulate fever because patients would experience fever that would wax and wane during the course of a day in a wavelike manner (i.e., in an undulating manner) and intermittent fever because the disease can become chronic with flare-ups occurring from time to time (i.e., intermittently).

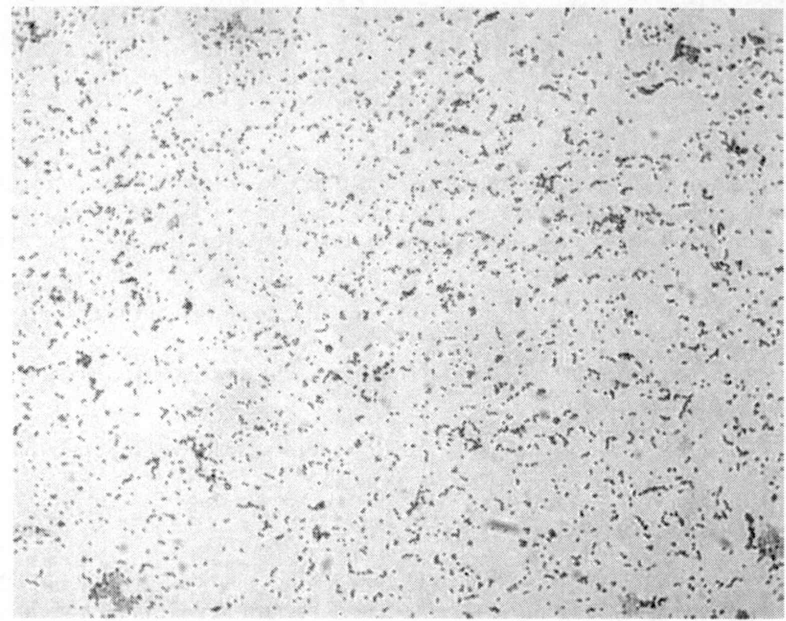

FIGURE 3.1 *Brucella* are small (0.5–1.5 μm in length), rod-shaped bacteria that can occur singly or in small clusters. (Courtesy of the CDC.)

TABLE 3.1
Animal Species Affected by Different Species of *Brucella* Are Denoted by a "Yes," "Rare," or "Possible"

Animal Host	*B. abortus*	*B. melitensis*	*B. suis*	*B. canis*	*B. ovis*
Cattle	Yes	Yes	Rare	—	—
Horse	Yes	Rare	Rare	—	—
Sheep	Rare	Yes	Possible	—	Yes
Goat	Rare	Yes	—	—	—
Hog	Rare	Rare	Yes	—	—
Dog	Yes[a]	Yes	Rare	Yes	—
Caribou	—	—	Yes	—	—
Water buffalo	Yes	Yes	—	—	—
Camel	Rare	Yes	—	—	—
Elk	Yes	—	—	—	—
Bison	Yes	—	—	—	—

Source: WHO, *World Health Organization*, 2006(7), 1, 2006.

[a] Dogs in North America are not infected with *B. abortus*.

A British doctor, David Bruce, who was treating soldiers in Malta, was the first to isolate *Brucella* from humans during 1887. He recovered the pathogen from the spleen of a soldier who had died of the disease and was able to demonstrate that the pathogen was the causative agent for the disease. The bacterium and the human disease were named *Brucella* in honor of Dr. Bruce.

Farmers realized centuries ago that abortion rates varied among livestock herds and that herds with high abortion rates were clustered together in both time and space. They also noted that the problem was contagious—introducing an animal from a herd or flock with a high abortion rate into a new one increased the abortion rate in the new herd. The disease was called infectious abortion, contagious abortion, and slinking of calves. But the cause of the illness was not recognized until 1897 when Bernhard Bang demonstrated that *Brucella* caused the disease in cows. The livestock disease was subsequently named Bang's disease to honor him. However, it was not until 1918 that scientists realized that the pathogens responsible for brucellosis in humans and the one responsible for Bang's disease in livestock were closely related and perhaps the same.

Currently, brucellosis is a worldwide disease of humans. By 2005, only 17 countries were considered to be brucellosis-free; most of them were in Europe and included Finland, Sweden, Norway, Denmark, Great Britain, Luxembourg, the Netherlands, Switzerland, Czechoslovakia, and Romania. In the New World, only the Virgin Islands are free of brucellosis. The disease is most common among people in countries where unpasteurized milk is consumed and in countries that lack effective public health programs or good health programs for livestock or pets. Areas where there was an enhanced risk of contracting brucellosis during 2012 included some countries in Africa, Asia, South America, Central America, and southern Europe (Madkour 2005, WHO 2006, CDC 2010).

In the United States, 100–200 people are diagnosed with brucellosis annually with over half of the cases occurring in just three states: Florida, Texas, and California (Figure 3.2). During 2011, there were only 79 reported cases in the United States. This is down considerably from the peak of 6321 reported cases in the United States during 1947 when drinking unpasteurized milk was more common (Figure 3.3). Although brucellosis is rare in the United States, it is one of the most common zoonotic diseases in the world (CDC 2010, 2013).

3.2 SYMPTOMS IN HUMANS

Brucellosis in humans has a range of symptoms, many of which are flu like: fever, chills, excessive sweating, fatigue, muscle pains, backaches, and headaches (Figure 3.4). In about half of the patients, there are no symptoms of brucellosis for several weeks after infection. Instead, the first signs of brucellosis occur weeks or months later. Often, brucellosis patients feel better during the morning, but their fever and other symptoms worsen during the day and peak during the afternoon or early evening (FAO 2003, WHO 2006). The Mayo Clinic (2010) recommends seeking medical attention if a person experiences a fever that is persistent or rises rapidly. People should also seek medical attention if they have any risk factors for the disease and unusual weakness or muscle aches.

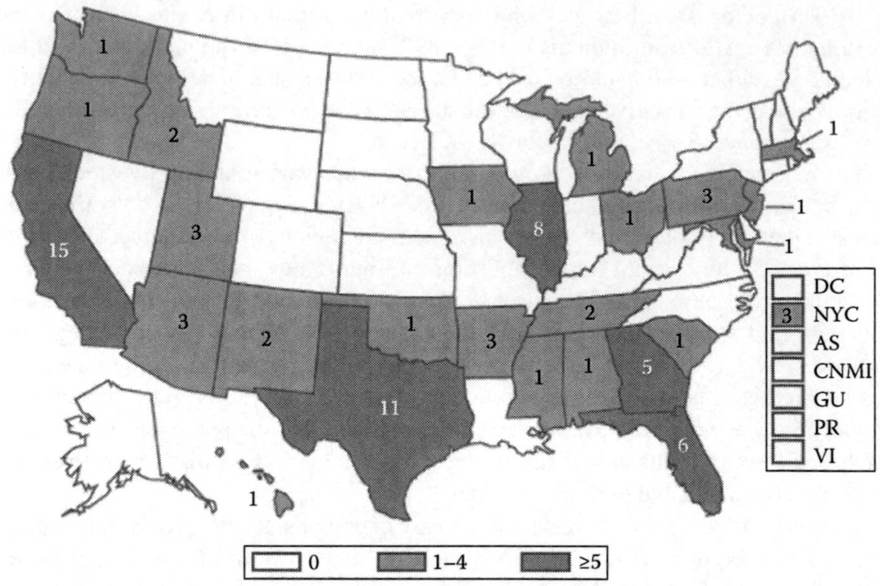

FIGURE 3.2 Distribution of human brucellosis cases that occurred in the United States during 2011. (From CDC, *Morbid. Mortal. Wkly. Rep.*, 60(53), 1, 2013; courtesy of the CDC.)

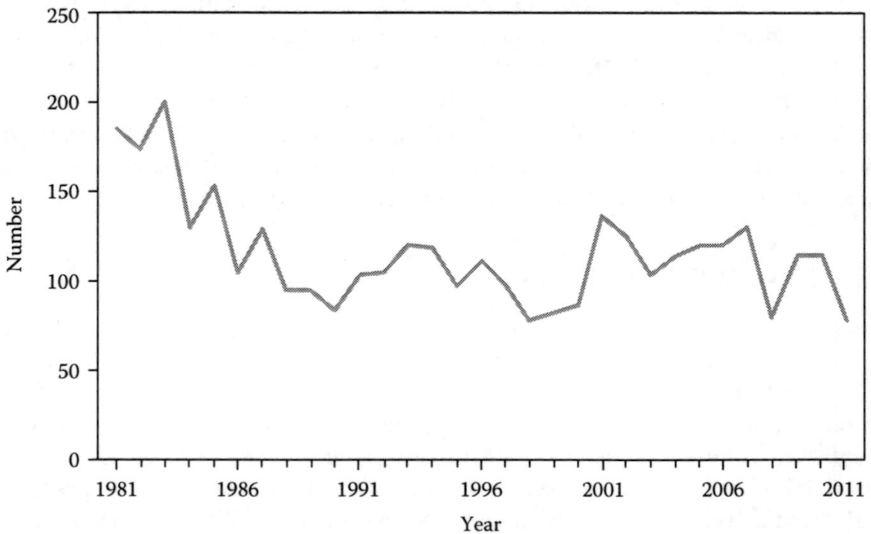

FIGURE 3.3 Number of brucellosis cases reported annually to the CDC. (From CDC, *Morb. Mort. Week. Rep.*, 60(53), 56, 2013; courtesy of the CDC.)

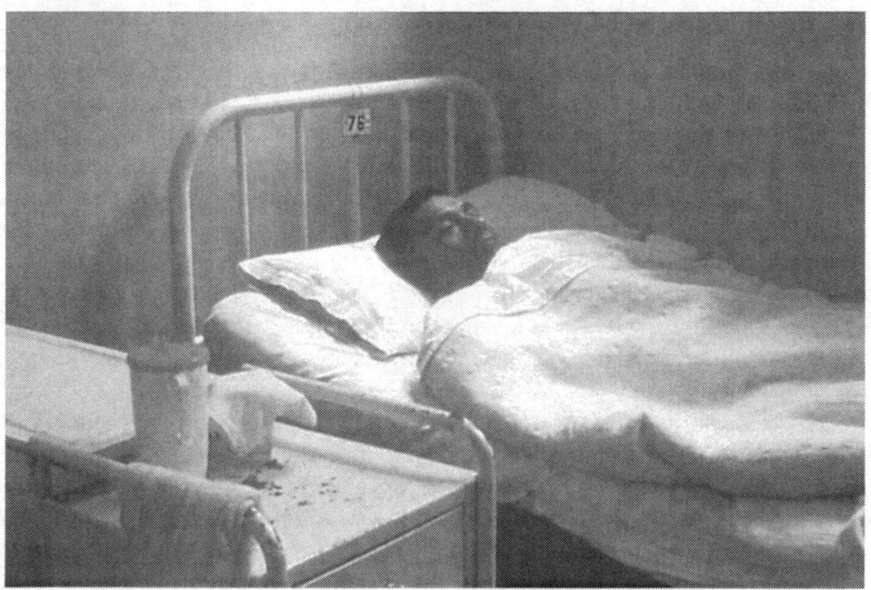

FIGURE 3.4 Patient being treated for brucellosis. (Courtesy of the CDC.)

In about 2% of patients, brucellosis develops into a serious condition because the infection spreads to the inner lining of the heart (endocarditis). If left untreated, endocarditis can damage heart valves, and this complication is the leading cause of mortality among brucellosis patients (Mayo Clinic 2010). In about 5% of patients infected with *B. melitensis*, the pathogen infects the central nervous system. These patients can develop meningitis causing both acute and chronic problems. Less than 1% of brucellosis patients die from the disease (WHO 2006).

Brucellosis can develop into a chronic infection (long-lasting illness) with recurring fever, pain in the joints, and fatigue. *Brucella* can infect any organ in the body; in males, it often inflects the testes, prostate gland, and kidneys. If this happens, the pain can be excruciating. *Brucella* can infect the joints and result in osteoarthritis; this is the most common complication of brucellosis; 40% of patients experience this (WHO 2006). Symptoms of osteoarthritis include pain and stiffness due to swelling of the joints. Once *Brucella* has invaded the joints or central nervous system, the pathogen can be particularly difficult to treat successfully. For this reason, prompt medical attention is advised whenever someone may have brucellosis (Mayo Clinic 2010). When a pregnant woman contracts brucellosis, there is a risk to her fetus, but her risk can be greatly reduced with antibiotics.

Blood tests can be conducted to determine if a person has developed antibodies against *Brucella* (i.e., the person is seropositive). Being seropositive indicates that the person either has or has had brucellosis or has been exposed to *Brucella*. It does not indicate that the person currently has brucellosis. A diagnosis (i.e., confirmation of a patient's disease) of an active case of brucellosis is made when *Brucella* is detected in samples of the patient's blood, bone marrow, or other tissue.

3.3 *BRUCELLA* INFECTIONS IN ANIMALS

Several free-ranging ungulates including elk and bison can serve as a reservoir host for *B. abortus*, and feral hogs and caribou are reservoir hosts for *B. suis*. Brucellosis is a disease of domestic sheep, goats, cattle, and hogs; it causes great economic losses for livestock producers. *Brucella* infects the reproductive system of animals, resulting in high abortion rates, premature births, retained placentas, and infertility. Infections usually spread among animals in a herd by inhalation or ingestion of the pathogen located in infected birth fluids and tissues. Females can pass along *Brucella* to their offspring through their milk. *B. ovis* is found in semen, and infected males can spread the pathogen to females through mating (Figure 3.5).

The most common mode of interspecific (between species) transmission of *Brucella* is through contact with the afterbirth or aborted fetuses from an infected animal. This happens because afterbirths are relatively novel in the environment and perhaps nutritious, so animals often smell, lick, or eat the afterbirth of other animals. *Brucella* can survive for long periods in feces, afterbirths, aborted fetuses, and in water. For example, *B. abortus* has survived 114 days in freshwater; 66 days in cool, wet soil; and 53 days in manure (WHO 2006). The ability of *Brucella* to survive for long periods of time after being shed by infected animals greatly enhances the probability that another animal will become infected. In Alaska, antibodies against *Brucella* antigens (i.e., chemicals on pathogens that provoke immune systems to develop antibodies against them) were detected in 10% of caribou, 3% of moose, and 5% of grizzly bears (Zemke 1983). Eliminating *Brucella* is very difficult once the pathogen has spread to wildlife species.

A national program to eradicate bovine brucellosis in the United States began during 1934, when 11% of adult cattle tested positive for *B. abortus*. The program was expensive, costing more than $3.5 billion, but it was effective in virtually eradicating bovine brucellosis from cattle herds everywhere in the United States except for the Greater Yellowstone Ecosystem where bison and elk within Yellowstone National Park (YNP) serve as reservoir hosts for the pathogen (Figure 3.6). Under

FIGURE 3.5 If bison or elk are infected with *Brucella* when nursing, their offspring can become infected by the pathogen. (Courtesy of Bill West and the USFWS.)

FIGURE 3.6 Bison grazing in YNP. (Courtesy of C.R. Madsen.)

the eradication program, an epidemiological investigation is required whenever a reproductively intact bovine tests positive for brucellosis to identify the infected animal's herd or origin and all cattle herds that may have contracted it. The infected herd is then destroyed or subjected to a testing program that takes over a year to complete. Any other herd that came into contact with the infected one is also quarantined. Owners are compensated if their herd is destroyed but not if it is quarantined. Still, an owner of an infected herd can incur losses of $35,000–$200,000 in uncompensated costs (Schumaker et al. 2012).

3.4 HOW HUMANS CONTRACT BRUCELLOSIS

People contract brucellosis after consuming or inhaling something that is contaminated with *Brucella* or by having the bacteria enter the body through a skin wound or mucous membranes around the eyes. Most people contract brucellosis by drinking unpasteurized milk or eating cheese, ice cream, or other dairy products made from unpasteurized milk. Milk is pasteurized by heating it to 161°F (72°C) for 15–20 seconds. This process kills *Brucella* and most other bacteria that might be in milk. People who have an elevated risk of contracting brucellosis include those who work in slaughter yards or meat packing plants, veterinarians, dairy farmers, ranchers, and wildlife biologists. Hunters may be infected when they harvest an elk or wild pig that has been infected with *Brucella*. In these cases, hunters may be infected when cleaning the carcass if the pathogen is inhaled or enters via a skin wound. For this reason, hunters, herdsmen, and others should wear rubber gloves when cleaning carcasses or handling viscera from wild animals or livestock. People can also be infected from consuming raw or undercooked meat from a game animal or other infected animal.

Besides consuming unpasteurized dairy products, people mostly contract brucellosis from exposure to cattle, sheep, and goats. Dogs can be infected with *B. canis* but few pet owners are at risk of contracting brucellosis from their dog because *Brucella* are concentrated in blood, semen, and placental fluids of dogs and few pet owners come into contact with these fluids. People with compromised immune systems are at risk of contracting brucellosis and should avoid dogs known to be infected with *B. canis* (Mayo Clinic 2010).

Brucellosis was a disappearing disease among Australians due to an effective eradication program of *Brucella* among cattle. However, the disease is now reemerging because of the popularity of hunting feral hogs (Sidebar 3.1). In Queensland, over 40% of people diagnosed with brucellosis were infected with *B. suis*, which is associated with free-ranging and domestic hogs. All of the patients infected with *B. suis* had been involved with the hunting, killing, or slaughter of feral pigs during the prior year. Given the increasing feral hog population in the United States and

SIDEBAR 3.1 BRUCELLOSIS AMONG FERAL HOG HUNTERS (GIURGIUTIU ET AL. 2009)

On July 14, 2008, a 37-year-old man from Pennsylvania went to the emergency room of the local hospital after experiencing chills, muscle aches, shortness of breath, night sweats, and weight loss for a week. Medical doctors initially thought the patient had a viral infection, but *B. suis* was isolated from his blood. The patient reported that he and his brother had hunted feral hogs in Florida during December 2007. They had shot four feral hogs, and both men had participated in field dressing and butchering the hogs without wearing personal protective equipment. The meat had been brought back to Pennsylvania and had been consumed by other members of the family. The meat was tested and found to contain *B. suis*.

According to the patient, all of the meat had been adequately cooked (internal temperature of 71°C or 160°F) and had been eaten by himself, his wife, and children. Fortunately, other family members did not exhibit symptoms of brucellosis. However, the patient reported that his brother, who had gone hunting with him, had become ill during April 2008 with similar symptoms to his own. Medical authorities contacted the brother and discovered that he had been ill but had not sought medical treatment. A blood test showed that the brother had antibodies for *Brucella*. The first patient recovered after 6 weeks of treatment with the antibiotics: rifampicin or doxycycline. The two brothers are believed to have been infected when they field dressed and butchered the feral hogs. The CDC recommends that patients be tested for brucellosis when they have symptoms consistent with brucellosis and a history of feral swine hunting.

the popularity of feral hog hunting, it is likely that brucellosis caused by *B. suis* will become more prevalent among U.S. hunters (Robson et al. 1993).

People can contract brucellosis directly from another person, but this is uncommon. Nursing babies can contract the disease from their mothers through breastfeeding. Transmission between sexual partners can occur, but this is rare.

3.5 MEDICAL TREATMENT

3.5.1 Treating Brucellosis in Humans

Treatment of brucellosis in humans is intended to relieve its acute and chronic symptoms and to eliminate *Brucella* from the body to prevent a relapse of the disease. Brucellosis is treated in humans by prescribing antibiotics that are effective against *Brucella*, usually doxycycline or rifampicin. Acute infections (i.e., an infection of short duration) of brucellosis respond well to appropriate antibiotic treatment, and most patients make a full recovery. If the pathogen invades the central nervous system, treatment options are more limited because many drugs do not readily cross the blood–brain barrier. Therefore, drugs that can cross it, such as rifampicin, are included in the drug regimen. When treatment is delayed or when the disease is severe, brucellosis can be difficult to treat and recovery may take several months. Brucellosis in humans can be misdiagnosed because the first symptoms of the disease are similar to those caused by flu. This may result in a delay in administering the correct treatment for the disease and increase the danger that the patient will develop serious complications. A person's susceptibility to brucellosis depends upon the strength of his or her immune system, the amount of *Brucella* to which the person was exposed, and the route of infection. The species of *Brucella* also influences severity of symptoms with *B. melitensis* and *B. suis* being the most virulent (i.e., capable of causing disease) for humans (WHO 2006, CDC 2010).

3.5.2 Treating Brucellosis in Animals

There is no approved treatment for brucellosis in livestock; instead, the emphasis is placed on preventing the spread of the disease to uninfected animals. While there is no vaccine for humans to protect themselves from brucellosis, there is a vaccine for livestock. With livestock, diagnosis and control of brucellosis must be made at the herd level rather than treating only those individuals that are infected. Hence, the identification of a single infected animal is sufficient to classify the entire herd as infected. This is necessary because brucellosis has a long incubation period (i.e., time between when a patient became infected with a pathogen and the onset of the first symptoms of illness), and some individuals in the herd may be asymptomatic (i.e., without symptoms) yet harbor *Brucella* and have the ability to pass the pathogen on to others. Infected herds in the United States are quarantined by state and federal regulations until the herd is proven free of *Brucella*. By quarantining infected animals and through vaccination programs, brucellosis has been eliminated from most U.S. states.

Unfortunately, free-ranging elk and bison can serve as a reservoir for *Brucella*, and eliminating the pathogen in wild animals is much more challenging than in livestock herds. Brucellosis in wildlife is a serious problem because wild animals can reinfect brucellosis-free livestock. For example, bison and elk from YNP are infected with *Brucella*, and ranchers around the park fear that infected bison will spread the disease to their livestock (Sidebar 3.2).

SIDEBAR 3.2 BRUCELLOSIS AROUND YELLOWSTONE NATIONAL PARK (YNP) AND THE SURROUNDING AREA: MANAGEMENT OF A ZOONOTIC DISEASE WHEN BOTH WILDLIFE AND LIVESTOCK ARE INVOLVED (SCHUMAKER ET AL. 2012, KAUFMAN ET AL. 2012)

The last known host reservoirs of bovine brucellosis (*B. abortus*) in the United States are the wild bison and elk living within YNP and the surrounding area. Bovine brucellosis was first discovered in YNP during 1917 among cattle kept there for park employees. It subsequently spread to park's free-ranging bison and elk. Despite numerous efforts over the last century, bovine brucellosis persists in YNP today. Half of the bison in YNP are seropositive (i.e., have specific antibodies for antigens unique to the pathogen) for *B. abortus* because they have been infected in the past, and up to 45% of seropositive bison are still infected. Cattle on neighboring ranches occasionally contract brucellosis from elk. Brucellosis was detected in 13 cattle herds in the area around YNP (eight in Wyoming, three in Idaho, and two in Montana) between 2004 and early 2011. When this happens, federal law requires that infected herds be destroyed or quarantined and tested multiple times. Furthermore, any herd that has had contact with the infected herd must be quarantined and tested. These restrictions impose significant costs on individual farmers and health agencies. States carefully guard their status as brucellosis-free because loss of this status triggers mandatory state-wide testing of any reproductively intact cattle being sold or moved across state lines. Risk of transmission of brucellosis from elk and bison to livestock also imposes costs on state wildlife agencies and YNP and exposes them to great political pressure to resolve the problem. But potential management actions, such as hazing wildlife, vaccinating them, test-and-slaughter, or reducing elk and bison populations, are controversial.

One wildlife management practice that enhances the spread of *B. abortus* among wild animals is providing hay and pelleted alfalfa during the winter at 23 feeding grounds in the Greater Yellowstone area (Figures 3.7 and 3.8). Approximately 80% of the 23,000 elk in the southern part of the Greater Yellowstone area take advantage of the feed that helps them survive the winter, but it also concentrates elk around these feeding grounds. Elk infected with brucellosis may abort their fetuses while at these feeding grounds, and the great numbers of elk concentrated at these feeding grounds greatly increase the risk that elk will become infected with *B. abortus*. During 2009, 22% of elk on feed grounds were seropositive to this pathogen. Yet feeding elk during the winter is popular with the public because it helps maintain elk populations, makes elk more visible for tourists, and keeps them off of private property. There is also concern that if feed grounds were closed, then hungry elk would disperse to private agricultural land and may increase the risk of spreading *B. abortus* to livestock.

Instead of stopping the winter feeding of elk, Wyoming Game and Fish Department shortened the feeding season, changed its procedure of providing hay in continuous lines to providing discrete and dispersed piles, and used biobullets to vaccinate elk calves located at feed grounds. Still management of brucellosis in the Greater Yellowstone area will remain controversial because of the important roles that elk, bison, and cattle play in both the ecology and culture of Yellowstone.

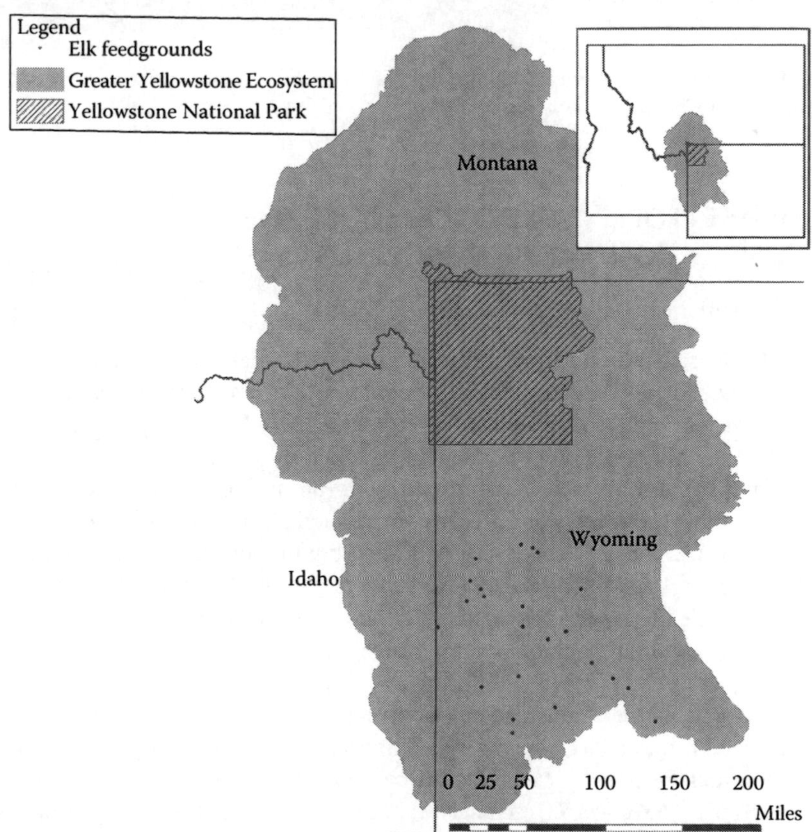

FIGURE 3.7 Map of the Greater Yellowstone area showing the location where elk are fed during the winter. (Courtesy of Brant Schumaker and *Human–Wildlife Interactions*.)

FIGURE 3.8 Elk on the National Elk Refuge, Jackson Hole, Wyoming. (Courtesy of Brant Schumaker and *Human–Wildlife Interactions*.)

3.6 WHAT PEOPLE CAN DO TO REDUCE THEIR RISK OF CONTRACTING BRUCELLOSIS

3.6.1 Preventing Brucellosis in Humans

China and Russia initiated programs to prevent brucellosis in humans during the 1950s, which involved vaccinating people with a live, attenuated strain (i.e., a strain that has reduced virulence) of *Brucella*. The Russian vaccine provided immunity for only 6–12 months. Hence, vaccination programs had to be timed so that they began only a few months prior to an anticipated disease outbreak. The vaccination program was effective in reducing the risk of morbidity (i.e., prevalence or incidence of disease in a human population) five- to tenfold among people who contracted brucellosis. However, a major problem with the vaccine was that some people became hypersensitive to the vaccine after repeated vaccinations. During 2012, a human vaccine against brucellosis was unavailable (WHO 2006, CDC 2010).

Most people in the United States contract brucellosis from consumption of unpasteurized milk or foods made from unpasteurized milk. People should avoid consuming these products. Meat should be cooked until its internal temperature is 145°F–165°F (63°C–74°C). This is especially important when cooking wild game or any meat that might contain *Brucella*. Hence, educating people about food safety is one of the best ways to reduce the number of people who contract brucellosis.

People at risk for brucellosis should inform their medical doctors of this if they become ill with symptoms similar to brucellosis. People at risk include those who interact with wild ungulates or livestock, consume unpasteurized milk, or have traveled to a foreign country where brucellosis is prevalent.

People should wear adequate protective clothing when carrying out high-risk procedures, such as making contact with infected animals or their carcasses. Protective clothing should include rubber gloves, boots, aprons, and face shields or goggles to protect the eyes. Care needs to be taken to make sure that *Brucella* is not transmitted outside of the work area. The infectious nature of this pathogen prompted military forces to consider its use as a weapon of war. Brucellosis in humans can never be eradicated as long as it occurs in livestock.

3.6.2 Preventing Brucellosis in Livestock

Brucellosis can be prevented in livestock through careful herd management and hygiene. Whenever an animal has had an abortion or premature birth, the placenta, dead fetus, and any body fluids should be disposed by burial or burning. Contaminated areas need to be thoroughly disinfected.

The risk of spreading *Brucella* to brucellosis-free herds and flocks can be reduced by selecting replacement animals from *Brucella*-free areas. Replacement animals should be tested for exposure to *Brucella* and also isolated for at least 30 days prior to introducing them into the herd. Herds should be tested for brucellosis on a regular basis; animals that test positive should be isolated or euthanized. This test-and-slaughter approach is effective when less than 2% of the herds or flocks are infected.

The most successful method for preventing brucellosis in livestock is through vaccination. In many countries, vaccination is the only economical and practical means of controlling brucellosis among livestock. Mass vaccination of all females (adverse side effects have been reported in bulls) also is recommended in areas where more than 10% of the flocks or herds are infected. Vaccinating adult sheep or hogs may result in some abortions so a safer approach may be to only vaccinate all 3–4-month-old replacement animals. This approach should result in the entire population being vaccinated within a decade. Cattle should be vaccinated with *B. abortus* strains 19 or RB-51; sheep and goats should be vaccinated with *B. melitensis* strain Rev-1. The RB-51 vaccine has a major advantage over strain-19 vaccine in that it does not cause false-positives when livestock are tested for brucellosis, thus preventing vaccinated animals from being mistaken for infected ones (Schumaker et al. 2012). Research to develop more effective vaccines is ongoing.

3.6.3 Controlling Brucellosis in Free-Ranging Wildlife

Efforts have been made to capture and test elk and bison for exposure to *Brucella* and to kill any infected animal. Wyoming Game and Fish Department conducted an experimental test-and-slaughter program from 2006 to 2010 by trapping elk at feeding grounds and killing any females that were seropositive for *Brucella*. During the program, 2226 elk were captured, 1286 cow elk were tested, and 197 seropositive elk were killed. At one feed ground, the program reduced seroprevalence (proportion of animals that were seropositive) from 37% to 5%, but the program was discontinued after it has been proven to be too expensive ($1.3 million) and controversial (Fenichel et al. 2010, Kauffman et al. 2012).

Brucella is most likely to spread among deer, elk, and bison when the animals are concentrated, such as when they are provided hay during the winter (Sidebar 3.2). For this reason, some wildlife agencies have stopped feeding wildlife. Efforts to stop the winter feeding of wildlife around YNP, however, are controversial because feeding programs make wildlife more accessible to the public, and hunters fear that ending the practice will cause deer and elk populations to decline.

Another approach to prevent infected elk and bison from spreading brucellosis to livestock is to keep infected wildlife separated from uninfected livestock. For years, YNP has tried to keep bison inside the park during the winter by hazing or killing any bison that migrated outside it. Unfortunately, it has proven more difficult to keep cattle and elk apart in the Greater Yellowstone Ecosystem (Schumaker et al. 2012).

In some areas where *B. suis* is endemic, programs to kill free-ranging hogs were implemented to reduce their density, but these programs often are unsuccessful due to the high reproductive rates of hogs (Figure 3.9). Other programs were aimed at vaccinating free-ranging hogs to reduce the number of free-ranging hogs that were susceptible to the disease. Still another approach is to confine domestic hogs behind double fences or other barriers that prevent domestic hogs and free-ranging hogs from coming into contact with each other. Food supplies for domestic hogs also should be kept in buildings or grain bins that are inaccessible to free-ranging hogs. Hog farmers need to realize that free-ranging hogs are usually hungry and will go to great lengths to gain entry to areas where food is available.

FIGURE 3.9 Feral hogs are a reservoir host for *B. suis*. (Courtesy of Tyler Campbell.)

Cattle herds that are free of brucellosis are most likely to be reinfected with *Brucella* from wildlife when coming into contact with the placenta or aborted fetus from an infected animal. Hence, livestock should be kept out of areas where elk or bison give birth.

3.7 ERADICATING BRUCELLOSIS FROM A COUNTRY

The complete elimination or eradication of *Brucella* from a country requires a highly organized effort. Eradication programs are more likely to be successful in areas where (1) *Brucella* is not endemic among local wildlife; (2) widespread mixing of herds does not occur; (3) there is widespread support for the eradication program; and (4) there is a willingness among ranchers, herders, and other livestock owners to comply with restrictions on the movement of livestock. A long-term surveillance program is also needed to detect outbreaks or reintroductions.

LITERATURE CITED

Capasso, L. 2002. Bacteria in two-millennia-old cheese, and related epizoonoses in Roman populations. *Journal of Infection* 46:122–127.

CDC. 2010. Brucellosis: General information. http://www.cdc.gov/nczved/divisions/dfbmd/diseases/brucellosis (accessed February 16, 2012).

CDC. 2013. Summary of notifiable diseases—United States, 2011. *Morbidity and Mortality Weekly Report* 60(53):1–118.

FAO. 2003. Guidelines for coordinated human and animal brucellosis surveillance. *FAO Animal Produce and Health Paper* 156:1–46.

Fenichel, E. P., R. D. Horan, and G. J. Hickling. 2010. Management of infectious wildlife diseases: Bridging conventional and bioeconomic approaches. *Ecological Applications* 20:903–914.

Giurgiutiu, D., C. Banis, E. Hunt, J. Mincer et al. 2009. *Brucella suis* infection associated with feral swine hunting—Three states, 2007–2008. *Morbidity and Mortality Weekly Report* 58:618–621.

Kauffman, M. E., B. S. Rashford, and D. E. Peck. 2012. Unintended consequences of bovine brucellosis management on demand for elk hunting in northwest Wyoming. *Human–Wildlife Interactions* 6:12–29.

Madkour, M. M. 2005. Brucellosis. In: D. A. Warnell, T. M. Cox, and J. D. Firth, editors. *Oxford Textbook of Medicine*. Oxford University Press. Oxford, U.K., pp. 543–545.

Massey, P. D., B. G. Polkinghorne, D. N. Durrheim, T. Lower, and R. Speare. 2011. Blood, guts and knife cuts: Reducing the risk of swine brucellosis in feral pig hunters in north-west New South Wales, Australia. *Rural and Remote Health* 11(1793):1–9.

Mayo Clinic. 2010. Brucellosis. http://www.mayoclinic.org/health/brucellosis (accessed February 21, 2012).

Robson, J. M., M. W. Harrison, R. N. Wood, M. H. Tilse et al. 1993. Brucellosis: Re-emergence and changing epidemiology in Queensland. *Medical Journal of Australia* 159:153–158.

Schumaker, B. A., D. E. Peck, and M. E. Kauffman. 2012. Brucellosis in the Greater Yellowstone area: Disease management at the wildlife-livestock interface. *Human–Wildlife Interactions* 6:48–63.

WHO. 2006. Brucellosis in humans and animals. *World Health Organization* 2006(7):1–90.

Zemke, R. L. 1983. Serologic survey for selected microbial pathogens in Alaska wildlife. *Journal of Wildlife Diseases* 19:324–329.

4 Tuberculosis

In Nineteenth Century Europe, the physical wasting away of the body from tuberculosis was believed to lead to euphoric release of passion and creativity of the soul. The illness in many poets exemplified genius bursting forth from ordinary talent as the body burned from fever. A subsequent decline in the arts was blamed on decreased incidence of tuberculosis.

Morens (2002)

A town meeting was held outside of Riding Mountain National Park (Manitoba, Canada) to discuss the problem of bovine tuberculosis (TB) in local elk. One participant remarked: "You need to kill off the wild elk herd. TB is a real problem and something drastic needs to be done." When government officials suggested that elk eradication was not feasible, participants became frustrated. One participant remarked bluntly: "We have given you the solution, but you don't have the guts to use it. Kill the elk."

Paraphrased from Brook (2009)

We ask you to remember not just the 5000 badgers we're talking about culling, but the 38,000 cattle slaughtered when herds were found to be infected with bovine tuberculosis.

Peter Kendall, president of the U.K. National Farmers' Union
Cited by Alan Cowell, *New York Times*, March 28, 2013

4.1 INTRODUCTION AND HISTORY

Tuberculosis (or TB) is a bacterial disease caused by several species of the genus *Mycobacterium*. Under the microscope, these bacteria appear as slender rods (Figure 4.1). They are not capable of movement, aerobic (i.e., requiring oxygen to survive), and Gram positive. Their cell membranes have a high concentration of lipids, and this waxy shell provides protection to the pathogen. The bacteria grow slowly, dividing about once a day versus other bacteria that divide about once an hour. They can survive outside of a host for weeks, even in a dry state. Most TB patients are infected with *M. tuberculosis*, and the source of their infection was another person. Humans serve as a reservoir host for these bacteria, but humans can transmit *M. tuberculosis* to other species. Other species of *Mycobacterium* can cause TB in humans. *M. bovis* is a disease of cattle, deer, and elk, but humans can become an accidental host. Most people contract *M. bovis* by drinking raw milk, and pasteurization of milk has greatly reduced this disease. *M. africanum* and *M. canetti* cause approximately half of the cases of human pulmonary TB in West Africa, but they are not a concern elsewhere. *M. microti* is primarily a risk to people with weakened immune systems (de Jong et al. 2010).

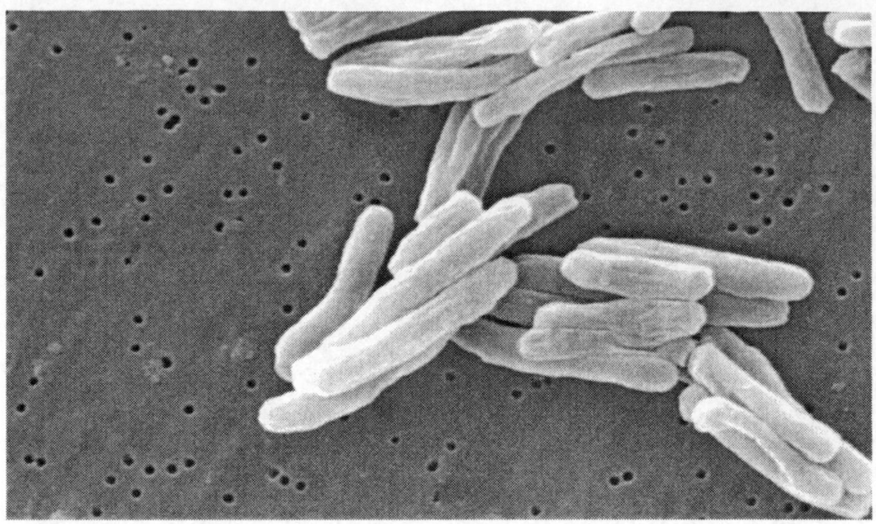

FIGURE 4.1 A photo of *M. tuberculosis* taken with a scanning electron microscope. (Courtesy of Ray Butler, Janice Carr, and the CDC.)

TB is an ancient disease of humans; signs of TB have been found in skeletal remains of a Neolithic man dating back to 8000 BC and in Egyptian mummies, who lived 3700–1000 BC. In South America, it has been detected in the bones of North American Indians who lived more than 1000 years ago. TB was known to the ancient Greeks, and Hippocrates described it as a widespread disease that involved the coughing up of blood and was usually fatal.

During the 1800s, TB was a disease of the urban poor. During 1815 in England, TB was responsible for 25% of all mortalities. A hundred years later, TB was still a major killer in Europe and was responsible for 17% of all mortalities in France. The causative agent of TB was first isolated during 1882 by Dr. Robert Koch, who received the 1905 Nobel Prize in medicine for his discovery.

Sanatoriums for TB patients became popular during the 1800s. Other patients moved to areas with warmer and drier climates in hopes of improving their health. During the 1830s and 1840s, a number of TB patients moved into Mammoth Cave, Kentucky, believing that the cave's constant air temperature and pure air might help them. Instead, many patients died of TB, and the experiment was abandoned (McCarthy 2001).

Early attempts to develop an effective vaccine against TB were unsuccessful, but development of drug therapies greatly reduced the morbidity rate from TB. In 1943, Selman Waksman discovered the antibiotic streptomycin and later received a Nobel Prize for his discovery. A short time later, William Feldman and Alfred Karlson showed that streptomycin was effective against *Mycobacterium*. Two other drugs providing an effective treatment for TB were isoniazid, which was introduced into the United States during 1952, and rifampin, introduced during 1971.

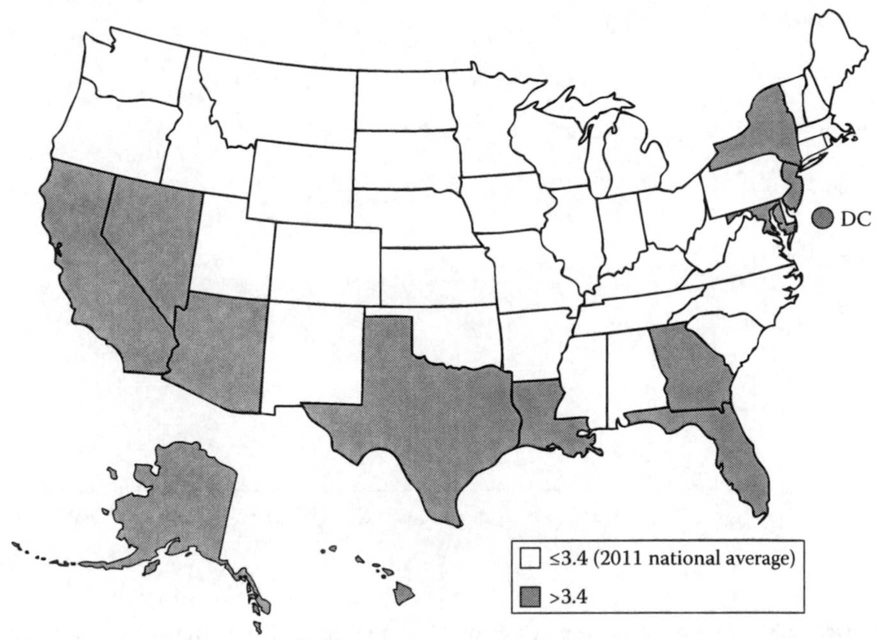

FIGURE 4.2 Incidence of TB among people living in different states during 2011. (Courtesy of the CDC, 2012.)

In the United States, there are about 10 million people infected with *Mycobacterium*; 9951 people from this infected group developed TB during 2012. Half of these cases occurred in just four states (California, Texas, New York, and Florida), but Hawaii had the highest incidence rate (number of new cases annually per 100,000 people) of TB at 8.8 (Figure 4.2). People predisposed to TB include alcoholics, intravenous drug users, people confined in nursing homes or prisons, people undergoing chemotherapy, and people with HIV or diabetes. The number of reported cases of TB in the United States ranged from 20,000 to 25,000 during the 1980s but decreased to 10,000 to 15,000 during 2000–2010 (Figure 4.3). Between 500 and 800 people die annually from TB in the United States (CDC 2013a,b).

Levels of human infection by *M. tuberculosis* vary considerably among countries, even between neighboring countries (Figure 4.4). This variation results from differences in the quality of public health systems and vigilance in identifying and treating patients with TB. Worldwide, there were 8.8 million new cases of TB in humans and 1.45 million deaths during 2010 (WHO 2011). Most of these were in Africa and Asia where 80% of humans tested positive during skin tests for *M. tuberculosis*; in comparison, less than 10% of the U.S. population tested positive. The spread of HIV among humans caused a worldwide surge of TB, starting during the 1980s. This global trend reversed more recently, with the incident rate falling since 2002; the absolute number of TB cases has been dropping since 2006 (WHO, 2010, 2011, Mayo Clinic 2011).

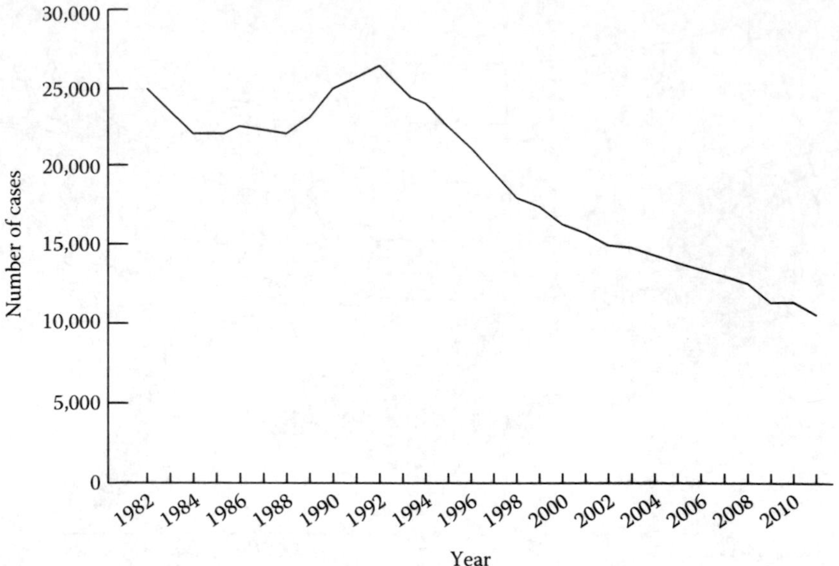

FIGURE 4.3 Number of reported cases of TB in the United States from 1982 until 2010. (Data courtesy of the CDC and obtained in 2012.)

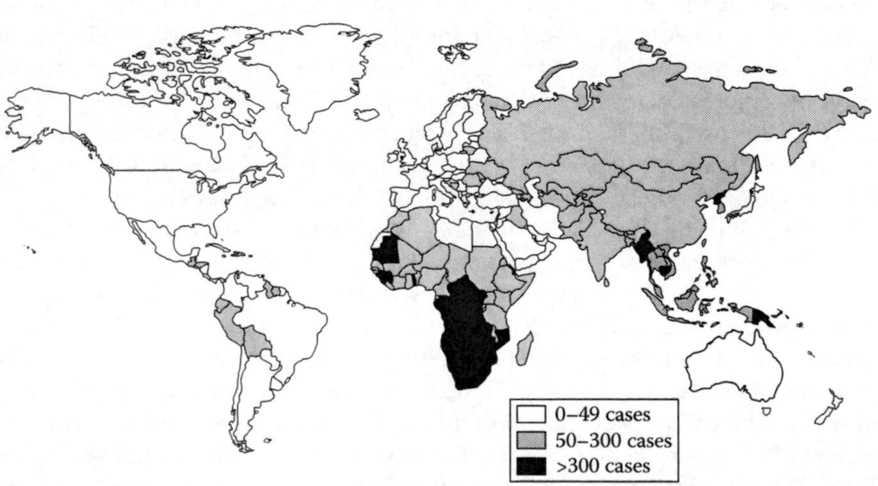

FIGURE 4.4 Incidence (number of new cases per 100,000 population) of TB among people living in different countries during 2010. (Data from WHO, *Global Tuberculosis Control 2011*, World Health Organization, Geneva, Switzerland, 2011.)

4.2 SYMPTOMS IN HUMANS

TB is a common infectious disease of humans with about one-third of the world's human population being infected. Most people infected with *M. tuberculosis* are asymptomatic, but about 10% of infected people develop TB. It is primarily a disease of the lungs (pulmonary TB) with symptoms including a chronic cough with bloody sputum (i.e., mucous produced by the airways or lungs and often mixed with saliva), chest pains, fever, excessive sweating at night, fatigue, and weight loss. This last symptom produced one of the disease's historical names, "consumption." In response to a pulmonary infection of *M. tuberculosis*, the body attempts to wall off the infection by forming a small nodule around it, which is called a granuloma. These can be seen on chest x-rays, but granulomas are not unique to TB patients. Therefore, other tests must be conducted before a diagnosis of TB can be made. If left untreated, most people with pulmonary TB will be killed by the disease. *M. tuberculosis* can also infect the central nervous system, lymphatic system, urogenital system, bones, and joints; because of this, TB can produce many symptoms. For example, TB in the spine causes back pain, and TB in the urogenital system produces blood in the urine. When *M. tuberculosis* has spread to other organs besides the lungs, it is called disseminated TB. Children are at greater risk of developing disseminated TB than adults.

Symptoms of TB caused by *M. bovis* or *M. tuberculosis* are similar but not identical. Most (65%) of the patients infected with *M. bovis* had an infection outside of the pulmonary system, as did only 18% of patients infected with *M. tuberculosis*—the reason being that *M. tuberculosis* is more likely to be inhaled while *M. bovis* is more likely to be ingested. When ingested, the pathogen is more likely to infect the gastrointestinal tract, spleen, kidneys, and liver than the lungs (WHO 2006, Hlavsa et al. 2008, Mayo Clinic 2011).

A person with a latent infection (i.e., an infection that produces no disease symptoms in the patient) may develop TB years after the initial infection. This happens when *M. tuberculosis* is able to overwhelm the patient's immune system and begins to multiply. Worldwide, TB kills more young people and adults than any other known infectious disease. The Mayo Clinic (2011) recommends a visit to the doctor when a person has a fever, unexplained weight loss, drenching night sweats, or a persistent cough; these are all symptoms of TB.

4.3 *MYCOBACTERIUM* INFECTIONS IN ANIMALS

While *M. tuberculosis* and *M. africanum* primarily infect humans, infected people can spread them to animals, especially to other primates. Other species of *Mycobacterium* can infect humans but primarily infect other mammal species. Wildlife species that serve as reservoir hosts for *M. bovis*, the causative agent for bovine TB, include the brush-tailed possum in New Zealand, the white-tailed deer in North America, and the European badger in the United Kingdom (Tweddle and Livingstone 1994, Campbell and VerCauteren 2011). Clinical signs of bovine TB include weight loss, swollen lymph nodes, discharging lymph nodes, abscesses, persistent coughing, and the presence of granulomas in the lungs and lymph nodes

(Campbell and VerCauteren 2011). It can develop into a disseminated illness and spread from the lungs to other organs. Bovine TB causes great economic losses for dairy farmers and cattle producers. Milk production is reduced by 10% and meat production by 6%–12% when cows become infected. For this reason, most countries try to eradicate bovine TB if possible and control its spread where eradication is not feasible (Sidebar 4.1).

In Manitoba, Canada, bovine TB was endemic in cattle until the 1970s when a campaign was initiated to control the disease. Manitoba was declared free of the disease during 1986, but the disease-free status did not last long. During 1991, seven cattle herds tested positive, and the suspected source of the pathogen was Riding Mountain National Park (Figure 4.5) when an elk from the park tested positive for TB the following year. Many people surrounding the park became angry and wanted elk in the park eradicated. Others wanted more elk and fewer cattle (Figure 4.6). To resolve the conflict, a bovine TB task force was created with representatives from the federal and provincial government and key stakeholders. The task force was successful in enhancing dialogue among the groups, but considerable conflict remained (Brook 2009).

SIDEBAR 4.1 ERADICATION OF BOVINE TB FROM AUSTRALIA (TWEDDLE AND LIVINGSTONE 1994, RADUNZ 2006, MORE 2009)

Cattle were introduced to Australia about 200 years ago by English settlers. By the 1990s, there were 23 million cattle in the country, distributed into 160,000 herds. With the importation of cattle came bovine TB, and by the twentieth century, the disease was a serious public health problem, especially in children who contracted the disease mainly by drinking unpasteurized milk. In 1970, Australia launched a compulsory campaign to eradicate bovine TB from cattle. Under the program, infected herds were quarantined until subsequent testing revealed no signs of infection over a long period of time. Herds and regions were labeled as infected, restricted, provisionally clear, and confirmed free. Movement of cattle among regions of the country was only allowed if the animals were going directly to slaughter or to a region with a similar or lower infection status. These controls created disruption of traditional trading patterns in Australia. For example, cattle normally moved from northern to southern Australia for fattening, but this was no longer allowed because bovine TB disease eradication was first achieved in the south. Another problem for Australia was that in remote parts of northern and central Australia, cattle were raised in semiferal operations that made it difficult to test cattle for *M. bovis*.

The Asiatic water buffalo was the only feral animal in Australia that served as a reservoir host for *M. bovis*. As part of the program to eradicate bovine TB, the water buffalo population was reduced from 340,000 during 1985 to 106,000 5 years later. Fortunately, Australia's eradication campaign was successful. The "last" case of bovine TB was detected during 2000, and the country was classified as free of bovine TB during 2008.

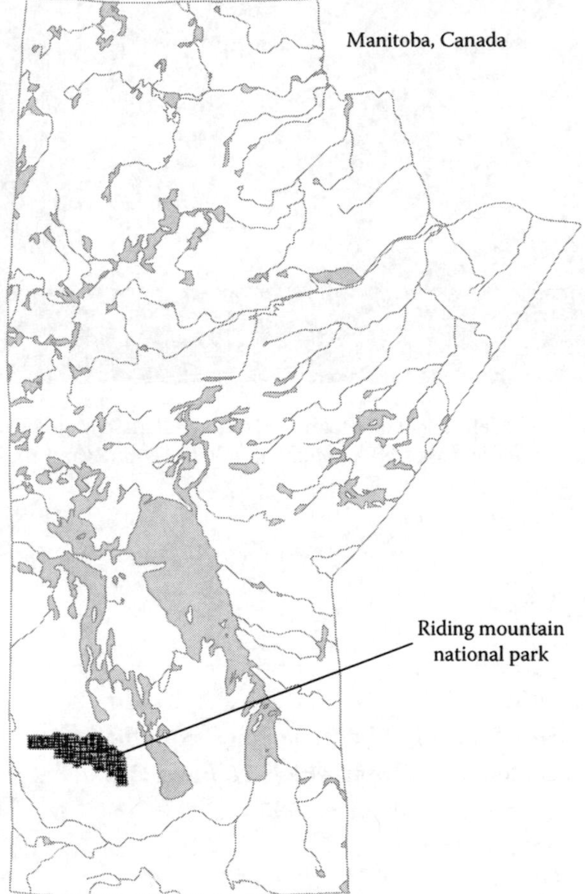

FIGURE 4.5 Map of Manitoba, Canada, showing the location of Riding Mountain National Park where bovine TB is endemic in the park's elk population. (From Brook, R.K., *Hum. Wildlife Conflicts*, 3, 72, 2009.)

Since 1995, *M. bovis* has been isolated from 15 wildlife species in the United States; most of these species are ungulates or predators (Table 4.1). In the United Kingdom and Europe, *M. bovis* has also been detected in European badgers, red deer, and fallow deer (Sidebar 4.2). In Africa, cattle likely spread the pathogen to native ungulates, which then infected native carnivores. The pathogen has been reported in lechwe antelope, topi antelope, warthog, wildebeest, African buffalo, kudu, African elephant, giraffe, and primates. Epizootic outbreaks (i.e., disease outbreaks among animals) of TB have occurred in lions, tigers, and leopards. Elsewhere, *M. bovis* has been detected in Arabian oryx, feral hogs, one-humped camels, Asiatic water buffalo, llamas, and alpacas (Thoen et al. 1992, Campbell and VerCauteren 2011, USDA APHIS 2011).

When *M. bovis* infects wildlife species, eradicating *M. bovis* or preventing its spread among dairy farms and livestock herds is more difficult. For instance,

FIGURE 4.6 A group of elk at Riding Mountain National Park in Manitoba, Canada, where bovine TB is endemic in the park's elk population. (Courtesy of Ryan Brooks and *Human–Wildlife Interactions*.)

TABLE 4.1
Free-Ranging Wildlife Species from the United States from which *M. bovis* Has Been Detected since 1995

White-tailed deer
Mule deer
Elk
Bison
Moose
Opossum
Raccoon
Feral hog
Coyote
Gray fox
Red fox
Wolf
Black bear
Feral cat
Bobcat

Source: USDA APHIS, *Guidelines for Surveillance of Bovine Tuberculosis in Wildlife*, USDA Animal Plant Health Inspection Service, Washington, DC, 2011.

SIDEBAR 4.2 REINTRODUCTION OF BOVINE TB INTO SWEDEN (WAHLSTRÖM AND ENGLUND 2006)

In Sweden, TB in any animal species must be reported to the government. If confirmed in cattle, the entire herd must be killed, and the farmer receives full compensation for any economic losses. These efforts were successful, and Sweden was declared free of bovine TB in 1958. After that, less than 10 human cases of *M. bovis* were diagnosed annually in Sweden, and most of these occurred in elderly people who were infected during their youth before bovine TB was eradicated.

Unfortunately, this situation changed in 1991 when bovine TB was rediscovered in Sweden, this time among farm-raised fallow deer. An investigation found that the source of the infection was 168 fallow deer that were imported into Sweden during 1987. Bovine TB had spread to 13 deer herds by 2005, and all these herds had direct or indirect contact with the imported deer. To avoid further spread, a compulsory control program was implemented where only owners of TB-free herds were allowed to sell live deer. A herd was classified as TB-free after three consecutive tests of the entire herd failed to detect any signs of *M. bovis*.

A problem developed with the whole-herd testing program for *M. bovis* because some herds were so large and spread over such a large area that it was impossible to collect the entire herd for testing. An alternate method was developed to determine if these herds were free of *M. bovis* in which at least 20% of the herd must be slaughtered annually for 15 years and the carcasses inspected for the pathogen. Sweden's experience shows that new species of livestock (e.g., fallow deer) and new husbandry techniques often challenge public health agencies and require new methods of disease surveillance and control of disease.

the prevalence (i.e., proportion of a population that has a disease or is infected or) of TB in Ireland is only 0.4% of cattle versus 45% among European badgers within *M. bovis* hotspots. Thus, badger-to-cattle transmission is of greater importance than cattle-to-cattle transmission. Because badgers act as a natural reservoir for *M. bovis*, Ireland has not been able to eradicate the disease (More 2009). At present, an oral vaccine is unavailable to prevent bovine TB in wildlife, but scientists are trying to develop one that would reduce the disease in European badgers.

M. microti causes TB in bank voles, field voles, wood mice, and shrews in the United Kingdom and Europe. It is most common in field voles where prevalence has ranged from 2% to 31%. It infects llamas, pigs, domestic cats, and ferrets in Europe and rock hyraxes from South Africa. *M. microti* have been considered unimportant as a causative agent for disease in humans, but they have infected a small number of patients from Europe who had weakened immune systems (Kremer and van Soolingen 1998, Cavanagh et al. 2002).

4.4 HOW HUMANS CONTRACT TUBERCULOSIS

TB is a contagious disease and is primarily spread from person to person rather than from animal to human. Transmission occurs when *M. tuberculosis* is expelled by the coughing and sneezing of an infected person, and the pathogen is inhaled by another person. Transmission can also occur through the consumption of contaminated food or water.

Human infections of *M. bovis* are often associated with outbreaks of bovine TB in dairy herds, other domesticated animals, or farm-reared wildlife. People contract bovine TB by drinking raw, unpasteurized milk or eating cheese, ice cream, or other dairy products made from unpasteurized milk. Milk is pasteurized by heating it to 161°F (72°C) for 15–20 seconds. This process kills most of the bacteria in milk, including *M. bovis*. For this reason, TB is more common in countries where most milk is not pasteurized. People can also contract *M. bovis* by inhaling aerosolized bacteria. People with an enhanced risk of contracting *M. bovis* include wildlife biologists, dairy farmers, game farm employees, veterinarians, zoo keepers, and others who work with species susceptible to TB.

4.5 MEDICAL TREATMENT

Tuberculin skin tests are often used for routine screening of individuals who may have been exposed to *M. tuberculosis*. A positive reaction often occurs if the patient is infected with *M. tuberculosis*, had an infection in the past, or has been inoculated against TB. Other tests are also available that are specific for *M. tuberculosis*. Chest x-rays are usually taken after a positive tuberculin skin test to determine if granuloma occur in the lungs.

A TB vaccine is effective against the disseminated TB that some children experience, but it does not provide consistent protection from pulmonary TB. The vaccine is widely used in countries where TB is rampant and provides some immunity for about 10 years. Vaccines are generally not recommended for people in Canada and the United States where TB is less common, excepting those with an elevated risk of becoming infected. One problem with the vaccine is that it produces a false-positive result to the tuberculin skin test.

Antibiotics are normally prescribed for people infected with *M. tuberculosis* to kill the pathogen; however, finding an effective treatment can be a challenge. The waxy cell membrane of *M. tuberculosis* makes many antibiotics ineffective. Further, strains of *M. tuberculosis* have developed resistance to some antibiotics that have been prescribed in the past. During 2010, a combination of the antibiotics, including isoniazid, rifampicin, ethambutol, and pyrazinamide, has often been used to treat TB patients (Mayo Clinic 2011). The drug treatment requires 6–9 months to be effective. Patients with latent infections of *M. tuberculosis* are often treated with an antibiotic to reduce the risk of the infection developing into TB later in life. Patients with latent infections are not contagious, and patients with active infections (i.e., the pathogen is causing illness) are not contagious after several weeks of an antibiotic treatment (WHO 2006). The risk of a latent infection becoming an active infection increases for patients with HIV, a weakened immune system, and chronic smokers.

When TB reoccurs in a patient, *M. tuberculosis* is isolated from that particular patient and tested with different drugs to determine which one(s) works best to kill that specific strain. This individual testing of patients allows for a cocktail of drugs to be administered that is both effective against *M. tuberculosis* and is less likely to allow the bacterium to develop drug resistance.

4.6 WHAT PEOPLE CAN DO TO REDUCE THEIR RISK OF CONTRACTING TUBERCULOSIS

M. tuberculosis is an infectious disease of humans; that is, humans are the reservoir host for this bacterium. People traveling to countries in Asia, Africa, and other areas where it is prevalent are at higher risk of infection. The risk of mortality increases for patients who do not seek or obtain medical treatment, are infected with a strain of *M. tuberculosis* that is resistant to antibodies, or have a weakened immune system. Efforts to eradicate *M. tuberculosis* from large areas have been unsuccessful, but its prevalence can be reduced in countries that have public health programs that can identify and treat TB patients (WHO 2006).

Programs to eradicate *M. bovis* have been successful in some countries. These programs involve the regular testing of cattle herds and surveillance for the disease at slaughterhouses. Cattle that test positive for *M. bovis* are killed, and infected herds are isolated. Eradication efforts in some countries are hindered because *M. bovis* is endemic in a local wildlife species. In New Zealand, brush-tailed possums serve as both vector and reservoir host for the disease, and the country has developed a program to eradicate brush-tailed possums where possible and to monitor possums for *M. bovis* elsewhere (Sidebar 4.3).

Many African countries lack the resources to fund programs to eliminate bovine TB in livestock and wildlife. Hence, the emphasis in Africa is on the prevention of *M. bovis* being transmitted from animals to humans. This includes the pasteurization of milk, which has been accomplished in most urban areas but not in rural areas.

In Ireland, European badgers both serve as reservoir host of *M. bovis* and transmit the bacteria to cattle (Figure 4.8). In Ireland, the prevalence of TB is only 0.4% of cattle versus 45% among European badgers within *M. bovis* hotspots. Hence, badger-to-cattle transmission is of greater importance than cattle-to-cattle transmission. In response, Ireland started a program to kill badgers, hoping that a reduction in badger densities would help control the disease in livestock. In England, badgers have been killed for decades in an effort to limit the spread of *M. bovis*. Despite this, TB in cattle has increased in both frequency and geographic range in Great Britain (Krebs et al. 1997). Not surprisingly, the program has become very controversial with some people wanting badgers to be saved and others wanting their eradication (Ireland Department of Agriculture and Food 2006, Cowell 2013). Contributing to this controversy are three research studies that reached different conclusions—one showed that badger culling (reducing an animal's population through selective slaughter) reduces the incidence of TB in cattle; one reported that badger culling increases bovine TB; one found that badger culling decreased TB in cattle within the treatment area but increased in adjacent areas. Donnelly et al. (2006) speculated that an increase in contact between cattle

SIDEBAR 4.3 NEW ZEALAND'S EFFORTS TO ERADICATE BOVINE TB (TWEDDLE AND LIVINGSTONE 1994)

In New Zealand, there are about 8.4 million cattle and 1.2 million captive deer that are raised in farm operations. Most (85%) captive deer are red deer that originated in Europe and the rest are fallow deer. Bovine TB probably arrived in New Zealand more than a century ago. A mandatory program of testing all dairy herds began during 1945; mandatory testing of beef herds began in 1968. New Zealand launched a campaign to eradicate *M. bovis* from the country in 1970, but the campaign was set back during 1978 when bovine TB was diagnosed in captive deer. Mandatory testing of captive deer began in 1989 with deer farmers paying for the testing.

New Zealand's compulsory test and slaughter program quickly restricted bovine TB in dairy cows to a few parts of the country but failed to resolve the problem in beef cattle. In some areas, cattle herds became reinfected years after the last confirmed case. Research confirmed that an introduced marsupial, the brush-tailed possum, was a reservoir host for *M. bovis* and spread the bacteria to both cattle and deer (Figure 4.7). In response, lethal control of possums was initiated during 1979. This program was so successful in reducing the incidence rate of TB in cattle that government funding for the possum control program was substantially reduced. Not surprisingly, the possum population rebounded, as did the number of cattle herds that tested positive for *M. bovis*.

FIGURE 4.7 A brush-tailed possum. (Courtesy of Weihong Ji and *Human–Wildlife Interactions*.)

FIGURE 4.8 A European badger. (Courtesy of British Wildlife Centre, Lingfield, Surrey, U.K.)

and infected badgers in areas adjacent to culling sites was the culprit because badgers travel more and have larger home ranges when their densities have been artificially reduced through culling. At present, an oral vaccine is unavailable to prevent bovine TB in wildlife, but scientists are trying to develop one that would reduce the disease in European badgers (Donnelly et al. 2006).

In North America, the primary reservoir host for *M. bovis* is the white-tailed deer and elk (Campbell and VerCauteren 2011). Whenever *M. bovis* is detected in domestic livestock or farm-raised wildlife, it is important to ascertain the pathogen's occurrence in local wildlife populations. If the bacteria are present in wildlife, a TB management plan is usually developed through joint efforts of the state wildlife agency and the U.S. Animal and Plant Health Inspection Service (APHIS) to eliminate *M. bovis* from free-ranging wildlife and domestic livestock. At most Michigan farms where bovine TB was found in the livestock, *M. bovis* was also detected in local raccoons or opossums. Many management plans recommend that resident raccoons and opossums be trapped and euthanized after the detection of *M. bovis* at a farm. Trapping should continue until no raccoons or opossums are captured for five consecutive nights. Trapping should then continue at regular intervals for at least 1 year after the last detection of *M. bovis* in either free-ranging wildlife or livestock (USDA APHIS 2011). Routine sampling of coyotes from a broad area is recommended to determine the extent of bovine TB across the landscape.

Bovine TB has not been detected among feral hogs in the continental United States but has elsewhere. The USDA APHIS advises that when *M. bovis* has been detected on a farm, all feral hogs living within 10 miles (16 km) be euthanized (USDA APHIS 2011).

M. bovis poses a risk to the health and safety of dairy farmers, wildlife biologists, and veterinarians who work with live animals or carcasses that might be infected. Detailed recommendations for personal safety guidelines and personal protective equipment are available at USDA APHIS (2011).

4.7 ERADICATING TUBERCULOSIS FROM A COUNTRY

M. bovis has proven difficult to eradicate in areas where the pathogen is endemic in local wildlife populations. Humans, and not animals, are the reservoir host for *M. tuberculosis*, and millions of people around the world are infected with the pathogen. Some strains of the pathogen are resistant to many antibiotics. Hence, eradicating *M. tuberculosis* from a country will be difficult even though animals do not serve as reservoir hosts.

LITERATURE CITED

Brook, R. K. 2009. Historical review of elk-agriculture conflicts in and around Riding Mountain National Park, Manitoba, Canada. *Human–Wildlife Conflicts* 3:72–87.

Campbell, T. A. and K. C. VerCauteren. 2011. Diseases and parasites. In: D. G. Hewitt, editor. *Biology and Management of White-Tailed Deer*. CRC Press, Boca Raton, FL, pp. 219–249.

Cavanagh, R., M. Begon, M. Bennett, T. Ergon et al. 2002. *Mycobacterium microti* infection (vole tuberculosis) in wild rodent populations. *Journal of Clinical Microbiology* 40:3281–3285.

CDC. 2012. Trends in tuberculosis—United States, 2011. *Morbidity and Mortality Weekly Report* 61:181–185.

CDC. 2013a. Summary of notifiable diseases—United States, 2011. *Morbidity and Mortality Weekly Report* 60(53):1–117.

CDC. 2013b. Trends in tuberculosis—United States, 2012. *Morbidity and Mortality Weekly* 62:201–205.

Cowell, A. 2013. First shots are fired in Britain after debate on a badger cull. *New York Times* (August 28, 2013).

de Jong, B. C., M. Antonio, and S. Gagneux. 2010. *Mycobacterium africanum*—Review of an important cause of human tuberculosis in West Africa. *PLoS Neglected Tropical Diseases* 4(9):1–17.

Donnelly, C. A., R. Woodroffe, D. R. Cox, F. J. Bourne et al. 2006. Positive and negative effects of widespread badger culling on tuberculosis in cattle. *Nature* 439:843–846.

Hlavsa, M. C., P. K. Moonan, L. S. Cowan, T. R. Navin et al. 2008. Human tuberculosis due to *Mycobacterium bovis* in the United States, 1995–2005. *Clinical Infectious Diseases* 47:168–175.

Ireland Department of Agriculture and Food. 2006. Disease eradication schemes—Bovine tuberculosis and brucellosis. http://www.agriculture.ie/index.jsp/fileanimal_health/TB.xml (accessed March 3, 2012).

Krebs, J. R., R. Anderson, T. Clutton-Brock, W. I. Morrison et al. 1997. *Bovine Tuberculosis in Cattle and Badgers*. Independent Scientific Review Group, London, U.K.

Kremer, K. and D. van Soolingen. 1998. *Mycobacterium microti:* More widespread than previously thought. *Journal of Clinical Microbiology* 36:2793–2794.

Mayo Clinic. 2011. Tuberculosis. http://www.mayoclinic.org/health/DiseasesIndex/ (accessed March 21, 2012).

McCarthy, O. R. 2001. The key to the sanatoria. *Journal of the Royal Society of Medicine* 94:413–417.

More, S. J. 2009. What is needed to eradicate bovine tuberculosis successfully: An Ireland perspective. *Veterinary Journal* 180:275–278.

Morens, D. M. 2002. At the deathbed of consumptive art. *Emerging Infectious Diseases* 8:1353–1358.

Radunz, B. 2006. Surveillance and risk management during the latter stages of eradication: Experiences from Australia. *Veterinary Microbiology* 112:283–290.

Thoen, C. O., W. J. Quinn, L. D. Miller, L. L. Stackhouse et al. 1992. *Mycobacterium bovis* infection in North American elk (*Cervus elaphus*). *Journal of Veterinarian Diagnostic Investigations* 4:423–427.

Tweddle, N. E. and P. Livingstone. 1994. Bovine tuberculosis control and eradication programs in Australia and New Zealand. *Veterinary Microbiology* 40:23–39.

USDA APHIS. 2011. *Guidelines for Surveillance of Bovine Tuberculosis in Wildlife*. USDA Animal Plant Health Inspection Service, Washington, DC.

Wahlström, H. and L. Englund. 2006. Adopting control principles in a novel setting. *Veterinary Microbiology* 112:265–271.

WHO. 2006. *Brucellosis in humans and animals*. World Health Organization 2006(7):1–90.

WHO. 2010. *World Health Statistics: 2010*. World Health Organization, Geneva, Switzerland.

WHO. 2011. *Global Tuberculosis Control 2011*. World Health Organization, Geneva, Switzerland.

5 Tularemia

Francisella tularensis subspecies *tularensis* occurs in the USA. It is one of the most infectious pathogens known in human medicine.

An increase in prevalence of type B tularemia has been observed during times of war. During the Second World War, at least 100,000 cases occurred each year.

Anda et al. (2007a)

5.1 INTRODUCTION AND HISTORY

After a massive earthquake hit San Francisco in 1906, a new bacterial species was found in ground squirrels. The bacterium was named *Bacterium tularense* after the California county where the discovery was made—Tulare County. Tularemia was first described as a human disease during 1911 in Utah, where it was called deerfly fever. In 1919, Edward Francis linked the two by isolating *B. tularense* from the blood of patients with deerfly fever. Hence, the disease was named tularemia after the pathogen, and the bacterium itself was renamed *Francisella tularensis* in honor of Edward Francis. Other names for tularemia include rabbit fever, Ohara fever, and Francis fever.

F. tularensis are small (0.2–0.7 μm), rod-shaped bacteria that are aerobic and Gram negative (Figure 5.1). They have been isolated from numerous birds and mammals, arthropods, and biting insects, as well as from mud and water. *F. tularensis* is one of two species that make up the Francisellaceae family of bacteria; the other species is *F. philomiragia*, which is found in salt water of the Atlantic Ocean and Mediterranean Sea (Tärnvik and Berglund 2003, Broman et al. 2007).

There are four subspecies of *F. tularensis*: (1) *F. tularensis tularensis* causes type A tularemia. This pathogen is found throughout North America (Canada, United States, and Mexico) and is one of the most infectious human pathogens known in the world; a person can become infected when inhaling as few as 25 bacterial cells. Because of its infectious nature, there is concern that the bacteria could be used for biowarfare or bioterrorism. (2) *F. tularensis holarctica* causes type B tularemia and is less infectious than *F. tularensis tularensis*. It is found throughout the Northern Hemisphere including North America, but most tularemia patients in North America have type A tularemia. (3) *F. tularensis novicida* was first isolated from water samples in Utah. It has low virulence but can infect people who have impaired immune systems. It is transmitted through water and has been detected in the United States, Canada, Australia, and Spain. Until recently, *F. tularensis novicida* was considered a separate species (*F. novicida*), but it is now considered a subspecies of *F. tularensis*. (4) *F. tularensis mediasiatica* has

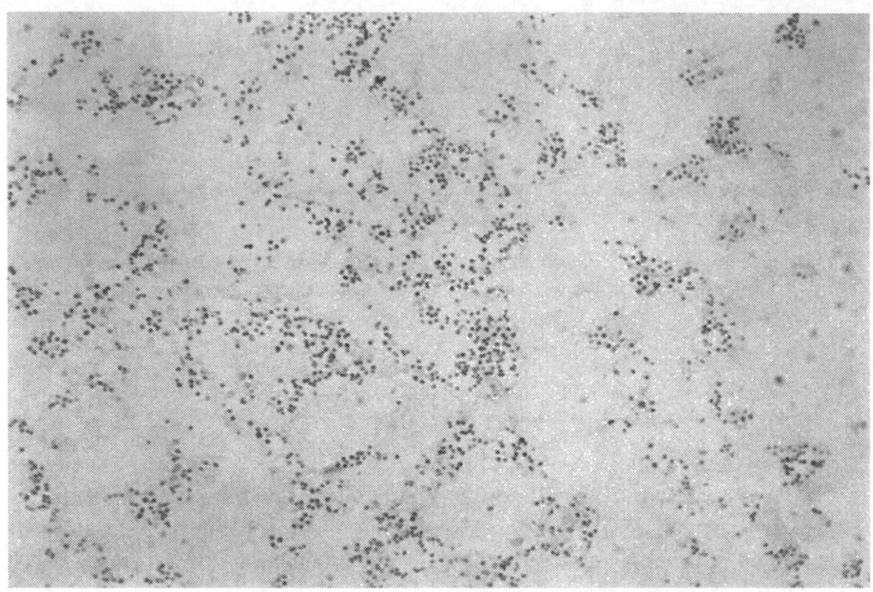

FIGURE 5.1 Photomicrograph of *F. tularensis*, the pathogen that causes tularemia. (From CDC, Tularemia, 2011, http://www.cdc.gov/tularemia/index.html, accessed November 1, 2012.)

only been detected in Kazakhstan and Turkmenistan and has virulence similar to *F. tularensis holarctica* (WHO 2007).

Incidence of tularemia varies considerably across years. In Sweden, the incidence rate ranges from just a few cases in most years to more than 2500 during 1967. During the winter of 1941, 67,000 people contracted tularemia around Rostov-on-Don near the Russian Sea of Azov—an area where tularemia was uncommon before then. An outbreak of tularemia in Spain occurred during 1996 when 585 people became infected. Before then, tularemia was unknown in Spain (Eliasson et al. 2002, Tärnvik and Berglund 2003, Broman et al. 2007).

In the United States, tularemia has been reported from all states except Hawaii, but it is more prevalent in Arkansas, Oklahoma, Missouri, and South Dakota (Figure 5.2). Historically, tularemia was a major health problem in the United States, with 14,000 cases from 1920 to 1945. From 2001 to 2011, there were between 90 and 154 reported cases annually in the United States. A decline in the incidence of tularemia from the early half of the twentieth century (Figure 5.3) may be due to a decrease in the number of people who hunt or consume rabbits, snowshoe hares, and jackrabbits (CDC 2011, 2013).

Tularemia occurs in many countries in the Northern Hemisphere (Figure 5.4). In Eurasia, the incidence of tularemia is higher in Finland, Sweden, and Russia than in central and southern countries; Great Britain may be free of tularemia. In Japan, the incidence of tularemia peaked during the 1950s and has since declined to less than a dozen cases annually. Human cases have not yet been reported in the Southern Hemisphere (Tärnvik and Berglund 2003).

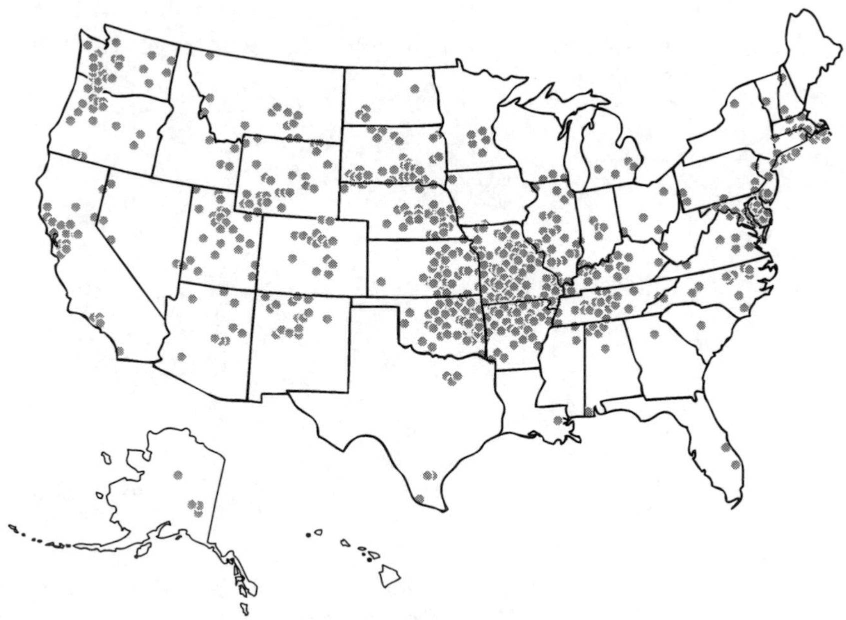

FIGURE 5.2 Distribution of reported cases of tularemia in the United States. (From CDC, Tularemia, 2011, http://www.cdc.gov/tularemia/index.html, accessed November 1, 2012.)

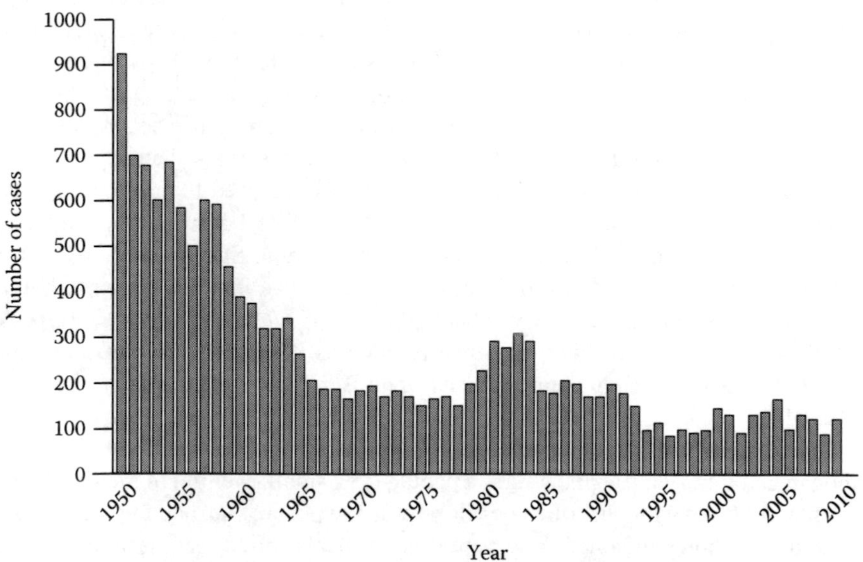

FIGURE 5.3 Annual changes in the number of reported cases of tularemia in the United States. (From CDC, Tularemia, 2011, http://www.cdc.gov/tularemia/index.html, accessed November 1, 2012.)

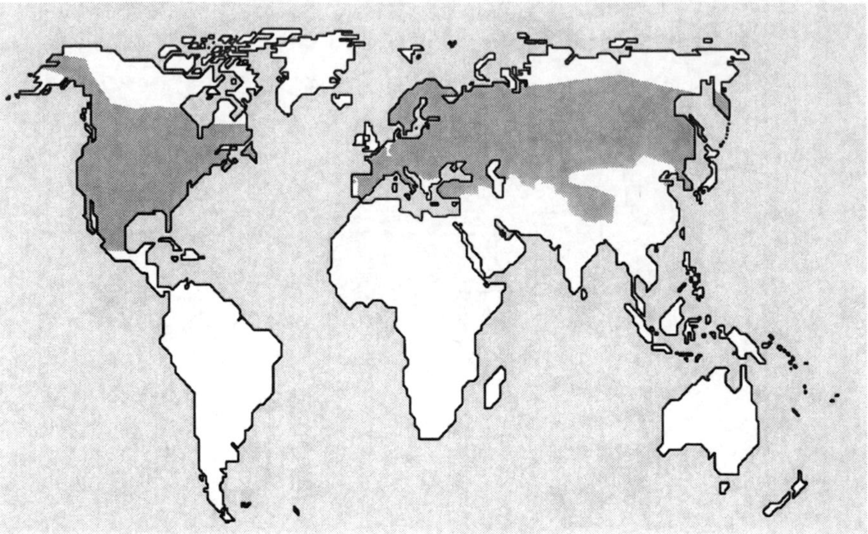

FIGURE 5.4 Regions of the world where tularemia has been reported are shaded on the map. (Data from WHO, *WHO Guidelines on Tularaemia*, World Health Organization, Geneva, Switzerland, 2007.)

5.2 SYMPTOMS IN HUMANS

The early symptoms of tularemia in humans are similar to other diseases, making it difficult to diagnose. The first symptoms usually occur 3–5 days after infection but can occur as early as 1 day or as late as 21 days after infection. Onset of disease is rapid, with symptoms including high fever, chills, loss of appetite, fatigue, headache, and muscle aches. A dry cough, sore throat, and chest pains may also occur. Infections by *F. tularensis tularensis* (type A tularemia) are more serious than those caused by *F. tularensis holarctica* (type B tularemia). Before the use of antibiotics, human fatality rates for type A tularemia ranged from 5% to 15% but have dropped to 2% with antibiotic treatment. Type B tularemia is rarely fatal. Survival in humans and animals depends on the individual developing an effective immune response before the pathogen numbers in the body reach lethal levels. *F. tularensis tularensis* (type A) multiplies more rapidly within a host than *F. tularensis holarctica* (type B), and this accounts for its more lethal nature (Tärnvik and Berglund 2003, Anda et al. 2007a, CDC 2011).

Clinical signs vary depending on how *F. tularensis* invades the body. When the pathogen gains entry through the skin, a papule (i.e., small swelling of the skin that is solid and not filled with pus) often develops at the exposure site, and the area around the papule becomes inflamed. The papule soon heals, often leaving a small scar. Lymph nodes can become swollen and tender, and the skin above them may redden.

F. tularensis can gain entry through the eye when the eye comes into contact with contaminated dust, aerosols, or fingers. When this happens, conjunctivitis (i.e., inflammation of the eye's surface) can result. Other symptoms include swelling of the eyelid, excessive watering of the eye, mucous discharge from the eye, photophobia (i.e., sensitivity to bright lights), and swelling of lymph nodes in the neck.

F. tularensis infections can be acquired through drinking contaminated water or eating contaminated food. In these cases, tularemia patients may experience a sore throat accompanied by redness and papules. Lymph nodes in the neck become swollen. This type of tularemia can be mistaken for a streptococcal infection resulting in the patient receiving penicillin, which is not effective against *F. tularensis* (Anda et al. 2007a).

When *F. tularensis* is inhaled, symptoms include a high fever, pneumonia, coughing, chest pain, vomiting, and a fast rate of respiration. Pulmonary tularemia is a serious condition, with a fatality rate of 30%–60% when antibiotics are not used (Feldman et al. 2001, CDC 2011). Pulmonary tularemia is much more common on Martha's Vineyard, Massachusetts, than elsewhere in the United States for reasons not entirely clear (Sidebar 5.1).

Complications can arise with the disease, especially with type A tularemia. These include blood poisoning, liver failure, kidney failure, endocarditis, and meningitis. Any of these complications can result in death (Anda et al. 2007a).

5.3 *FRANCISELLA TULARENSIS* INFECTIONS IN ANIMALS

More than 250 species can become infected with *F. tularensis*, but none of these animals are reservoir hosts, either because the animals do not survive the infection or their immune systems eradicate the pathogen from their bodies. Animals can, however, serve as amplifying hosts and are often the source of the pathogen when humans contract tularemia. In North America, the main amplifying hosts are rodents, rabbits, hares, beavers, and muskrats. Unfortunately, despite decades of research, uncertainty exists regarding how *F. tularensis* is able to maintain itself in North America (Tärnvik and Berglund 2003, Telford and Goethert 2011).

F. tularensis holarctica is associated with surface water (e.g., ponds and streams) that has been contaminated by infected beavers or muskrats. In Russia, the Eurasian water vole, common vole, red-backed vole, and house mouse are amplifying hosts and are the main sources of *F. tularensis holarctica* when humans develop type B tularemia. However, none of these animals are reservoir hosts, leaving uncertainty as to how *F. tularensis holarctica* is able to maintain itself in Eurasia. It is possible that an aquatic invertebrate may serve as the reservoir host (Tärnvik and Berglund 2003, Sjöstedt 2007, Telford and Goethert 2011).

F. tularensis tularensis causes illness in muskrats, house mice, lemmings, beavers, eastern cottontail rabbits, hares, black-tailed jackrabbits, white-tailed jackrabbits, squirrels, prairie dogs, and shrews. When rabbits and mice inhale this pathogen, lesions develop in the bronchi, lung, spleen, liver, and lymph nodes. Infected prairie dogs become dehydrated due to diarrhea, lethargic, and their lymph nodes swell (La Regina et al. 1986, Petersen et al. 2004, Sjöstedt 2007).

Some rodent populations are cyclic in temperate regions, undergoing large swings in population. In some species, outbreaks of *F. tularensis* play a role in causing a population crash. Evidence for this came from an 8,500 mile^2 (22,000 km^2) region of Saskatchewan, Canada, where numbers of deer mice soared in 2005. During the

SIDEBAR 5.1 TULAREMIA ON MARTHA'S VINEYARD (MATYAS ET AL. 2007)

Martha's Vineyard is a small island (100 square miles or 26,000 hectares) that is located south of Cape Cod and is part of Massachusetts (Figure 5.5). With 15,000–20,000 year-round residents, its population swells to 100,000 during the summer tourist season. There have been only two outbreaks of pneumonic tularemia in the United States, and both have occurred on this small island. Why both happened on the same island has created a medical mystery.

Pneumonic tularemia occurs when the pathogen *F. tularensis* is inhaled rather than acquired by other means. The first outbreak of pneumonic tularemia occurred during 1978 when there were 15 cases of tularemia, 12 of the pneumonic form, and all patients recovered. One of the pneumonic patients was a sheep shearer, two were gardeners, and six had spent a week in August at the same cottage. All recovered, and an investigation was unable to identify a single source of the outbreak other than all patients who had spent time at Martha's Vineyard. A second outbreak started in 2000 when there were 15 confirmed cases of tularemia on the island, and 11 of the patients had the pneumonic form. Unfortunately, one of the patients, a 43-year-old male, succumbed. Half of the patients were landscapers, and 80% reported using a powered brush cutter or lawnmower during the 2 weeks prior to the onset of illness. From 2000 to 2006, there were 59 cases of tularemia on Martha's Vineyard, with more than 60% being pneumonic tularemia.

Several studies have tried to identify why the pneumonic form of tularemia is common on Martha's Vineyard and rare elsewhere in the United States. Rabbits, beavers, muskrats, white-tailed deer, voles, skunks, raccoons, and rodents are common on the island, but these species are also common throughout most of northeastern states and Canadian provinces. Dog ticks and blacklegged ticks occur on Martha's Vineyard and are vectors for *F. tularensis*; the same arthropods, however, occur throughout the eastern United States and Canada. Despite much research, it remains an enigma why pneumonic tularemia is common on Martha's Vineyard but rare elsewhere.

subsequent population crash, several dead mice were collected and tested positive for *F. tularensis*. Similar results were found following population crashes of voles in California and Oregon (Wobeser et al. 2007).

Livestock can become ill when infected with *F. tularensis*. Horses, pigs, and sheep are especially susceptible (up to 15% of infected lambs may die); in contrast, cattle appear resistant. Dogs and cats can become ill after biting or consuming an infected animal. In all of these domesticated animals, symptoms include a high fever, lethargy, reduced mobility, muscle stiffness, coughing, diarrhea, and frequent urination. The liver, spleen, and lymph nodes often become enlarged and develop small lesions. In some animals, death may occur within a few hours or days after infection while other animals are asymptomatic. *F. philomiragia* rarely causes

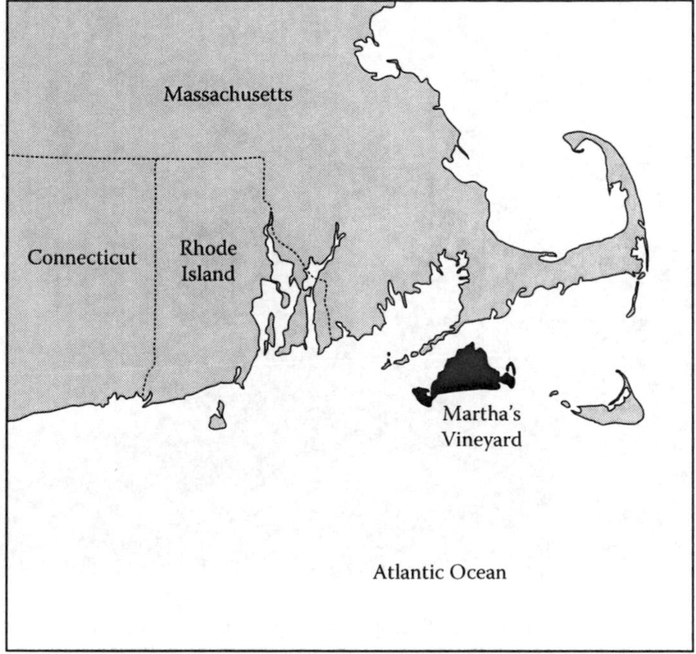

FIGURE 5.5 The island of Martha's Vineyard is blackened in this map.

human illness but some people have become infected after inhaling seawater when they nearly drowned (Anda et al. 2007a, Merck 2010).

5.4 HOW HUMANS CONTRACT TULAREMIA

There are several ways that humans can become infected with *F. tularensis*, including through skin contact with an infected animal or consumption of one. The pathogen can be transmitted through the skin in the absence of visible skin lesions or abrasions, but cuts greatly increase the risk of becoming infected (Figure 5.6). In the United States and Canada, people, who trap beavers and muskrats or hunt and eat rabbits, have an elevated risk of contracting tularemia. In Oklahoma, the annual variation in the incidence of tularemia correlates closely with the number of rabbits harvested annually by hunters. Wild gallinaceous birds, especially grouse, are a source of human infection. Other wildlife species, which are possible sources of human tularemia in Canada, include ground squirrels and coyotes (Martin et al. 1982, Rohrbach et al. 1991, Hopla and Hopla 1994).

Biting arthropods, such as the American dog tick, Rocky Mountain wood tick, and lone star tick, can spread *F. tularensis* to humans and animals. The rabbit tick also transmits *F. tularensis* among North American animals, but this tick rarely bites people and is not known to transmit the pathogen to humans. The rabbit tick's main hosts are rabbits, hares, rodents, and grouse, and it plays a major role in spreading the pathogen among them. The rabbit tick also feeds on migratory birds,

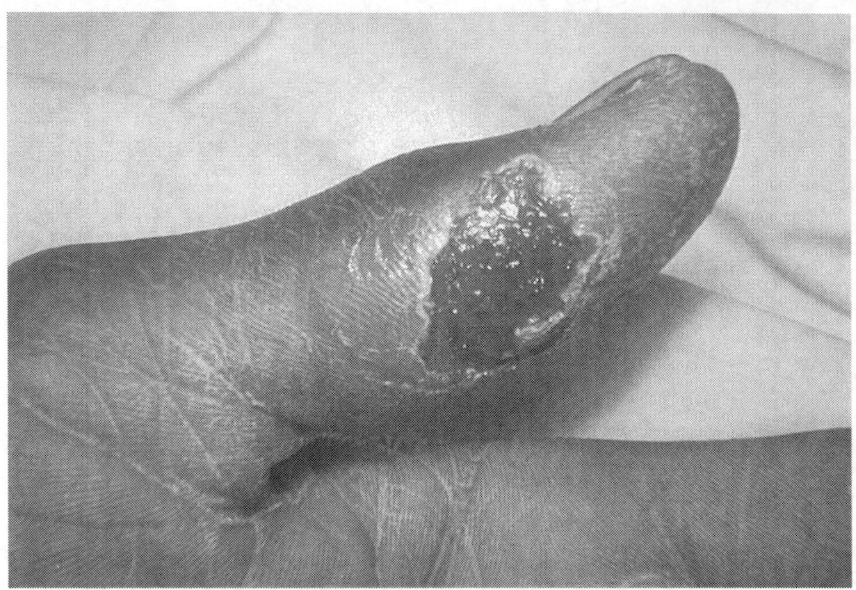

FIGURE 5.6 An eschar on a person's finger at the site where the person was infected by *F. tularensis*. (Courtesy of the CDC, Emory University, and Dr. Sellers.)

which can carry the infected ticks over large areas and spread *F. tularensis* to new areas (Telford and Goethert 2011). In Utah and other Rocky Mountain states, deerflies and other biting flies are important vectors in spreading the pathogen among animals and to humans. In Sweden, mosquitoes are the main vector (Eliasson et al. 2002); in Russia, both mosquitoes and ticks serve as vectors (Sjöstedt 2007).

F. tularensis does not form spores or encapsulate, but the pathogen can still survive up to 12 weeks in stored feed, 14 weeks in water or mud, and 26 weeks in straw. Ponds, streams, and other water sources can be contaminated when infected beavers or muskrats inhabit the water body. Voles and rodents also shed the pathogen in their feces, and people can acquire tularemia when they consume or touch contaminated items or inhale the pathogen. For example, field voles live in European hay fields and haystacks; farmers contract tularemia by inhaling dust from contaminated hay (Parker et al. 1951, Feldman et al. 2001).

Human cases of tularemia often occur during peak populations of small mammals. In Russia, outbreaks of tularemia in humans correspond with outbreaks of *F. tularensis* among Eurasian water voles (Mörner 1992) or with high densities of water rats. Eurasian water vole populations are cyclic with peaks occurring every 5–7 years. Incidence of tularemia in humans often peak soon after local populations of Eurasian water voles reached their zenith (Efimov et al. 2003). In Sweden, the incidence of tularemia corresponded with vole and hare populations during the 1950s and 1960s, but this pattern has not been apparent in recent years (Sjöstedt 2007). In North America, the seasonal pattern of tularemia is bimodal. One peak occurs in late spring and summer when people are likely to be bitten by ticks, and a second peak occurs during the fall hunting season for rabbits (Figure 5.7).

FIGURE 5.7 Photo of a jackrabbit, which is a host for *F. tularensis* in the United States and Canada. (Courtesy of Fred Knowlton.)

Worldwide, the incidence of type B tularemia is much more common than type A. Type B increases during wars and natural disasters, most notably during World War II and during the War of Kosovo (Sidebar 5.2). Transmission of *F. tularensis* from one human to another is not known to occur, but people can contract tularemia from their pets, either by petting them or being bitten. Domestic cats acquire the pathogen from mouthing or eating an infected rodent (Figure 5.9).

5.5 MEDICAL TREATMENT

People who suspect that they have tularemia should seek medical attention immediately because the disease can be serious or fatal if not treated with antibiotics. Patients with tularemia are usually treated with streptomycin, but gentamicin is also an acceptable treatment. Doses should continue for at least 10 days. Most patients make a complete recovery, though the illness can last for several weeks (Anda et al. 2007b, CDC 2011).

5.6 WHAT PEOPLE CAN DO TO REDUCE THEIR RISK OF CONTRACTING TULAREMIA

People who have had tularemia develop a long-lasting immunity to the disease; very few people have tularemia more than once during their lives. This suggests that a vaccine against tularemia should be effective. Gaiskii and El'bert in Russia were the first to develop an effective vaccine against tularemia using a live but attenuated strain of

SIDEBAR 5.2 TULAREMIA OUTBREAK IN A WAR-TORN COUNTRY (REINTJES ET AL. 2002)

Outbreaks of tularemia often occur during wars, including the War in Kosovo. This small country in eastern Europe has a population of about two million people (Figure 5.8). During the 1990s, civil war resulted in the displacement of a large number of people, economic collapse, and the disruption of public sanitation and hygiene. Prior to the war, there had been no known cases of tularemia in the country, but soon afterward, a public health physician reported a cluster of patients with tularemia. By late June 2000, almost a thousand suspected cases of tularemia were identified, and more than three hundred cases were confirmed as tularemia. Many of the patients were ethnic Albanians who had to flee their rural farming villages during the war and became refugees. They reported that when they returned home after the war, they found their village overrun by striped field mice and black rats.

To determine if there was a linkage between rodents and human cases of tularemia, medical investigators compared patient households with a randomly selected group of households. The analyses demonstrated that patient households were more likely to have rodent feces in food preparation and storage areas and to have water sources (mostly crude, open wells) that rodents could easily access. *F. tularensis* was also isolated from the feces of both black rats and striped field mice. Investigators concluded that the population explosion of rodents and the lack of safe drinking water and food supplies were responsible for the outbreak of tularemia in Kosovo.

F. tularensis. Russia implemented a national vaccination program starting in 1946, which was successful in reducing the incidence of tularemia. The United States used the Russian strain to develop its own live-strain vaccine, which reduced the incidence of respiratory tularemia but not tularemia transmitted by arthropods. Unfortunately, the vaccine did not produce consistent immunity, and the vaccine is no longer available in Western countries (Tärnvik and Berglund 2003).

To decrease the chance of contracting tularemia, people should avoid touching a sick or dead animal with their bare hands and instead should wear plastic or rubber gloves and a face mask, which should be thrown away after use. Grouse and rabbit hunters and fur trappers should wear rubber or plastic gloves when cleaning, dressing, or skinning game; knives and other equipment used in the process should be disinfected after use. Hunters should not shoot animals that appear ill. Game meat should be thoroughly cooked. When outdoors, people should not drink untreated surface water regardless of how clean the water appears.

Pets, especially rodents, rabbits, hamsters, or hares, should be taken to a veterinarian if they exhibit signs indicative of an *F. tularensis* infection. Tularemia outbreaks among people have occurred when pet stores sold infected rabbits or hamsters (Sidebar 5.3). Preventative steps to reduce the incidence of infection among livestock are limited to reducing tick populations.

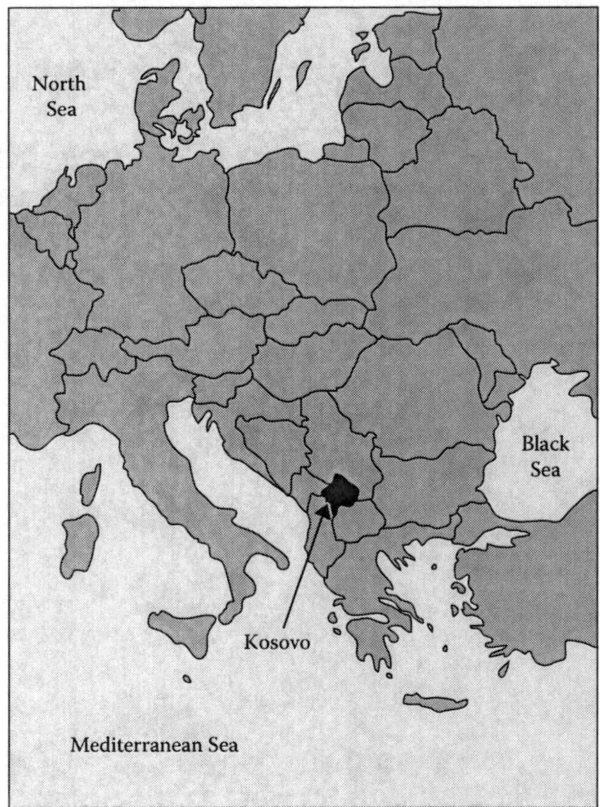

FIGURE 5.8 Map showing the location of Kosovo in Europe.

Dead animals or rabbit nests should not be run over with lawn mowers or chopped up with weed whackers. In Tennessee, two boys contracted tularemia after doing just that. In Martha's Vineyard, Massachusetts, professional landscapers who used power blowers or powered weed whackers were more likely to be seropositive for *F. tularensis* than landscapers who used hand tools. The power equipment caused an increase in the amount of contaminated dust in the air, therefore increasing the risk of pneumonic tularemia (McCarthy and Murphy 1990, Feldman et al. 2003).

Additional precautions are warranted during a local outbreak of tularemia. Trapping, hunting, or consumption of beaver, muskrats, rabbits, jackrabbits, and hares should be curtailed. Domestic cats should be kept indoors to minimize their risk of consuming an infected mammal. Protective gloves should be worn when in contact with wild animals, and respiratory masks should be worn during activities that stir up dust in areas frequented by rodents. The risk of being bitten by a tick or deerfly can be reduced by wearing long-sleeved shirts and long pants. Insect repellents containing DEET should also be used. More information on how to avoid tick bites is provided in Chapter 13 on Lyme disease.

FIGURE 5.9 Cats can become infected with *F. tularensis* after consuming infected animals. (Courtesy of the Vertebrate Pest Conference.)

SIDEBAR 5.3 ARE PRAIRIE DOGS LOVABLE PETS OR TROJAN HORSES? (AVASHIA ET AL. 2004, AZAD 2004)

Many people enjoy having unusual pets, and this has created a thriving demand for pet prairie dogs in the United States and abroad. Thousands of wild prairie dogs have been captured to meet this demand each year (Figure 5.10). Some of these animals may be infected with zoonotic diseases and pose a risk of spreading the disease to humans.

In 2002, there was a die-off of captive prairie dogs at a commercial distributor of unusual pets. Investigators soon determined that most of the dead prairie dogs tested positive for *F. tularensis* (type B). They also discovered that during the prior 6 months, there had been a 10-fold increase in prairie dog deaths at the facility, and 3600 prairie dogs had been sold to people in several states and countries. One of the animal handlers at the facility was diagnosed as having had tularemia. These results indicate that it may be safer not to select a prairie dog for a pet.

5.7 ERADICATING TULAREMIA FROM A COUNTRY

It is difficult to eradicate tularemia from an endemic area because many avian and mammalian species can be infected. Migratory birds carry ticks infected with *F. tularensis* and can reintroduce the pathogen into areas where it has been eradicated. Efforts toward eradication are hampered because of a lack of knowledge regarding

FIGURE 5.10 Photo of a prairie dog. (Courtesy of Claire Dobert and the USFWS.)

how *F. tularensis* is able to maintain itself in the environment. Focus should instead be made on trying to prevent the spread of type A tularemia or type B tularemia into new areas.

LITERATURE CITED

Anda, P., A. Pearson, and A. Tärnvik. 2007a. Chapter 4. Clinical expression in humans. In: A. Tärnvik, editor. *WHO Guidelines on Tularaemia*. World Health Organization, Geneva, Switzerland, pp. 11–20.

Anda, P., A. Pearson, and A. Tärnvik. 2007b. Chapter 5. Treatment. In: A. Tärnvik, editor. *WHO Guidelines on Tularaemia*. World Health Organization, Geneva, Switzerland, pp. 21–26.

Avashia, S. B., J. M. Petersen, C. M. Lindley, M. E. Schriefer et al. 2004. First reported prairie dog-to-human tularemia transmission, Texas, 2002. *Emerging Infectious Diseases* 10:483–486.

Azad, A. F. 2004. Prairie dog: Cuddly pet or Trojan horse? *Emerging Infectious Diseases* 10:542–543.

Broman, T., M. Forsman, J. Petersen, and A. Sjöstedt. 2007. Chapter 2. The infectious agent. In: A. Tärnvik, editor. *WHO Guidelines on Tularaemia.* World Health Organization, Geneva, Switzerland, pp. 3–4.

CDC. 2011. Tularemia. http://www.cdc.gov/tularemia/index.html (accessed November 1, 2012).

CDC. 2013. Summary of notifiable diseases—United States, 2011. *Morbidity and Mortality Weekly Report* 60(30):1–118.

Efimov, V. M., Y. K. Galaktionov, and T. A. Galaktionova. 2003. Reconstruction and prognosis of water vole population dynamics on the basis of tularemia morbidity among Novosibirsk oblast residents. *Doklady Biological Sciences* 388:59–61.

Eliasson, H., J. Lindbäck, J. P. Nuorti, M. Arneborn et al. 2002. The 2000 tularemia outbreak: A case-control study of risk factors in disease-endemic and emergent areas, Sweden. *Emerging Infectious Diseases* 8:956–960.

Feldman, K. A., R. E. Enscore, S. L. Lathrop, B. T. Matyas et al. 2001. An outbreak of primary pneumonic tularemia on Martha's Vineyard. *New England Journal of Medicine* 345:1601–1606.

Feldman, K. A., D. Stiles-Enos, K. Julian, B. T. Matyas et al. 2003. Tularemia on Martha's Vineyard: Seroprevalence and occupational risk. *Emerging Infectious Diseases* 9:350–354.

Hopla, C. E. and A. K. Hopla. 1994. Tularemia. In: G. W. Beran, editor. *Handbook of Zoonoses*, 2nd edition. CRC Press, Boca Raton, FL, pp. 113–126.

La Regina, M., J. Lonigro, and M. Wallace. 1986. *Francisella tularensis* infection in captive, wild caught prairie dogs. *Laboratory Animal Science* 36:178–180.

Martin, T., I. H. Holmes, G. A. Wobeser, R. F. Anthony, and I. Greefkes. 1982. Tularemia in Canada with a focus on Saskatchewan. *Canadian Medical Association Journal* 127:279–282.

Matyas, B. T., H. S. Nieder, and S. R. Telford III. 2007. Pneumonic tularemia on Martha's Vineyard: Clinical, epidemiologic, and ecological characteristics. *Annals of the New York Academy of Science* 1105:351–377.

McCarthy, V. P. and M. D. Murphy. 1990. Lawnmower tularemia. *Pediatric Infectious Disease Journal* 9:298–300.

Merck. 2010. *Merck Veterinary Manual,* 10th edition. Merck, Whitehouse Station, NJ.

Mörner, T. 1992. The ecology of tularaemia. *Revue Scientifique et Technique* 11:1123–1130.

Parker, R. R., E. A. Steinhaus, G. M. Kohls, and W. L. Jellison. 1951. Contamination of natural waters and mud with *Pasteurella tularensis* and tularemia in beavers and muskrats in the northwestern United States. *National Institutes of Health Bulletin* 193:1–61.

Petersen, J. M., M. E. Schriefer, L. G. Carter, Y. Zhou et al. 2004. Laboratory analysis of tularemia in wild-trapped, commercially traded prairie dogs, Texas, 2002. *Emerging Infectious Diseases* 10:419–425.

Reintjes, R., I. Dedushaj, A. Gjini, T. R. Jorgensen et al. 2002. Tularemia outbreak investigation in Kosovo: Case control and environmental studies. *Emerging Infectious Diseases* 8:69–73.

Rohrbach, B. W., E. Westerman, and G. R. Istre. 1991. Epidemiology and clinical characteristics of tularemia in Oklahoma, 1979 to 1985. *Southern Medical Journal* 84:1091–1096.

Sjöstedt, A. 2007. Chapter 3: Epidemiology. In: A. Tärnvik, editor. *WHO Guidelines on Tularaemia.* World Health Organization, Geneva, Switzerland, pp. 5–10.

Tärnvik, A. and L. Berglund. 2003. Tularaemia. *European Respiratory Journal* 21:361–373.

Telford III, S. R. and H. K. Goethert. 2011. Toward an understanding of the perpetuation of the agent of tularemia. *Frontiers in Microbiology* 1:150.

WHO. 2007. *WHO Guidelines on Tularaemia.* World Health Organization, Geneva, Switzerland.

Wobeser, G., M. Ngeleka, G. Appleyard, L. Bryden, and M. R. Mulvey. 2007. Tularemia in deer mice (*Peromyscus maniculatus*) during a population irruption in Saskatchewan, Canada. *Journal of Wildlife Diseases* 43:23–31.

6 Leprosy

Command the children of Israel, that they put out of the camp every leper, and every one that hath an issue, and whosoever is defiled by the dead.

Numbers 5:2, King James Version of the Bible

And behold a leper came to him [Jesus] and knelt before him, saying 'Lord, if you will, you can make me clean.' And he stretched out his hand clear and touched him, saying 'I will; be clean.' And immediately his leprosy was cleansed.

Matthew 8:2–3, King James Version of the Bible

Heal the sick, cleanse the lepers, raise the dead, cast out devils; freely ye have received, freely give.

Matthew 10:8, King James Version of the Bible

6.1 INTRODUCTION AND HISTORY

Like TB, leprosy is a disease caused by *Mycobacterium*; this time, the infectious agent is *Mycobacterium leprae*, but recently, a second *Mycobacterium* was identified (*M. lepromatosis*) that also causes leprosy in humans. Both are rod-shaped bacteria and surrounded by a waxy membrane (Figure 6.1). Neither species can survive outside of their living hosts and cannot be cultured in the lab, making the study of them difficult.

Leprosy is an ancient disease of humans. The disease evolved either in South Asia or East Africa during the Pleistocene and from there spread throughout the world. Examination of human skeletons indicated that it had reached India by 2000 BC, Egypt and Italy by 300 BC, and Britain by 400 AD (Reader 1974, Mariotti et al. 2005, Robbins et al. 2009). It was a scourge of the ancient civilizations of China, India, Egypt, and Rome. Originally, lepers were forcibly segregated from the rest of the society and confined to leper colonies where conditions were often appalling. In the Old Testament of the Bible, lepers were viewed as unclean and were ostracized. In the New Testament, attitudes toward lepers changed, and some of the miracles of Jesus involved the healing of lepers (see previous texts). During the Roman Empire, leprosy was called elephantiasis and was described by Pliny the Elder. Although leprosy reached Europe by 300 BC, it did not become widespread until the development of large cities during the medieval period (Robbins et al. 2009).

Leprosy is also called Hansen's disease, named after Gerhard Hansen who first observed *M. leprae* in skin tissue taken from a leprosy patient during 1873. The first effective drug treatment for leprosy began during the 1930s with the discovery of

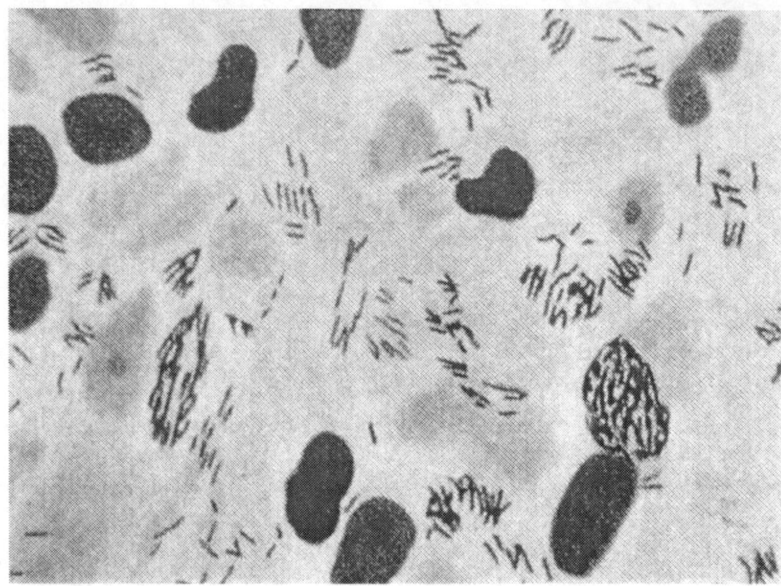

FIGURE 6.1 The small rods in this micrograph are *M. leprae*. (Courtesy of the CDC.)

the antibiotic, dapsone. Due to its overuse, strains of *M. leprae* evolved that were resistant to dapsone. To circumvent the problem of drug resistance, other drugs have been combined with dapsone since the 1980s; this multidrug treatment has proven effective against *M. leprae*.

Leprosy is endemic in 122 countries across the world, but since 2010, it is no longer considered a public health problem in 119 of them, which is defined as a prevalence rate (i.e., proportion of the human population that has a particular disease) of more than one case per 10,000 humans. Since 1985, 15 million patients have been cured of leprosy due to the development of an effective multidrug treatment; and the prevalence of leprosy has been reduced by more than 90% worldwide. At the beginning of 2010, there were 245,000 new cases of leprosy worldwide (WHO 2010). Most of these new cases were in Asia, Africa, and South America (Figure 6.2). In the United States, about 150 people contract leprosy annually.

6.2 SYMPTOMS IN HUMANS

Leprosy is a communicable disease that mainly affects the skin and nerves. Clinical signs include skin patches that can appear anywhere on the body. The patches may be pale, reddish, or copper. They are raised or flat with the skin surrounding them and do not quickly appear, disappear, or spread. They differ from other skin rashes in that they do not itch or hurt and lack sensation to heat, touch, or pain. This last characteristic of leprosy results from the bacteria destroying the underlying nerves. Leprosy can also cause pain and discharge from the eyes and a blurring of vision.

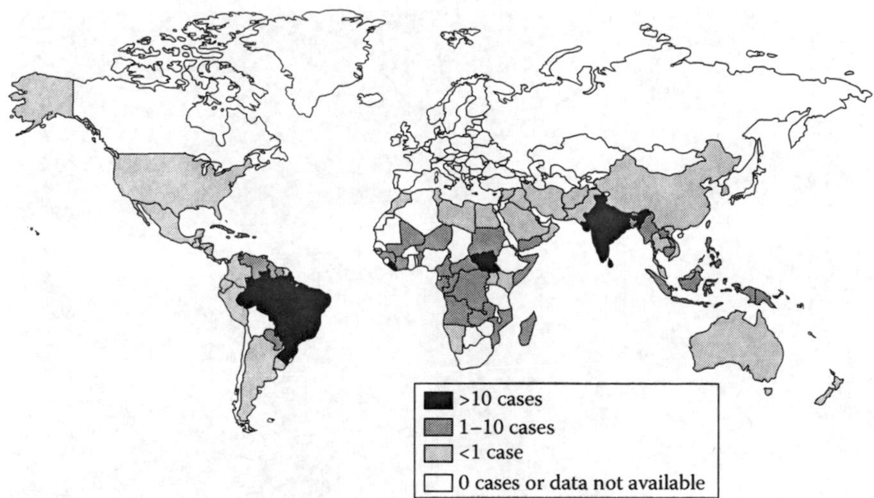

FIGURE 6.2 Incidence of leprosy (number of new cases per 100,000 people) in different countries as of January 2012 based on data reported to the WHO.

If left untreated, leprosy can develop into painful reddish nodules on the skin, swelling of the hands and feet, fever, muscle weakness, and nerve damage resulting in a loss of sensation. Loss of sensation of the hands and feet can result in patients injuring themselves without realizing it, and these repeated injuries and infections cause some patients to lose the use of their hands and feet (WHO 2000).

Similar to other *Mycobacterium*, *M. leprae* grow slowly. Hence like TB, leprosy has a long incubation period, averaging 3 years (WHO 2000), but the incubation period can extend beyond 30 years. *M. leprae* infects the cooler parts of the human body: testes, skin, upper respiratory tract, peripheral parts of the eye, and nerves near the skin, owing to *M. leprae* having an optimal temperature that is less than normal body temperature for humans (98.6°F or 37°C).

Most people (more than 90%) are resistant to *M. leprae* and are not at risk of acquiring leprosy. Physicians have observed that leprosy is often clustered in particular families suggesting a genetic basis for susceptibility. This was confirmed when genes responsible for resistance were located on the sixth chromosome (Buschman and Skamene 2004).

6.3 *MYCOBACTERIUM LEPRAE* INFECTIONS IN ANIMALS

Other than humans, nine-banded armadillos are the only known animals that serve as a reservoir host for *M. leprae* (Figure 6.3). These armadillos occur in the New World, and they contracted leprosy from infected humans sometime since European settlement. The current range of armadillos in the United States includes all of the southern states and extends northward to Colorado and North Carolina. Within the United States, the highest concentrations of armadillos occur in Texas and Louisiana. Armadillos are highly susceptible to *M. leprae*; perhaps one

FIGURE 6.3 A photo of a nine-banded armadillo. (Courtesy of the USFWS.)

reason for this is because armadillos have a low body temperature of 90°F–95°F (32°C–35°C), which is more optimal for growth of *M. leprae*. Armadillos infected with leprosy occur mostly west of the Mississippi River (Howerth et al. 1990). In eastern Texas, 4%–8% of armadillos have leprosy (Thoen and Williams 1994). Armadillos in the United States came from two sources. Some crossed the Rio Grande River into Texas during the 1820s; since then, this population has been expanding farther north and east. A second source originated from captive armadillos released in Florida during the 1920s; this population has been spreading north and west. The two populations met in eastern Mississippi or western Alabama during the 1980s, forming a continuous population across the south (Loughry et al. 2009). Only the western population was infected with *M. leprae*, which raises the question of whether the bacteria are spreading eastward into the uninfected eastern population of armadillos. There is some evidence that this might be occurring with the discovery of leprosy in armadillos from eastern sites previously considered disease-free (Loughry et al. 2009). *M. leprae* has been detected in armadillo populations in Mexico and Brazil but not in Columbia and Paraguay (Deps et al. 2007, Pedrini et al. 2010). Also, it has not been detected in other species of armadillos.

While humans are the only primate that serves as a reservoir host for *M. leprae*, other primates can become infected. *M. leprae* has been isolated from sooty mangabey monkeys, cynomolgus macaques, and common chimpanzees (Thoen and Williams 1994, Rojas-Espinosa and Lovik 2001). Scientists also have experimentally infected other primate species with the disease agent but *M. leprae* cannot maintain itself in any primate species except for humans.

6.4 HOW HUMANS CONTRACT LEPROSY

Leprosy is a contagious disease in that most patients contract *M. leprae* from another person. Yet, it is not very contagious; only about 5% of spouses that live with a patient contract the disease (Binford et al. 1982). Despite this, people with leprosy have been quarantined for a millennium (Sidebar 6.1). It is not clear how *M. leprae* is transmitted from person to person. Nasal mucus of patients contains large number of *M. leprae* and inhalation of nasal droplets containing the disease agent may be the primary way that the disease agent is passed from one person to another. Infected skin patches of patients contain large numbers of bacteria so the pathogen may spread through skin-to-skin contact. Insects may be able to transmit *M. leprae*, but this has not been proven.

Almost all leprosy patients contracted the pathogen from another person, but in North America, some people have contracted leprosy from armadillos (Truman et al. 2011). By comparing DNA from *M. leprae* isolated from infected patients in the southeastern United States to *M. leprae* isolated from local armadillos, scientists have confirmed that about a third of the U.S. patients contracted leprosy from armadillos (Harris 2011).

6.5 MEDICAL TREATMENT

Leprosy responds well to a multidrug treatment, which is available free from the WHO. The 6-month treatment involves monthly doses of rifampicin and daily doses of dapsone. For more severe cases (i.e., patients with more than five skin patches of

SIDEBAR 6.1 A HELL IN PARADISE (GRANGE 2012)

Within a few decades of Captain James Cook's discovery of the Hawaiian Islands in 1778, European sailors accidently brought *M. leprae* to the Islands. Leprosy soon spread to the native Hawaiians and caused such fear that the local government passed the "Act to Prevent the Spread of Leprosy." This law allowed the use of force to confine leprosy patients to a place of treatment or isolation. A year later, the first patients were taken by ship to a peninsula on the island of Molokai that was isolated from the rest of the island by steep cliffs that prevented anyone from escaping. The patients were told that they would never be able to leave their tropical prison.

A Catholic priest, Father Damien, arrived during 1873 to what he called "a living cemetery." Aided by other missionaries, he grew food for the patients and built homes, schools, and a church. At its peak, over 1000 patients were confined to the peninsula. Father Damien's greatest accomplishment was to provide dignity, love, and hope to the patients (Figure 6.4). Father Damien contracted leprosy but was not dismayed to share the same fate as his parishioners. He died in 1889 and was beatified by the Pope in 2009.

FIGURE 6.4 A photo of Father Damien and some of the leper patients he attended. (Courtesy of the Hawaiian State Archives.)

leprosy), treatment lasts a year and involves monthly doses of rifampicin and clofazimine and daily doses of clofazimine and dapsone (WHO 2000). After 1 month of treatment, patients are no longer in danger of spreading *M. leprae* to others.

6.6 WHAT PEOPLE CAN DO TO REDUCE THEIR RISK OF CONTRACTING LEPROSY

The incidence of leprosy in the southern United States could be reduced if the prevalence of *M. leprae* in free-ranging armadillos could be reduced, but it is unknown how to achieve this. There is no evidence that reducing armadillo populations would reduce the prevalence of *M. leprae* in armadillos. Moreover, it would be hard to reduce armadillo densities given their high productivity and large numbers in much of their range.

M. leprae are fragile bacteria and do not grow outside of its hosts: humans and armadillos; it can survive for only a few days in moist soil. Scientists have not been able to get it to grow in petri dishes; instead, they must use infected armadillos to study the disease. It is not known how the disease agent is transmitted from armadillos to humans, but *M. leprae*'s fragile nature suggests that transmission is through direct contact with an infected armadillo. Many people consider armadillos a pest and kill them; others find dead armadillos near their homes perhaps ones that have been killed by vehicle collisions. In both cases, many people pick up the carcass to dispose of it. Other people like the taste of armadillos and consume them.

People can reduce their risk of contracting leprosy from an armadillo by not touching one with bare hands or consuming one.

Most leprosy patients in the United States were infected with *M. leprae* from another person, often when living or traveling to countries where leprosy is more common. Leprosy responds well to drugs if treated early. One problem is that patients may ignore the symptoms. Sometimes, medical doctors may not consider a diagnosis of leprosy owing to its rarity. Hence, people, who handle or consume armadillos or who have lived in countries where leprosy is common, should provide this information to their medical providers if they develop symptoms similar to leprosy.

6.7 ERADICATING LEPROSY FROM A COUNTRY

The WHO has launched a program to eradicate leprosy from the world. This is feasible because people with leprosy can be cured with drugs, and only humans are infected by *M. leprae* in most parts of the world. The program involves trying to identify anyone with leprosy and treating them before they can spread the disease agent to others (WHO 2000). The eradication program is concentrated in those countries where leprosy is most common. Eradicating leprosy from the United States will be particularly difficult if not impossible because it is endemic in free-ranging armadillos.

LITERATURE CITED

Binford, C. H., W. M. Meyers, and G. P. Walsh. 1982. Leprosy. *Journal of the American Medical Society* 247:2283–2292.

Buschman, E., and E. Skamene. 2004. Linkage of leprosy susceptibility to Parkinson's disease genes. *International Journal of Leprosy* 72:169–170.

Deps, P. D., J. M. A. P. Antunes, and J. Tomimori-Yamashita. 2007. Detection of *Mycobacterium leprae* infection in wild nine-banded armadillos (*Dasypus novemcinctus*) using the rapid ML Flow test. *Revista da Sociedade Brasileira de Medicina Tropical* 40:86–87.

Grange, K. 2012. Exiled to paradise. *National Parks* 86(3):62–63.

Harris, G. 2011. Armadillos can transmit leprosy to humans, federal researchers confirm. *New York Times* (April 27, 2011).

Howerth, E. W., D. E. Stallknecht, W. R. Davidson, and E. J. Wentworth. 1990. Survey for leprosy in nine-banded armadillos (*Dasypus novemcinctus*) from the southeastern United States. *Journal of Wildlife Diseases* 26:112–115.

Loughry, W. J., R. W. Truman, C. M. McDonough, M. Tilak et al. 2009. Is leprosy spreading among nine-banded armadillos in the southeastern United States? *Journal of Wildlife Diseases* 45:144–152.

Mariotti, V., O. Dutour, M. G. Belcastro, F. Facchini, and P. Brasili. 2005. Probable early presence of leprosy in Europe in a Celtic skeleton of the 4th–3rd century BC (Casalecchio di Reno, Bologna, Italy). *International Journal of Osteoarchaeology* 15:311–325.

Pedrini, S. C. B., P. S. Rosa, I. M. Medri, G. Mourão et al. 2010. Search for *Mycobacterium leprae* in wild mammals. *Brazilian Journal of Infectious Diseases* 14:47–53.

Reader, R. 1974. New evidence for the antiquity of leprosy in early Britain. *Journal of Archaeological Science* 1:205–207.

Robbins, G., V. M. Tripathy, V. N. Misra, R. K. Mohanty et al. 2009. Ancient skeletal evidence for leprosy in India (2000 B.C.). *PLoS ONE* 4(5) E5669:1–8.

Rojas-Espinosa, O. and M. Lovik. 2001. *Mycobacterium leprae* and *Mycobacterium lepraemurium* infections in domestic and wild animals. *Revue Scientifique et Technique* 20:219–251.

Thoen, C. O. and D. E. Williams. 1994. Tuberculosis, tuberculoidoses, and other mycobacterial infections. In: G. W. Beran, editor. *Handbook of Zoonoses*. CRC Press, Boca Raton, FL, pp. 41–59.

Truman, R. W., P. Singh, R. Sharma, P. Busso et al. 2011. Probable zoonotic leprosy in the southern United States. *New England Journal of Medicine* 364:1626–1633.

WHO. 2000. *Guide to Eliminating Leprosy as a Public Health Problem*. World Health Organization, Geneva, Switzerland.

WHO. 2010. Global Health Observatory (GHO). Leprosy. http://www.who.int/gho/neglected_diseases/leprosy/en/index.html (accessed May 1, 2012).

7 Anthrax

Although anthrax has been recognized for centuries, little is known about the disease, and among the most basic questions frequently asked, but not yet answered, is how precisely do grazing and browsing animals acquire it.

WHO (2008)

7.1 INTRODUCTION AND HISTORY

Anthrax is a disease caused by *Bacillus anthracis*, which are small (0.4–1.0 μm in length), rod-shaped bacteria. These pathogens are Gram positive and not capable of movement. When grown in blood or tissue samples, they appear as short chains of a few cells and have a square end similar to a railroad box car (Figure 7.1). Anthrax is mainly a disease of mammalian herbivores, but it can infect all warm-blooded animals, including humans. The bacteria are obligate pathogens and cannot multiply outside of a living animal host. The bacteria start producing spores when exposed to the atmosphere: conditions occur after an infected animal has died and body fluids leak from the carcass or when the carcass is opened by scavengers (hereafter, the carcasses of animals killed by anthrax will be called anthrax carcasses). It is these spores that allow *B. anthracis* to infect new hosts. Half the spore's volume is taken up by the spore's coat, which protects the spore from environmental extremes of temperature, pH, chemicals, and desiccation. Spores can survive in the environment until conditions are suitable for germination. These conditions, which would occur inside a mammal's body, include high humidity, temperatures between 46°F and 113°F (8°C and 45°C), pH between 5 and 9, and the presence of amino acids (WHO 2008). The ability of anthrax spores to survive for long periods of time is astounding. Spores prepared by Dr. Louis Pasteur were still viable when retested 68 years later. Viable spores were also recovered from horse hair used more than a century earlier to bind plaster in London buildings (WHO 2008). Spore survival is enhanced in areas where the soil is dry, alkaline (pH greater than 8), and rich in calcium. The spores can germinate in less than 16 minutes when exposed to organic chemicals (e.g., L-alanine and ribosides).

Anthrax has been studied as a potential weapon in biological warfare. Anthrax spores have been used in bioterrorism, most notably during 2001 when several important politicians and government officials in the United States received letters and packages containing the spores. Anthrax vaccines have been developed for humans and stockpiled to combat the threat of using anthrax spores as a biological weapon.

Anthrax has been a scourge of humans since ancient times and may have been one of the plagues of Egypt during the time of Moses. *B. anthracis* evolved in the Middle East or Africa, and humans accidently introduced the pathogen to

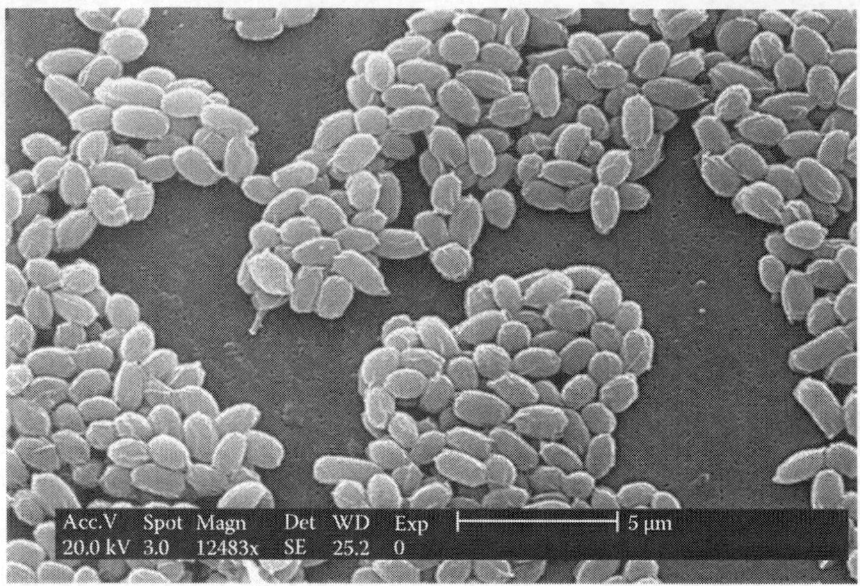

FIGURE 7.1 A highly magnified photo of spores from *B. anthracis*. (Courtesy of Janice H. Carr and the CDC.)

the New World. It reached Haiti by 1770, when 15,000 people succumbed, and the United States by at least the 1830s, when widespread outbreaks of anthrax occurred in Louisiana and Texas. The large cattle drives following the American Civil War helped spread anthrax northward from Texas into the prairie states and Canada. Currently, anthrax is uncommon among people in developed countries, including the United States and Canada. In the United States, the annual incidence rate has dropped from 130 cases at the beginning of the twentieth century to one or two cases by the beginning of the next century (Figure 7.2). There was one confirmed case of inhalation anthrax in the United States during 2011—a Florida resident who became ill while vacationing in Minnesota. The source of the infection could not be located, but the patient had handled antlers and driven through dust clouds stirred up by animal herds during the prior month (Shadomy and Smith 2008, Baille and Huwar 2011, CDC 2013).

Anthrax is more common in less developed countries, such as India, Chad, Ethiopia, Zambia, and Zimbabwe. During the 1970s, about 7000 human cases were reported annually across the world, but the incidence of anthrax among humans has steadily decreased since then, owing to better management of anthrax in livestock and better disposal of anthrax carcasses. Current incidence rates are unknown, in part, because mild cases are often not diagnosed. Anthrax incidence rates in humans vary considerably from 1 year to the next, depending upon whether an outbreak has occurred. In Zimbabwe, more than 10,000 people contracted anthrax during a civil war that disrupted public health and veterinarian services. During a 2000 outbreak

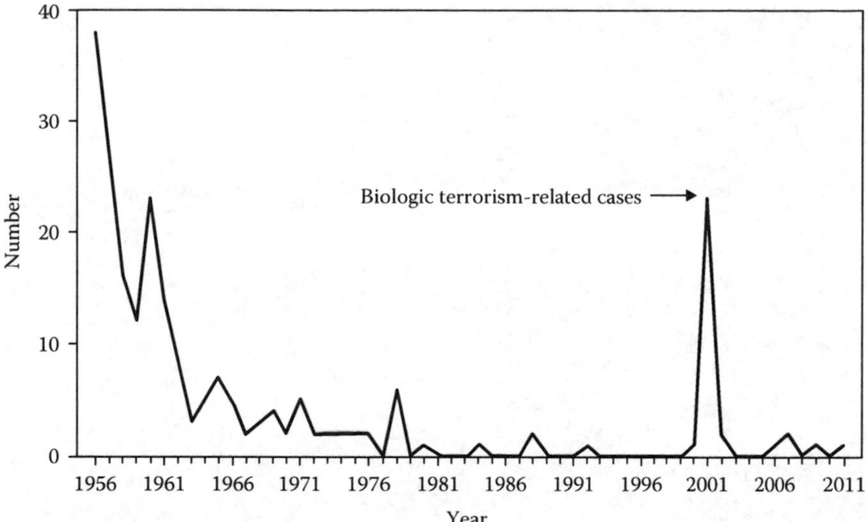

FIGURE 7.2 Annual number of reported cases of anthrax in the United States from 1956 to 2011. (Courtesy of the CDC.)

in Ethiopia, hundreds of people contracted anthrax after handling or consuming meat from contaminated animals. Fifty people in Zambia died of anthrax during the epidemic of 1990–1991 (Nass 1992, WHO 2008).

7.2 SYMPTOMS IN HUMANS

B. anthracis produces a toxin complex consisting of three proteins that act in concert. Due to the presence of these toxins, symptoms of anthrax can continue or worsen even after antibiotics have killed all of the pathogens within a patient. Anthrax produces a variety of symptoms in humans depending on whether *B. anthracis* was inhaled (inhalational anthrax), swallowed (enteric anthrax), or invaded the body via the skin (cutaneous anthrax). Worldwide, more than 95% of human cases of anthrax involve cutaneous anthrax that results from skin contact with contaminated material.

Cutaneous anthrax normally occurs on exposed skin, such as the face, neck, arms, and hands, usually where there is a cut or abrasion in the skin that allows entry to anthrax spores. Symptoms normally start to develop 5–7 days after exposure. The first clinical sign often is a small, painless pimple. In a day or two, the pimple expands to a 0.4–1.1 in. (1–3 cm) round lesion surrounded by fluid-filled skin blisters. Within a week, the lesion has developed a characteristic hard, black crust or scab (Figure 7.3). Unless there is a secondary infection, the lesion does not develop pus and is not painful. There is usually swelling that extends some distance from the lesion; low-grade fever, weakness, and headaches may occur. Anthrax lesions may take several weeks to heal. Complications can arise if the pathogen spreads to the

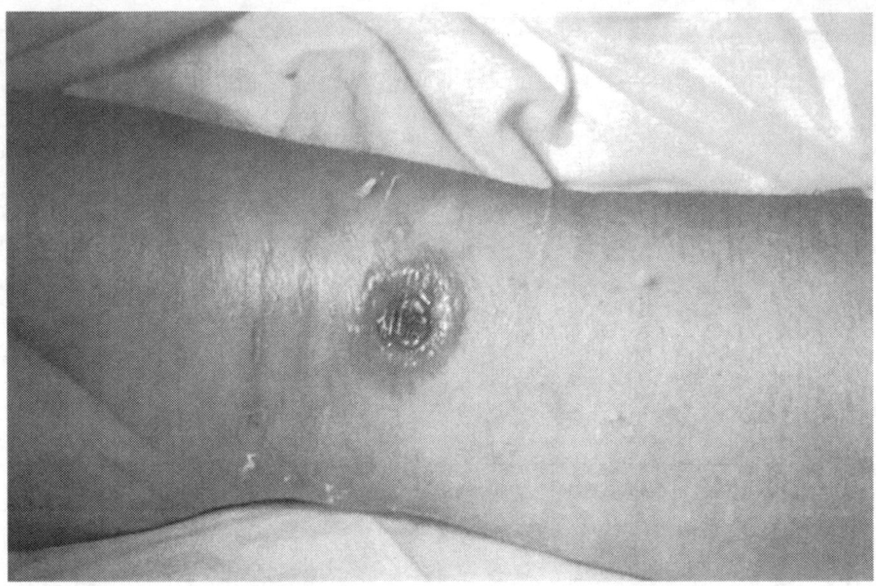

FIGURE 7.3 A cutaneous anthrax lesion on the arm. (Photo courtesy of Phil S. Brachman and the CDC.)

blood stream or if the lesion is located on the neck and swelling hinders the ability to breathe. Without medical attention, the fatality rate of people with cutaneous anthrax is approximately 20%, but with medical attention, the fatality rate drops to less than 1% (Shadomy and Smith 2008, WHO 2008).

Inhalational anthrax is often fatal because it goes unrecognized until it is too late. Surprisingly, inhalational anthrax does not start in the lungs; rather, the patient's macrophages carry the inhaled spores to the lymph nodes where the infection begins. Anthrax spores germinate in the macrophages. The vegetative forms of *B. anthracis* kill the macrophages and are released into the bloodstream where they multiply and cause blood poisoning. Headaches, muscle aches, fever, chills, sweating, fatigue, and nausea are the initial clinical signs of inhalational anthrax (Mayo Clinic 2010). Two or three days later, these mild clinical signs may be followed by labored breathing and a blue or purple appearance of the skin due to the lack of oxygen. Disorientation, coma, and death may quickly follow. During the twentieth century, 18 cases of naturally acquired inhalational anthrax have occurred in the United States, resulting in 16 deaths (90% fatality rate); 74 of 75 (99%) inhalational anthrax cases in England and Wales were fatal (WHO 2008).

When anthrax spores are swallowed, *B. anthracis* cannot enter the body and cause infection unless there are abrasions or cuts somewhere in the digestive system. When anthrax infections occur in the mouth, tongue, tonsils, or throat, the pathogen causes a sore that is 2–3 cm in diameter, covered by a gray membrane, and surrounded by swollen tissue. Patients experience a sore throat and swollen and painful lymph nodes on the side of the neck where the lesion is located. The neck may swell to the point where the airway is obstructed, requiring a tracheotomy (i.e., surgery to cut open the trachea so the patient can breathe). Even with treatment, these symptoms are often followed by shock, coma, and death (WHO 2008).

Swallowed anthrax spores may also cause an infection in the stomach or intestines. Symptoms of gastrointestinal anthrax begin 1–6 days after exposure and are nonspecific; they include a loss of appetite, nausea, fever, headache, and fainting. These symptoms may be followed by swelling of the abdomen, perforation of the infected digestive organ, massive hemorrhaging, and blood poisoning. Death often follows 2–5 days after clinical signs appear. Without medical attention, more than half of the humans diagnosed with gastrointestinal anthrax will die; with it, the fatality rate drops to less than 40% (Shadomy and Smith 2008).

Despite the route of infection, *B. anthracis* infections tend to become systemic and spread throughout the body; shock, coma, and death can develop rapidly. The patient can develop meningitis, a condition that is usually fatal. The Mayo Clinic (2010) recommends that people seek immediate medical attention if they have symptoms similar to anthrax and might have been exposed to anthrax (e.g., people who have been in a country where anthrax is endemic or exposed to animals or animal parts from such areas).

7.3 *BACILLUS ANTHRACIS* INFECTIONS IN ANIMALS

Anthrax is primarily a disease of large mammalian herbivores with each animal species having a characteristic pattern of illness. Death occurs when the toxin produced by anthrax reaches a critical level in the blood; the amount of toxin in the blood is related to the amount of *B. anthracis* in the body. Anthrax causes a high fatality rate among mammalian herbivores but some mammal species, such as carnivores, are resistant to anthrax, as are birds (Good et al. 2008).

An outbreak of anthrax is usually detected following sudden death of wildlife or livestock. Prior to that, infected animals may stop eating, have difficulty breathing, and have a swelling beneath the jaw. In susceptible species, only a few hours may exist between when these clinical signs are noted and when the animal goes into a coma and dies. In other species, the illness is more protracted. In cattle, death usually occurs 2–3 days after the animal first exhibits signs of illness, which include shivering and cramps. Anthrax should be considered a potential diagnosis when a mammalian herbivore dies suddenly and bloody fluids leak from the carcass's orifices. Pigs are more resistant to anthrax than other livestock. Most pigs that swallow anthrax spores survive but can be ill for a week with fever, blood in feces, lethargy, and anorexia (i.e., unwillingness to eat). Infected pigs also exhibit swelling in the throat and in the lymph glands (Redmond et al. 1997, WHO 2008).

Anthrax was a problem for both livestock and wildlife throughout the first half of the twentieth century. During 1915, anthrax was reported in 157 counties located in 21 states. By the 1940s, the endemic area had increased to 405 counties within 37 states. Since then, anthrax outbreaks became less common in North America due to better disposal of anthrax carcasses and treatment of livestock exposed to *B. anthracis* spores with antibiotics or vaccinations to break the cycle of infection.

FIGURE 7.4 *B. anthracis* is endemic in bison in parts of Canada. (Courtesy of Jack Dykinga and the USDA Agricultural Research Service.)

Still, numerous anthrax outbreaks among livestock and wild herbivores (deer, elk, and bison) have occurred in North America since 1950. Most outbreaks occur in a broad swath of land that starts in Texas and Louisiana and extends north through Nebraska, South Dakota, Alberta, Saskatchewan, Manitoba, and the Northwest Territories (Blackburn et al. 2007). In northern Canada, anthrax can maintain itself by infecting cattle and bison (Figure 7.4). Anthrax outbreaks occurred during 1999 in Alberta, during 2005 in North Dakota, and during 2006 in Canada's Northwest Territories and Saskatchewan (Sidebar 7.1).

Sick bison are indifferent to activities around them, walk with difficulty, and appear swollen. Anthrax is often fatal in bison and can occur suddenly. Death is often preceded by a bloody discharge from the animal's mouth, nose, and anus that continues after the animal has died. This leakage of body fluid results from the anthrax toxin destroying the lining of blood capillaries and is crucial to the life cycle of *B. anthracis* because the bacteria will only form spores once exposed to the air. The carcass bloats extensively, causing its legs to assume a sawhorse position (Gainer and Saunders 1989, Gates et al. 1995, Dragon and Elkin 2001, Parkinson et al. 2003, Nishi et al. 2007, Himsworth and Argue 2008, Shadomy and Smith 2008).

Among wild herbivores in Africa, anthrax has been reported in black rhinos, African buffalo, common eland, elephants, hippos, impala, hartebeest, springbok, wildebeest, and zebra. An anthrax outbreak in Tanzania's Lake Manyara National Park killed more than 90% of the impala. In the Luangwa Valley of Zambia, more than 4000 hippos died during a 1987 outbreak. In the Malilangwe Wildlife Reserve of Zimbabwe, an outbreak during 2004 killed almost all of the reserve's kudu and

SIDEBAR 7.1 ANTHRAX OUTBREAKS AMONG BISON IN NORTHERN CANADA (DRAGON ET AL. 1999, KENEFIC ET AL. 2010)

Anthrax was first detected among bison in the Slave River Lowlands of the Northwest Territories during 1962. Out of a herd of 1300 bison, 281 anthrax carcasses were found and buried, but many more bison probably died without being located. The origin of this outbreak is unknown, but the bison may have contracted anthrax from local livestock. Over time, *B. anthracis* spread southward and reached the Mackenzie Bison Sanctuary and the Woods Buffalo National Park where there were at least nine outbreaks among bison from 1962 to 1993, killing more than 1300 bison. *B. anthracis* is now endemic in these areas (Figure 7.5), and anthrax epidemics occurred during 2000, 2001, and 2006.

Until 1967, dead bison killed by anthrax were buried under mounds of dirt due to a high water table; this practice, however, exposed other wildlife species to the anthrax spores. The dead bison attracted mammalian scavengers, including wolves. The mounds of dirt in an otherwise flat plain attracted the attention of burrowing mammals and people who liked to camp on them. Ants also liked to build their nests on the mounds and were constantly bringing anthrax up to the surface where the spores would pose a danger to wildlife.

more than 40% of its nyala, bushbuck, waterbuck, and roan antelope (Prins and Weyerhaeuser 1987, Turnbull et al. 1991, Clegg et al. 2007).

Animals that have some resistance to anthrax may survive *B. anthracis* infection. Most bison males (87%) had antibodies against *B. anthracis* 8 months after an anthrax outbreak within Canada's Mackenzie Bison Sanctuary, indicating that some bison survived anthrax. Most bison that died of anthrax were adult males (Gates et al. 1995, Dragon and Elkin 2001). It is not clear what accounts for this sex bias, but several hypotheses have been proposed (WHO 2008). Bulls use wallows for dominance displays and their rutting behavior kicks up spore-contaminated dust. They also graze closer to the ground than females or calves. The general health of bulls declines when they become dominant, owing to time and energy expended in fights to maintain dominance. All of these behaviors may increase the probability of a bull becoming infected contracting anthrax (Fox et al. 1977).

Not all herbivore species are equally susceptible to anthrax. Cattle and other animals that pull up plants when foraging are more likely to consume or inhale anthrax spores than other herbivores, such as deer, that forage on leaves located above the ground. Among wild herbivores, species that are vulnerable to anthrax may differ from one region to another. In Namibia's Etosha Nation Park, 50 zebra were killed by anthrax for every kudu that was killed by it (Lindeque and Turnbull 1994). Yet kudus are more vulnerable to anthrax than any other wildlife species in South Africa's Kruger National Park (de Vos 1990).

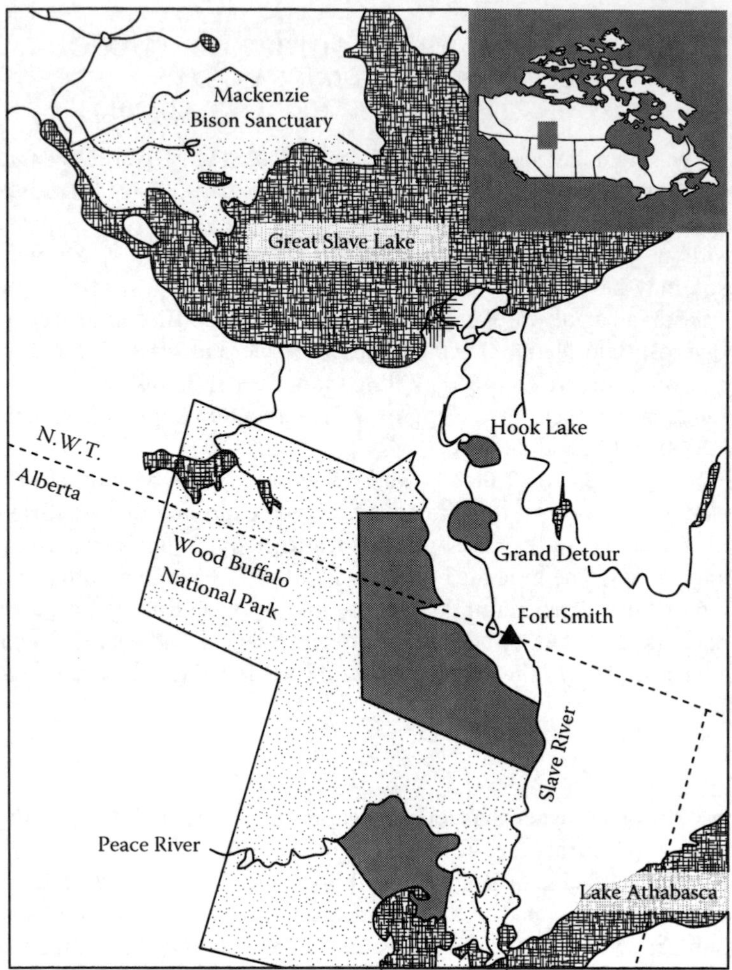

FIGURE 7.5 Map showing the location of Mackenzie Bison Sanctuary and the Woods Buffalo National Park in Canada.

Carnivores are more resistant than herbivores to anthrax, but they too can die from the disease. Cheetahs are one carnivore that is susceptible to anthrax (Figure 7.6). Cheetahs do not scavenge the kills of other animals and may not have evolved immunity to *B. anthracis* (Good et al. 2008).

In some areas, outbreaks among herbivores often occur during or immediately after a drought. Perhaps this pattern occurs because animals are forced during droughts to consume tough vegetation containing thorns or spikes, and this poor diet produces the lesions in the mouth or digestive system that the pathogen needs to invade the body. Another theory is that during droughts, animals are forced to graze plants closer to the ground and are more likely to inhale dry dust and anthrax spores (McCulloch 1961, Bell and Laing 1977, Fox et al. 1977).

FIGURE 7.6 Cheetahs are more susceptible to *B. anthracis* than other carnivores. (Courtesy of Gary Stolz and the USFWS.)

Anthrax outbreaks can occur far from endemic areas. One reason for this is the worldwide movement of animal products, some of which may be contaminated. Anthrax outbreaks have also occurred among cattle pastured downstream from tanneries or when cattle feed was supplemented with bone meal or meat byproducts obtained from a diseased animal. Strong wind storms can carry anthrax spores and initiate a new outbreak among wildlife or livestock in areas that were previously free of anthrax. For example, an accidental release of anthrax spores in the former Soviet Union sickened sheep at a village 37 miles (60 km) downwind (Henry 1936, Meselson et al. 1994, WHO 2008). Anthrax spores have been found in the feather and feces of birds that forage on carcasses such as gulls, ravens, and vultures (Figure 7.7). Hence, these birds may play a role in spreading anthrax spores over long distances (Dragon et al. 1999, Saggese et al. 2007). Vultures and other scavengers are suspected of spreading *B. anthracis* in southern Africa (Lindeque and Turnbull 1994, Lindeque et al. 1996).

B. anthracis can be spread from an infected animal to a healthy one by mosquitoes (*Aedes aegypti* and *A. taeniorhynchus*), horseflies (*Tabanus rubidus*), and biting stable flies (*Stomoxys calcitrans*; Davies 1983). An anthrax epidemic in Zimbabwe during the 1970s may have been facilitated by horseflies (Davies 1983), and another outbreak among hippos may have been spread by biting flies (WHO 2008). *B. anthracis* can be transmitted from a dead animal to a healthy one by blowflies and bluebottle flies. These insects feed on the body fluids of anthrax carcasses and then deposit contaminated feces or vomit on leaves that are then consumed by healthy animals (Braack and deVos 1990, Shadomy and Smith 2008).

FIGURE 7.7 Vultures and other birds that scavenge carcasses can transport *B. anthracis* long distances in Africa. (Courtesy of Chris Fallows and the U.S. Public Library of Science.)

7.4 HOW HUMANS CONTRACT ANTHRAX

People normally contract anthrax from infected animals or anthrax carcasses. High-risk professions include farmers, veterinarians, wildlife biologists, meat packers, and butchers. Also at risk are those people who work in industries involving wool, mohair, horse hair, or animal hides. During 2009 and 2010, 273 people in Bangladesh contracted cutaneous anthrax, and 90% of them were involved with butchering animals or handling uncooked meat or animal skins (Chakraborty et al. 2012). From 2005 to 2010, four people in the United Kingdom and United States contracted anthrax from either making or playing drums constructed from contaminated animal hides (CDC 2010). Lastly, a few people have contracted anthrax from spores deliberately released by acts of bioterrorism (Sidebar 7.2).

The risk of human-to-human transmission of *B. anthracis* is remote, provided that precautions are taken. Anthrax lesions on the skin should be covered, and disposable gloves should be worn by health workers when treating the patient. Disinfection, decontamination, and correct disposal of contaminated material are important.

Humans are moderately resistant to anthrax. Despite exposure to anthrax carcasses, few wildlife biologists have contracted the disease. In Namibia's Etosha National Park, there have been only two unconfirmed cases among park employees despite 30 years of collecting and disposing of anthrax carcasses. In Zimbabwe's Malilangwe Trust, two employees involved with anthrax carcass disposal developed anthrax lesions on their hands. In Canada's Wood Buffalo National Park, two people developed cutaneous anthrax after performing autopsies on the first carcasses to

SIDEBAR 7.2 THE ANTHRAX LETTERS: USING A ZOONOTIC DISEASE FOR BIOTERRORISM (INGLESBY ET AL. 2002, JERNIGAN ET AL. 2002, WHO 2008)

During 2001, letters and packages containing anthrax spores were mailed to U.S. congressional leaders and news media companies. Twenty-two people in the eastern United States were either diagnosed or suspected of having anthrax, 20 of whom were either mail handlers or people working in buildings where the contaminated mail was received or processed. Half of the victims developed inhalational anthrax; the other half had cutaneous anthrax. Five victims died, all of whom had the more deadly inhalational anthrax. It is unknown how many people were exposed to the anthrax spores, but it is likely that only a small portion of people exposed to the spores developed anthrax. People known to have been exposed to the spores were given antibiotics, and this reduced the morbidity and mortality rates.

Media coverage of the incident was intense, raising alarm among people who feared they may have been exposed. Thousands of people sought medical attention for possible exposure to anthrax. Thousands of others, fearing they had received mail containing anthrax spores, forwarded the suspicious mail to the police or emergency management agencies. This incident showed that a terrorist does not have to kill many people to disrupt normal activities.

determine the cause of death, and a backhoe operator developed inhalational anthrax after burying anthrax carcasses (WHO 2008).

7.5 MEDICAL TREATMENT

Anthrax responds well to penicillin and other broad-spectrum antibiotics, and the prompt use of these drugs greatly reduces the mortality rate in both humans and livestock. Antibiotics should be administered as soon as possible to stop the production of toxin. Penicillin still remains the drug of choice in those parts of the world where the disease is common. In mild cases of cutaneous anthrax, intramuscular injections of penicillin are recommended, usually one or two daily for 3–7 days. Amoxicillin can also be used to treat anthrax. Antibiotics usually kill *B. anthracis* within 24 hours, reduce swelling, and limit the size of the skin lesion, but several weeks or months may be required before the skin lesions fully heal. Anthrax patients are often given penicillin or other antibiotic(s) intravenously. Treatment for shock can be life saving because death is often caused by toxin-induced shock. Patients may be placed on a respirator if they have difficulty breathing (Shadomy and Smith 2008, WHO 2008).

Administration of antibiotics to prevent anthrax from developing is not warranted unless the person has had a substantial exposure to *B. anthracis* (e.g., the consumption of raw or undercooked meat from an animal that died of anthrax). But, people who have potentially been exposed to *B. anthracis* should be watched carefully and given antibiotics if they develop a flu-like illness or a pimple or boil-like lesion within 4 weeks of the exposure (WHO 2008).

7.6 WHAT PEOPLE CAN DO TO REDUCE THEIR RISK OF CONTRACTING ANTHRAX

Anthrax is easily treated if diagnosed at an early stage of infection. Wildlife biologists or other people who might have been exposed to an anthrax carcass should alert their health-care providers if they develop symptoms similar to anthrax.

An anthrax vaccine has been developed for humans and is recommended for people in high-risk groups, such as veterinarians, wildlife biologists, and people who work with animal hides or furs. The primary immunization series consists of five injections given at day 0 and months 1, 4, 6, 12, and 18. This is followed by a booster shot that is administered annually. In the past, military personnel were vaccinated against anthrax to protect them from the use of anthrax as a biological weapon, a use that has become controversial because it was unclear if the remote threat may not justify the risk involved with vaccinating large numbers of troops (WHO 2008, CDC 2010).

Control of anthrax in humans is largely dependent upon preventing it in livestock (Figure 7.8). Cattle that may have anthrax should be separated from the herd and treated with antibiotics. Preferred antibiotics for infected animals include intravenous injections of sodium benzylpenicillin followed by intramuscular injections of a long-acting antibiotic such as benethamine penicillin or amoxicillin. The rest of the

FIGURE 7.8 The best method to prevent anthrax in humans is to prevent it in cattle. (Courtesy of Keith Weller and the USDA Agricultural Research Service.)

herd should be checked at least three times a day for 2 weeks to detect other animals with anthrax. When this is not possible, the entire herd should be treated with antibiotics (WHO 2008).

The site where an animal died of anthrax should be decontaminated. Infected herds are often quarantined for at least 3 weeks after the last mortality. Within endemic areas, disease outbreaks often occur after soil disturbance (e.g., ditch digging, laying water lines, and bulldozing) exposes forgotten sites where anthrax-killed livestock were buried.

B. anthracis is much more likely to spread by its production of spores than by its vegetative cells, and spore formation requires oxygen. For this reason, steps should be taken not to open up a carcass or to conduct an autopsy on an animal that may have died of anthrax. The preferred method to dispose of an anthrax carcass is incineration where it died. The carcass should be raised off the ground to ensure that the carcass is completely incinerated. There is some concern that *B. anthracis* spores may survive the incineration and become airborne in the updraft. Some viable spores have been recovered in smoke plumes, but the number of viable spores that become airborne is low enough that the probability of an infection spreading this way is remote. The soil beneath the carcass should also be incinerated to a depth of 0.7 ft (21 cm). If this is not possible, a solution of 10% formaldehyde should be injected into the soil with the injections spaced 0.5 yards (0.5 m) apart horizontally throughout the contaminated surface area. Spreading lime has little effect because lime does not kill the spores (WHO 2008).

All anthrax vaccines for animals (but not vaccines for humans) utilize live, attenuated strains of *B. anthracis*, but these strains can still produce toxin. Because of this, vaccinated livestock should not be slaughtered for human consumption within 6 weeks after the vaccination date. A single dose of the vaccine provides immunity for about 1 year, and annual boosters can be used to lengthen the period of immunity (WHO 2008). Vaccinated animals gain protection from anthrax within 1 or 2 weeks of being vaccinated. Antibiotics and vaccines using live strains of *B. anthracis* should not be given together because the antibiotic will interfere with the vaccine's effectiveness. Losses of livestock exposed to *B. anthracis* may be minimized by first administering antibiotics to treat any infected animals, followed by a vaccination program. Anthrax spores can last in the soil for years, and new outbreaks of anthrax have been reported in animals grazing in contaminated pastures 3 years after the prior outbreak. Thus, livestock should be revaccinated annually for at least 3 years after the initial outbreak to prevent future problems (WHO 2008).

Wildlife agencies often view anthrax as a natural form of mortality in wildlife and may only intervene if an endangered species is at risk (Turnbull et al. 2004). While a hands-off approach may be suitable for large and remote national parks, this approach is less appropriate when an anthrax outbreak spreads to game animals during the hunting season or to livestock. Wildlife managers confront a difficult problem when anthrax occurs among wildlife because of the inability to locate and treat more than a small portion of a wildlife population. Still, there have been some successful interventions to stop an anthrax outbreak among wildlife. WHO (2008) reported that one outbreak among wildlife in Tanzania was controlled by using a dart to administer antibiotics to roan, sable antelope, and kudu. Anthrax vaccines have been used to immunize wildlife including zebra, black rhinoceros, elephants, and cheetahs by vaccinating them twice, spaced 2 months apart (Turnbull et al. 2004). In some anthrax-endemic areas, the risk of anthrax spreading from wildlife to livestock has been minimized by constructing wildlife-proof fences to minimize co-mingling of livestock and wildlife.

7.7 ERADICATING ANTHRAX FROM A COUNTRY

The complete elimination or eradication of *B. anthracis* from a country or large area is possible, providing that no wildlife species serve as a reservoir host for the bacteria. Eradication and surveillance programs must be sustained for many years to be effective against a pathogen that can persist for such long periods of time.

Due to successful programs to combat anthrax, the disease rarely occurs in northern Europe and Russia but still occurs in countries around the Mediterranean Sea. It is also absent from New Zealand and from Caribbean countries with the exception of Haiti. In the United States, anthrax has been confined to a few areas located in Texas, Nevada, Nebraska, Minnesota, and the Dakotas. In Canada, it has been restricted to the area around the Wood Bison National Park located in northern Alberta and in a few isolated areas within southern Manitoba, Alberta, and Saskatchewan (WHO 2008).

LITERATURE CITED

Baille, L. and T. Huwar. 2011. Anthrax. In: S. R. Palmer, L. Soulsby, P. R. Torgerson, and D. W. G. Brown, editors. *Oxford Textbook of Zoonoses*, 2nd edition. Oxford University Press, Oxford, U.K.

Bell, W. J. and P. W. Laing. 1977. Pulmonary anthrax in cattle. *Veterinary Record* 100:573–574.

Blackburn, J. K., K. M. McNyset, A. Curtis, and M. E. Hugh-Jones. 2007. Modeling the geographic distribution of *Bacillus anthracis*, the causative agent of anthrax disease, for the contiguous United States using predictive ecologic niche modeling. *American Journal of Tropical Medicine and Hygiene* 77:1103–1110.

Braack, L. E. and V. de Vos. 1990. Feeding habits and flight range of blow-flies (*Chrysomyia* spp.) in relation to anthrax transmission in the Kruger National Park, South Africa. *Onderstepoort Journal of Veterinary Research* 57:141–142.

CDC. 2010. Anthrax. http://www.cdc.gov/agent/anthrax (accessed May 16, 2012).

CDC. 2013. Summary of notifiable diseases—United States, 2011. *Morbidity and Mortality Weekly Report* 60(53):1–118.

Chakraborty, A., S. U. Khan, M. A. Hasnat, S. Parveen et al. 2012. Anthrax outbreaks in Bangladesh, 2009–2010. *American Journal of Tropical Medicine and Hygiene* 86:703–710.

Clegg, S. B., P. C. B. Turnbull, C. M. Foggin, and P. M. Lindeque. 2007. Massive outbreak of anthrax in wildlife in the Malilangwe Wildlife Reserve, Zimbabwe. *Veterinary Record* 160:113–118.

Davies, J. C. 1983. A major epidemic of anthrax in Zimbabwe. 2. Distribution of cutaneous lesions. *Central African Journal of Medicine* 29:8–12.

de Vos, V. 1990. The ecology of anthrax in the Kruger National Park, South Africa. *Salisbury Medical Bulletin* 68S:19–23.

Dragon, D. C. and B. T. Elkin. 2001. An overview of early anthrax outbreaks in northern Canada: Field reports of the Health of Animals Branch, Agriculture Canada, 1962–71. *Arctic Institute of North America* 54:32–40.

Dragon, D. C., B. T. Elkin, J. S. Nishi, and T. R. Ellsworth. 1999. A review of anthrax in Canada and implications for research on the disease in northern bison. *Journal of Applied Microbiology* 87:208–213.

Fox, M. D., J. M. Boyce, A. F. Kaufmann, J. B. Young, and H. W. Whitford. 1977. An epizootiologic study of anthrax in Falls County, Texas. *Journal of the American Veterinary Medical Society* 170:327–333.

Gainer, R. S. and J. R. Saunders. 1989. Aspects of the epidemiology of anthrax in Wood Buffalo National Park and environs. *Canadian Veterinary Journal* 30:953–956.

Gates, C. C., B. T. Elkin, and D. C. Dragon. 1995. Investigation, control and epizootiology of anthrax in a geographically isolated, free-roaming bison population in northern Canada. *Canadian Journal of Veterinary Research* 59:256–264.

Good, K. M., A. M. Houser, L. Arntzen, and P. C. B. Turnbull. 2008. Naturally acquired anthrax antibodies in a cheetah (*Acinonyx jubatus*) in Botswana. *Journal of Wildlife Diseases* 44:721–723.

Henry, M. 1936. A further note on the incidence of anthrax in stock in Australia. *Australian Veterinary Journal* 12:235–239.

Himsworth, C. G. and C. K. Argue. 2008. Anthrax in Saskatchewan 2006: An outbreak overview. *Canadian Veterinary Journal* 49:235–237.

Inglesby, T. V., T. O'Toole, D. A. Henderson, J. G. Bartlett et al. 2002. Anthrax as a biological weapon, 2002: Updated recommendations of management. *Journal of the American Medical Association* 287:2236–2252.

Jernigan, D. B., P. L. Raghunathan, B. P. Bell, R. Brechner et al. 2002. Investigation of bioterrorism-related anthrax, United States, 2001: Epidemiologic findings. *Emerging Infectious Diseases* 8:1019–1028.

Kenefic, L. J., R. T. Okinaka, and P. Keim. 2010. Population genetics of *Bacillus*: Phylogeography of anthrax in North America. In: D. A. Robinson, D. Falush, and E. J. Feil, editors. *Bacterial Population Genetics in Infectious Disease*. John Wiley & Sons, New York, pp. 169–180.

Lindeque, P. M., C. Brain, and P. C. B. Turnbull. 1996. A review of anthrax in the Etosha National Park. *Salisbury Medical Bulletin* 87(S):24–26.

Lindeque, P. M. and P. C. B. Turnbull. 1994. Ecology and epidemiology of anthrax in the Etosha National Park, Namibia. *Onderstepoort Journal of Veterinary Research* 61:71–83.

Mayo Clinic. 2010. Anthrax. http://www.mayoclinic.org/health/anthrax (accessed February 20, 2013).

McCulloch, B. 1961. Pulmonary anthrax in cattle. *Veterinary Record* 73:805.

Meselson, M., J. Guillemin, M. Hugh-Jones, A. Langmuir et al. 1994. The Sverdlovsk anthrax outbreak of 1979. *Science* 266:1202–1208.

Nass, M. 1992. Anthrax epizootic in Zimbabwe, 1978–1980: Due to deliberate spread? *Physicians for Social Responsibility Quarterly* 2:198–209.

Nishi, J. S., T. R. Ellsworth, N. Lee, D. Dewar et al. 2007. An outbreak of anthrax (*Bacillus anthracis*) in free-roaming bison in the Northwest Territories, June-July 2006. *Canadian Veterinary Journal* 48:37–38.

Parkinson, R., A. Rajic, and C. Jenson. 2003. Investigation of an anthrax outbreak in Alberta in 1999 using a geographic information system. *Canadian Veterinary Journal* 44:315–318.

Prins, H. H. T. and F. J. Weyerhaeuser. 1987. Epidemics in populations of wild ruminants: Anthrax and impala, rinderpest and buffalo in Lake Manyara National Park, Tanzania. *Oikos* 49:28–38.

Redmond, C., G. A. Hall, P. C. B. Turnbull, and J. S. Gillgan. 1997. Experimentally assessed public health risks associated with pigs from farms experiencing anthrax. *Veterinary Record* 141:244–247.

Saggese, M. D., R. P. Noseda, M. M. Uhart, S. L. Deem et al. 2007. First detection of *Bacillus anthracis* in feces of free-ranging raptors from central Argentina. *Journal of Wildlife Diseases* 43:136–141.

Shadomy, S. V. and T. L. Smith. 2008. Zoonosis update: Anthrax. *Journal of the American Veterinary Medical Association* 233:63–72.

Turnbull, P. C. B., R. H. Bell, K. Saigawa, F. E. Munyenyembe et al. 1991. Anthrax in wildlife in the Luangwa Valley, Zambia. *Veterinary Record* 128:399–403.

Turnbull, P. C. B., B. W. Tindall, J. D. Coetzee, C. M. Conradie et al. 2004. Vaccine-induced protection against anthrax in cheetah (*Acinonyx jubatus*) and black rhinoceros (*Diceros bicornis*). *Vaccine* 22:3340–3347.

WHO. 2008. *Anthrax in Humans and Animals*, vol. 2006(7), 4th edition. World Health Organization, Geneva, Switzerland, pp. 1–90.

8 Rat-Bite Fever

The rise of rats: a growing pediatric issue [due to rat-bite fever].

Khatchadourian et al. (2010)

Rat-bite fever was once considered an infection exclusive to children living in poverty; however, dense urban housing and changing pet-keeping practices may be altering this profile.

Schachter et al. (2006)

Another name for rat-bite fever is "sodoku, which is a Japanese word meaning rat (so) poison (doku)."

Paraphrased from Will (1994)

8.1 INTRODUCTION AND HISTORY

For many centuries, rat-bite fever has been recognized in patients who became ill after being bitten by a rat. Wagabhatt, who lived in India 2300 years ago, reported that many people bitten by rats develop a skin rash. The first medical description of rat-bite fever in North America was made by Eli Ives during 1831, in Europe by Millot-Carpentier a half century later, and in Japan by Miyaki during 1900 (Will 1994, Gaastra et al. 2009).

Rat-bite fever is caused by two different species of bacteria, both of which are transmitted to humans through rat bites. These bacteria produce two diseases in humans that are similar but not identical. Streptobacillary rat-bite fever (SRBF) is caused by *Streptobacillus moniliformis*. Spirillary rat-bite fever, or sodoku, is caused by *Spirillum minus*. Both diseases occur worldwide but SRBF is more common in North America and Europe while sodoku is more common in Asia.

Streptobacillus are Gram-negative, anaerobic (i.e., not requiring oxygen to survive) bacteria that lack the ability to move. The cells are 1–5 μm in length and less than 1 μm in diameter. *Streptobacillus* is a combination of a Greek word "streptos" meaning curved and the Latin word "bacillus" meaning small rod. Cells form long chains or filaments of up to 150 μm in length. These filaments have swellings irregularly spaced along them, which explains the species name of this bacterium; *moniliformis* is Latin and means "in the form of a necklace" (Figure 8.1). *Spirillum minus* is a spirochete or spiral-shaped bacterium with two or three twists, and it is capable of movement. "Spirillum" means coiled in Latin, and "minus" translates as minor or small. It is also an anaerobic bacterium, 3–5 μm in length and 0.2 μm in diameter (Gaastra et al. 2009).

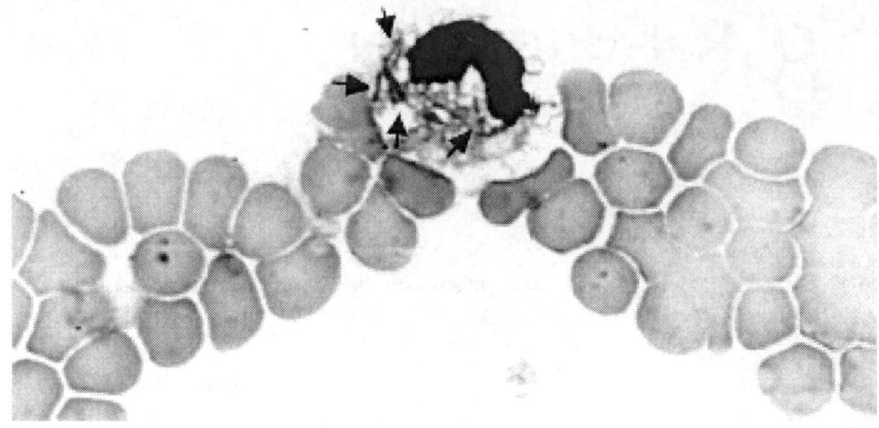

FIGURE 8.1 Photo of *S. moniliformis*. (Courtesy of CDC.)

People develop SRBF after being bitten by an infected animal or consuming food or drink contaminated by *S. moniliformis*. Disease symptoms differ based on how the pathogen gains entry into the body; the disease is called Haverhill fever when the pathogen is ingested. The name comes from Haverhill, Massachusetts, where an epidemic occurred in 1926 that sickened 86 people who had consumed unpasteurized milk. A year earlier, 400–600 people around Chester, Pennsylvania, became ill during another outbreak. During 1983, 304 of 700 children at a rural girls' boarding school in Essex, England, developed Haverhill fever. Several of the girls required hospitalization for weeks. The source of infection was a water system that had been contaminated by the feces and urine of rats (Will 1994).

SRBF occurs in both the United States and Canada, but reported human cases are uncommon because it is not a notifiable disease in either country. In Baltimore, Maryland, 11% of people bitten by rats developed SRBF (Will 1994). Historically, most SRBF patients were poor children, possibly because they were more likely to be bitten by a wild rat. Although this is less true today, half of patients with confirmed cases of SRBF in California were less than 10 years old (Graves and Janda 2001). There was a twofold increase in the incidence of SRBF in California during the 1990s, and there is a concern that the disease may be an emerging zoonotic disease (Schachter et al. 2006).

8.2 SYMPTOMS IN HUMANS

When people develop SRBF, the first signs of illness usually occur 3–10 days after exposure to the bacteria but can be delayed as long as 3 weeks. By then, patients may no longer remember being bitten or scratched by a rodent, and any wound would have had time to heal. Fever, vomiting, and muscle and joint aches are early symptoms. A few days later, 75% of patients develop a red rash with numerous small bumps on their limbs, including on the palms of hands and soles of the feet. In half of patients, multiple joints may become inflamed, swollen, and painful.

If left untreated, SRBF can become a recurrent illness that can last weeks or months. Serious complications can result if the infection spreads to the heart (resulting in endocarditis or myocarditis), brain (meningitis), lungs (pneumonia), or other internal organs (multiple organ failure). A presumptive diagnosis of SRBF can be made when patients exhibit clinical signs consistent with the disease and have a history of exposure to rodents. The diagnosis is confirmed through isolating the pathogen from the blood, urine, lymph nodes, or lesions. SRBF can rapidly become fatal (Sidebar 8.1) with a mortality rate among untreated patients of 7%–15% (Gaastra et al. 2009, CDC 2011). People with rheumatic heart disease have a higher risk of developing rat-bite endocarditis, which has a high mortality rate (53%); half of the survivors of endocarditis suffer damage to their heart valves (Rupp 1992).

The first signs of sodoku begin after a 7- to 21-day incubation period (Table 8.1). By then, the bite wound usually has healed, but it will reopen later. The area around the wound becomes swollen. In about half of the patients, a distinctive macular rash (i.e., a skin rash consisting of red spots that are flat with the surrounding skin) develops at the infection site, spreading rapidly to other parts of the body, especially to skin located over joints. Lymph nodes draining the infection site become swollen and tender. Fever is common; without treatment, fever may last 3–5 days, temporarily disappearing for several days, and then returning along with the macular rash. In untreated patients, the illness may recur for months. With sodoku, the central and peripheral nervous system can become infected, resulting in headaches, restlessness, nervousness, and sudden pain. Some patients can develop meningoencephalitis (a medical condition that resembles both meningitis and encephalitis) marked by twitching, convulsions, and unconsciousness (Will 1994). Other complications of sodoku can include hepatitis, inflammation of the kidney or spleen, and myocarditis

SIDEBAR 8.1 SRBF AMONG THE OWNERS OF PET RATS (POLLOCK ET AL. 2005)

If your children are begging for a pet rat and you are seeking reasons to say no, this sidebar may be helpful (Figure 8.2). A 19-year-old woman in Washington died before reaching the local hospital in November 2003. During the prior 3 days, she had experienced fever, headache, muscle aches, weakness, and jaundice (i.e., a yellowing of the skin and eyes caused when the liver is not functioning correctly). Immediately before transport to the hospital, the patient exhibited confusion, anxiety, and labored breathing. An autopsy was conducted, and SRBF was diagnosed after *S. moniliformis* was isolated from her liver and kidney tissue.

The patient lived in an apartment with nine pet rats. Two of them had recently experienced respiratory problems and were prescribed doxycycline by a veterinarian. It is unknown if the rats were infected with *S. moniliformis* because they were not tested. The patient had no known animal bites during the 2 weeks prior to her illness.

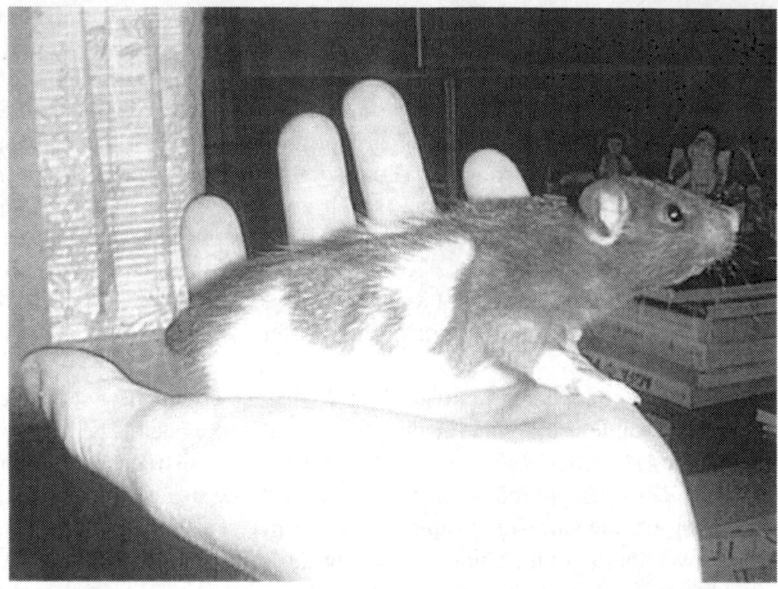

FIGURE 8.2 Photo of a pet rat. (Courtesy of Waldemar Zboralski.)

TABLE 8.1
Characteristics of SRBF and Sodoku

Characteristics	SRBF	Sodoku
Pathogen	*S. moniliformis*	*Spirillum minus*
Geographic range	North America and Europe	Asia
Incubation period	Less than 2 weeks	1–4 weeks
Infection site	Wound heals	Reopens and is inflamed
Lymph nodes	Not swollen	Swollen and tender
Arthritis	Common in multiple joints	Uncommon
Rash	Rash on hands and feet	Usually none

Source: Booth, C.M. et al., *Can. J. Infect. Dis.*, 13, 269, 2002.

(i.e., inflammation of the heart muscle). The mortality rate among untreated patients is 6%. In North America, sodoku is mainly seen among travelers returning from Asia. Symptoms of sodoku are similar enough to SRBF that it can be difficult to distinguish between them. Fortunately, the same antibodies are effective against both, so it is not necessary to differentiate between them before starting antibiotics (Will 1994, Gaastra et al. 2009, CDC 2011).

Haverhill fever results when *S. moniliformis* is swallowed. The symptoms differ from those caused through rat bites and scratches because the digestive system is involved. Symptoms include a sore throat and severe vomiting. There is some evidence suggesting that Haverhill fever is caused by a different strain of *S. moniliformis* rather than the strains causing SRBF (Costas and Owen 1987, CDC 2011).

8.3 *STREPTOBACILLUS MONILIFORMIS* AND *SPIRILLUM MINUS* INFECTIONS IN ANIMALS

Black rats and brown rats are the reservoir hosts of both *S. moniliformis* and *Spirillum minus*. Both pathogens are part of the common bacterial flora that live in the nose, throat, and trachea (i.e., thin-walled tube made of cartilage extending from the nose and mouth to the lungs) of healthy rats. Neither pathogen causes illness in rats. Among wild rats, 25% may be harboring *Spirillum minus* and 50% harboring *S. moniliformis*; infected rats excrete the bacteria in their urine or feces. *S. moniliformis* also live in the intestines of mice and chickens. Despite its presence in the intestines of house mice, *S. moniliformis* can sicken or kill mice if the pathogen invades their bodies, causing spleen and liver abscesses, arthritis in multiple joints, kidney infections, and blood poisoning. Chronic infections can continue for months. Abortions and arrested pregnancies can result when pregnant mice become infected (Taylor et al. 1994, Will 1994, Wullenweber 1994, Gaastra et al. 2009).

When domesticated guinea pigs, turkeys, koala, rhesus monkeys, and red-bellied titi monkeys are infected with *S. moniliformis*, they exhibit clinical signs similar to rat-bite fever in humans. Predators of rats including dogs, cats, and ferrets can become infected after eating an infected rat or having one in their mouth. The presence of DNA from *S. moniliformis* was found in 15% of dogs that had access to rats, yet few dogs ever become ill. If illness does result, dogs have clinical signs that include diarrhea, vomiting, anorexia, arthritis, and death. However, this pathogen rarely causes illness in most animals (Will 1994, Valverde et al. 2002, Gaastra et al. 2009).

8.4 HOW HUMANS CONTRACT RAT-BITE FEVER

Most people with rat-bite fever became infected after being bitten by a black rat or brown rat or less commonly by the bite of mice, gerbils, domesticated guinea pigs, or squirrels. People can also acquire rat-bite fever while handling infected rodents or through exposure to predators (usually domestic cats and dogs) that have preyed on infected rodents. In California, over 80% of patients with rat-bite fever had either recent exposure to a rodent or had been bitten by one. The pathogen can also be spread through saliva, nasal discharge, feces, and urine. Haverhill fever, a form of rat-bite fever, occurs from consuming food or water that has been contaminated by infected rodents (Graves and Janda 2001).

Rat-bite fever is not contagious; there are no known cases where it spread from one person to another. Given the presence of *S. moniliformis* and *Spirillum minus* in the nose, mouth, and throat of numerous rats, it is surprising that rat-bite fever is not more common in humans.

8.5 MEDICAL TREATMENT

The CDC (2011) recommends that anyone exhibiting symptoms of rat-bite fever after being exposed to rats or other rodents should seek medical attention immediately. Most people with rat-bite fever make a complete recovery within 2 weeks, even without antibiotic treatment. However, serious complications and death can result if antibiotics are not administered. Death can occur within 3 days of the onset of symptoms (Pollock et al. 2005). Hence, a person infected with *S. moniliformis* should not delay in seeking medical help. Rat-bite fever is a rare disease, and a doctor may not initially consider it as a possible diagnosis. Hence, patients should inform their physician if they have been bitten by a rodent or exposed to them during the previous 3 weeks.

According to Schachter et al. (2006), treatment for rat-bite fever is penicillin G given intravenously for 7–10 days, followed by penicillin V administered orally for 7 days. Tetracycline is considered the best alternative for patients who are allergic to penicillin (Freunek et al. 1997, Gaastra et al. 2009).

SIDEBAR 8.2 KEEPING RODENTS OUT OF BUILDINGS (CONOVER 2002)

Most people realize that their home contains rats or mice when they see the animals scurrying about or notice their feces in drawers, cabinets, pantries, or under furniture. Adult mice are small, 3–3.5 in. (8–9 cm) in length not counting their 2.5–3.5 in. (6–9 cm) tail; their feces are about 0.2 in. (0.5 cm) in length (Figures 8.2 and 8.3). Adult rats are twice as big; their body lengths are 7–10 in. (18–25 cm), not counting their 5–8 in. (13–20 cm) tail. Likewise, rat feces are twice as large as those of mice.

Rats and mice often infest buildings where they can obtain food. Rodents eat foods containing seeds and grains, including nuts, cereals, crackers, bread, and cookies. Thus, these foods should be removed or stored in rodent-proof containers (e.g., metal, glass, or plastic containers with a tight lid). As long as rodents have access to food, getting rid of them will be difficult. For the same reason, trash can lids should fit tightly, areas around garbage dumpsters should be kept clean, and dumpster drains should be equipped with rodent screening.

Rodents are resourceful at finding ways into rooms and buildings. Common entry points include vents, drains, pipes, spaces at the base of doors or windows, and openings in walls for electric, water, or gas lines. Rats can enter into buildings through holes as small as 0.5 in. (1.3 cm), and mice can enter through holes as small as 0.25 in. (0.6 cm). Hence, all holes greater than 0.25 in. should be sealed. When access points are blocked, rodents will often attempt to reopen them by gnawing, especially if they can see light along an edge. For this reason, patches should be larger than the hole and placed securely on the surface so no cracks remain. Vigilance is required to make sure rodents do not find or create new access points. Vents should be covered with 0.25 × 0.25 in. (0.6 cm) galvanized hardware cloth. Lastly, domestic cats can reduce rodent populations in buildings. However, they cannot keep a building free of rodents.

8.6 WHAT PEOPLE CAN DO TO REDUCE THEIR RISK OF CONTRACTING RAT-BITE FEVER

People who have rats, mice, gerbils, or other rodents as pets or who work in pet stores have an increased risk of contracting rat-bite fever. The risk is also higher for people who live in rat-infested buildings, technicians who work with laboratory rats, and wildlife biologists who work with free-ranging rodents (Sidebar 8.2).

People can reduce their risk of contracting rat-bite fever by avoiding contact with rodents and avoiding places where rodents are abundant. Any rodents living inside rooms or buildings should be trapped and holes providing access to rodents should be sealed. Rooms or buildings contaminated with rodent feces or urine should be decontaminated (Sidebar 8.3). The prevalence of rat-bite fever and urban rat populations can be reduced through better garbage disposal, public sanitation, and rat control.

(a)

(b)

FIGURE 8.3 Photos of (a) brown rat and (b) a house mouse. These rodents are commonly found in homes, barns, and other buildings. (Rat photo courtesy of CDC; mouse photo courtesy of Lucille K. Georg and the CDC.)

SIDEBAR 8.3 CDC (2011) RECOMMENDED STEPS FOR CLEANING UP AFTER RODENTS

1. All rodents should be removed by trapping them; trapping should continue for at least a week after the last rodent was captured. Trapping rodents is better than using poisons because the bodies of trapped rodents are easily removed but poisoned rodents can die in inaccessible locations where carcasses can produce a stench and serve as a continued source of infection.
2. All windows and doors should be opened 30 minutes prior to cleaning to allow fresh air to enter the area. When possible, window and doors on different sides of the room should be opened to increase airflow through cross ventilation.
3. Wear rubber, latex, or vinyl gloves when cleaning surfaces contaminated with rodent urine or feces.
4. Spray rodent urine, feces, or nests with a disinfectant and let soak for at least 5 minutes.
5. Do not stir up dust by sweeping or vacuuming up rodent feces and urine. Instead use a paper towel to wipe them up and place them in a covered garbage can.
6. After removal of rodent feces and urine, items in the room should be disinfected. Floors should be mopped, countertops and drawers cleaned with a disinfectant, and any bedding or clothing washed with laundry detergent and hot water. Upholstered furniture and carpets should be steam cleaned or washed with a commercial disinfectant.
7. When finished, gloves need to be sterilized with a disinfectant. Hands should be washed thoroughly with soap and water.

When handling dead or living rodents, people should wear protective gloves. People should avoid drinking unpasteurized milk or water from sources that are not protected from rodents. People should also not consume food contaminated by rodents. The risk of Haverhill fever can be reduced through chlorination of water supplies, pasteurization of milk, and protection of food supplies from rodents. There are no vaccines against rat-bite fever.

The risk of a person contracting rat-bite fever after being bitten by a rat is unknown, but the risk is small. However, it is important to remember that rats are reservoir hosts for many zoonotic diseases and that 10% of people bitten by a rat develop some type of infection. Hence, anyone bitten by a wild or pet rat should be alert to signs of an infection. Antibiotics are often prescribed after a rat bite to prevent rat-bite fever from developing (Khatchadourian et al. 2010).

8.7 ERADICATING RAT-BITE FEVER FROM A COUNTRY

The pathogens responsible for rat-bite fever occur in the nose, throat, and trachea of free-ranging black rats and brown rats and in the intestines of house mice. These rodents have spread to all corners of the world. It will be difficult to eradicate rat-bite fever in any country where these rodents are endemic. Most humans contract rat-bite fever after being bitten by a rat, often a family pet. For this reason, the most efficacious method to reduce the incidence of rat-bite fever is to ban the sale of rats at pet stores.

LITERATURE CITED

Booth, C. M., K. C. Katz, and J. Brunton. 2002. Fever and a rat bite. *Canadian Journal of Infectious Diseases* 13:269–272.

CDC. 2011. Rat-bite fever. http://www.cdc.gov/rat-bite-fever/symptoms/index.html (accessed December 1, 2012).

Conover, M. R. 2002. *Resolving Human–Wildlife Conflicts: The Science of Wildlife Damage Management*. CRC Press, Boca Raton, FL.

Costas, M. and R. J. Owen. 1987. Numerical analysis of electrophoretic protein patterns of *Streptobacillus moniliformis* strains from human, murine and avian infections. *Journal of Medical Microbiology* 23:303–311.

Freunek, K., A. Turnwald-Maschler, and J. Pannenbecker. 1997. Rattenbissfieber Infektion mit *Streptobacillus moniliformis* (rat-bite fever infection with *Streptobacillus moniliformis*). *Monatsschrift Kinderheilkunde* 145:473–476.

Gaastra, W., R. Boot, H. T. K. Ho, and L. J. A. Lipman. 2009. Rat-bite fever. *Veterinary Microbiology* 133:211–228.

Graves, M. H. and J. M. Janda. 2001. Rat-bite fever (*Streptobacillus moniliformis*): A potential emerging disease. *International Journal of Infectious Diseases* 5:151–155.

Khatchadourian, K., P. Ovetchkine, P. Minodier, V. Lamarre et al. 2010. The rise of the rats: A growing paediatric issue. *Paediatrics and Child Health* 15:131–134.

Pollock, W. J., J. Lanza, S. Buck, P. A. Williams et al. 2005. Fatal rat-bite fever—Florida and Washington, 2003. *Morbidity and Mortality Weekly Report* 53:1198–1202.

Rupp, M. E. 1992. *Streptobacillus moniliformis* endocarditis: Case report and review. *Clinical Infectious Diseases* 14:769–772.

Schachter, M. E., L. Wilcox, N. Rau, D. Yamamura et al. 2006. Rat-bite fever, Canada. *Emerging Infectious Diseases* 12:1301–1302.

Taylor, J. D., C. P. Stephens, R. G. Duncan, and G. R. Singleton. 1994. Polyarthritis in wild mice (*Mus musculus*) caused by *Streptobacillus moniliformis*. *Australian Veterinary Journal* 71:143–145.

Valverde, C. R., L. J. Lowenstine, C. E. Young, R. P. Tarara, and J. A. Roberts. 2002. Spontaneous rat bite fever in non-human primates: A review of two cases. *Journal of Medical Primatology* 31:345–349.

Will, L. A. 1994. Rat-bite fevers. In: G. W. Beran, editor. *Handbook of Zoonoses*, 2nd edition. CRC Press, Boca Raton, FL, pp. 231–244.

Wullenweber, M. 1994. *Streptobacillus moniliformis*—A zoonotic pathogen. Taxonomic considerations, host species, diagnosis, therapy, geographical distribution. *Laboratory Animals* 29:1–15.

9 Salmonellosis

Humpty Dumpty lay on the ground a crushed and broken fella, no one wanted to put him together, cause he had salmonella.

Edwards (1999)

Every year, approximately 40,000 cases of salmonellosis are reported in the United States. Because many milder cases are not diagnosed or reported, the actual number of infections may be thirty or more times greater.

CDC (2013a)

9.1 INTRODUCTION AND HISTORY

Salmonellosis is caused by bacteria of the genus *Salmonella* and is one of the most widespread and dangerous foodborne diseases in the United States. The genus was named after Daniel Salmon, an American veterinary surgeon and bacteriologist who first isolated the bacteria in 1885. These bacteria are rod shaped and Gram negative (Figure 9.1). Most *Salmonella* species are capable of movement and can survive either with or without oxygen.

Human infections are caused primarily by *S. enterica*, a pathogen that produces three different human diseases. One of them is typhoid fever (also called typhoid) caused by *S. enterica* serovar Typhi and another is paratyphoid fever caused by *S. enterica* serovar Paratyphi (bacteria are considered to be of the same serovar if they have the same antigenic makeup). Neither disease should be confused with typhus, which is the subject of Chapter 17. Both *S. enterica* serovar Typhi and *S. enterica* serovar Paratyphi are able to penetrate intestinal walls; suppress the host's defenses; invade blood, lymph nodes, and internal organs (e.g., spleen and liver); and multiply there. The pathogens then reinfect the intestines, and the bacteria are shed in feces.

Typhoid fever is a serious disease in humans with clinical signs including fever, vomiting, rash, distended abdomen, delirium, exhausted stupor, and sometimes death. The disease is prevalent in countries that lack adequate sanitation and safe drinking water. The highest incidence rate of typhoid fever occurs in some of the developing countries of Asia, Africa, Latin America, and the Caribbean. Worldwide, typhoid fever sickened an estimated 21 million people and caused 216,000 fatalities in 2000 (Crump et al. 2004). During 2010, an estimated 27 million people contracted typhoid fever (Buckle et al. 2012). Typhoid fever is less common in the United States, Canada, and Europe. Between 2000 and 2006, there were approximately 1821 cases of typhoid fever annually in the United States but few deaths (Scallan et al. 2011). Typhoid fever is not a zoonotic disease because the pathogen responsible for typhoid fever (*S. enterica* Typhi) only infects humans. Paratyphoid fever is similar to typhoid fever but has milder symptoms, does not last as long, and has a lower mortality rate.

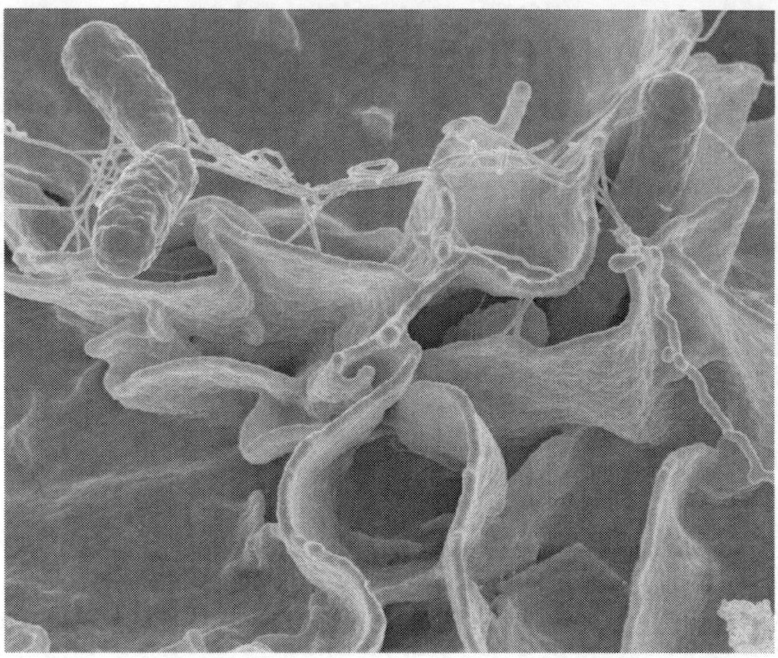

FIGURE 9.1 Photo of several rod-shaped *Salmonella* invading a culture of human cells. (Courtesy of Rocky Mountain Laboratories, National Institute of Allergy and Infectious Diseases, National Institutes of Health.)

Worldwide, paratyphoid fever sickens over five million people annually (Crump et al. 2004). Neither typhoid nor paratyphoid will be discussed further in this book.

Another disease caused by *S. enterica* is salmonellosis, which results in enteritis (i.e., inflammation of the small intestine) with symptoms of diarrhea and abdominal cramps. While only one serovar of *S. enterica* causes typhoid fever, about 2000 different serovars cause salmonellosis in humans or animals. Only three serovars (Enteritidis, Typhimurium, and Newport) are responsible for about half of all human cases of salmonellosis in the United States. The serovars that cause salmonellosis infect the digestive system, especially the intestines, but usually do not penetrate intestinal walls and do not cause a systemic infection. During 2011, out of an estimated 1,028,000 people in the United States stricken with salmonellosis, 19,300 of them required hospitalization, and 378 died. The incidence of confirmed cases of salmonellosis has changed little in the United States from 1996 to 2011 (Scallan et al. 2011; CDC 2012, 2013b).

An outbreak of salmonellosis occurs when several people become ill from the same source and from the same strain of *S. enterica*. The latter is determined by examining the pathogen's DNA. The CDC (2012) reported 12 outbreaks of salmonellosis in the United States caused by 14 different serotypes of *S. enterica* in 2012. Three of the outbreaks involved contaminated poultry; other outbreaks resulted from contaminated peanut butter, ground beef, dry dog food, raw ground

SIDEBAR 9.1 OUTBREAKS OF SALMONELLOSIS LINKED TO HEDGEHOGS (HANDELAND ET AL. 2002, CDC 2013)

From December 2011 through August 2012, the CDC identified 14 people who had been infected by *S. enterica* Typhimurium. The patients were located in five states in the Midwest and Washington. Half of the patients were less than 10 years old. Three patients required hospitalization, but no deaths were reported.

When interviewed, all patients reported keeping hedgehogs as pets or having had contact with someone else's hedgehog. Some patients specifically noted contact with African pygmy hedgehogs. The same strain of *S. enterica* Typhimurium found in patients was isolated from hedgehogs that they had handled. Because the hedgehogs were purchased from multiple breeders located in several different states, it was unclear if there had been a single source of infection. The CDC noted that hedgehogs are a known source of several infections in humans. Infected hedgehogs shed *S. enterica* in their feces. Furthermore, the skin of hedgehogs and the areas they frequent can easily become infected with the pathogen. People should carefully wash their hands immediately after touching a hedgehog or areas where they live or roam.

An outbreak of salmonellosis in humans occurred during 1996 in Moss, Norway, and during 2000 in the Norwegian municipalities of Askøy, Bergen, and Os (Figure 9.2). Epidemiology investigations failed to find a single source of infection. In both areas, there were large populations of wild European hedgehogs (Figure 9.3). Some of these animals had habituated to humans and came into gardens and parks seeking food handouts. To determine if they were the source of the infection, several hedgehogs were collected, and 40% of those from endemic areas were infected with the same strain of *S. enterica* Typhimurium as the patients. Hedgehogs collected elsewhere in Norway were uninfected. Where people fed them, 71% of hedgehogs were infected, but infection rates dropped to 25% away from feeding sites. Handeland et al. (2002) concluded that hedgehogs were the source of the pathogen that caused the outbreak of salmonellosis in humans.

tuna, a restaurant chain, mangoes, cantaloupes, pet hedgehogs (Sidebar 9.1), and pet turtles. Sometimes, salmonellosis outbreaks come from unusual sources; one outbreak in 2010 was a result of frozen rodents that were in demand as food for pet snakes.

9.2 SYMPTOMS IN HUMANS

The first signs of salmonellosis begin after an incubation period of 12–72 hours and include fever, diarrhea, and abdominal cramps; these usually last 4–7 days, and most patients recover without the use of antibiotics. However, diarrhea can become so

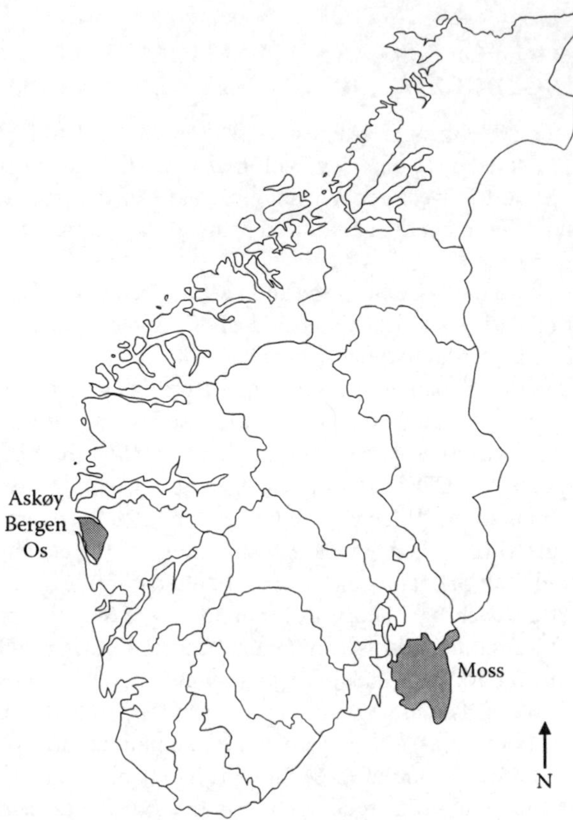

FIGURE 9.2 Map of southern Norway showing the location of Moss and Bergen where outbreaks of salmonellosis in humans were linked to wild hedgehogs. (From Handeland, K. et al., *Epidemiol. Infect.*, 128, 523, 2002.)

severe that patients become dehydrated and need hospitalization so that fluids can be given intravenously. Patients should drink plenty of fluids, and young children can be given an oral rehydration solution, such as Pedialyte®. Warning signs of dehydration include dry mouth and tongue, sunken eyes, inability to produce tears, and a decrease in urine (Mayo Clinic 2011).

Some salmonellosis patients develop a complication called Reiter's syndrome in which the body tries to mount a defense against a pathogen. Symptoms include painful joints, eye irritation, and pain when urinating. Reiter's syndrome is difficult to treat, can persist for months, and can result in chronic arthritis (Mayo Clinic 2011). Infants and the elderly have a higher risk of developing a serious illness if they contract salmonellosis.

Because many different diseases cause diarrhea, fever, and abdominal pain, salmonellosis is diagnosed by isolating *S. enterica* from a patient's fecal sample. However, lab tests to confirm a diagnosis of salmonellosis and identify the serotype of *S. enterica* are conducted on only 5% of U.S. patients suspected of having salmonellosis.

FIGURE 9.3 Photo of a European hedgehog. (Courtesy of the British Wildlife Centre, Lingfield, Surrey, U.K.)

9.3 *SALMONELLA ENTERICA* INFECTIONS IN ANIMALS

While typhoid fever occurs only in humans, other mammalian species experience a similar disease called salmonellosis with septicemia. Only a few serovars (bacteria are considered to be of the same serovar if they have the same antigenic makeup) cause this disease, and the serovars that produce it vary among animal species (Table 9.1). This disease starts as an infection of the intestines, resulting in diarrhea. As the disease continues, localized infections can occur in the brain (resulting in meningoencephalitis), uterus of pregnant animals (resulting in abortions), and in the end of limbs, ears, and tails (resulting in gangrene). Salmonellosis with septicemia is especially prevalent among newborn calves, piglets, foals, and lambs; their mortality rates can approach 100%. The disease is less lethal after animals have lived long enough to acquire a protective intestinal microflora (Merck 2010).

The serovar *S. enterica* Dublin is endemic at many sheep farms and cattle ranches but does not cause illness among livestock; these animals shed the pathogen, posing a risk to other animals. The tonsils and lymph nodes of animals can have an infection of *S. enterica*, without the animals exhibiting illness or excreting the pathogen. Later, the animals can develop salmonellosis or start shedding the pathogen, especially when stressed by moving to a new location, crowding, or calving. In contrast, the serovar *S. enterica* Typhimurium can cause an explosive outbreak of enteritis when first introduced into a herd. Outbreaks often occur when infected calves are brought in from another farm (Merck 2010). Clinical signs of salmonellosis in cattle include high fevers of 105°F–107°F (40.5°C–41.5°C), severe

TABLE 9.1
More Common *S. enterica* Serovars that Cause Enteritis in Domesticated Animals

Animal	Common Serovars of *S. enterica*
Horses	Typhimurium, Anatum, Newport, Enteritidis
Cattle	Typhimurium, Dublin, Newport
Sheep	Typhimurium, Dublin, Abortusovis, Anatum, Montevideo
Goats	Typhimurium, Dublin, Abortusovis, Anatum, Montevideo
Hogs	Typhimurium, Choleraesuis
Poultry	Typhimurium, Gallinarum, Pullorum
Wild birds	Typhimurium
Wild ungulates	Typhimurium
Domestic cats	Typhimurium, Heidelberg

Sources: Hall, A.J. and Saito, E.K., *J. Wildl. Dis.*, 44, 585, 2008; Philbey, A.W. et al., *Vet. Rec.*, 164, 120, 2009; Merck, *Merck Veterinary Manual*, 10th edn., Merck, Whitehouse Station, NJ, 2010.

diarrhea, and a decline in milk production in dairy cows. If salmonellosis becomes chronic, pregnant animals may abort (Tauni and Österlund 2000, Merck 2010).

Almost all wild mammals can become infected by *S. enterica.* In California, the pathogen has been isolated from Virginia opossum, coyote, mule deer, elk, feral hog, and skunk (Gorski et al. 2011). *S. enterica* was isolated from 1% of free-ranging white-tailed deer shot by hunters in Nebraska and 8% in Texas (Branham et al. 2005, Renter et al. 2006); four *S. enterica* serovars isolated from deer in Nebraska can cause human disease (Dessau, Enteritidis, Infantis, and Litchfield). *S. enterica* was not detected in wild ungulates in Finland, Norway, Sweden, or Poland, but it has been found in ungulates from southern European countries (Pagano et al. 1985, Wahlström et al. 2003, Paulsen et al. 2012). In Spain, the pathogen was isolated from 41% of wild reptiles; furthermore, 38% of the *S. enterica* isolates from wild reptiles were known to cause human disease (Briones et al. 2004).

Infection rates can be high in wild carnivores that become infected from consuming infected prey. European investigations have reported *S. enterica* was isolated from fecal material from 7% to 18% of European badgers and 7% of red foxes. Among feral hogs, the pathogen was found in fecal samples from 47% of feral hogs in Slovenia, 22% in Portugal, and 7% in Spain (Gaffuri and Holmes 2012).

The pathogen is less prevalent in deer than in carnivores and feral hogs. In Italy, *S. enterica* was isolated from 19% of wild boars harvested by hunters but only 2% of European roe deer, 1% of red deer, and 0% of alpine chamois (Magnino et al. 2011). The low infection rates in wild deer result from deer living in small groups, limiting opportunities for the pathogen to spread from one individual to another.

Most wild mammals infected with *S. enterica* are asymptomatic, but the pathogen produces serious illness or death in some wildlife, such as feral hogs, elk, Florida Key deer, red deer, and sika deer (McAllum et al. 1978, Sato et al. 2000,

Foreyt et al. 2001, Nettles et al. 2002, Paulsen et al. 2012). The risk of illness is greatest for very young animals. The risk of a wild animal becoming ill increases when it is fed by humans or when it resides near farms, ranches, or cities. Hence, humans and domesticated animals are more likely to spread *S. enterica* to wildlife than vice versa. For example, an outbreak of salmonellosis in wild alpine chamois occurred when they foraged in a pasture containing infected cattle (Paulsen et al. 2012).

In birds, *S. enterica* infections first start among the cells lining the cecum, esophagus, and crop. All avian species can be infected by *S. enterica*, but in healthy adult birds, the pathogen does not usually cause illness. Clinical signs of salmonellosis in birds include ruffled feathers, diarrhea, rapid breathing, lesions in the esophagus that appear as yellow nodules, and lesions in the liver that appear as nodules or spots. Most birds that exhibit signs of salmonellosis die from the infection (Hall and Saito 2008, Gaffuri and Holmes 2012).

A significant number of wild birds die each year from salmonellosis, particularly if infected by the serovar *S. enterica* Typhimurium (Figure 9.4). In the United States, there were 186 reported outbreaks of salmonellosis among birds from 1985 to 2004. Salmonellosis is a significant source of mortality for evening grosbeak, common redpoll, pine siskin, American goldfinch, and northern cardinal (Hall and Saito 2008). In Great Britain, Pennycott et al. (2006) examined 779 wild birds that were found dead by the public; salmonellosis was responsible for more than half of all deaths in house sparrows and goldfinches (Table 9.2). In Norway, *S. enterica* was isolated from half of all Eurasian bullfinches, European greenfinches, Eurasian siskins, and common redpolls found dead by the public (Table 9.3).

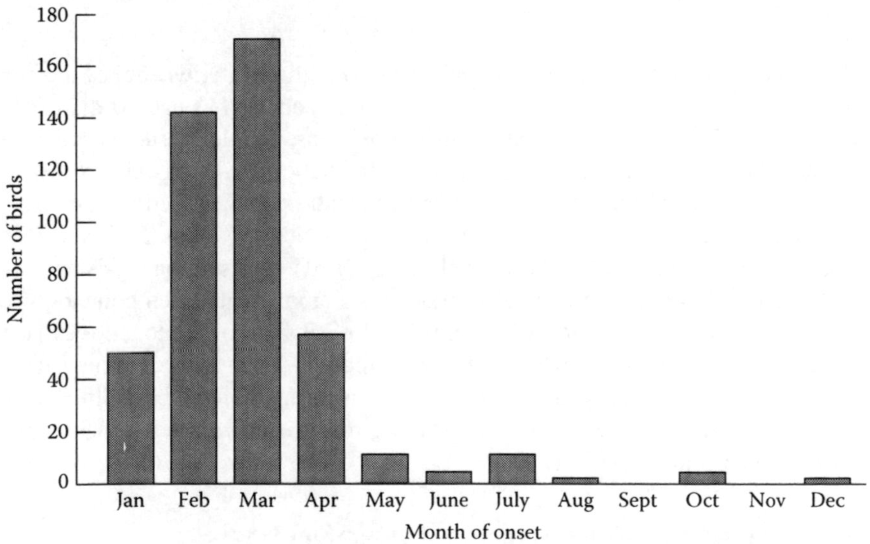

FIGURE 9.4 Monthly variation in the number of dead birds from which *S. enterica* were isolated. Birds were found dead by the public in Norway from 1969 to 2000 and turned into the National Veterinary Institute. (From Refsum, T. et al., *Appl. Environ. Microbiol.*, 68, 5595, 2002.)

TABLE 9.2
Percent of Dead Birds Found by the Public in Great Britain from which *S. enterica* Were Isolated

	% Birds with *S. enterica*	No. of Birds Sampled
Greenfinch	72	190
House sparrow	66	50
Gull (immature)	7	41
Goldfinch	60	10
Herring gull (adult)	0	32
Rook	11	19
Feral pigeon	9	53
European starling	0	49
Mute swan	0	23
Eurasian blackbird	0	22
Carrion crow	0	21
Wood pigeon	0	19
Eurasian collared dove	0	15
Lesser black-backed gull (adult)	0	12
Eurasian jackdaw	0	11

Note: The authors of the study (Pennycott et al. 2006) hypothesized that birds infected with this pathogen likely died from it.

Many colonial birds contract salmonellosis during the winter when they are concentrated at roosting sites. In North America, salmonellosis has caused die-offs of colonial waterbirds including American white pelicans, double-crested cormorants, and ring-billed gulls (Hall and Saito 2008). Salmonellosis kills passerines primarily during the winter because birds are often concentrated when feeding, especially at bird feeders, allowing *S. enterica* to spread easily among them (Hall and Saito 2008, Pennycott et al. 2010, Velarde et al. 2012). Birds with salmonellosis are easy prey for domestic cats; outbreaks of salmonellosis among cats often coincide with outbreaks among passerine birds (Sidebar 9.2). Clinical signs of salmonellosis in cats include lethargy, anorexia, and diarrhea. Other predators, such as birds of prey, target sick birds and can develop salmonellosis after consuming infected birds. In Europe, *S. enterica* has been isolated from common buzzards, griffon vultures, and little owls (Molina-Lopez et al. 2011).

9.4 HOW HUMANS CONTRACT SALMONELLOSIS

People become infected from eating food or drinking liquids contaminated with minute amounts of fecal material from infected humans or animals. Although food can become contaminated, most salmonellosis patients become ill after consuming beef, pork, chicken, fish, milk, or eggs. In Germany, public health authorities

TABLE 9.3
Percent of Dead Birds Found by the Public in Norway from which *S. enterica* Were Isolated, 1969–2000

Avian Species	% Birds with *S. enterica*	No. of Birds Sampled
Eurasian bullfinch	81	242
European greenfinch	80	71
Common redpoll	78	41
Eurasian siskin	75	69
House sparrow	23	31
Great tit	21	14
Blue tit	13	14
Yellow hammer	13	15
Common chaffinch	10	21
Brambling	6	16
Rock dove	4	72
Black-billed magpie	3	40
Hooded crow	2	52

Source: Refsum, T. et al., *Appl. Environ. Microbiol.*, 68, 5595, 2002.

SIDEBAR 9.2 ROLE OF BACKYARD BIRD FEEDERS AS SITES WHERE SALMONELLOSIS CAN SPREAD AMONG BIRDS (GIOVANNINI ET AL. 2012)

Winter can concentrate some avian species at roosts and foraging sites, such as bird feeders in backyards. Fecal material often accumulates at these sites, providing an opportunity for *S. enterica* to spread among birds and cause avian die-offs. One such event occurred in Switzerland during February and March 2010 when many Eurasian siskins (Figure 9.5) were found ill or dead in backyards with bird feeders. That winter was particularly harsh, and the density of Eurasian siskins was unusually high at bird feeders. Postmortem examinations of the Eurasian siskins showed that they were in poor health and had lesions on their liver and spleen. *S. enterica* Typhimurium was isolated from the intestines of most birds.

Concomitantly, local veterinarians reported an increase in the number of domestic cats with fever, anorexia, and vomiting. Most of the ill cats spent time outdoors and had a history of hunting wild birds. Rectal samples were collected from eight cats, and all contained *S. enterica* Typhimurium. The sick birds were easy targets for domestic cats and other predators, which became ill after consuming them.

FIGURE 9.5 Photo of a Eurasian siskin.

regularly sample meat for *S. enterica*. The pathogen was isolated in 0.5% of beef, 2.8% of game meat, and 3.3% of pork (Paulsen et al. 2012). The pathogen also occurs in meat from waterfowl, pheasants, and hares. Game meat is more likely to be contaminated with *S. enterica* if the intestines were perforated when the animal was shot or being cleaned. People can also contract *S. enterica* when cleaning wild game (Paulsen et al. 2012).

Foods can be contaminated during cleaning, packing, and processing if washed in contaminated water, handled by infected workers, or infected rodents occupied the facilities. Thorough cooking kills *S. enterica*, but cooked items can become contaminated again if the cook and food handler have not adequately washed their hands. In one outbreak, more than 220,000 people in the United States became infected with *S. enterica* from ice cream that was contaminated following pasteurization. The source of the infection was traced back to tanker trucks used to haul the ice cream premix. The same trucks had been used earlier to transport unpasteurized liquid eggs. The egg residue then contaminated the ice cream (Hennessy et al. 1996).

Fruits and vegetables can become infected with *S. enterica* when grown in fields visited by infected wild animals or irrigated with contaminated water. Preventing such outbreaks is challenging because there are many environmental sources of *S. enterica*, and the pathogen can survive for months in pond water, soil, or dried feces (Gaffuri and Holmes 2012). In one part of California where fresh vegetables are grown, *S. enterica* was isolated from samples of water, soil, and wildlife, including birds, mule deer, elk, feral hogs, and skunks (Gorski et al. 2011).

SIDEBAR 9.3 PET TURTLE SALES IN WISCONSIN LEAD TO SALMONELLOSIS IN YOUNG CHILDREN (CDC 2005)

In the mid-1970s, the Food and Drug Administration (FDA) banned commercial distribution of small pet turtles in the United States (with shells less than four inches) due to concerns over salmonellosis (Figure 9.6). However, pet turtles continued to be sold illegally in pet stores and souvenir shops in many states. In the summer of 2004, three separate incidences of salmonellosis were identified in children under the age of 10 whose families had recently purchased small turtles from souvenir shops throughout Wisconsin. Patients exhibited clinical signs including diarrhea, fever, and vomiting. In all three patients, cultures of stool samples revealed a rare serotype *S. enterica* Pomona. That same summer, the Wisconsin Division of Public Health (WDPH) became aware of a resurgence in illegal turtle sales or giveaways among souvenir shops near the state's tourist destinations. When informed of the ban on small turtle distribution by local health departments, most Wisconsin shops discontinued the sale of turtles and returned them to their suppliers. One retailer refused to comply, forcing an emergency order from health officials to stop the sale of turtles. The retailer sent a sample supply of turtles to a laboratory to be tested for salmonellosis. Test results indicated that one turtle from the sample group carried *S. enterica* Pomona.

Many salmonella patients became infected from wild birds that shed *S. enterica* in their feces. In Norway, an outbreak of *S. enterica* Typhimurium sickened 349 people; concomitantly, there was an outbreak of the same pathogen among wild passerines. In one British hospital, 160 people developed salmonellosis after house sparrows got into the kitchen (Gaffuri and Holmes 2012).

People can contract the pathogen from their pets, including cats, reptiles, and birds. Pet turtles (Sidebar 9.3) and young birds are particularly likely to harbor the pathogen. Petting animals at farms, zoos, wildlife centers, and fairs increases children's risk of contracting salmonellosis. This risk can be mitigated if parents require their children to wash their hands carefully after touching the animals (Friedman et al. 1998).

9.5 MEDICAL TREATMENT

Cases of salmonellosis usually resolve themselves within 5–7 days without treatment. Hence, antibiotics are often unnecessary unless the pathogen spreads from the intestines, and the infection becomes systemic. Patients should drink plenty of fluids to avoid becoming dehydrated; those with severe diarrhea may need to be provided fluid intravenously.

Some *S. enterica* strains have become resistant to antibiotics. Some strains of *S. enterica* Typhimurium are now resistant to at least five antibiotics, strains of *S. enterica* Newport are resistant to at least seven antibiotics, and the strain of *S. enterica*

FIGURE 9.6 Red slider turtles are often sold for pets but can serve as a reservoir for *Salmonella*. (Courtesy of Eric Grafman and the CDC.)

Heidelberg that sickened 136 people during a 2011 outbreak in the United States is resistant to several commonly prescribed antibiotics (Institute of Medicine 2012). This increasing resistance of *S. enterica* strains to antibiotics is a major concern among medical doctors who are left with fewer options to treat infections (CDC 2013a).

9.6 WHAT PEOPLE CAN DO TO REDUCE THEIR RISK OF CONTRACTING SALMONELLOSIS

No human vaccine exists that can prevent salmonellosis, but humans have some natural defenses against *S. enterica* infections, such as stomach acid; however, taking antacids reduces this defense by lowering the stomach's acidity. Another defense is provided by the normal bacteria that populate healthy intestines, which can prevent the establishment of *S. enterica*. Use of oral antibiotics can alter the normal bacterial flora found in the intestines and may increase, rather than decrease, the risk of salmonellosis (Mayo Clinic 2011).

Hens can infect their eggs with *S. enterica* even before the shell is created. Hence, an egg's shell is not a barrier to the pathogen. Eggs should be cooked thoroughly, and people should avoid foods containing raw eggs. In 2010, eggs from two poultry farms were responsible for an outbreak of *S. enterica* Enteritidis that sickened more than 1900 people in the United States. Both poultry farms bought chicken feed from the same feed mill. Tests showed the feed was contaminated with the pathogen. The outbreak led to the voluntary recall of 500 million eggs (Institute of Medicine 2012).

Care should be taken so that *S. enterica* from eggs, poultry, seafood, and meat are not accidentally spread to other foods that are normally eaten raw, such as fruits and vegetables. Cooks and food handlers need to wash their hands, cooking utensils, and work surfaces immediately after they have been in contact with eggs, raw meat, or chicken. Food should not be placed on unwashed plates that previously held meat. It is wise to have two cutting boards in the kitchen—one for raw meat and another for fruits, vegetables, and cooked meat.

Public health agencies investigate confirmed cases of salmonellosis to determine the specific type of *S. enterica* and its source. Once the source has been identified, public health authorities can take corrective steps to prevent future illness. As one example, the United States banned the sale of pet turtles in 1975 after several people acquired salmonellosis from them (Sidebar 9.3). After an outbreak of *S. enterica* Heidelberg was linked to a turkey processing plant, 36 million pounds (16 million kg) of turkey products were recalled, and the processing plant was disassembled, steam cleaned, washed with an antiseptic solution, and equipped with a modern system to detect pathogens (Institute of Medicine 2012, CDC 2011).

The risk of salmonellosis in livestock can be reduced by purchasing animals from disease-free farms and isolating new animals for at least a week while their health is monitored. Feed should only be purchased from reliable sources. Use of pelleted feed can reduce the risk of *S. enterica* in livestock because the heat required to form the pellet is usually sufficient to kill the pathogen. Efforts to reduce the prevalence of *S. enterica* at poultry and livestock operations are more likely to succeed if rodents and birds can be eliminated at farms (Figure 9.8; Sidebar 9.4). European governments have implemented programs to reduce levels of *S. enterica* in livestock and poultry. The program includes the compulsory survey of pigs and poultry for *S. enterica* (Merck 2010, Gaffuri and Holmes 2012). Wildlife raised in European game farms must be inspected and treated just like any livestock except that wildlife may be killed on-site instead of at a slaughterhouse. A trained inspector must examine the carcass and inner organs of game animals for signs of infection. If this examination raises concerns, a veterinarian has to examine the carcass.

The incidence of salmonellosis in humans can be lowered by reducing the number of livestock and poultry that are infected with *S. enterica*. Rodents and birds can be vectors of *S. enterica* among ranches, dairy farms, and feedlots (Figure 9.8). *S. enterica* have been detected in birds that are concentrated during the winter at roosts or feeding sites.

Once an outbreak of salmonellosis occurs at a farm, a strict disease management program should be implemented that includes protecting water and feed from fecal material. Movement of livestock needs to be restricted to minimize the risk of infection spreading to a larger group of animals. Infected animals need to be identified and then culled or isolated and treated. Contaminated straw or other material must be destroyed, and contaminated buildings must be cleaned and disinfected. A vaccine has been licensed for use in pigs that uses a live but attenuated (i.e., weakened) strain of *S. enterica* (Merck 2010).

People can contract *S. enterica* by consuming wild game meat. Hunters should avoid shooting animals that appear ill. When hunting ungulates or feral hogs, a shot should not be taken if the bullet might perforate the digestive system. Game animals

SIDEBAR 9.4 CAN CONTROLLING NUMBERS OF EUROPEAN STARLINGS AT A FEEDLOT REDUCE THE PREVALENCE OF *S. ENTERICA* IN CATTLE? (CARLSON ET AL. 2011, 2012)

One way to reduce the incidence of salmonellosis in humans is to reduce the number of cattle that are infected with *S. enterica*, but this can be difficult when infected European starlings are abundant at dairy farms and feedlots. To determine if reducing starling numbers at cattle feedlots can reduce the risk of infection, Carlson et al. (2011, 2012) compared *S. enterica* levels at one feedlot, where starling populations were reduced 66% by using a toxicant, to another feedlot, where no toxicants were used and starling populations remained constant (reference feedlot). Before the toxicant was used (pretreatment period), 33% of water samples from watering troughs contained *S. enterica*, as did 8% of cattle feed samples (Figures 9.7 and 9.8). After the toxicant was applied (posttreatment period), *S. enterica* was found in only 5% of water samples and 0% of cattle feed samples. Concomitantly, the percent of water samples and cattle feed samples that contained the pathogen increased ninefold and threefold, respectively, at the reference feedlot. *S. enterica* was also isolated from 15% of cattle fecal samples from the treatment feedlot after the toxicant was applied, compared to 50% of cattle fecal samples from the reference feedlot.

FIGURE 9.7 Starlings congregate in large numbers at cattle feedlots. (Courtesy of George Linz and the USDA Wildlife Services.)

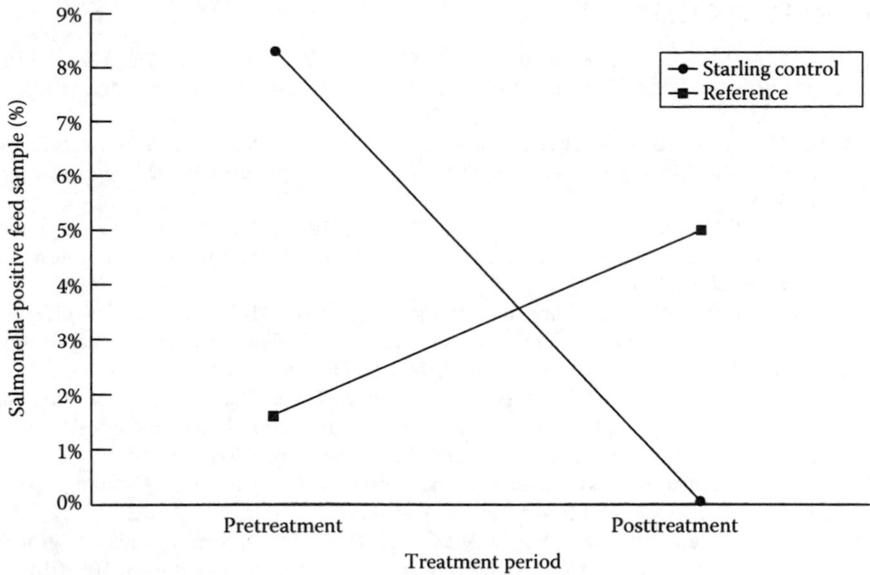

FIGURE 9.8 Percentage of cattle feed samples that tested positive for *S. enterica* at one feedlot where starling numbers were reduced (starling control) and at a nearby feedlot where starling numbers were not reduced (reference). The figure shows results both before starling numbers were reduced (pretreatment period) and after their numbers were reduced (posttreatment period). (Courtesy of James Carlson and George Linz.)

should be cleaned as soon as possible after they are shot; care should be taken not to contaminate the carcass with intestinal material. The meat should be cooled to below 44°F (7°C) as soon as possible (Paulsen et al. 2012).

Salmonellosis in wild birds is often associated with bird feeders. The risk of infection to birds can be reduced by regularly cleaning and disinfecting bird feeders and water baths. Careful placement of feeders away from branches or perch sites prevents perching birds from defecating into feeders and water. Accumulation of fecal material should be removed, and feeders and water baths should be relocated to new sites as needed.

Children under the age of five are five times more likely to become ill from salmonellosis than older children or adults. For this reason, reptiles, amphibians, baby chicks, or ducklings are not appropriate pets for young children. Everyone, but especially children, should wash hands with soap after handling reptiles, birds, their cages, bird baths, or bird feeders (CDC 2013a).

9.7 ERADICATING SALMONELLOSIS FROM A COUNTRY

S. enterica is so ubiquitous in livestock, poultry, and wildlife that it is impossible to eradicate it. Still, incidence of salmonellosis in humans can be lowered by reducing the number of infected animals.

LITERATURE CITED

Branham, L. A., M. A. Carr, C. B. Scott, and T. R. Callaway. 2005. *E. coli* 0157 and *Salmonella* spp. in white-tailed deer and livestock. *Current Issues in Intestinal Microbiology* 6:25–29.

Briones, V., S. Téllez, J. Goyache, C. Ballesteros et al. 2004. *Salmonella* diversity associated with wild reptiles and amphibians in Spain. *Environmental Microbiology* 6:868–871.

Buckle, G. C., C. L. Fischer Walker, and R. E. Black. 2012. Typhoid fever and paratyphoid fever: Systematic review to estimate global morbidity and mortality for 2010. *Journal of Global Health* 2(10401):1–9.

Carlson, J. C., J. W. Ellis, S. K. Tupper, A. B. Franklin, and G. M. Linz. 2012. The effect of European starlings and ambient air temperature on *Salmonella enterica* contamination within cattle feed bunks. *Human–Wildlife Interactions* 6:64–71.

Carlson, J. C., R. M. Engeman, D. R. Hyatt, R. L. Gilliland et al. 2011. Efficacy of European starling control to reduce *Salmonella enterica* contamination in a concentrated animal feeding operation in the Texas Panhandle. *BMC Veterinary Research* 7:9.

CDC. 2005. Salmonellosis associated with pet turtles—Wisconsin and Wyoming, 2004. *Morbidity and Mortality Weekly Report* 54:223–226.

CDC. 2011. Multistate outbreak of human *Salmonella* Heidelberg infections linked to ground turkey. http://www.cdc.gov/salmonella/heidelberg/092911 (accessed April 10, 2014).

CDC. 2012. CDC estimates of food borne illnesses in the United States. http://www.cdc.gov/foodborneburden/2011-foodborne-estimates.html (accessed December 10, 2012).

CDC. 2013a. Salmonella. http://www.cdc.gov/salmonella/ (accessed July 7, 2013).

CDC. 2013b. Summary of notifiable diseases—United States, 2011. *Morbidity and Mortality Weekly Report* 60(53):1–117.

Crump, J. A., S. P. Luby, and E. D. Mintz. 2004. The global burden of typhoid fever. *Bulletin of the World Health Organization* 82:346–353.

Edwards, R. W. 1999. Humpty dumpty. http://www.familypoet.com/?s=humpty+dumpty (accessed April 6, 2013).

Foreyt, W. J., T. E. Besser, and S. M. Lonning. 2001. Mortality in captive elk from salmonellosis. *Journal of Wildlife Diseases* 37:399–402.

Friedman, C. R., C. Torigian, P. J. Shillam, R. E. Hoffman et al. 1998. An outbreak of salmonellosis among children attending a reptile exhibit at a zoo. *Journal of Pediatrics* 132:802–807.

Gaffuri, A. and J. P. Holmes. 2012. Salmonella infection in wild mammals. In: D. Gavier-Widén, J. P. Duff, and A. Meredith, editors. *Infectious Diseases of Wild Mammals and Birds in Europe*. Wiley-Blackwell, Oxford, U.K., pp. 386–397.

Giovannini, S., M. Pewsner, D. Hüssy, H. Hächler et al. 2012. Epidemic of salmonellosis in passerine birds in Switzerland with spillover to domestic cats. *Veterinary Pathology* 50:597–606.

Gorski, L., C. T. Parker, A. Liang, M. B. Cooley et al. 2011. Prevalence, distribution, and diversity of *Salmonella enterica* in a major produce region of California. *Applied and Environmental Microbiology* 77:2734–2748.

Hall, A. J. and E. K. Saito. 2008. Avian wildlife mortality events due to salmonellosis in the United States, 1985–2004. *Journal of Wildlife Diseases* 44:585–593.

Handeland, K., T. Refsum, B. S. Johansen, G. Holstad et al. 2002. Prevalence of *Salmonella* Typhimurium infection in Norwegian hedgehog populations associated with two human disease outbreaks. *Epidemiology and Infection* 128:523–527.

Hennessy, T. W., C. W. Hedberg, L. Slutsker, K. E. White et al. 1996. A national outbreak of *Salmonella enteritidis* infections from ice cream. *New England Journal of Medicine* 334:1281–1286.

Institute of Medicine. 2012. *Improving Food Safety through a One Health Approach.* National Academies Press, Washington, DC.

Magnino, S., M. Frasnelli, M. Fabbi, A. Bianchi et al. 2011. The monitoring of selected zoonotic diseases of wildlife in Lombardy and Emilia-Romagna, northern Italy. In: P. Paulsen, A. Bauer, M. Vodnansky, R. Winkelmayer et al., editors. *Game Meat Hygiene in Focus: Microbiology, Epidemiology, Risk Analysis and Quality Assurance.* Wageningen Academic Publishers, Wageningen, Netherlands, pp. 223–244.

Mayo Clinic. 2011. Salmonella infection. http://www.mayoclinic.org/health/salmonella/DS00926.html (accessed December 12, 2012).

McAllum, H. J., A. S. Familton, R. A. Brown, and P. Hemmingsen. 1978. Salmonellosis in red deer calves (*Cervus elaphus*). *New Zealand Veterinary Journal* 26:130–131.

Merck. 2010. *Merck Veterinary Manual*, 10th edition. Merck, Whitehouse Station, New Jersey.

Molina-Lopez, R. A., N. Valverdú, M. Martin, E. Mateu et al. 2011. Wild raptors as carriers of antimicrobial-resistant *Salmonella* and *Campylobacter* strains. *Veterinary Record* 168:21565.

Nettles, V. F., C. F. Quist, R. R. Lopez, T. J. Wilmers et al. 2002. Morbidity and mortality factors in Key deer (*Odocoileus virginianus clavium*). *Journal of Wildlife Diseases* 38:685–692.

Pagano, A., G. Nardi, C. Bonaccorso, V. Falbo et al. 1985. Faecal bacteria of wild ruminants and the alpine marmot. *Veterinary Research Communications* 9:227–232.

Paulsen, P., F. J. M. Smulders, and F. Hilbert. 2012. *Salmonella* in meat from hunted game: A central European perspective. *Food Research International* 45:609–616.

Pennycott, T. W., H. A. Mather, B. Bennett, and G. Foster. 2010. Salmonellosis in garden birds in Scotland, 1995 to 2008: Geographic region, *Salmonella enterica* phage type and bird species. *Veterinary Record* 166:419–421.

Pennycott, T. W., A. Park, and H. A. Mather. 2006. Isolation of different serovars of *Salmonella enterica* from wild birds in Great Britain between 1995 and 2003. *Veterinary Record* 158:817–820.

Philbey, A. W., F. M. Brown, H. A. Mather, J. E. Coia, and D. J. Taylor. 2009. Salmonellosis in cats in the United Kingdom: 1955 to 2007. *Veterinary Record* 164:120–122.

Refsum, T., K. Handeland, D. L. Baggesen, G. Holstad, and G. Kapperud. 2002. Salmonellae in avian wildlife in Norway from 1969 to 2000. *Applied and Environmental Microbiology* 68:5595–5599.

Renter, D. G., D. P. Gnad, J. M. Sargeant, and S. E. Hygnstrom. 2006. Prevalence and serovars of *Salmonella* in the feces of free-ranging white-tailed deer (*Odocoileus virginianus*) in Nebraska. *Journal of Wildlife Diseases* 42:699–703.

Sato, Y., C. Kobayashi, K. Ichikawa, R. Kuwamoto et al. 2000. An occurrence of *Salmonella* Typhimurium infection in sika deer (*Cervus nippon*). *Journal of Veterinary Medical Science* 62:313–315.

Scallan, E., R. M. Hoekstra, F. J. Angulo, R. V. Tauxe et al. 2011. Foodborne illness acquired in the United States—Major pathogens. *Emerging Infectious Diseases* 17:7–15.

Tauni, M. A. and A. Österlund. 2000. Outbreak of *Salmonella typhimurium* in cats and humans associated with infection in wild birds. *Journal of Small Animal Practice* 41:339–341.

Velarde, R., M. C. Porrero, E. Serrano, I. Marco et al. 2012. Septicemic salmonellosis caused by *Salmonella* Hessarek in wintering and migrating song thrushes (*Turdus philomelos*) in Spain. *Journal of Wildlife Diseases* 48:113–121.

Wahlström, H., E. Tysén, E. O. Engvall, B. Brändstrom et al. 2003. Survey of *Campylobacter* species, VTEC O157 and *Salmonella* species in Swedish wildlife. *Veterinary Record* 153:74–80.

10 *Escherichia coli* and Other Foodborne Diseases

The incident [an outbreak of *E. coli* O157:H7 in spinach] led to increased efforts to promote food safety across the entire "farm to fork" supply chain.

Langholz and Jay-Russell (2013)

Current federal standards for food safety certification incentivize farmers nationwide to remove wildlife.

Langholz and Jay-Russell (2013)

Approximately 48 million cases of food-borne illness occur annually in the United States—one for every six residents.

Institute of Medicine (2012)

10.1 INTRODUCTION AND HISTORY

The CDC (2012) estimates that 48 million people in the United States are stricken annually with gastroenteritis (i.e., inflammation of the stomach and/or small intestine, which results in vomiting, diarrhea, abdominal pains, and cramps) caused by an unspecified agent. The CDC defines an unspecified agent either as (1) an agent not yet identified or (2) identified microbes, chemicals, or other substances known to be in food but with an unknown ability to cause human illness. These unspecified agents are responsible for an estimated 72,000 hospitalizations and 1700 deaths annually in the United States.

Additionally, 9.4 million people in the United States are stricken annually with a foodborne illness caused by a known pathogen or toxin. Of these, 39% were caused by bacteria, 59% by viruses, and 2% by parasites. Each year, more than 225,000 patients in the United States are hospitalized, and 2500 die due to a foodborne illness caused by a known pathogen (Table 10.1). Bacteria were responsible for 64% of the U.S. deaths, parasites for 25%, and viruses for 12%. The leading causes of death were *Salmonella enterica* (nontyphoid fever, 28%), *Toxoplasma gondii* (24%), *Listeria monocytogenes* (19%), and norovirus (11%). Worldwide, more than one billion people become ill with a foodborne disease each year.

Escherichia coli bacteria live in the intestines of all humans, birds, and mammals (Figure 10.1). Most *E. coli* are harmless or beneficial in helping to maintain a healthy intestinal tract. However, a few serotypes of *E. coli* cause illness in humans, and they are the main focus of this chapter. One virulent group of *E. coli* produces Shiga toxin and is called Shiga toxin-producing *E. coli* (STEC). The serotype of STEC

TABLE 10.1
Estimated Annual Number of Foodborne Illnesses (Both Zoonotic and Nonzoonotic Diseases) Acquired in the United States, Including the Percent of Illnesses Acquired from Consuming Food or Drink, Annual Numbers of Foodborne Illnesses, Those Requiring Hospitalization, and Mortalities

Pathogen	% Foodborne	Number of Illnesses	Number Hospitalized	Number of Mortalities
Bacteria				
Bacillus cereus[a]	100	63,400	20	0
Brucella spp.	50	839	55	1
Campylobacter spp.	80	845,024	8,463	76
C. botulinum[a]	100	55	42	9
C. perfringens[a]	100	965,958	438	26
E. coli O157 (STEC)	68	63,153	2,138	20
E. coli non-O157 (STEC)	82	112,752	271	0
E. coli (ETEC)[a]	100	17,894	12	0
E. coli but not STEC or ETEC	30	11,982	8	0
L. monocytogenes	99	1,591	1,455	255
Mycobacterium bovis	95	60	31	3
S. enterica (nontyphoid)	94	1,027,561	19,336	378
S. enterica (typhoid)	96	1,821	197	0
Shigella spp.	31	131,254	1,456	10
S. aureus[a]	100	241,148	1,064	6
Streptococcus spp. (Group A)[a]	100	11,217	1	0
Vibrio cholerae (toxigenic)	100	84	2	0
Vibrio parahaemolyticus	86	36,664	100	4
Vibrio vulnificus	47	96	93	36
Vibrio spp. (other species)	57	17,564	83	8
Y. enterocolitica	90	97,656	533	29
Subtotal for bacteria		3,645,773	35,796	861
Viruses				
Astroviruses	<1	15,433	87	0
Hepatitis A virus	7	1,566	99	7
Noroviruses	26	5,461,731	14,663	149
Rotaviruses	<1	15,433	348	0
Sapoviruses	<1	15,433	87	0
Subtotal for viruses		5,509,597	15,284	156
Parasites				
Cryptosporidium spp.	8	57,616	210	4
Cyclospora cayetanensis	99	11,407	11	0
Giardia intestinalis	7	76,840	225	2
Toxoplasma gondii	50	86,666	4,428	327

(*Continued*)

TABLE 10.1 (*Continued*)

Estimated Annual Number of Foodborne Illnesses (Both Zoonotic and Nonzoonotic Diseases) Acquired in the United States, Including the Percent of Illnesses Acquired from Consuming Food or Drink, Annual Numbers of Foodborne Illnesses, Those Requiring Hospitalization, and Mortalities

Pathogen	% Foodborne	Number of Illnesses	Number Hospitalized	Number of Mortalities
Trichinella spp.	100	156	6	0
Subtotal for parasites		232,705	4,881	333
Total		9,388,075	55,961	1,351

Source: Based on Scallan, E. et al., *Emerg. Infect. Dis.*, 17, 7, 2011.

Note: The table includes both zoonotic diseases and diseases that are not zoonotic.

STEC, Shiga toxin-producing *E. coli*; ETEC, enterotoxigenic *E. coli*.

[a] Only foodborne illnesses were included for this pathogen.

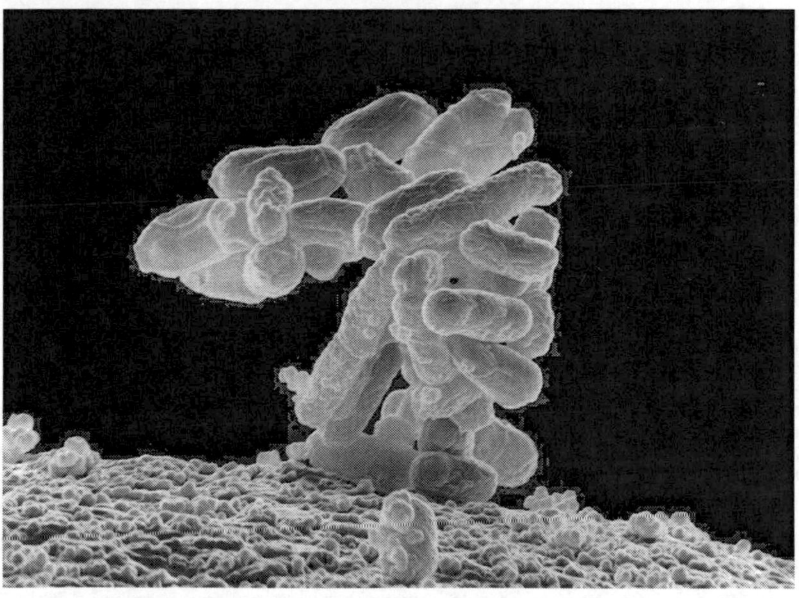

FIGURE 10.1 Photo of a cluster of *E. coli* taken with a scanning electron microscope. (Courtesy of Eric Erbe and the USDA Agricultural Research Service, Washington, DC.)

that causes the most problems in the United States is *E. coli* O157. Other virulent serotypes of STEC that cause human disease are collectively referred to as *E. coli* non-O157 and include *E. coli* O26, O45, O103, O111, O121, and O145. In the United States, there are 63,000 illnesses attributable to *E. coli* O157 and 112,000 attributable to *E. coli* non-O157 (Table 10.1). One large outbreak of *E. coli* non-O157 occurred during 2011 in Germany. This outbreak sickened at least 4321 people and killed 43. The outbreak was linked to the consumption of plant sprouts contaminated with *E. coli* O104:H4. Ultimately, the pathogen was traced back to a shipment of fenugreek seeds from Egypt (Institute of Medicine 2012, Muniesa et al. 2012).

Each *E. coli* O157 serotype can be further subdivided into genotypes based on its DNA "fingerprint." This allows the source of an outbreak to be identified because the genotype of the pathogen will be the same in the victims and the source.

Another group of *E. coli* that produce enterotoxins is collectively referred to as enterotoxigenic *E. coli* (ETEC). These pathogens have external hairlike fibers that allow them to attach to the cells that line the intestines. ETEC cannot normally penetrate cells and do not cause bloody diarrhea, but they produce watery diarrhea and dehydrate many ETEC patients. ETEC infections are deadly in countries where patients have limited access to intravenous fluids to replace their fluid loss. In developing countries, 210 million people are stricken with ETEC annually, and 170,000 die; most victims are children (CDC 2012). In the United States, there are an estimated 18,000 people infected annually by ETEC, but few if any deaths occur because dehydrated patients can be provided with intravenous fluids (Table 10.1).

Humans are the reservoir host of ETEC, while livestock, poultry, and wildlife are the reservoir hosts of STEC. In the United States, STEC outbreaks are usually associated with consumption of meat and fresh produce. STEC outbreaks are more common during summer than winter, perhaps because hamburgers, the source of many STEC infections, are frequently consumed during the summer or because hamburgers prepared on outdoor grills are more likely to be undercooked. The largest STEC outbreak in the United States happened during the 1990s when more than 700 people became ill after consuming undercooked hamburgers at a chain of fast-food restaurants (Boyce et al. 1995).

10.2 SYMPTOMS IN HUMANS

After being ingested, STEC adheres to mucus membranes of the intestines and produces one or more toxins that result in local and/or general injury to the digestive system. Symptoms usually begin 3–5 days after infection, but the incubation period can be as short as 1 day or as long as 8. Symptoms usually begin slowly with mild cramps, vomiting, and diarrhea, which worsen over the course of a few days. Often there is no fever or only a mild one (less than 101°F or 38.5°C). In STEC infections produced by *E. coli* O157, 39% of patients experienced bloody diarrhea compared to 7% of patients with *E. coli* non-O157 (Pennington 2010). Most STEC patients recover within 5–7 days of the onset of symptoms even without the use of antibiotics.

In 5%–10% of STEC cases, inflammatory compounds gain access to the circulatory system and damage the kidneys; this condition is known as hemolytic-uremic syndrome (HUS) and can become life threatening. HUS begins about a week after the first signs of STEC, with the waning of diarrhea. The symptoms of HUS include weakness, a decrease in urination, and loss of pink color in the cheeks and inside the eyelids. Most HUS patients make a full recovery within a few weeks, but the disease can be fatal or result in permanent kidney damage. For this reason, HUS patients require hospitalization. The risk of kidney failure can be reduced by quickly providing fluids intravenously. Children under the age of 5 have the highest risk of developing HUS; this disease is the most common cause of acute kidney failure among children who live in the United States and Europe (Boyce et al. 1995, CDC 2011a, 2012). Among 180 HUS patients in Scotland, 13% had renal (i.e., kidneys and urinary system) impairment, 7% became dependent on dialysis, 4% developed neurological problems, and 4% died (Pennington 2010).

10.3 *ESCHERICHIA COLI* INFECTIONS IN ANIMALS

E. coli O157 are commonly found in the feces of livestock, including cattle, pigs, sheep, and goats (Table 10.2). In Great Britain, 4.7% of cattle, 0.7% of sheep, and 0.3%

TABLE 10.2
Prevalence (% of Samples that Were Positive) of *E. coli* O157 in the Feces of Domesticated Animals

Animal Source	Location	Prevalence (%)
Cattle	Colorado, Nebraska	24.7
Cattle	California	33.8
Cattle	Mexico	1.2
Cattle	United Kingdom	13.2
Cattle	Switzerland	4.2
Cattle	Norway	7.0
Cattle at feedlot	United States	10.2
Cattle, beef	United States	4.7
Cattle, beef at feedlot	Canada	1.9
Cattle, beef	Spain	1.6
Cattle, dairy	United States	3.9
Cattle, dairy	Ohio	0.7
Cattle, dairy	Spain	7.0
Cattle, dairy	Norway	0.0
Chicken	Korea	0.0
Chicken	United States	2.7
Chicken, turkey	United States	2.7
Pig	United States	8.8
Pig	Mexico	2.1
Pig	South Korea	0.3
Sheep	Norway	17.1
Sheep	Britain	20.8
Sheep	Spain	8.7

Source: Erickson, M.C. and Doyle, M.P., Plant food safety issues: Linking production agriculture with One Health. In: Institute of Medicine, editor. Improving Food Safety through a One Health Approach, National Academies Press, Washington, DC, pp. 140–175, 2012.

TABLE 10.3
Prevalence of *E. coli* O157 (% of Samples that Tested Positive) in Feces from Different Species of Wild Birds

Avian Species	Location	Sample Size	Prevalence of *E. coli* O157 (%)
Duck	Washington	40	3
Canada goose	Northeast United States	360	0
	Colorado	397	0
	Washington	121	0
	Sweden	105	0
Trumpeter swan	Washington	67	0
Tundra swan	Alaska	100	0
Gull	Washington	150	0
	Sweden	161	0
	England	400	1
Pigeon	Wisconsin	99	1
	Czech Republic	50	0
European starling	Kansas	434	0
	Ohio	430	1
	Washington	124	0
	Denmark	244	1
Wild turkey	Washington	83	0

Source: Langholz, J.A. and Jay-Russell, M.T., *Human–Wildlife Interact.*, 7, 140, 2013.

of hogs sent to slaughter during 2003 had the pathogen in their intestines (Milnes et al. 2008). *E. coli* O157 are also found in the feces of wild birds (Table 10.3), wild game species (Table 10.4), and other wild mammals (Table 10.5), though its prevalence in wildlife is much lower than in livestock (Table 10.2). Feral hogs, deer, elk, rodents, and other wildlife species that forage in pastures containing cattle and sheep can acquire the pathogen from the livestock. This close association allows *E. coli* O157 and non-O157 to spread between livestock and wildlife (Langholz and Jay-Russell 2013). In Spain, *E. coli* O157 was isolated from 3.3% of wild hogs killed during the hunting season, compared to *E. coli* non-O157 isolated from 5.2% of wild hogs (Sánchez et al. 2010).

Prevalence of *E. coli* O157 among cattle at individual farms and ranches can vary considerably with periods of high prevalence spaced among longer periods when *E. coli* O157 is absent. Why the pathogen reappears is unclear, but there is concern that European starlings may be spreading the pathogen from one dairy farm or animal feedlot to another. European starlings carry *E. coli* O157 in their intestines and shed it in their feces. Large numbers of starlings congregate at farms and feedlots during the winter and often defecate in the water troughs and feeding areas used by livestock (Figure 10.2). Williams et al. (2011) found that both cattle and starlings at the same feedlot were infected with same strain of *E. coli* O157:H7.

TABLE 10.4
Prevalence of *E. coli* O157 in Feces from Wild Game Mammals

Avian Species	Location	Sample Size	Percentage with *E. coli* O157
Deer, white-tailed	Washington	630	0.8
	Nebraska	1608	0.2
	Kansas	122	2.4
	Georgia	919	0.3
	Louisiana	338	0.3
Deer, black-tailed	Oregon	32	9.4
	California	9	11.1
Deer, roe	Sweden	195	0.0
Deer, red	Spain	470	1.3
Elk	Washington	244	0.0
Bison	Washington	57	0.0
Bighorn sheep	Washington	32	0.0
Moose	Sweden	84	0.0
Feral hogs	California	87	14.9
	Sweden	68	1.5
	Spain	474	1.7

Source: Langholz, J.A. and Jay-Russell, M.T., *Human–Wildlife Interact.*, 7, 140, 2013.

TABLE 10.5
Prevalence of *E. coli* O157 in Feces from Wild Mammals

Mammal Species	Location	Sample Size	Percentage with *E. coli* O157
Coyote	Kansas and Nebraska	100	0
Red fox	Ireland	124	0
	Spain	260	0
Raccoon	Kansas and Nebraska	230	0
Opossum	Kansas and Nebraska	25	4
Bat	Trinidad	377	0
Hare	Sweden	125	0
European rabbit, European	England	138	20
Rodents	Washington	300	0

Source: Langholz, J.A. and Jay-Russell, M.T., *Human–Wildlife Interact.*, 7, 140, 2013.

FIGURE 10.2 Starling flocks numbering in the tens of thousands feed during the fall and winter at dairy farms and cattle feedlots. (Courtesy of George Linz.)

Outbreaks of *E. coli* O157 are rarely linked to wildlife, but there have been a few exceptions where the same strain of a pathogen responsible for the outbreak was also isolated from wildlife feces. A large outbreak in the United States that claimed at least three lives was traced to a spinach field frequented by feral hogs (Sidebar 10.1). During an outbreak of *E. coli* O157:H7 in the state of Washington,

SIDEBAR 10.1 OUTBREAK OF *E. COLI* O157:H7 IN THE UNITED STATES (LANGHOLZ AND JAY-RUSSELL 2013)

In the United States, 200 people were sickened and at least 3 died from an infection of *E. coli* O157:H7 during 2006. An intensive investigation by public health authorities tracked the outbreak to bagged spinach from a single field in California. The grower reported that he had observed feral hogs foraging in the field prior to harvest (Figure 10.3). Both feral hogs and cattle had direct access to the surface water source on the farm used for irrigation. *E. coli* O157:H7 was found in 34% of cattle feces, 15% of feral hog feces, 8% of soil samples, and 0% of water samples. Eight samples from feral hogs and 15 from cattle contained the same strain of *E. coli* O157:H7 that was responsible for the outbreak. How the pathogen was transmitted to the spinach was unclear because the contamination occurred prior to the start of the epidemiological investigation. Investigators concluded that cattle and feral hogs were among the potential sources of the pathogen that caused the outbreak.

FIGURE 10.3 Photo of a feral hog. (Courtesy of the USFWS, Washington, DC.)

37 children became ill, 8 were hospitalized, and 8 developed HUS. The source of the infection was located at Battle Ground Lake, where the children had gone swimming. The same isolate of *E. coli* O157:H7 responsible for the outbreak was isolated in water from the lake and from duck feces (Samadpour et al. 2002, Bruce et al. 2003). During 2011, another outbreak was traced to strawberries at an Oregon farm and to the deer feces recovered from the field (Laidler et al. 2013, Langholz and Jay-Russell 2013). In 1995, 11 people in an extended family in Oregon were infected with *E. coli* O157:H7 after consuming homemade venison jerky (Keene et al. 1997).

10.4 HOW HUMANS CONTRACT DISEASES CAUSED BY *ESCHERICHIA COLI*

People contract STEC when they consume meat, fresh produce, water, unpasteurized milk, or apple cider contaminated with the pathogen or when they visit petting zoos, swallow lake water while swimming, or consume food prepared by food handlers or cooks who failed to wash their hands properly. While the risk of STEC cannot be eliminated, it can be reduced through improved hygiene. The largest known outbreak of STEC occurred in Sakai City, Japan, when 7966 people became ill from eating radish sprouts served in school lunches (Pennington 2010).

During 2011 and 2012, STEC outbreaks in the United States were linked to organic spinach, raw clover sprouts, romaine lettuce, bologna, and in-shell hazelnuts (CDC 2012). In the United States, most STEC patients (52%) were infected after eating contaminated food. Some people (14%) were infected from person to person, 9% from water-related transmissions, 3% from physical contact with an infected animal, and 21% from an unknown transmission route. Water-related transmissions usually occurred during a recreational activity in a lake or pond (68%), and the remaining transmissions (32%) involved drinking water (Rangel et al. 2005). From 1994 to 2003, 40% of *E. coli* O157 outbreaks in Scotland were foodborne, 54% were environmental, and 6% involved both transmission routes (Pennington 2010). Environmental transmission often involved people who became infected after visiting pastures or farms. For example, an outbreak of *E. coli* O157:H7 occurred among visitors at a dairy farm in Pennsylvania (Crump et al. 2002). Water-related outbreaks of *E. coli* O157 are often associated with heavy rains. In many of these outbreaks, water from spring or private wells was the source of the pathogen. In Walkerton, Canada, an outbreak of *E. coli* O157:H7 during 2000 sickened 2300 and killed 7. The source was the drinking water supply, which had been contaminated by cattle feces (WHO 2000). People have also become infected from eating venison (Rabatsky-Ehr et al. 2002).

Children infected with STEC shed the pathogen for an average of 13 days (ranging from 2 to 62 days) in their feces, while children with HUS shed the pathogen, on average for 21 days (ranging from 5 to 124 days). While *E. coli* O157 can spread from one person to another, it cannot maintain itself by just infecting people (Pennington 2010).

10.5 MEDICAL TREATMENT

The CDC (2011) recommends that people with diarrhea seek medical attention if it becomes bloody, lasts more than 3 days, and is accompanied with a fever of more than 101.5°F (38.6°C). People with an enhanced risk of developing a foodborne illness include those with chronic diseases, older adults, pregnant woman, infants, and young children. A serious complication of foodborne illnesses is severe dehydration, which results when loss of water and essential salts and minerals from vomiting and diarrhea exceed the amount replaced by drinking. In such cases, hospitalization and intravenous fluids may be required.

Doctors seldom prescribe antibiotics to treat STEC infections because most patients improve in 2–3 days, regardless of whether antibiotics are prescribed. Moreover, taking antibiotics or antidiarrhea agents can increase the risk of HUS. Instead of these medical treatments, patients should be made comfortable and encouraged to drink plenty of liquids to prevent dehydration (CDC 2012). STEC patients should be monitored for signs of HUS.

Most STEC patients usually stop shedding the pathogen in their feces by the time they recover. Some people, nonetheless, continue to do so for several more weeks, especially children. Therefore, proper hand washing is important for people who have had STEC. In some areas, public health laws prohibit STEC patients from attending schools or workplaces. Other schools and workplaces have policies that discourage STEC patients from attending.

10.6 WHAT PEOPLE CAN DO TO REDUCE THEIR RISK OF CONTRACTING DISEASES CAUSED BY *ESCHERICHIA COLI*

Public health agencies are mandated to safeguard food while being produced on a farm, harvested, distributed, cooked, or served. The intentions are to reduce the risk of contamination at each step of food processing and to identify the pathogenic source when people become sick. California and Arizona, which grow much of the fresh produce in the United States, have adopted a Leafy Green Marketing Agreement that specifies food safety practices. Among other requirements, the agreement prohibits harvesting plants within 5 ft (1.5 m) of locations where animals defecated (Langholz and Jay-Russell 2013).

There are no vaccines against *E. coli* O157 that are approved for human use. However, the United States and Canada have approved two *E. coli* O157 vaccines for use in cattle. Both vaccines are effective in reducing the prevalence of the pathogen in the digestive system of cattle (Snedeker et al. 2012).

People who live in urban areas have a lower risk of contracting STEC than people who live in rural areas, especially those living where high densities of cattle and sheep occur (Pennington 2010). People with an elevated risk of being infected with STEC include those who spend time in cattle or sheep pastures, even if these pastures had not contained livestock for weeks (Sidebar 10.2). *E. coli* O157 can survive up to 234 days in water and 179 days in soil (Duffitt et al. 2011).

To reduce the risk of infections by *E. coli* and foodborne pathogens, food should be refrigerated and not kept at room temperature. Frozen food should be defrosted quickly in a microwave or have cold water run over it. When uncertain if a food is safe to eat, do not taste it; rather, just discard the food. The Mayo Clinic (2011a) recommends that poultry be cooked to 165°F (74°C), hamburger or pork to 160°F (71°C),

SIDEBAR 10.2 *E. COLI* O157 OUTBREAK AT A BOY SCOUT CAMP (HOWIE ET AL. 2003)

In 2000, 223 boys and 104 adults attended a Boy Scout camp in New Deer, Scotland. Soon thereafter, three boys visited doctors complaining of diarrhea. *E. coli* O157 was isolated from the feces of one boy, while the other two boys had probable infections of the same pathogen. Upon learning that other campers had similar illnesses, public health authorities declared an outbreak of *E. coli* O157. Of 288 campers that were contacted by health authorities, 70 reported diarrhea or vomiting. Forty-eight of those 70 campers supplied fecal samples, and 20 of them (42%) were confirmed as having an infection of *E. coli* O157. An epidemiological study was initiated to determine the source of the pathogen. None of the samples of drinking water, milk, and food contained *E. coli* O157. A week prior to the camp, 300 sheep were turned loose in the 20 acre (8 ha) campsite to graze down the tall grass. *E. coli* O157 was later isolated from fecal samples from 14 of 25 of these sheep and from soil samples at the campsite. All isolates of *E. coli* O157 from the sheep, soil, and patients were identical. Heavy rains during the camp were believed to have washed the pathogen from the sheep feces into the upper layers of the soil. Health authorities concluded that sheep had been the source of the pathogen and that the outbreak was due to environmental exposure.

and steaks, roasts, or fish to 145°F (63°C). A thermometer is the best gauge to determine when meat is thoroughly cooked because judging solely by the meat's color is unreliable. Steps should be taken to avoid contaminating food preparation areas by carefully cleaning utensils, cutting boards, and kitchen surfaces after they have come into contact with raw meat. Unpasteurized milk, fruit juice, and the products containing them should not be ingested. People should avoid swallowing water when swimming in lakes, ponds, streams, swimming pools, and backyard wading pools. Personal hygiene is also important. During one outbreak of *E. coli* O157, people who failed to wash their hands before eating were nine times more likely to become infected than those who did. Likewise, people who ate with their fingers were seven times more likely to become ill than those who used eating utensils (Howie et al. 2003).

10.7 OTHER PATHOGENS THAT CAUSE FOODBORNE DISEASES IN THE UNITED STATES

Most foodborne diseases that infect humans are contagious; most patients are infected by pathogens obtained from another person, rather than directly from an animal. Many of these same pathogens infect both domestic and wild animals, and animals are often implicated as the original source of the pathogen that caused an epidemic. This is especially true with wildlife because they are usually no longer present by the time a disease outbreak has begun. The bacteria and viruses that cause many of the more common or serious foodborne illnesses in the United States and Canada are discussed next.

10.7.1 *CAMPYLOBACTER* SPP.

These bacteria are Gram negative and appear under the microscope as curved spirals or corkscrews; the word campylobacter means "twisted bacteria" in Latin (Figure 10.4). These bacteria can move by twisting around their central axis. There are 14 recognized species in this genus, all producing gastroenteritis in humans or animals including wild birds and wild mammals. Three species are recognized as causing human diseases (*Campylobacter jejuni*, *C. fetus*, and *C. coli*). However, a single species (*C. jejuni*) is responsible for almost all of the 2.4 million human cases of campylobacteriosis (the name of the disease caused by *Campylobacter*) in the United States annually.

Campylobacteriosis begins 2–10 days after infection and lasts 7–14 days (Table 10.6). Symptoms include diarrhea that is often watery and/or bloody, fever, abdominal cramps, and nausea. Most campylobacteriosis patients are infants and young children. In developing countries, almost all children have been exposed to the pathogen, and many children die from the diarrhea caused by *Campylobacter* (WHO 2000, 2009).

Some campylobacteriosis patients develop reactive arthritis (i.e., joint pain and swelling due to an infection elsewhere in the body), and others (less than one in a thousand) exhibit the Guillain–Barré syndrome, which results when a person's immune system attacks the nerves (WHO 2009, Gardner et al. 2011, CDC 2012). Early symptoms of this syndrome include numbness and tingling sensations that

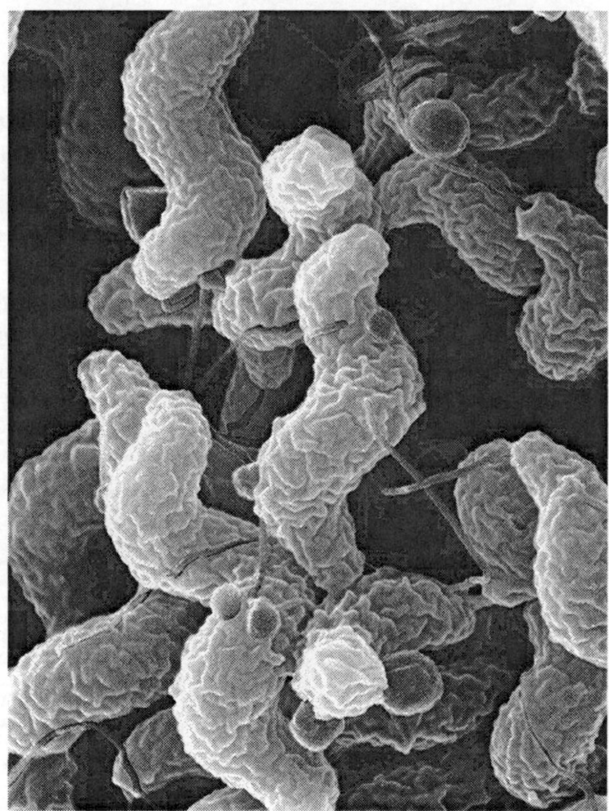

FIGURE 10.4 Scanning electron microscope photo showing the characteristic corkscrew shape of *C. jejuni* cells. (Courtesy of DeWood, Pooley, and the USDA, Agricultural Research Service, Electron Microscopy Unit, Washington, DC.)

start in the feet and lower legs and spread to the hands, upper arms, and upper body. Patients may also have difficulty with eye or facial movements, walking, swallowing, chewing, or speaking. Guillain–Barré syndrome is a serious medical problem because symptoms can rapidly progress to paralysis of the legs, arms, and breathing muscles. The Mayo Clinic (2011a) recommends seeking emergency medical attention if any of the following symptoms develop: tingling sensation that begins in the feet or toes and ascends through the body, tingling or weakness that spreads rapidly, tingling involving both feet or both hands, shortness of breath, and choking on saliva.

Campylobacter are found in cattle, pigs, sheep, and poultry (Table 10.7). Domestic chickens, turkeys, ducks, and geese are especially important because they are reservoir hosts for *C. jejuni*. Most campylobacteriosis patients became infected from eating undercooked chicken or other food contaminated with juices that drip from uncooked poultry. Humans, rodents, wild birds, and houseflies can transmit these pathogens from one poultry flock to another (Merck 2010, CDC 2012).

TABLE 10.6
Incubation Period (i.e., Time between When a Patient Became Infected with a Pathogen and the Onset of the First Symptoms of Illness) and the Foods or Liquids that Commonly Are Contaminated by the Pathogen

Pathogen	Incubation Period	How Transmitted to Humans
B. cereus	1–5 hours	Meat, pasta, rice, vegetables
Campylobacter	2–10 days	Meat, poultry, milk, fresh produce, water
C. botulinum	12–72 hours	Home-canned foods, unrefrigerated foods
C. perfringens	8–16 hours	Meat, stews, poultry
E. coli O157	1–8 days	Hamburger, meat, milk, apple cider, fresh produce
Listeria	9–48 hours	Luncheon meats, milk, fresh produce, water
Shigella	1–2 days	Fish, shellfish, fresh produce
S. aureus	1–6 hours	Meat, cream sauces, cream-filled pastries, salads
V. vulnificus	1–7 days	Oysters, mussels, clams, scallops
Y. enterocolitica	1–10 days	Meat, milk
Astroviruses	1–3 days	Food, water
Hepatitis A virus	1 month	Shellfish, contaminated water, drug needles
Noroviruses	12–48 hours	Shellfish, fresh produce
Rotavirus	1–3 days	Fresh produce

Sources: Olsen, S.J. et al., *Morb. Mortal. Wkly Rep.*, 49(SS01), 1, 2000; Mayo Clinic, Food poisoning, 2011, http://mayoclinic.com/health/food-poisoning/DS00981, accessed February 1, 2013.

Campylobacter have been isolated from many free-ranging mammals and birds, and people can develop campylobacteriosis from consuming infected game meat (Table 10.8). *Campylobacter* were isolated from samples of meat from wild European roe deer, red deer, and feral hogs that were shot in Europe (Atanassova et al. 2008). Game meat is more likely to be contaminated if the animal was shot through the body cavity, allowing material from the intestines to contaminate meat. In Sweden, Wahlström et al. (2003) detected *Campylobacter* in fecal samples from 4% of roe deer and 12% of wild hogs.

Many wild birds, including gulls, ducks, pigeons, and crows, serve as reservoir hosts for *C. jejuni*. These pathogens occur in their intestines; infected birds are generally asymptomatic, but there are exceptions. *C. jejuni* causes dysentery (i.e., inflammation of the large intestine) in some finches, canaries, and parrots: an infection can be fatal in these birds. *C. jejuni* also causes dysentery in dogs, cats, cattle, sheep, pigs, mink, ferrets, and primates.

People can acquire campylobacteriosis through contact with infected pets, but the vast majority of people become infected after consuming contaminated food or drinking contaminated water. During 2008, a *C. jejuni* outbreak in Alaska sickened at least 98 people, most of whom reported consuming raw peas. The contaminated

TABLE 10.7
Prevalence (% of Samples that Were Positive) of *Campylobacter* spp. in Feces of Livestock and Poultry based on Studies Summarized by Erickson and Doyle (2012)

Animal Source	Location	Prevalence
Cattle	United Kingdom	13%
Cattle	Britain	10%–13%
Cattle	Ireland	25%
Cattle	Australia	94%
Dairy cattle	New Zealand	64%
Pig	United Kingdom	13%
Pig	Britain	10%–13%
Sheep	United Kingdom	21%
Sheep	Britain	11%–21%
Chicken	United Kingdom	19%
Poultry	Britain	8%–19%
Broilers	Sweden	47% of ceca samples
Duck, mallard	United Kingdom	93%–100%

TABLE 10.8
List of Wild Mammals from Europe that Are Known to Be Infected with, or Are Hosts of, *Campylobacter coli*, *C. jejuni*, *C. lari*, or *C. hyointestinalis*

	Infected by			
Mammal Species	***C. coli***	***C. jejuni***	***C. lari***	***C. hyointestinalis***
Hare	Y	Y		
Rabbit			Y	
Roe deer		Y		Y
Moose		Y		
Eurasian red squirrel		Y		
European hedgehog		Y		
Red fox		Y		
Badger		Y	Y	

Source: Speck, S., *Campylobacter* infections. In: D. Gavier-Widén et al., eds., *Infectious Diseases of Wild Mammals and Birds in Europe*, Wiley-Blackwell, Chichester, U.K., pp. 398–401, 2012.

peas were traced to a single Alaskan farm. Sandhill cranes had been foraging in the field daily when the crop was being harvested (Figure 10.5). Sandhill crane feces were collected from the pea field, and all contained *C. jejuni*. The fecal samples and pea samples from the farm contained the same strain of *C. jejuni* that produced the outbreak (Gardner et al. 2011).

FIGURE 10.5 Sandhill cranes were identified as a source of *C. jejuni* that sickened 98 people during one outbreak in Alaska. (Courtesy of the USFWS, Washington, DC.)

At present, there are no human vaccines that are effective against *Campylobacter*, but research to develop one is progressing. The risk of people contracting campylobacteriosis can be reduced through improved washing of chicken carcasses at meat processing plants. Chemical disinfectants or gamma irradiation can be used to eliminate the pathogen from chicken carcasses. Poultry should be thoroughly cooked so the core temperature will reach 165°F (74°C); raw meat should be handled carefully so it does not contaminate other foods (Merck 2010).

C. fetus also causes human illness, but unlike *C. jejuni*, this pathogen rarely causes gastroenteritis in humans. *C. fetus* infections in humans can become serious and result in blood poisoning, meningitis, and endocarditis. People most at risk for developing serious complications are the elderly or those with immune deficiencies. Among pregnant women, infections by *C. fetus* can cause abortions or preterm births (Merck 2010).

C. fetus causes genital campylobacteriosis in both sexes of cattle and sheep, can be sexually transmitted, and results in infertility, embryonic death, and abortions. Strict animal hygiene is necessary to stop an outbreak in a cattle herd or sheep flock; use of tetracyclines can reduce the risk of animals aborting. A vaccine is available for use in livestock to prevent the disease. In infected herds, both infected and uninfected animals should be vaccinated. Artificial insemination can be used to reduce the risk of the pathogen spreading, providing that the semen does not come from an infected male.

10.7.2 *Clostridium perfringens*

The genus *Clostridium* contains a group of anaerobic, Gram-positive species of bacteria (Figure 10.6). These bacteria normally live in the soil, aquatic environments, and intestines of animals across the world. They are rod-shaped, incapable of movement, and produce spores that can survive in the environment for long periods of time. The spores are widely distributed in the environment but are most prevalent in soil, mud, and aquatic and marine sediment. *Clostridium* spp. produce toxins and can cause illness in humans through three different modes. *C. tetani* and *C. perfringens* can reproduce within human or animal tissue if they

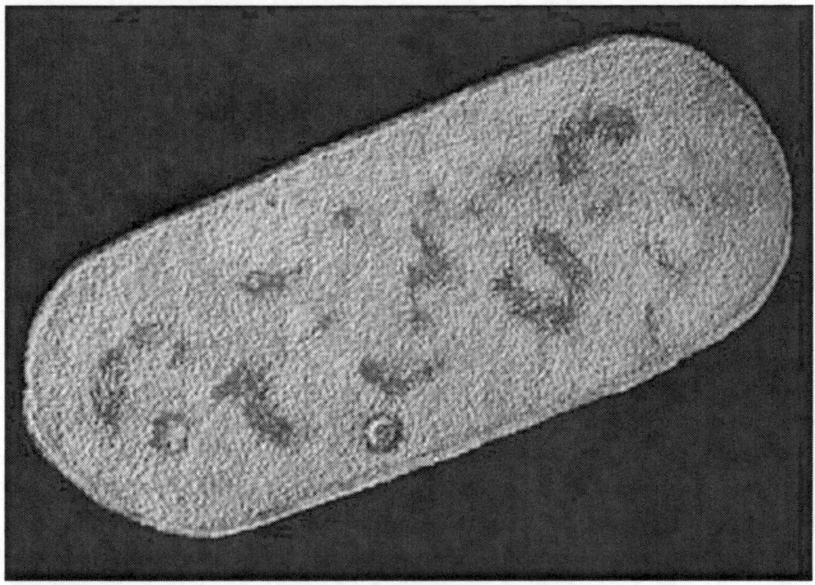

FIGURE 10.6 Photo of *C. perfringens* cell. (From Institute of Medicine, *Improving Food Safety through a One Health Approach*, National Academies Press, Washington, DC, p. 25, 2012.)

gain entry through a puncture wound and cause tetanus and gas gangrene, respectively. *C. botulinum* causes botulism and is discussed later.

C. perfringens occurs in raw meat and poultry, and most people are infected by eating undercooked meat and poultry. Spores of *C. perfringens* can survive the high temperatures used to cook meat and then germinate after the meat has cooled. Vegetative forms of the bacteria can proliferate in meat kept above 68°F (20°C). If the meat is then consumed, the bacteria can infect the intestines and cause dysentery. There are almost a million people in the United States who become ill from *C. perfringens* infections each year (Table 10.1). Fortunately, only a small fraction of them require hospitalization (CDC 2012, Institute of Medicine 2012).

There are five different serotypes of *C. perfringens* (types A, B, C, D, and E) that differ in the toxins they each produce and the animal species they infect (Table 10.9). Most human infections are caused by type A, which causes diarrhea, abdominal cramps, and vomiting. Types B and D rarely cause human disease. Type C is less common in humans than type A but causes necrotizing enteritis (Iowa State University 2004).

All five types of *C. perfringens* produce illness in animals. Type A causes necrotic enteritis (enteritis with tissue death) in birds and dogs, inflammation of the intestine in horses, and diarrhea in pigs. Types B and C produce dysentery, toxemia, and mortality in lambs, calves, piglets, and foals. Type C also causes enterotoxemia in adult cattle, sheep, and goats. Type D often is responsible for enterotoxemia in sheep and on occasion in goats and cattle. Type E causes the same disease in calves, lambs, and rabbits (Merck 2010).

TABLE 10.9
Toxins Produced by Different Strains of *C. perfringens*

Strain	Toxins
Type A	Alpha
Type B	Alpha, beta, epsilon
Type C	Alpha, beta
Type D	Alpha, epsilon
Type E	Alpha, iota

Source: Iowa State University, Epsilon toxin of *Clostridium perfringens*, Center for Food Security and Public Health, Iowa State University, Ames, IA, 2004.

Among wild birds, infections by *C. perfringens* types A and C are the most serious because they cause necrotic enteritis. This disease results when the pathogen proliferates in the intestines and produces toxins, which in turn cause intestinal ulcers, bleeding, and tissue death. From there, the pathogen and toxins can spread to the liver. Infected birds can be uncoordinated, lack the ability to fly, and die. The disease primarily occurs in ducks, geese, and swans. It also has been reported in grouse, including greater sage grouse in the United States and western capercaillies in Europe (Wobeser and Rainnie 1987, Hagen and Bildfell 2007). Wild birds are rarely the source of human infections.

10.7.3 *Clostridium botulinum*

Like other species of the *Clostridium* genus, *C. botulinum* grows in meat, sometimes in plant material, and produces toxins that contaminate the food. When animals or humans consume the contaminated food, they consume the toxin; it is the toxin, not the pathogen, that produces illness. Some of the first patients identified as suffering from botulism had eaten contaminated sausage; it was called sausage poisoning after "botulus," the Latin word for sausage.

Botulinum toxins are neurotoxins that are among the most toxic chemicals known to man. When ingested, they are absorbed and attached to nerve–muscle junctions where they inhibit the release of acetylcholine, preventing the transmission of nerve impulses. As a result, patients have difficulty contracting their muscles. Botulism poisoning should be suspected if someone has consumed food during the previous week that could potentially be contaminated with the toxin (e.g., home-canned food or unrefrigerated meat) and if they exhibit three or more of the following symptoms: (1) nausea or vomiting, (2) double vision, (3) dilated pupils, (4) trouble swallowing, and (5) dry mouth or throat (Horowitz 2010). The cranial nerve is usually one of the first nerves impacted by botulinum toxins. Clinical signs can include drooping eyelids; inability to move eyes left, right, or downward; and difficulty saying "ah," smiling, and moving the tongue from side to side or biting down. Clinical signs are usually bilateral (both eyes or both sides of the body are affected). Paralysis can progress to the arms and upper body and descends through the body (i.e., descending paralysis). There is usually no fever because botulism results from a toxin and not from an infection (Horowitz 2010).

Antitoxins have been developed for botulinum toxins, but death can result if they are not provided in time. Recovery from botulism can take more than a year because the binding of botulinum toxins to the nerve membranes is irreversible, requiring nerve regeneration for recovery. Human cases of botulism are uncommon in the United States, but it is a serious disease with a high rate of mortality. Humans can be exposed to the toxin by ingestion, *C. botulinum* wound infection, or infant botulism. In most years, there are 10–40 people in the United States diagnosed with ingestion botulism, 20–40 with wound botulism, and 60–100 with infant botulism (Figure 10.7). Botulism in infants is associated with certain foods, particularly honey. For this reason, infants should not be given honey until they are at least 1 year old (Mayo Clinic 2012).

C. botulinum spores are common in the environment and are found in soils worldwide. Once the spore germinates, the vegetative form of bacteria requires anaerobic conditions and a protein-rich environment. Such conditions occur in rotting carcasses, unrefrigerated meat, and home-canned foods. It is the vegetative form of the bacteria that produces botulinum toxins. *C. botulinum* spores cannot germinate in adult humans except in those who lack an intact gastrointestinal tract. Instead, adults develop botulism from eating food that is contaminated with botulinum toxins from vegetative forms of *C. botulinum* that have been growing within the food. However, spores can germinate inside the intestines of infants and can produce toxins internally (Mayo Clinic 2012).

There are seven types of *C. botulinum* (A, B, C, D, E, F, G) based on the type of neurotoxin each produces. Outbreaks of botulism in wild birds are usually caused by types C and E. Botulism in humans is usually caused by types A, B, and E. In the United States, type A is more common in the eastern United States, while type B is more prevalent in the west. Type E is often associated with soils in the tidal zone, fish, and fish-eating birds. People can develop type E botulism from consuming stored or fermented fish or other seafood that were improperly prepared. Botulism in humans can also be caused by other species of *Clostridium*: type E can also be caused by *C. butyricum*, type F by *C. baratii*, and type G by *C. argentinense*, but botulism patients were rarely infected by these species of *Clostridium* (Fenicia et al. 1999, Horowitz 2010).

All avian species are susceptible to type C toxin with the possible exception of some carrion-feeding birds such as vultures and buzzards. But the lethal dose varies considerably among avian species with ducks being more susceptible than many other birds. Outbreaks of type C botulism have resulted in massive die-offs involving thousands of ducks, especially in the western United States and Canada (Sidebar 10.3). Outbreaks occur in duck populations when *C. botulinum* and maggots both grow in the same carcass. The pathogen produces type C toxin, and maggots accumulate the toxin as they feed on the carcass. A single maggot can accumulate enough toxin to kill a duck, but the maggots are immune to the toxin. A single duck carcass can produce hundreds of maggots, which can appear on the outside of the

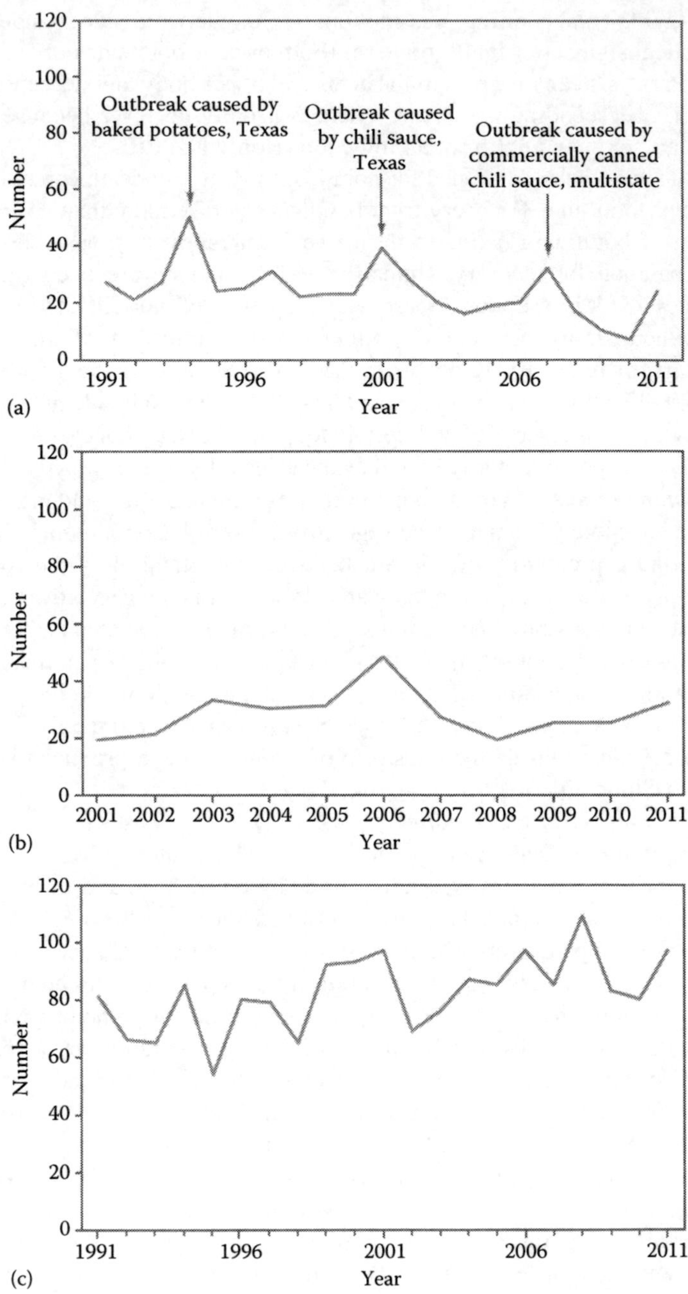

FIGURE 10.7 Annual number of confirmed cases of (a) ingestion botulism, (b) wound botulism, and (c) infant botulism in the United States. (From CDC, *Morb. Mortal. Wkly Rep.*, 60(53), 1, 2013.)

SIDEBAR 10.3 AVIAN BOTULISM AT UTAH'S GREAT SALT LAKE (BARRAS AND KADLEC 2000, KADLEC 2002)

The Great Salt Lake is a large hypersaline lake located in the western United States, which attracts millions of waterfowl (Figure 10.8). Around its margins, freshwater inflow is diked to produce shallow impoundments for waterfowl and shorebirds. The most famous of these is the Bear River Migratory Bird Refuge (BRMBR). These shallow-water impoundments attract millions of waterfowl and shorebirds during the late summer and fall. Different species of *Clostridium*, including *C. botulinum*, are abundant in these marshes, especially in the top layer of mud, where they use fermentation to break complex carbohydrates and proteins into simpler organic compounds. In some years, large outbreaks of *C. botulinum* type C kill thousands of ducks and shorebirds foraging in these impoundments, but during other years, no outbreaks occur. For example, an estimated 400,000 birds died during 1997. It is unclear why these impoundments are so conducive to outbreaks or why they occur during some years but not others. Anecdotal observations suggest that outbreaks are more likely to occur during wet years when water inflow into the marshes is high. They also are more likely after mudflats are flooded or wind events stir the water–soil interface. Kadlec (2002) speculated that botulism outbreaks occur when conditions are conducive to the growth of *C. botulinum* and to the production of toxin by the pathogen. Abundant invertebrates in the impoundments also are required because they accumulate the toxin and transfer it to the birds when consumed. Once a botulism outbreak begins and starts killing ducks, the outbreak becomes self-perpetuating because maggot populations explode from the duck carcasses; more ducks then feed on maggots, and they too die, continuing the cycle.

carcass within 3 days of the duck's death (Cliplef and Wobeser 1993). Ducks consume these maggots and receive a lethal dose of the toxin. Hence, maggots from a single carcass can kill several ducks, and a single outbreak can kill thousands of ducks. These dead ducks become infected with both *C. botulinum* and maggots, and the cycle is repeated. Management efforts to stop an outbreak involve the removal of dead birds, but these efforts are often unsuccessful because most carcasses are never located (Cliplef and Wobeser 1993).

Birds suffering from botulism poisoning exhibit a flaccid paralysis called limber neck because the toxin prevents birds from being able to contract their neck muscles. Stricken birds often are prostrate; their wings droop away from their body, similar to how birds hold their wings when brooding their young. Botulinum toxins cannot cross the blood–brain barrier, so paralyzed birds often appear alert. Death can result from lack of oxygen if the paralysis spreads to the respiratory muscles, drowning if birds are paralyzed while on water, dehydration, or starvation. Ducks with limber neck are also easy targets for predators (Neimanis and Speck 2012).

While type C botulism dominates in ducks, type E is responsible for most outbreaks in fish-eating birds. On Lake Michigan, outbreaks of type E botulism have

FIGURE 10.8 Photo of ducks on the Great Salt Lake. (Courtesy of the USFWS, Washington, DC.)

killed hundreds of common loons and ring-billed gulls, but outbreaks do not occur every year (Brand et al. 1983). Fish are susceptible to type E toxin, and waterbirds have an easy time catching fish that have been intoxicated and killed by the toxin. During an outbreak of type E botulism in France, thousands of black-headed gulls and herring gulls died after consuming contaminated fish waste in the Bay of Canche (Gourreau et al. 1998). Elsewhere in Europe, outbreaks of type C or E botulism have killed numerous waterbirds in Sweden, the Netherlands, the United Kingdom, and the Czech Republic (Neimanis and Speck 2012).

Optimal temperatures for vegetative growth of *C. botulinum* type C are 86°F–99°F (30°C–37°C). Hence, most outbreaks of type C botulism among waterfowl occur during summer and are more likely to occur when summers are hot (Neimanis and Speck 2012). Optimal temperatures for the vegetation growth of *C. botulinum* type E are below 86°F (30°C). Hence, outbreaks of type E botulism generally occur during the cooler weather of fall.

Botulism outbreaks among waterfowl and shorebirds usually do not spread to livestock or other domestic animals. However, cattle have developed botulism after grazing and drinking in wetlands during a botulism outbreak (Wobeser et al. 1997). Outbreaks of botulism type C among birds or other wildlife species pose little danger to humans because type C neurotoxin does not affect humans; outbreaks of type E botulism among wildlife, however, pose a risk to humans, especially to people handling sick or dead animals (Neimanis and Speck 2012).

Type C and D botulism are called forage poisoning when occurring in horses and shaker foal syndrome when occurring in foals. Botulism has been observed in cattle that ingest food or water contaminated with botulinum toxin C or D. This poisoning has resulted after the incorporation of dead birds into the feed rations of cattle, use of poultry litter and carcasses to fertilize cow pastures, or allowing cattle to forage in marshes where botulism outbreaks had killed waterbirds (Wobeser et al. 1997). The first clinical signs of botulism in cattle are difficulty chewing and swallowing caused by paralysis of the tongue and jaw. This can progress to general paralysis and death.

The Mayo Clinic (2012) recommends that people seek urgent medical attention any time botulism is suspected. An antitoxin has been developed that is effective against types A, B, and E. It can neutralize the toxin that remains in the bloodstream, but it cannot reverse nerve damage that has already occurred. For this reason, the antitoxin needs to be injected as soon as possible. If the toxin was ingested, doctors may attempt to remove it from the digestive system by inducing vomiting and bowel movements. A mechanical ventilator may be required if the toxin has impaired the patient's breathing muscles. Many patients make full recoveries from botulism, but it may take months for the nerves to regenerate (Mayo Clinic 2012).

10.7.4 *Listeria monocytogenes*

The genus *Listeria* was named after the famous English doctor, Joseph Lister. These bacteria are Gram positive and rod-shaped and often form short chains; they do not form spores (Figure 10.9). The species *L. monocytogenes* causes a gastrointestinal illness called listeriosis. Compared to other foodborne diseases, listeriosis is

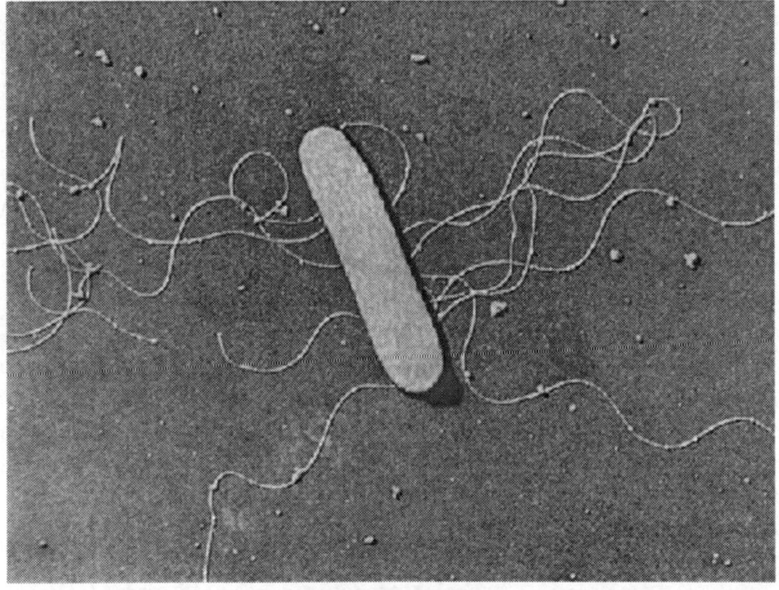

FIGURE 10.9 Photo of *L. monocytogenes* cell taken with an electron microscope. (From Institute of Medicine, *Improving Food Safety through a One Health Approach*, National Academies Press, Washington, DC, p. 27, 2012.)

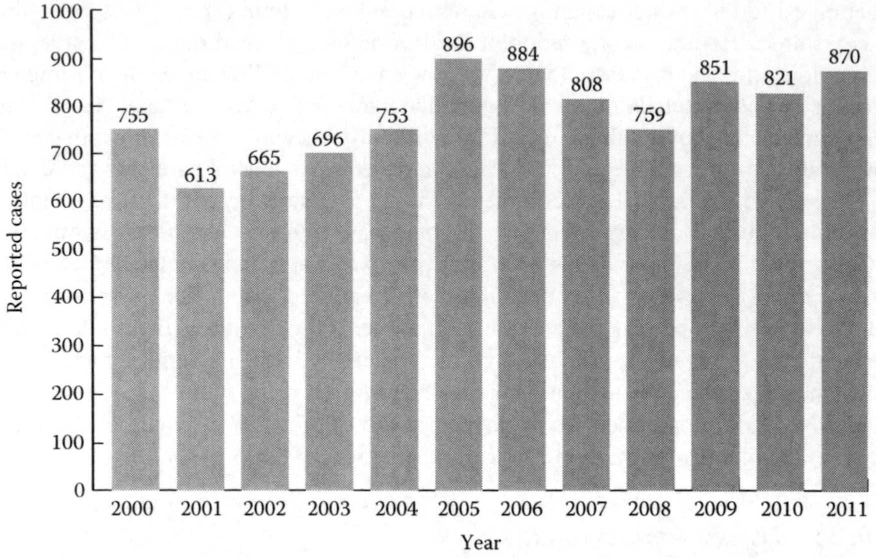

FIGURE 10.10 Number of human cases of listeriosis reported annually in the United States. (Courtesy of the CDC, Atlanta, GA.)

uncommon in humans; less than 2000 people in the United States develop listeriosis annually, but it is a serious disease. Most patients require hospitalization, and the case fatality rate in the United States exceeds 15% (Table 10.1). Older people, those with weakened immune systems, and pregnant women have a higher risk of contracting the disease or developing a serious complication. Since 2000, there have been between 600 and 800 confirmed cases of listeriosis annually in the United States (Figure 10.10).

Listeriosis usually starts as an infection of the digestive system; initial clinical signs include diarrhea and nausea. In most patients, *L. monocytogenes* can penetrate the intestines and cause a systemic infection resulting in fever, headaches, and muscle aches. Complications include blood poisoning and meningitis. The Mayo Clinic (2011b) recommends that people seek emergency medical care if symptoms include high fever, stiff neck, confusion, severe headaches, and photophobia because these symptoms may indicate meningitis. Listeriosis among pregnant women can have serious consequences for their fetuses, resulting in miscarriages, stillbirths, and premature births. Fetuses that survive the infection can experience delayed development or neurological damage after birth (Mayo Clinic 2011b).

L. monocytogenes occurs in soil, surface water, sewage, plants, fish, and crustaceans. The pathogen has been isolated from more than 40 species of domestic and wild mammals and 17 species of poultry and wild birds. It is common in cattle, sheep, and goats and also occurs in cats and dogs, although less often. Infected animals shed the pathogen in their feces, milk, and uterine discharges (Ferroglio 2012, Institute of Medicine 2012).

Animals usually become infected with *L. monocytogenes* after ingesting the pathogen. Many infected animals are asymptomatic, but *L. monocytogenes* can cause encephalitis (i.e., inflammation of the brain) and septicemia in moose, white-tailed deer, European roe deer, fallow deer, giraffes, and llamas. Animals suffering from encephalitis caused by the pathogen sometimes walk in circles; this behavior accounts for the disease's common name: circling disease. Other clinical signs include leaning, uncoordinated movements, and lethargy. Circling disease can be fatal in animals, but most survive if antibiotics are administered in a timely manner. This disease is more likely to occur when bacteria infect the mouth and throat; from there, they can invade the trigeminal nerve and follow it into the brain, resulting in encephalitis. Primates and carnivores infected with *L. monocytogenes* can develop encephalitis and septicemia. In some rodents and hares, *L. monocytogenes* causes septicemia; uterine infections have been observed in European hares and mountain hares, but mortality rates in these species are low (Mörner 2001, Ferroglio 2012).

Humans become infected when they consume food contaminated with *L. monocytogenes*. Contaminated foods can include hot dogs, poultry, other meats, fresh produce, milk, and milk products (CDC 2011b, Claiborn 2011, Institute of Medicine 2012). During 2011, the largest outbreak of listeriosis in the U.S. history was linked to contaminated cantaloupe (Sidebar 10.4). The pathogen can grow at low temperatures; this ability increases the risk that refrigerated food may become contaminated.

Wildlife species are rarely the source of human infections of *L. monocytogenes*, but people can develop listeriosis when they consume game meat contaminated with the pathogen. Among freshly shot wildlife in Germany, Atanassova et al. (2008) found *Listeria* spp. in 7 of 127 feral hogs, 4 of 95 European roe deer, and 3 of 67 red deer. In other studies cited by Atanassova et al. (2008), *L. monocytogenes* were found in 3%–24% of meat samples from feral hogs shot in Europe. Game meat is

SIDEBAR 10.4 OUTBREAK OF LISTERIOSIS LINKED TO CANTALOUPES (CDC 2011B, 2013, CLAIBORN 2011, INSTITUTE OF MEDICINE 2012)

A deadly outbreak of listeriosis occurred in the United States during the summer of 2011. Before it ended, 147 people had become ill with the same strain of L. *monocytogenes*, 143 were hospitalized, and 33 patients died. Epidemiologists traced the pathogen to cantaloupes grown by a producer in Colorado (Figure 10.11). The producer announced a voluntary recall of 1.5–4.5 million cantaloupes harvested between July 29 and September 10. Federal officials found four different strains of L. *monocytogenes* in a packing shed. The producer was cited for (1) having old and corroded equipment that was difficult to clean, (2) allowing standing pools of water in the shed, and (3) not adequately cooling the melons after harvest. *L. monocytogenes* was not found on melons growing in the fields or in the soil. Hence, epidemiologists were unable to identify the original source of the pathogen.

FIGURE 10.11 Cantaloupe field. (Courtesy of the USDA Agricultural Research Service, Washington, DC.)

more likely to be contaminated when the animal was shot in the abdomen. Wild animals also play a role in transmitting the pathogen among farms and livestock (Webster and Macdonald 1995) (Figure 10.12).

The risk of contracting listeriosis can be reduced by thoroughly washing fruit and vegetables and thoroughly cooking all meat. Uncooked meat should be separated from other foods so any pathogens in it will not be accidentally transferred to foods that are eaten raw, such as fruit. Smoked fish and soft cheese made from unpasteurized milk should be avoided. Foods should be refrigerated at 40°F (4°C).

FIGURE 10.12 Photo of European roe deer foraging. (Courtesy of Stig Nordskilde.)

10.7.5 *Staphylococcus aureus*

These bacteria are Gram positive, are not capable of movement, and do not form spores. They often appear as irregular grapelike clusters (*Staphylococcus* in Greek means grape cluster). They are part of the normal microbial flora of the skin and mucus membranes of the nose and throat of healthy humans and animals; 30% of people in the United States have *S. aureus* in their nasal passages (Graham et al. 2006). This pathogen causes a human disease called staphylococcosis, or staph infections, which usually are minor infections of skin wounds. However, staph infections can become serious if they become systemic and spread to vital body organs. *S. aureus* can also cause gastroenteritis with vomiting and diarrhea. This occurs when the pathogen grows on food and produces toxins. When consumed, the contaminated food results in staphylococcal food intoxication, also called food poisoning. Inadequate personal hygiene can contaminate foods such as poultry, ham, potato salad, and egg salad. At room temperature, these foods provide a fertile environment for *S. aureus* to grow and produce toxins. Each year in the United States, 240,000 people develop staphylococcal food poisoning; of these, 19,000 require hospitalization, and about 6 people die (Table 10.1).

S. aureus has been isolated from a diverse group of wild birds and mammals, including voles, mice, shrews, rabbits, feral hogs, deer, seals, and killer whales. Infections are asymptomatic in many of these species, but sometimes the pathogen can cause enteritis, conjunctivitis, liver lesions, meningitis, abortions, tonsillitis, urinary tract infections, and pneumonia. The disease is more likely to be fatal in hares and rabbits than in other wildlife species (WHO 2000, 2009, Atanassova et al. 2008, Merck 2010).

Humans can be infected with *S. aureus* from contact with livestock, companion animals, and wildlife; animals can also be infected by humans. For example, the Spanish imperial eagle is the most endangered bird of prey in Europe, and wildlife biologists band eaglets to track their survival. Ferrer and Hiraldo (1995) observed that several chicks developed wartlike lesions on their talons and around their eyes caused by *Staphylococcus* infections. The authors hypothesized that humans were the source of the infection. Their hypothesis was supported by the finding that 44% of chicks were infected when handled by barehanded humans, but this rate dropped to 4% when people wore disposable gloves.

Some strains of *S. aureus* have developed resistance to the antibiotic methicillin. These resistant strains have been called superbugs because of the difficulty of treating patients who have the misfortune of being infected by them. Most patients became infected with a resistant strain during a hospital stay, and approximately 18,000 patients in the United States die annually from the infection. These methicillin-resistant strains probably evolved in hospitals where antibiotics are commonly used to treat infections. Resistant strains did not stay there; they have been found in domestic pigs, cattle, horses, dogs, and cats. In these cases, the animals probably acquired the pathogen from their human caretakers. Methicillin-resistant strains were first detected in wildlife during 2008 when they were isolated from black rats and brown rats on European pig farms, from two eastern cottontail rabbits captured in Iowa, and from a migratory shorebird, the lesser yellowlegs. It is unknown how these wild mammals and birds became infected or if wildlife can spread these superbugs to other animals or humans. This spread of superbugs to wildlife is alarming (van de Giessen et al. 2009, Adams 2012, Wardyn et al. 2012).

10.7.6 *Yersinia* spp.

This genus of Gram-negative bacteria contains seven species. *Y. pestis* causes plague and is the subject of Chapter 2. Two other species that cause human illness are *Y. enterocolitica* and *Y. pseudotuberculosis*. Both are foodborne pathogens, which produce a disease known as yersiniosis. *Y. enterocolitica* also has a global distribution. *Y. pseudotuberculosis* evolved in Europe but has now spread worldwide. It is most prevalent in northern Europe and Scandinavia (Najdenski and Speck 2012).

Both *Y. pseudotuberculosis* and *Y. enterocolitica* cause gastroenteritis in humans with fever, diarrhea, and abdominal pain. The pain is often intense on the right side of the abdomen where it can be mistaken for appendicitis. Unfortunately, yersiniosis has resulted in many unnecessary appendectomies. After being ingested, both pathogens attach to the intestinal mucosa; from there, they cross the gut epithelium and invade the underlying tissue, particularly the Peyer's patches. From the Peyer's patches, the pathogens can invade the lymph nodes and eventually the liver, spleen, and lungs (Najdenski and Speck 2012). These pathogens can also cause reactive arthritis.

Most yersiniosis patients became infected from ingesting contaminated meat, salads, vegetables, milk, or dairy products that have been contaminated with fecal material from infected animals or humans. Humans can also acquire the pathogens by drinking contaminated water or by direct exposure to infected animals. In Great Britain, *Y. enterocolitica* was found in the intestines of livestock sent to slaughter during 2002—4% of cattle, 8% of sheep, and 10% of hogs (Milnes et al. 2008). House mice, black rats, brown rats, and birds, especially wood pigeons, are infected with *Y. enterocolitica* and *Y. pseudotuberculosis* and can transmit the pathogens from one pig or chicken farm to another (Mair 1973, Backhans and Fellström 2012). Livestock feed contaminated with the feces of infected rodents or birds produced outbreaks among livestock. Feral pigeons and rats often test positive for *Yersinia* when trapped at farms where there are outbreaks of yersiniosis (Gasper and Watson 2001). In Europe, 26 different strains of *Y. pseudotuberculosis* were isolated from brown rats and black rats trapped at pig farms (Kapperud 1975).

During 1998, an outbreak of *Y. pseudotuberculosis* was traced to iceberg lettuce grown in Norwegian fields, where there were unusually high densities of roe deer feces (Sidebar 10.5). During 2004, another outbreak of *Y. pseudotuberculosis* at a Finnish school sickened 53 children and staff. It was linked to raw carrots served at lunch. The strain of *Y. pseudotuberculosis* responsible for the outbreak was also isolated from the intestines of common shrews collected at the farm where the carrots were produced (WHO 2000, Kangas et al. 2008).

Both *Y. pseudotuberculosis* and *Y. enterocolitica* infect sheep, goats, pigs, dogs, cats, wild mammals, wild birds, and reptiles (Merck 2010). However, pigs are the reservoir hosts for strains of *Y. enterocolitica* that cause most human diseases; rodents and small animals are the reservoir hosts for virulent strains of *Y. pseudotuberculosis*. Both pathogens can tolerate temperatures ranging from 41°F to 107°F

SIDEBAR 10.5 OUTBREAK OF *Y. PSEUDOTUBERCULOSIS* FROM ICEBERG LETTUCE (NUORTI ET AL. 2004)

During an outbreak of yersiniosis in Finland in 1998, 47 people were infected with *Y. pseudotuberculosis* O:3. Of the 47 patients, 16 were hospitalized, 5 had appendectomies before the diagnosis of yersiniosis was made, and 1 patient died. An epidemiological study traced the source to iceberg lettuce supplied by a single shipping company. Four farms in the southwest archipelago of Finland were identified as the source of the contaminated lettuce. Two of the farms were irrigated from man-made ponds that collected surface runoff, and the two farms were irrigated by a ditch leading to a nearby lake. Fields and water sources used for irrigation were not fenced to exclude wildlife, and large quantities of feces from European roe deer were found in all lettuce fields and around the water sources (Figure 10.12). Finnish health authorities were unable to determine the exact mechanism that contaminated the lettuce but noted that deer and other wildlife served as a reservoir for *Y. pseudotuberculosis*. Health authorities concluded that the outbreak likely resulted when lettuce growing in the fields was contaminated with animal fecal material. They recommended that wildlife-proof fences be erected to protect the fields and water supplies from wildlife.

(5°C to 42°C) and can survive in soil and aquatic environments for months or years (Najdenski and Speck 2012).

Animals usually become infected with *Yersinia* from ingesting food or water that has been contaminated with feces from infected animals, but predators can also become infected by consuming infected prey. In New York, Shayegani et al. (1986) sampled 1426 wild animals for *Y. enterocolitica* and isolated the pathogen from 25% of Virginia opossum, 19% of white-tailed deer, 14% of red fox, 13% of black bear, 14% of North American porcupine, 12% of gray fox, 12% of eastern gray squirrel, 11% of striped skunk, 6% of beaver, 5% of muskrat, and 4% of woodchuck. Among birds, *Y. enterocolitica* were isolated from 25% of red-winged blackbirds, 12% of common grackles, and 10% of rugged grouse. The pathogen was also isolated from less than 10% of Canada geese, wood ducks, mallards, canvasbacks, ring-billed gulls, great horned owls, long-eared owls, red-tailed hawks, American kestrels, wild turkeys, brown-headed cowbirds, and European starlings. One hundred forty-eight serotypes of *Y. enterocolitica* were collected from these wild mammal and birds; most of the serotypes do not cause yersiniosis in humans. In Bulgaria, pathogenic strains of *Yersinia* were isolated from European hare, feral hog, mouflon, golden jackal, red fox, European otter, beech marten, European polecat, and wildcat (Nikolova et al. 2001). In Sweden, *Y. enterocolitica* was found in 6% of migratory birds including barnacle geese, dunlin, rough-legged hawk, common redshank, Eurasian blackbird, common redstart, and Eurasian blackcap (Niskanen et al. 2003).

In the United Kingdom, *Y. pseudotuberculosis* has been isolated from 7 species of wild mammals (i.e., house mouse, field vole, nutria, European rabbit, mountain hare, European hare, and red fox) and 21 species of wild birds. On the European continent, the pathogen infects mink, martens, moles, European hedgehogs, marmots, European roe deer, and fallow deer (Najdenski and Speck 2012).

Most animals infected with *Y. enterocolitica* and *Y. pseudotuberculosis* are asymptomatic, but illness becomes more likely when animals are cold, wet, overcrowded, malnourished, or in poor condition. Acute infections of *Y. pseudotuberculosis* in animals cause gastroenteritis with clinical signs of anorexia, lethargy, and diarrhea. Acute infections can progress to respiratory distress, blood poisoning, and death; chronic infections can result in weight loss, listlessness, and muscle weakness. Hares and beavers are particularly susceptible to diseases caused by *Y. pseudotuberculosis* or *Y. enterocolitica.* In contrast, most mice and rats are asymptomatic when infected with *Y. pseudotuberculosis.* In the United States, Canada, Australia, and New Zealand, yersiniosis is one of the most significant diseases among farm-reared elk and fallow deer, especially among young animals (Gasper and Watson 2001). Both pathogens cause abortions, mastitis, and gastrointestinal diseases in cervids. McLean et al. (1991) reported that yersiniosis was the leading cause of death among wild muskox on Banks Island, Northwest Territories, Canada. *Yersinia* produces illness in many avian species and is often fatal in birds with clinical signs including diarrhea and lameness (Mair 1973, Hacking and Sileo 1974).

The prevalence of *Y. pseudotuberculosis* and *Y. enterocolitica* at farms can be reduced through good hygiene, regular use of disinfectants, and effective programs to control rodents and exclude birds from the premises. Countries should restrict the importation or movements of animals that may be infected to prevent the introduction of new serotypes into an area.

10.7.7 Hepatitis E Viruses

Hepatitis is an inflammation of the liver and is mainly caused by single-stranded RNA viruses known as hepatitis viruses. There are five main serotypes (A, B, C, D, and E) that cause human illness and death throughout the world. Hepatitis A and E infections result from ingesting contaminated food or water; hepatitis B, C, and D are transmitted through human body fluids (e.g., blood). Only hepatitis E will be described because the others are not considered zoonotic diseases (WHO 2001).

Each year, the hepatitis E virus infects 20 million people across the globe, 3 million of whom become ill, and 70,000 die. Infected adults are at higher risk of becoming ill than children. Symptoms of hepatitis include nausea, abdominal cramps, fatigue, fever, loss of appetite, dark urine, and jaundice. The dark urine and jaundice result from an impairment of liver function. Hospitalization is usually not required, and the disease is self-limiting in most patients. In a few patients, hepatitis E causes liver failure and death. In some parts of the world (e.g., northern India and Pakistan), pregnant women who become infected during their third trimester have an increased risk of death and miscarriage. Immunocompromised patients have a higher risk of developing a chronic case of hepatitis E. During 2011, the first vaccine for hepatitis E was approved for use in China. At the time this book was being written, it was unavailable elsewhere (Pavio et al. 2010, WHO 2012).

People infected with hepatitis E shed the virus in their feces; most patients become ill from drinking water or eating food contaminated with fecal material containing the virus. In developing countries, large epidemics of hepatitis E result from contaminated water and are more likely to occur during the rainy season when flooding causes water supplies to be contaminated with human sewage. In Europe and North America, hepatitis E occurs sporadically rather than in large epidemics.

Some people have contracted hepatitis E from eating meat from infected animals including venison, pork, or shellfish. People can also become infected by working with infected animals. In China, pig farmers and people living downstream from pig farms have an increased risk of being infected with hepatitis E virus. In the United States, Canada, and elsewhere, antibodies for hepatitis E virus are more prevalent among veterinarians who work with pigs than other people (Meng et al. 2002, Pelosi and Clarke 2008, Widén 2012).

Hepatitis E virus infects pigs across the world. Antibodies against the hepatitis E virus have been detected in at least one pig at more than 90% of pig farms tested in the United States, Mexico, Spain, and New Zealand. Antibodies were found in 0%–71% of wild hogs collected in Japan and Spain (Pavio et al. 2010). Most pigs infected with hepatitis E have a mild disease or none at all. Antibodies against hepatitis E virus have been found in pigs, horses, cattle, goats, rabbits, dogs, cats, rats, and monkeys from several different countries (Pelosi and Clarke 2008, Widén 2012). They have also been found in 2%–35% of wild sika deer from Japan and 34% of roe deer from Hungary (Sonoda et al. 2004, Reuter et al. 2009, Tomiyama et al. 2009).

In many parts of the world, black rats, brown rats, and Pacific rats have antibodies against the hepatitis E virus. In the United States, antibodies against the virus have been discovered in 44% of brown rats from Louisiana, 77% from Maryland, and 90% from Hawaii (Kabrane-Lazizi et al. 1999). Likewise, Favorov et al. (2000) have detected antibodies against hepatitis E virus in most rats (woodrats, rice rats, hispid cotton rats, and brown rats) collected from across the United States. The same study reports hepatitis E antibodies in several species of mice and voles. Hence, rodents may play an important role as a vector or amplifying host for the virus (Pelosi and Clarke 2008, Pavio et al. 2010).

There are four genotypes (1, 2, 3, and 4) of the hepatitis E virus that can infect humans and other mammals. Genotypes 1 and 2 infect only humans and often cause waterborne epidemics. Genotypes 3 and 4 mainly infect domestic and wild pigs and also infect humans. Genotypes 3 and 4 also sporadically infect humans rather than produce large epidemics and are less virulent in humans than genotypes 1 and 2 (Pelosi and Clarke 2008, Pavio et al. 2010). Genotype 1 occurs mainly in Asia and Africa; genotype 2 in Mexico and Africa; genotype 3 in Europe, Russia, Australia, New Zealand, Argentina, and North America; and genotype 4 in India, China, and Southeast Asia (Pelosi and Clarke 2008, Pavio et al. 2010). In addition to the four genotypes of mammalian hepatitis, there also is an avian hepatitis E genotype that infects only birds. Poultry infected with this virus may develop an enlarged liver and spleen (Pelosi and Clarke 2008, Pavio et al. 2010).

10.8 ERADICATING *ESCHERICHIA COLI* AND FOODBORNE DISEASES FROM A COUNTRY

Most foodborne pathogens are ubiquitous in both humans and wildlife so there is little prospect of being able to eradicate them from large areas. Migratory birds are infected with some of these pathogens and can reintroduce them into areas where the pathogens have been eradicated (Hamasaki et al. 1989, Niskanen et al. 2003).

LITERATURE CITED

Adams, J. U. 2012. Antibiotic-resistant bugs go wild. *Science Now*. http://news.sciencemag.org/plants-animals/2012/10/antibiotic-resistant-bugs-go-wild (accessed March 4, 2013).

Atanassova, V., J. Apelt, F. Reich, and G. Klein. 2008. Microbiological quality of freshly shot game in Germany. *Meat Science* 78:414–419.

Backhans, A. and C. Fellström. 2012. Rodents on pig and chicken farms—A potential threat to human and animal health. *Infection Ecology and Epidemiology* 2:17093.

Barras, S. C. and J. A. Kadlec. 2000. Abiotic predictors of avian botulism outbreaks in Utah. *Wildlife Society Bulletin* 28:724–729.

Boyce, T. G., D. L. Swerdlow, and P. M. Griffin. 1995. *Escherichia coli* O157:H7 and the hemolytic-uremic syndrome. *New England Journal of Medicine* 333:364–368.

Brand, C. J., R. M. Duncan, S. P. Garrow, D. Olson, and L. E. Schumann. 1983. Waterbird mortality from botulism type E in Lake Michigan: An update. *Wilson Bulletin* 95:269–275.

Bruce, M. G., M. B. Curtis, M. M. Payne, R. K. Gautom et al. 2003. Lake-associated outbreak of *Escherichia coli* O157:H7 in Clark County, Washington, August 1999. *Archives of Pediatrics and Adolescent Medicine* 157:1016–1021.

CDC. 2011a. Food safety. http://www.cdc.gov/foodsafety/facts.html (accessed January 1, 2013).

CDC. 2011b. Investigation update: Multistate outbreak of listeriosis linked to whole cantaloupes from Jensen Farms, Colorado. http://www.cdc.gov/listeria/outbreaks/cantaloupes-jensen-farms/092711/index.html (accessed February 2, 2013).

CDC. 2012. *E. coli* (*Escherichia coli*). http://www.cdc.gov/ecoli/general/index.html (accessed January 10, 2013).

CDC. 2013. Summary of notifiable diseases—United States, 2011. *Morbidity and Mortality Weekly Report* 60(53):1–117.

Claiborn, K. 2011. Update on the listeriosis outbreak. *Journal of Clinical Investigation* 121:4569.

Cliplef, D. J. and G. Wobeser. 1993. Observations on waterfowl carcasses during a botulism epizootic. *Journal of Wildlife Diseases* 29:8–14.

Crump, J. A., A. C. Sulka, A. J. Langer, C. Schaben et al. 2002. An outbreak of *Escherichia coli* O157:H7 infections among visitors to a dairy farm. *New England Journal of Medicine* 347:555–560.

Duffitt, A. D., R. T. Reber, A. Whipple, and C. Chauret. 2011. Gene expression during survival of *Escherichia coli* O157:H7 in soil and water. *International Journal of Microbiology* 2011:340506.

Erickson, M. C. and M. P. Doyle. 2012. Plant food safety issues: Linking production agriculture with One Health. In: Institute of Medicine, editor. *Improving Food Safety through a One Health Approach*. National Academies Press, Washington, DC, pp. 140–175.

Favorov, M. O., M. Y. Kosoy, S. A. Tsarev, J. E. Childs, and H. S. Margolis. 2000. Prevalence of antibody to hepatitis E virus among rodents in the United States. *Journal of Infectious Diseases* 181:449–455.

Fenicia, L., G. Franciosa, M. Pourshaban, and P. Aureli. 1999. Intestinal toxemia botulism in two young people, caused by *Clostridium butyricum* Type E. *Clinical Infectious Diseases* 29:1381–1387.

Ferrer, M. and F. Hiraldo. 1995. Human-associated staphylococcal infection in Spanish imperial eagles. *Journal of Wildlife Diseases* 31:534–536.

Ferroglio, E. 2012. Listeria infections. In: D. Gavier-Widén, J. P. Duff, and A. Meredith, editors. *Infectious Diseases of Wild Mammals and Birds in Europe*. Wiley-Blackwell, Chichester, U.K., pp. 413–416.

Gardner, T. J., C. Fitzgerald, C. Xavier, R. Klein et al. 2011. Outbreak of campylobacteriosis associated with consumption of raw peas. *Clinical Infectious Diseases* 53:26–32.

Gasper, P. W. and R. P. Watson. 2001. Plague and yersiniosis. In: E. S. Williams and I. K. Barker, editors. *Infectious Diseases of Wild Mammals*, 3rd edition. Iowa State University Press, Ames, IA, pp. 313–329.

Gourreau, J. M., O. Debaère, P. Raevel, F. Lamarque et al. 1998. Étude d'un episode de botulisme de type E chez des mouettes rieuses (*Larus ridibundus*) et des goélands argentés (*Larus argentatus*) en baie de Canche (Pas-de-Calais). *Gibier Faune Sauvage* 15:357–363.

Graham III, P. L., S. X. Lin, and E. L. Larson. 2006. A U.S. population-based survey of *Staphylococcus aureus* colonization. *Annals of Internal Medicine* 144:318–325.

Hacking, M. A. and L. Sileo. 1974. *Yersinia enterocolitica* and *Yersinia pseudotuberculosis* from wildlife in Ontario. *Journal of Wildlife Diseases* 10:452–457.

Hagen, C. A. and R. J. Bildfell. 2007. An observation of *Clostridium perfringens* in greater sage-grouse. *Journal of Wildlife Diseases* 43:545–547.

Hamasaki, S., H. Hayashidani, K. Kaneko, M. Ogawa, and Y. Shigeta. 1989. A survey for *Yersinia pseudotuberculosis* in migratory birds in coastal Japan. *Journal of Wildlife Diseases* 25:401–403.

Horowitz, B. Z. 2010. Type E botulism. *Clinical Toxicology* 48:880–895.

Howie, H., A. Mukerjee, J. Cowden, J. Leith, and T. Reid. 2003. Investigation of an outbreak of *Escherichia coli* O157 infection caused by environmental exposure at a scout camp. *Epidemiology and Infection* 131:1063–1069.

Institute of Medicine. 2012. *Improving Food Safety through a One Health Approach.* National Academies Press, Washington, DC.

Iowa State University. 2004. Epsilon toxin of *Clostridium perfringens.* Center for Food Security and Public Health, Iowa State University, Ames, IA.

Kabrane-Lazizi, Y., J. B. Fine, J. Elm, G. E. Glass et al. 1999. Evidence for widespread infection of wild rats with hepatitis E virus in the United States. *American Journal of Tropical Medicine and Hygiene* 61:331–335.

Kadlec, J. A. 2002. Avian botulism in Great Salt Lake marshes: Perspectives and possible mechanisms. *Wildlife Society Bulletin* 30:983–989.

Kangas, S., J. Takkinen, M. Hakkinen, U. M. Nakari et al. 2008. *Yersinia pseudotuberculosis* O:1 traced to raw carrots, Finland. *Emerging Infectious Diseases* 14:1959–1961.

Kapperud, G. 1975. *Yersinia enterocolitica* in small rodents from Norway, Sweden, and Finland. *Acta Pathologica Microbiologica Scandinavica. Section B, Microbiology* 83:335–342.

Keene, W. E., E. Sazie, J. Kok, D. H. Rice et al. 1997. An outbreak of *Escherichia coli* O157:H7 infections traced to jerky made from deer meat. *Journal of the American Medical Association* 227:1229–1231.

Laidler, M. R., M. Tourdjman, G. L. Buser, T. Hostetler et al. 2013. *Escherichia coli* O157:H7 infections associated with consumption of locally grown strawberries contaminated by deer. *Clinical Infectious Diseases* 57:1129–1134.

Langholz, J. A. and M. T. Jay-Russell. 2013. Potential role of wildlife in pathogenic contamination of fresh produce. *Human–Wildlife Interactions* 7:140–157.

Mair, N. S. 1973. Yersiniosis in wildlife and its public health implications. *Journal of Wildlife Diseases* 9:64–71.

Mayo Clinic. 2011a. Food poisoning. http://mayoclinic.org/health/food-poisoning/DS00981 (accessed February 1, 2013).

Mayo Clinic. 2011b. Listeria infection. http://mayoclinic.org/health/food-poisoning/DS00963 (accessed January 25, 2013).

Mayo Clinic. 2012. Botulism. http://mayoclinic.org/health/food-poisoning/DS00657 (accessed March 25, 2013).

McLean, B. D., P. Fraser, and J. E. Blake. 1991. Yersiniosis in muskoxen on Banks Island, N.W.T., 1987–1990. *Rangifer* 13:65–66.

Meng, X. J., B. Wiseman, F. Elvinger, D. K. Guenette et al. 2002. Prevalence of antibodies to hepatitis E virus in veterinarians working with swine and in normal blood donors in the United States and other countries. *Journal of Clinical Microbiology* 40:117–122.

Merck. 2010. *Merck Veterinary Manual*, 10th edition. Merck, Whitehouse Station, NJ.

Milnes, A. S., I. Stewart, F. A. Clifton-Hadley, R. H. Davies et al. 2008. Intestinal carriage of verocytotoxigenic *Escherichia coli* O157, *Salmonella*, thermophilic *Campylobacter* and *Yersinia enterocolitica*, in cattle, sheep and pigs at slaughter in Great Britain during 2003. *Epidemiology and Infection* 136:739–751.

Mörner, T. 2001. Listeriosis. In: E. S. Williams and I. K. Barker, editors. *Infectious Diseases of Wild Mammals*, 3rd edition. Iowa State University Press, Ames, IA, pp. 502–505.

Muniesa, M., J. A. Hammerl, S. Hertwig, B. Appel, and H. Brüssow. 2012. Shiga toxin-producing *Escherichia coli* O104:H4: A new challenge for microbiology. *Applied and Environmental Microbiology* 78:4065–4073.

Najdenski, H. and S. Speck. 2012. Yersinia infections. In: D. Gavier-Widén, J. P. Duff, and A. Meredith, editors. *Infectious Diseases of Wild Mammals and Birds in Europe*. Wiley-Blackwell, Chichester, U.K., pp. 293–302.

Neimanis, A. and S. Speck. 2012. *Clostridium* species and botulism. In: D. Gavier-Widén, J. P. Duff, and A. Meredith, editors. *Infectious Diseases of Wild Mammals and Birds in Europe*. Wiley-Blackwell, Chichester, U.K., pp. 417–427.

Nikolova, S., Y. Tzvetkov, H. Najdenski, and A. Vesselinova. 2001. Isolation of pathogenic yersiniae from wild animals in Bulgaria. *Journal of Veterinary Medicine Series B* 48:203–209.

Niskanen, T., J. Waldenström, M. Fredriksson-Ahomaa, B. Olsen, and H. Korkeala. 2003. *virF*-positive *Yersinia pseudotuberculosis* and *Yersinia enterocolitica* found in migratory birds in Sweden. *Applied and Environmental Microbiology* 69:4670–4675.

Nuorti, J. P., T. Niskanen, S. Hallanvuo, J. Mikkola et al. 2004. A widespread outbreak of *Yersinia pseudotuberculosis* O:3 infection from iceberg lettuce. *Journal of Infectious Diseases* 189:766–774.

Olsen, S. J., L. C. MacKinon, J. S. Goulding, N. H. Bean, and L. Slutsker. 2000. Surveillance for foodborne-disease outbreaks—United States, 1993–1997. *Morbidity and Mortality Weekly Report* 49(SS01):1–51.

Pavio, N., X. J. Meng, and C. Renou. 2010. Zoonotic hepatitis E: Animal reservoirs and emerging risks. *Veterinary Research* 41:46.

Pelosi, E. and I. Clarke. 2008. Hepatitis E: A complex and global disease. *Emerging Health Threats Journal* 1:e8.

Pennington, H. 2010. *Escherichia coli* O157. *Lancet* 376:1428–1435.

Rabatsky-Ehr, T., D. Dingman, R. Marcus, R. Howard et al. 2002. Deer meat as the source for a sporadic case of *Escherichia coli* O157:H7 infection, Connecticut. *Emerging Infectious Diseases* 8:525–527.

Rangel, J. M., P. H. Sparling, C. Crowe, P. M. Griffin, and D. L. Swerdlow. 2005. Epidemiology of *Escherichia coli* O157:H7 outbreaks, United States, 1982–2002. *Emerging Infectious Diseases* 11:603–609.

Reuter, G., D. Fodor, P. Forgách, A. Kátai, and G. Szücs. 2009. Characterization and zoonotic potential of endemic hepatitis E virus (HEV) strains in humans and animals in Hungary. *Journal of Clinical Virology* 44:227–281.

Samadpour, M., J. Stewart, K. Steingart, C. Addy et al. 2002. Laboratory investigation of an *E. coli* O157:H7 outbreak associated with swimming in Battle Ground Lake, Vancouver, Washington. *Journal of Environmental Health* 64(10):16–26.

Sánchez, S., R. Martínez, A. García, D. Vidal et al. 2010. Detection and characterization of O157:H7 and non-O157 Shiga toxin-producing *Escherichia coli* in wild boars. *Veterinary Microbiology* 143:420–423.

Scallan, E., R. M. Hoekstra, F. J. Angulo, R. V. Tauxe et al. 2011. Foodborne illness acquired in the United States—Major pathogens. *Emerging Infectious Diseases* 17:7–15.

Shayegani, M., W. B. Stone, I. DeForge, T. Root et al. 1986. *Yersinia enterocolitica* and related species isolated from wildlife in New York state. *Applied and Environmental Microbiology* 52:420–424.

Snedeker, K. G., M. Campbell, and J. M. Sargeant. 2012. A systematic review of vaccinations to reduce the shedding of *Escherichia coli* O157 in the faeces of domestic ruminants. *Zoonoses and Public Health* 59:126–138.

Sonoda, H., M. Abe, T. Sugimoto, Y. Sato et al. 2004. Prevalence of hepatitis E virus (HEV) infection in wild boars and deer and genetic identification of a genotype 3 HEV from a boar in Japan. *Journal of Clinical Microbiology* 42:5371–5374.

Speck, S. 2012. *Campylobacter* infections. In: D. Gavier-Widén, J. P. Duff, and A. Meredith, editors. *Infectious Diseases of Wild Mammals and Birds in Europe*. Wiley-Blackwell, Chichester, U.K., pp. 398–401.

Tomiyama, D., E. Inoue, Y. Osawa, and K. Okazaki. 2009. Serological evidence of infection with hepatitis E virus among wild yezo-deer, *Cervus nippon yesoensis*, in Hokkaido, Japan. *Journal of Viral Hepatitis* 16:524–528.

van de Giessen, A. W., M. G. van Santen-Verheuvel, P. D. Hengeveld, T. Bosch et al. 2009. Occurrence of methicillin-resistant *Staphylococcus aureus* in rats living on pig farms. *Preventive Veterinary Medicine* 91:270–273.

Wahlström, H., E. Tysén, E. Olsson Engvall, B. Brändström et al. 2003. Survey of *Campylobacter* species, VTEC O157 and *Salmonella* species in Swedish wildlife. *Veterinary Record* 153:74–80.

Wardyn, S. E., L. K. Kauffman, and T. C. Smith. 2012. Methicillin-resistant *Staphylococcus aureus* in central Iowa wildlife. *Journal of Wildlife Diseases* 48:1069–1073.

Webster, J. P. and D. W. Macdonald. 1995. Parasites of wild brown rats (*Rattus norvegicus*) on UK farms. *Parasitology* 111:247–255.

WHO. 2000. *Guidelines for Drinking-Water Quality: Factsheets on Potential Waterborne Pathogens*. WHO, Geneva, Switzerland.

WHO. 2001. *Hepatitis Data*. WHO, Geneva, Switzerland.

WHO. 2009. Initiative for vaccine research (IVR): Diarrhoeal diseases. http://www.who.int/vaccine_research/diseases/diarrhoeal/en/index.html (accessed February 1, 2013).

WHO. 2012. Hepatitis. http://www.who.int/mediacentre/factsheets/fs328/en/index.html (accessed March 3, 2013).

Widén, F. 2012. Hepatitis E. In: D. Gavier-Widén, J. P. Duff, and A. Meredith, editors. *Infectious Diseases of Wild Mammals and Birds in Europe*. Wiley-Blackwell, Chichester, U.K., pp. 249–250.

Williams, M. L., D. L. Pearl, and J. T. LeJeune. 2011. Multiple-locus variable-nucleotide tandem repeat subtype analysis implicates European starlings as biological vectors for *Escherichia coli* O157:H7 in Ohio, USA. *Journal of Applied Microbiology* 111:982–988.

Wobeser, G., K. Baptiste, E. G. Clark, and A. W. Deyo. 1997. Type C botulism in cattle in association with a botulism die-off in waterfowl in Saskatchewan. *Canadian Veterinary Journal* 38:782.

Wobeser, G. and D. J. Rainnie. 1987. Epizootic necrotic enteritis in wild geese. *Journal of Wildlife Diseases* 23:376–385.

11 Psittacosis and Other Zoonotic Diseases Caused by *Chlamydia* Species

Close contact with psittacine birds and their products might lead to a disease which was described by Fra Bartolomeo, a Dominican sent to Peru in 1615, as 'a pest which is characterized by fever, malaise and drowsiness mainly occurred in the cold months, it affects and kills predominantly females especially pregnant ones.'

Pospischil (2009)

11.1 INTRODUCTION AND HISTORY

Psittacosis in humans and avian chlamydiosis in birds are both caused by the same pathogen: *Chlamydia psittaci*, which is an aerobic, Gram-negative bacterium (Figure 11.1). The pathogen originated in South America where it infected parrots and other species of psittacine birds. These birds were popular with the Native Americans in the area, who kept them as pets and used their feathers for decoration, but these practices sickened many people. A monk stationed in Peru first described the disease in 1615 (see quote earlier). The disease remained unknown in Europe until the 1880s when people started importing parrots from South America for household pets. In 1880, Jakob Ritter, a general practitioner, documented an unusual pneumonia-like disease in Switzerland that involved five members of his brother's household and two visitors; three of the seven succumbed. He realized that the family's newly imported parrots and finches were the source of the disease.

C. psittaci traveled to the United States soon after reaching Europe. The first cases in the United States occurred during 1904 when three family members were diagnosed with psittacosis; the apparent source was a sick pet parrot (Henry and Crossley 1986). During the winter of 1929–1930, a lethal pandemic of psittacosis swept through both Europe and the United States; there were 215 psittacosis patients and 45 deaths in Germany, 125 patients and 27 deaths in the United Kingdom, and 169 patients and 34 deaths in the United States (Pospischil 2009). The source of the pathogen in this pandemic was traced back to green Amazon parrots imported from Argentina; in response, public health authorities around the world prohibited the importation of parrots and other related birds (Vanrompay et al. 1995). Concomitantly, an outbreak of psittacosis struck in the Faroe Islands with lethal consequences for the local population (Sidebar 11.1).

During 1929, Bedson et al. (1930) isolated the pathogen that produced psittacosis. Initially, the pathogen was classified as a virus but was later identified as a bacterium

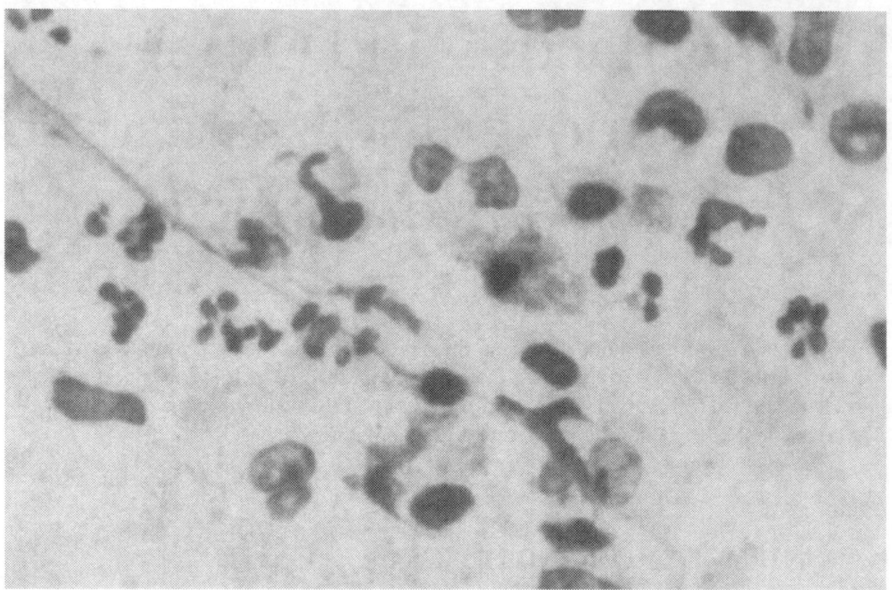

FIGURE 11.1 Photo of *C. psittaci*. (Courtesy of Dorothy Reese and the CDC.)

and named *C. psittaci* (Harkinezhad et al. 2009). The pathogen has two distinct life forms and alternates between them. One is a highly condensed infectious form called the elementary body. This nonmetabolic form allows the bacteria to survive outside of cells and to invade them. Once inside a cell, the elementary body changes into a reticulate body that is metabolically active and reproduces through cell division. The reticulate bodies then change back into elementary bodies, which erupt from the cell and complete the life cycle.

At various times in the past, *C. psittaci* has been called *Chlamydophila psittaci*, *Miyagawanella psittaci*, *Rickettsia psittaci*, *Bedsonia*, and *Psittacosis* virus (Kaleta and Taday 2003). The most common name for the human disease is psittacosis; it is a derivation of the Latin word for parrot (psittacus) because these birds were a common source of infection. The disease in humans has been called parrot fever, parrot disease, and ornithosis. In birds, the disease is commonly referred to as avian chlamydiosis. In this book, the human disease will be referred to as psittacosis and the disease in animals as avian chlamydiosis.

C. psittaci is a member of the Chlamydiaceae family. Some authors divide the nine species of this family into two genera *Chlamydophila* and *Chlamydia*; other authorities consider them all members of a single genus *Chlamydia* (Stephens et al. 2009). In this book, all species will be considered members of the *Chlamydia* genus. In addition to *C. psittaci*, this genus includes *C. pneumoniae* and *C. trachomatis*, which infect humans; *C. felis*, which infects cats; *C. caviae* (domesticated guinea pigs); *C. pecorum* (pigs and ruminants); *C. abortus* (cattle, sheep, and goats); *C. muridarum* (mice and hamsters); and *C. suis* (pigs). *C. psittaci* will be considered initially; other

SIDEBAR 11.1 AN OUTBREAK OF PSITTACOSIS ON THE FAROE ISLANDS (HERRMANN ET AL. 2006, POSPISCHIL 2009)

The Faroe Islands are in the North Atlantic halfway between Ireland and Iceland (Figure 11.2). Large numbers of northern fulmars nest on the islands, and the local people used to catch between 50,000 and 100,000 juvenile fulmars for food. During 1930, people were stricken with an influenza-like disease not seen before on the Faroe Islands. Over the next 8 years, 174 people out of a total population of 25,000–30,000 contracted the disease. The disease proved deadly, resulting in a human fatality rate of 20%. Pregnant women were especially susceptible to the disease; their fatality rate was 80%. The disease was soon identified as psittacosis. The source of the pathogen was traced to the northern fulmars; most patients had either hunted or consumed them. In response, the hunting and consumption of fulmars were banned, and the incidence of psittacosis declined in the Faroe Islands. However, *C. psittaci* persisted in northern fulmars; 10% of juvenile fulmars collected on the Faroe Islands during 1999 were infected.

The strain of *C. psittaci* isolated from these fulmars proved almost identical to a strain isolated from parakeets. How a South American disease had reached these isolated islands in the North Atlantic remains unclear. One hypothesis is that shippers transporting parrots and parakeets from South America to Europe during 1930 threw dead birds overboard, and fulmars fed on the floating carcasses. That year was also the start of a psittacosis outbreak in Europe, which was linked to the importation of infected birds from South America. If this hypothesis is correct, it provides a classic example of humans inadvertently spreading zoonotic diseases across the world. *C. psittaci* now infects waterbirds across the Arctic, and many people living in the Arctic who consume native birds have antibodies against *C. psittaci* (Wilt et al. 1959, Hildes et al. 1965).

species that cause zoonotic diseases (*C. abortus*, *C. felis*, and *C. pneumoniae*) will be discussed at the end of the chapter (Kerr et al. 2005).

There are six serovars of *C. psittaci* that infect birds (A, B, C, D, E, and F) with different serovars infecting different avian species. Serovar A primarily infects Psittaciformes, B infects pigeons, C infects ducks and geese, and D infects turkeys. Serovars E and F are associated with a more diverse array of species including ducks, pigeons, and Psittaciformes. All six avian serovars cause human illness. *C. psittaci* infections have also been reported in one or more species of scallops, fish, frogs, tortoises, and snakes (Vanrompay et al. 1995, Harkinezhad et al. 2009).

11.2 SYMPTOMS IN HUMANS

Psittacosis is an uncommon disease of humans that occurs sporadically or as an outbreak. It can range from a mild flu-like disease to being fatal. The first symptoms of

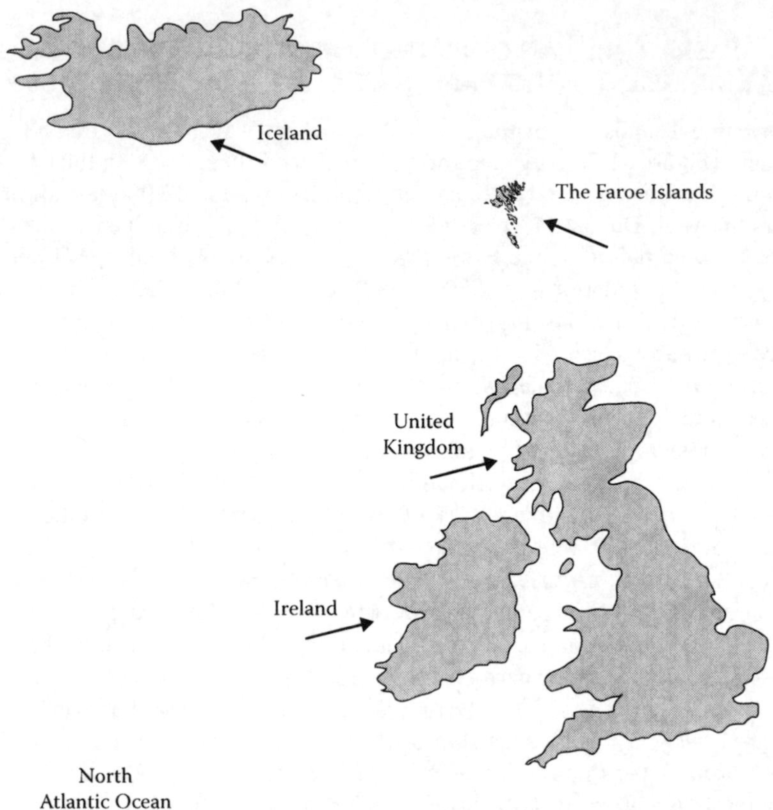

FIGURE 11.2 Map of the Faroe Islands showing their location in the North Atlantic.

psittacosis begin abruptly after a 5- to 15-day incubation period. Dry cough, muscle aches, headache, fever, and a skin rash are early symptoms. Patients may experience tightness in the chest or difficulty breathing; pneumonia can result. The infection can spread to the heart (resulting in endocarditis or myocarditis), central nervous system (encephalitis), liver (hepatitis), joints (arthritis), and eye (conjunctivitis). Most patients require hospitalization. Disease severity increases in the young, elderly, and pregnant women. Psittacosis during pregnancy can result in an abortion or premature delivery (Idu et al. 1998). Before the use of antibiotics, 15%–20% of psittacosis patients died, but mortality is rare (less than 1%) when antibiotics are available (Rodolakis and Mohamad 2010, U.S. National Association of State Public Health Veterinarians 2010).

From 1988 to 2003, 935 human cases of psittacosis in the United States were reported to the CDC (Smith et al. 2005). Vanrompay et al. (2007) estimated that there are about 100 psittacosis cases annually in the United States, resulting in one fatality. During 2004, the number of patients diagnosed with psittacosis was 239 in Australia, 62 in Great Britain, 40 in Japan, 33 in the Netherlands, 15 in Germany, 13 in Northern Ireland, 12 in Belgium, 8 in Denmark, 7 in Hungary, 7 in Sweden, and 4 in Croatia (Harkinezhad et al. 2009). Worldwide, it is unknown how many people with psittacosis fail to seek medical attention.

11.3 *CHLAMYDIA PSITTACI* INFECTIONS IN ANIMALS

C. psittaci infections in birds occur in the pharyngeal, cloacal, and respiratory epithelium and in the liver and spleen. Infected birds intermittently shed the pathogen in their feces and mucous discharges from their eyes and nose. Birds are more likely to develop avian chlamydiosis when young or stressed from overcrowding, a change in diet, and reproduction. Most birds become infected from inhaling the pathogen, but they can also become infected from drinking or bathing in contaminated water, consuming fecal material from an infected bird, or being bitten by lice, mites, or blood-sucking insects. The pathogen can be transmitted through the egg in some species, such as chickens and snow geese (Wittenbrink et al. 1993). More commonly, birds infect their young when they regurgitate food for them. Birds of prey and avian scavengers can become ill by consuming infected birds. The pathogen is found frequently on the plumage of infected birds because birds use their beaks to clean their feathers (Kaleta and Taday 2003, Smith et al. 2005).

Both captive and wild birds are the reservoir hosts for *C. psittaci*, especially birds from the order Psittaciformes (parakeets, canaries, lorikeets, and cockatoos; Figure 11.3) and Columbiformes (pigeons and doves). Infections occur in 14%–16%

(a)

(b)

FIGURE 11.3 *C. psittaci* probably evolved among Central and South American parrots, but the pathogen now has a worldwide distribution because parrots, such as the scarlet macaw (a) and the double yellow-headed Amazon parrot (b), were sent around the world for pets and many of them were infected with the pathogen. (Courtesy of Eric Grafman and the CDC.)

of Psittaciformes kept as pets in Europe (Vanrompay et al. 1995). Studies of urban pigeons found that 58% of feral pigeons in Belgium were seropositive, as were 56% in Switzerland, 48% in Italy, 30% in Japan, 26% in Bosnia and Herzegovina, and 19% in Macedonia (Henry et al. 1977, Chiba et al. 1984, Haag-Wackernagel and Moch 2004, Harkinezhad et al. 2009). By 2010, *C. psittaci* had been detected in over 450 avian species from 30 avian orders (Kaleta and Taday 2003).

Birds infected with *C. psittaci* can be asymptomatic or develop an acute or chronic illness depending on the bird's age, health, and species, the virulence of the *C. psittaci* strain, and infectious dose. In many avian species, adults infected with the pathogen are asymptomatic, while young birds experience acute infections. Lethargy, closed eyes, and ruffled feathers are clinical signs of avian chlamydiosis. Infected birds may have mucous discharges from their eyes or nose and may produce green or yellow feces. Birds with severe infections may become dehydrated, emaciated, and die. When autopsied, birds with avian chlamydiosis have an enlarged spleen or liver, and their lungs appear cloudy. *C. psittaci* infections may spread to the central nervous system, resulting in tremors, convulsions, and an inability to walk or fly. Chronically infected pigeons can be lame and have difficulty flying (Vanrompay et al. 1995, Harkinezhad et al. 2009).

Avian chlamydiosis is a serious economic problem for turkey and duck producers. The disease can be introduced into a flock by wild birds or by bringing in infected poultry. *C. psittaci* is prevalent in birds such as blackbirds, pigeons, and house sparrows that frequently forage at poultry farms, and they can be the source of infection. Clinical signs in domestic ducks include nasal discharges, conjunctivitis, diarrhea, and trembling. Mortality rates from avian chlamydiosis can approach 30% among domestic ducks. In turkeys, clinical signs include ruffling of the feathers, anorexia, diarrhea with yellow feces, coughing, nasal discharges, and difficulty breathing. Fatality rates in domestic turkeys can approach 40%. Chickens are more resistant to the pathogen than ducks and turkeys and usually do not become ill when infected (Vanrompay et al. 1995, Harkinezhad et al. 2009).

In mammals, *C. psittaci* produces diverse clinical signs, including conjunctivitis or abortions in horses, cattle, sheep, goats, pigs, and koalas. It can infect the testes of cattle and pigs; the spinal cord or brain of cattle and dogs; the joints of horses, cattle, sheep, and pigs; and the lungs of horses, cattle, sheep, goats, pigs, dogs, cats, and mice. In snowshoe hares and muskrats, the pathogen infects the intestines and causes diarrhea (Vanrompay et al. 1995).

11.4 HOW HUMANS CONTRACT PSITTACOSIS

C. psittaci is transmitted from birds to humans, but this transmission is difficult considering the frequency of avian chlamydiosis among pets and the infrequency of psittacosis among humans. Infected birds excrete *C. psittaci* in their feces and nasal discharges. An infected bird can shed the pathogen for months; the pathogen can survive for weeks within fecal material. Most people infected with *C. psittaci* acquired the pathogen from pet birds, usually by inhaling the pathogen as a result of aerosolized particles of dried feces, urine, or nasal discharges from infected birds. Humans can be infected from kissing pet birds (mouth-to-beak contact) or touching the plumage of infected birds.

Person-to-person transmission can occur but is rare. Infected birds do not shed the pathogen constantly but are more likely to do so when they are stressed (Harkinezhad et al. 2009, U.S. National Association of State Public Health Veterinarians 2010).

People who have an elevated risk of contracting psittacosis include poultry farmers, workers at poultry processing plants, veterinarians, wildlife rehabilitators, bird-banders, and wildlife biologists who handle wild birds (Sidebar 11.2). This risk is also enhanced for pigeon fanciers and people living in households with pet parrots, parakeets, and other psittacines (Figure 11.3). Outbreaks of psittacosis have stricken workers at turkey, duck, and chicken farms and poultry processing plants (Hinton et al. 1993, Gaede et al. 2008). Seroprevalence for antibodies against *C. psittaci* was 15% of humans working in 39 breeding facilities for Psittaciformes in Belgium (Vanrompay et al. 2007).

Urban populations of domestic pigeons often congregate in large groups where humans feed them, and *C. psittaci* can easily spread in such settings from one pigeon

SIDEBAR 11.2 TWO WATERFOWL BIOLOGISTS DEVELOP PSITTACOSIS (WOBESER AND BRAND 1982)

During December 1979, a wildlife biologist working for the USFWS sought medical attention for fever, lower back pain, headaches, and a loss of appetite. Influenza was suspected, and acetaminophen was prescribed. As symptoms worsened over the next few days, the patient was hospitalized with a suspected viral infection; chest x-rays revealed a lung infection, but tests for routine pathogens were negative. The patient was given several antibiotics intravenously, but 2 days later, the entire right lung and part of the left lung were infected, and the patient's fever rose to 106°F (41°C). The next day, the patient exhibited hypoxemia and confusion; he was transferred to the hospital's intensive care unit. Thereafter, a diagnosis of psittacosis was made, antibiotics effective against *C. psittaci* were employed, and the patient recovered.

During May 1980, a faculty member of the University of Saskatchewan sought medical attention for breathlessness, fever, headaches, muscle aches, and a dry cough. A physical exam revealed an enlarged spleen, rapid pulse, and infiltration in the right lung. A week later, the symptoms were worse, and the patient was hospitalized. Blood, urine, and throat cultures were negative for bacterial pathogens other than *C. psittaci*. The patient recovered after being placed on a tetracycline treatment. Antibodies against *C. psittaci* were detected later, and the final diagnosis was psittacosis.

The source of the pathogen was never identified in either case. Neither patient could recall any contact with poultry, psittacines, or feral pigeons during the prior month. The first patient had investigated the mortality of snow geese and sandhill cranes at the Aransas National Wildlife Refuge in Texas. The second patient had necropsied sandhill cranes, geese, and ducks that had died in Saskatchewan and Manitoba. The data suggested that both people had contracted psittacosis from working with wild birds.

FIGURE 11.4 Snow geese can be infected with *C. psittaci.* (Courtesy of Dave Menke and the USFWS.)

to another. Feral pigeons can transmit the pathogen to people who feed them, care for them, or live in the urban environment. Three members of a family in Minnesota were diagnosed with psittacosis after finding a sick feral pigeon outside their apartment and bringing it indoors to care for it (Henry and Crossley 1986, Haag-Wackernagel and Moch 2004). People can also become infected from environmental exposure. For example, mowing a lawn can cause feces or urine from infected birds to become airborne where they can be inhaled.

Several cases of psittacosis in humans can be traced to waterbirds, including shearwaters and gulls. An outbreak on the Faroe Islands, which had a case fatality rate of 23%, was linked to northern fulmars; an outbreak in Louisiana that killed eight people was linked to snowy egrets. One scientist, who drew blood from cormorants, developed psittacosis. A high proportion of Native Americans living in the Canadian Arctic have antibodies against *C. psittaci.* The source of the pathogen remains elusive, but snow geese (Figure 11.4) may be the culprit; a high proportion of these birds have antibodies against *C. psittaci* (Hildes et al. 1965, Wilt et al. 1972, Wobeser and Brand 1982).

11.5 MEDICAL TREATMENT

Tetracyclines are commonly used to treat psittacosis in humans. Oral doses of doxycycline or tetracycline hydrochloride are prescribed for mild cases, while doxycycline hyclate is given intravenously to patients who are seriously ill. Antibiotics should be continued 10–14 days after the fever subsides to prevent a relapse. Macrolide antibiotics are sometimes prescribed for patients who should not be given tetracyclines (e.g., pregnant women, young children, or people allergic to tetracyclines). People with pet birds should seek prompt medical attention if they experience influenza-like symptoms or develop respiratory problems (U.S. National Association of State Public Health Veterinarians 2010).

11.6 WHAT PEOPLE CAN DO TO REDUCE THE RISK OF CONTRACTING PSITTACOSIS

People should wear protective clothing when cleaning bird cages or handling birds that might be infected with *C. psittaci.* Protective clothing include gloves, eyewear, surgical hat, and a respirator with at least an N95 rating (U.S. National Association of State Public Health Veterinarians 2010). Prior infections by *C. psittaci* do not produce long-lasting immunity in humans or other mammals, so people can be infected more than once. A vaccine effective against *C. psittaci* has not been developed. Antibiotics are not routinely given to people who have been exposed to *C. psittaci* in the hope of preventing psittacosis (U.S. National Association of State Public Health Veterinarians 2010).

Psittacosis is rarely fatal in humans if diagnosed and treated quickly. Hence, awareness of the disease and its potential consequences is important for people who are at risk of infection and the physicians who treat them. Wobeser and Brand (1982) recommended that people working with domestic or wild birds alert their physician to the possibility of psittacosis if they experience persistent fever, persistent flu-like symptoms, unresolved pneumonia, or typhoid-like conditions.

One way to diminish the risk of psittacosis in humans is to reduce the prevalence of avian chlamydiosis in pet birds. This can be accomplished by taking sick birds to a veterinarian who may recommend the use of antibiotics for all birds in the household. Antibiotic treatment should continue for a minimum of 45 days (Association of Avian Veterinarians 2010). Birds should not be sold or purchased if they have been exposed to *C. psittaci* or are exhibiting clinical signs of avian chlamydiosis. Newly acquired birds should be quarantined for at least 30 days before they join other birds. Birds that come into contact with the public (e.g., those in schools, petting zoos, or public displays) should be routinely tested for *C. psittaci.* Bird cages, food bowls, and water bowls should be cleaned daily and disinfected regularly. Rooms and cages that housc infected birds should be cleaned and disinfected by workers wearing appropriate protective clothing. Disinfectants could include 1% Lysol® or a 1:32 dilution of household bleach (1/2 cup of bleach in a gallon of water). Rooms should be well ventilated when using disinfectants because they can irritate the lungs of people or birds if too much is inhaled (U.S. National Association of State Public Health Veterinarians 2010). Poultry and duck farms can become infected with *C. psittaci* if the facilities attract wild birds that are shedding the pathogen.

The U.S. National Association of State Public Health Veterinarians (2010) published a compendium of measures to reduce the risk of psittacosis in humans and avian chlamydiosis in birds. Their recommendations include as follows: (1) educate and protect people at risk, (2) test birds before they are boarded or sold, (3) avoid purchasing or selling birds that exhibit clinical signs of avian chlamydiosis, (4) maintain records of all bird-related transactions for at least a year so that sources of infected birds can be identified, (5) isolate all ill or newly acquired birds, (6) avoid mixing birds from multiple sources, (7) practice preventive husbandry of birds, (8) control the spread of infections, and (9) use disinfection measures.

Animal or public health authorities may issue an order to quarantine all birds that have been infected or exposed to *C. psittaci* to prevent the pathogen from spreading to other birds. The USDA APHIS regulates the importation of birds into the United States to prevent sick birds and avian diseases from entering the country. APHIS can also prohibit the movement of poultry, carcasses, or offal from any farm or other location where avian chlamydiosis occurs.

Breeding facilities for cockatoos, parrots, parakeets, and lories frequently use antibiotics to keep birds healthy. Nonetheless, 8% of birds at these facilities and 15% of people working in them were seropositive, indicating a prior infection by *C. psittaci*. The situation creates a serious risk of developing antibiotic-resistant strains of *C. psittaci*. The development of effective antibiotics against this pathogen reduced the mortality rate of psittacosis in humans from 15% to 20% to less than 1%. If strains of *C. psittaci* become resistant to antibiotics, human mortality rates may increase. This risk has led Vanrompay et al. (2007) to recommend restrictions of the use of antibiotics for captive psittacines and the development of a vaccine against psittacosis.

11.7 ERADICATING PSITTACOSIS FROM A COUNTRY

It will be difficult to eradicate psittacosis in countries where the *C. psittaci* have already spread to wild birds. The chances of success are greater on isolated islands where the pathogen is limited to domesticated animals.

11.8 ZOONOTIC DISEASES CAUSED BY OTHER CHLAMYDIAL SPECIES

11.8.1 *Chlamydia abortus*

This pathogen causes domestic sheep and goats to abort their fetuses, resulting in large economic losses to the European livestock industry. Infected females shed the pathogen through their fetuses, placentas, and uterine discharges. *C. abortus* then infects other sheep when they ingest the pathogen. Infected females may harbor the pathogen in their tonsils or lymph nodes and often are asymptomatic until the next lambing season when infected ewes will abort during the last month of their pregnancy.

When *C. abortus* first infects a sheep flock, up to 30% of pregnant ewes may abort or give birth to weak offspring that die soon thereafter. In subsequent years, 5%–15% of pregnant ewes in an infected flock may abort with most abortions occurring in young ewes. After aborting once, most ewes have enough immunity to the pathogen to prevent further abortions. Many of these ewes, however, are still infected and can continue to shed *C. abortus*. The pathogen also causes abortions in horses, cows, pigs, and deer but is less likely to do so than sheep and goats (Papp et al. 1994, Kerr et al. 2005, Wheelhouse and Longbottom 2012). Young sheep and goats infected from their mothers may suffer from arthritis, pneumonia, and conjunctivitis. Males can be infected by their mothers or by mating with infected females. *C. abortus* can infect the seminal vesicles where it can reduce male fertility. Infected males can also shed the pathogen with their sperm. Wild birds are rarely infected with *C. abortus*, but those that are can transport the pathogen over long distances (Rodolakis and Mohamad 2010).

A few people have been sickened by *C. abortus* in both Europe and the United States. Some pregnant women have experienced a spontaneous abortion after being exposed to infected sheep or goats when these animals were aborting or giving birth. The first signs of illness in a pregnant woman include fever, headache, malaise (i.e., a general feeling of not being well), vomiting, and lower abdominal pain. These can be followed by abortion (Pospischil et al. 2002, Rodolakis and Mohamad 2010). Pregnant women infected with *C. abortus* can develop life-threatening complications, but these are rare (Walder et al. 2005). Additionally, some people have developed a respiratory infection from inhaling *C. abortus*; symptoms are similar to psittacosis. Others experience a pelvic inflammatory disease (Barnes and Brainerd 1964, Walder et al. 2003).

Two vaccines for *C. abortus* are available for use in sheep and goats. The most efficient vaccine contains live, but attenuated bacteria. It produces immunity against all strains of *C. abortus* and *C. pecorum* but is ineffective against *C. psittaci* (Rodolakis and Mohamad 2010).

11.8.2 *Chlamydia felis*

Cats are the reservoir host for *C. felis*. This pathogen causes conjunctivitis in cats and swelling of their eyelids. Moreover, infections result in mucous discharges from the nose and eyes, sneezing, fever, and lameness. Cats can shed the pathogen in mucous discharges for 2 months and in feces or vaginal excretions for much longer (Sykes 2005). Cats that live in close association with other cats have an enhanced risk of becoming infected. In Japan, 17% of pet cats have antibodies against *Chlamydia*, as do 51% of stray cats (Yan et al. 2000).

C. felis can produce conjunctivitis in humans. Most patients with *C. felis* conjunctivitis acquired the pathogen from a domestic cat in their household; veterinarians who treat cats also are at risk of acquiring this disease. *C. felis* infections are common in cats but rare in people, indicating that the risk of someone contracting the disease is very low. The pathogen is suspected of causing pneumonia, endocarditis, and hepatitis in humans based on serological (i.e., pertaining to antibodies) investigations, but this has not been confirmed (Wheelhouse and Longbottom 2012). *C. felis* infection in cats is usually treated with doxycycline for a period of 2 weeks. All cats in a household should be treated simultaneously. Prompt treatment of infected cats is recommended to prevent human disease. Both live and inactivated vaccines have been approved for cats in the United States, Europe, and Japan (Sykes 2005).

11.8.3 *Chlamydia pneumoniae*

This pathogen causes respiratory infections in people across the world. Most patients become infected after inhaling it. After initially infecting the respiratory system, *C. pneumoniae* can spread throughout the human body; it has been detected in the brain, heart, joints, and vascular system. The ubiquitous distribution of the pathogen in the body suggests that it does not produce symptoms in most people. The DNA of *C. pneumoniae* is found in plaque located in blood arteries and veins, causing some speculation that infections might contribute to hardening of the arteries (Roulis et al. 2013).

Humans are the reservoir host for *C. pneumoniae*, but the pathogen has been detected in a diverse group of animals including horses and dogs in Europe; koalas, western barred bandicoots, Blue Mountains tree frogs, and great barred frogs in Australia; and green iguanas in Central America (Bodetti et al. 2002, Kutlin et al. 2007). In these animals, infections have been reported in the lungs, urinary system, genital tracts, and eyes (Wheelhouse and Longbottom 2012, Roulis et al. 2013).

LITERATURE CITED

Association of Avian Veterinarians. 2010. *Chlamydiosis in Birds*. Association of Avian Veterinarians, Denver, CO.

Barnes, M. G. and H. Brainerd. 1964. Pneumonitis with alveolar-capillary block in a cattle rancher exposed to epizootic bovine abortion. *New England Journal of Medicine* 271:981–985.

Bedson, S. P., G. T. Western, and S. L. Simpson. 1930. Observations on the aetiology of psittacosis. *Lancet* 215:345–346.

Bodetti, T. J., E. Jacobson, C. Wan, L. Hafner et al. 2002. Molecular evidence to support the expansion of the hostrange of *Chlamydophila pneumoniae* to include reptiles as well as humans, horses, koalas and amphibians. *Systematic and Applied Microbiology* 25:146–152.

Chiba, N., J. Arikawa, I. Takashima, and N. Hashimoto. 1984. Isolation and serological survey of chlamydiosis in feral pigeons and crows in Hokkaido. *Japanese Journal of Veterinary Science* 46:243–245.

Gaede, W., K.-F. Reckling, B. Dresenkamp, S. Kenklies et al. 2008. *Chlamydophila psittaci* infections in humans during an outbreak of psittacosis from poultry in Germany. *Zoonoses and Public Health* 55:184–188.

Haag-Wackernagel, D. and H. Moch. 2004. Health hazards posed by feral pigeons. *Journal of Infection* 48:307–313.

Harkinezhad, T., T. Geens, and D. Vanrompay. 2009. *Chlamydophila psittaci* infections in birds: A review with emphasis on zoonotic consequences. *Veterinary Microbiology* 135:68–77.

Henry, K. and K. Crossley. 1986. Wild-pigeon-related psittacosis in a family. *Chest* 90:708–710.

Henry, M. C., F. Hebrant, and J. B. Jadin. 1977. Importance and distribution of serological findings of ornithosis-psittacosis in semi-domestic pigeons. *Bulletin de la Société de Pathologie Exotique et de ses Filiales* 70:144–151.

Herrmann, B, H. Persson, J.-K. Jensen, H. D. Joensen et al. 2006. *Chlamydophila psittaci* in fulmars, the Faroe Islands. *Emerging Infectious Diseases* 12:330–332.

Hildes, J. A., W. L. Parker, A. Delaat, W. Stackiw, and J. C. Wilt. 1965. The elusive source of psittacosis in the Arctic. *Canadian Medical Association Journal* 93:1154–1155.

Hinton, D. G., A. Shipley, J. W. Galvin, J. T. Harkin, and R. A. Brunton. 1993. Chlamydiosis in workers at a duck farm and processing plant. *Australian Veterinary Journal* 70:174–176.

Idu, S. R., C. Zimmerman, L. Mulder, and J. F. Meis. 1998. A very serious course of psittacosis in pregnancy. *Nederlands Tijdschrift voor Geneeskunde* 142:2586–2589.

Kaleta, E. F. and E. M. A. Taday. 2003. Avian host range of *Chlamydophila* spp. based on isolation, antigen detection and serology. *Avian Pathology* 32:435–461.

Kerr, K., G. Entrican, D. McKeever, and D. Longbottom. 2005. Immunopathology of *Chlamydophila abortus* infection in sheep and mice. *Research in Veterinary Science* 78:1–7.

Kutlin, A., P. M. Roblin, S. Kumar, S. Kohlhoff et al. 2007. Molecular characterization of *Chlamydophila pneumoniae* isolates from western barred bandicoots. *Journal of Medical Microbiology* 56:407–417.

Papp, J. R., P. E. Shewen, and C. J. Gartley. 1994. Abortion and subsequent excretion of chlamydiae from the reproductive tract of sheep during estrus. *Infection and Immunity* 62:3786–3792.

Pospischil, A. 2009. From disease to etiology: Historical aspects of chlamydia-related diseases in animals and humans. *Drugs of Today (Supplement B)* 45:141–146.

Pospischil, A., R. Thoma, M. Hilbe, P. Grest, and J.-O. Gebbers. 2002. Abortion in woman caused by caprine *Chlamydophila abortus* (*Chlamydia psittaci* serovar 1). *Swiss Medical Weekly* 132:64–66.

Rodolakis, A. and K. Y. Mohamad. 2010. Zoonotic potential of *Chlamydophila*. *Veterinary Microbiology* 140:382–391.

Roulis, E., A. Polkinghorne, and P. Timms. 2013. *Chlamydia pneumoniae*: Modern insights into an ancient pathogen. *Trends in Microbiology* 21(3):120–128.

Smith, K. A., K. K. Bradley, M. G. Stobierski, and L. A. Tengelsen. 2005. Compendium of measures to control *Chlamydophila psittaci* (formerly *Chlamydia psittaci*) infection among humans (psittacosis) and pet birds, 2005. *Journal of the American Veterinary Medical Association* 226:532–539.

Stephens, R. S., G. Myers, M. Eppinger, and P. M. Bavoil. 2009. Divergence without difference: Phylogenetics and taxonomy of *Chlamydia* resolved. *FEMS Immunology and Medical Microbiology* 55:115–119.

Sykes, J. E. 2005. Feline chlamydiosis. *Clinical Techniques in Small Animal Practice* 20:129–134.

U.S. National Association of State Public Health Veterinarians. 2010. Compendium of measures to control *Chlamydophila psittaci* infection among humans (psittacosis) and pet birds, 2010. http://www.nasphv.org/documentsCompendiaPsittacosis.html (accessed March 23, 2013).

Vanrompay, D., R. Ducatelle, and F. Haesebrouck. 1995. *Chlamydia psittaci* infections: A review with emphasis on avian chlamydiosis. *Veterinary Microbiology* 45:93–119.

Vanrompay, D., T. Harkinezhad, M. van de Walle, D. Beeckman et al. 2007. *Chlamydophila psittaci* transmission from pet birds to humans. *Emerging Infectious Diseases* 13:1108–1110.

Walder, G., H. Hotzel, C. Brezinka, W. Gritsch et al. 2005. An unusual cause of sepsis during pregnancy: Recognizing infection with *Chlamydophila abortus*. *Obstetrics and Gynecology* 106:1215–1217.

Walder, G., H. Meusburger, H. Hotzel, A. Oehme et al. 2003. *Chlamydophila abortus* pelvic inflammatory disease. *Emerging Infectious Diseases* 9:1642–1644.

Wheelhouse, N. and D. Longbottom. 2012. Endemic and emerging chlamydial infections of animals and their zoonotic implications. *Transboundary and Emerging Diseases* 59:283–291.

Wilt, J. C., J. A. Hildes, and F. J. Stanfield. 1959. The prevalence of complement fixing antibodies against psittacosis in the Canadian Arctic. *Canadian Medical Association Journal* 81:731–733.

Wilt, P. C., N. Kordová, and J. C. Wilt. 1972. Preliminary characterization of a chlamydial agent isolated from embryonated snow goose eggs in northern Canada. *Canadian Journal of Microbiology* 18:1327–1332.

Wittenbrink, M. M., M. Mrozek, and W. Bisping. 1993. Isolation of *Chlamydia psittaci* from a chicken egg: Evidence of egg transmission. *Journal of Veterinary Medicine, Series B* 40:451–452.

Wobeser, G. and C. J. Brand. 1982. Chlamydiosis in two biologists investigating disease occurrences in wild waterfowl. *Wildlife Society Bulletin* 10:170–172.

Yan, C., H. Fukushi, H. Matsudate, K. Ishihara et al. 2000. Seroepidemiological investigation of feline chlamydiosis in cats and humans in Japan. *Microbiology and Immunology* 44:155–160.

Section II

Spirochetal Diseases

12 Leptospirosis

Leptospirosis has the widest geographic distribution of any zoonotic disease. Clinical and serologic evidence for the presence of this disease in humans and domestic animals has been recorded in almost every country around the globe.

Paraphrased from Torten and Marshall (1994)

12.1 INTRODUCTION AND HISTORY

Leptospirosis is a disease caused by aerobic bacteria of the genus *Leptospira* (hereafter, these bacteria are called leptospires), which resemble both Gram-negative and Gram-positive bacteria (Vijayachari et al. 2008). *Leptospira* are flexible microorganisms that have the appearance of a tightly wound screw (Figure 12.1). They are thin (less than 0.2 μm wide) and 10–20 μm in length. They can move by spinning around their axis or by wiggling their bodies from side to side; their mobility increases their chances of locating a host and moving through tissues once inside a host. Virulent leptospires have a hooklike structure on one or both ends of their bodies. They survive by infecting renal tubules of host animals. There are also nonvirulent leptospires that live in seawater, freshwater, and moist soil. Based on DNA, the genus *Leptospira* is divided into 17 species, but most infections are caused by *L. interrogans* (Bharti et al. 2003). There are more than 300 serovars of leptospires, 200 of which are known to produce disease (Victoriano et al. 2009).

Leptospirosis is an ancient disease of humans. It plagued Napoleon Bonaparte's troops during their invasion of Egypt and U.S. forces during the American Civil War when they occupied the South. Leptospirosis may have been responsible for the collapse of the Native American population in New England when English pilgrims established the Plymouth Bay colony during 1620 (Sidebar 12.1).

The disease leptospirosis was first described by Adolf Weil in 1886, and a serious form of leptospirosis, Weil's disease, is named after him. At that time, only the severe cases of leptospirosis could be diagnosed clinically. This led to the assumption that leptospirosis causes jaundice and kidney failure (Weil's disease). The bacteria, *Leptospira*, were isolated and named during the 1910s. With improved methods to detect leptospire infections, medical doctors realized that patients have quite variable symptoms.

Leptospirosis is a common zoonotic disease throughout the world and the most widely distributed. This explains why there are so many common names for the disease that reflect either where it was found (Fort Bragg fever, Stuttgart disease, Japanese autumnal fever, Japanese seven-day fever, swamp fever, mud fever, field fever, and water fever) or who was likely to become ill (fish handler's disease, haymaker's disease, pea picker's disease, rat catcher's yellows, rice field worker's

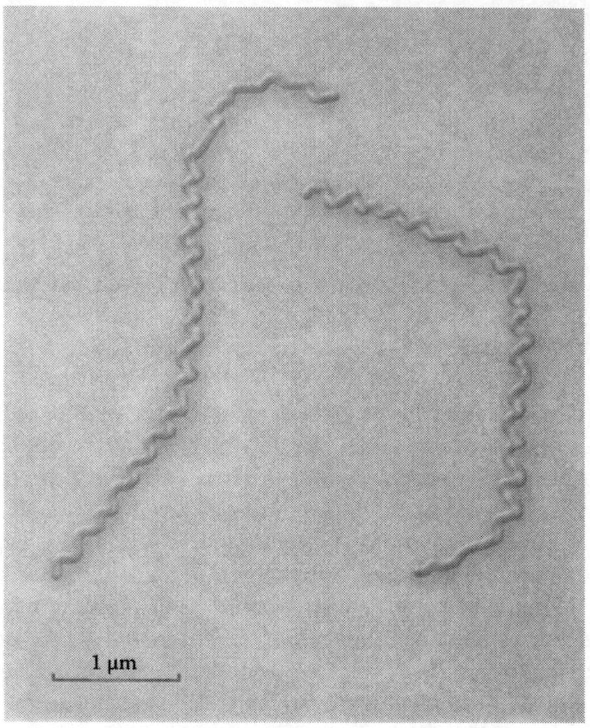

FIGURE 12.1 Leptospires are tightly coiled spirochetes that are thin enough to pass through filters that remove most bacteria. (Courtesy of the Leptospirosis Information Center, Leptospirosis, 2009, http://www.leptospirosis.org, accessed November 15, 2012.)

disease, sewer man's flu, and swine herder's disease). Leptospirosis is more prevalent in tropical and subtropical countries than temperate countries and more common during the rainy season or after floods than during droughts. Many U.S. soldiers contracted the disease while serving in Southeast Asia during the Vietnam War. People become infected when they come into contact with urine from an infected animal or water that has been contaminated with infected urine. Leptospirosis was initially considered a rural disease, but it is also prevalent in urban areas where there are high densities of rodents.

Leptospirosis is rare among people in the continental United States but more common in Hawaii due to the state's tropical climate. In Hawaii, the reported annual incidence rate is 3.3 per 100,000 people, but the actual incidence rate in Hawaii is probably twice as high because many people with leptospirosis do not seek medical attention (Ellis et al. 2008). Each year, there are between 100 and 200 reported cases of leptospirosis in the United States. For example, two outbreaks of leptospirosis in the United States were reported to the CDC during 2005 (Meites et al. 2004). One outbreak involved three people in California who had contact with a slow-flowing stream, and the other involved 43 people who participated in an adventure race in Florida (Sidebar 12.2).

SIDEBAR 12.1 WHICH DISEASE DECIMATED NATIVE AMERICA POPULATIONS PRIOR TO THE ARRIVAL OF THE ENGLISH PILGRIMS AT THE PLYMOUTH BAY? (MARR AND CATHEY 2010)

One of the first permanent English settlements in the New World occurred during 1620 when pilgrims established the Plymouth Colony in what is now Massachusetts. In the years immediately prior to the pilgrims' arrival, the local population of Native Americans crashed, owing to an introduced disease to which they lacked immunity. The disease probably was introduced by European fishermen, whale hunters, or traders who were in the area at the time. Eyewitnesses described the disease as causing fever, headaches, nose bleeds, jaundice, and skin lesions. Fatality rates among the Native Americans are unknown but probably ranged from 33% to 90%. While the epidemic was horrific for Native Americans, English pilgrims benefitted from it because the few surviving Native Americans around Plymouth Bay were unable to resist the English settlement. Many historians believe that the pilgrim settlement would have been unsuccessful if local populations of Native Americans were at full strength.

Speculation as to the disease has included YF, plague, smallpox, chicken pox, typhus, and typhoid fever, but the observed symptoms do not match well with the symptoms caused by these diseases except for YF. However, the first known epidemic of YF in the New World did not occur until 1648, occurring in Yucatan, far from the cold lands of New England. YF was carried by slave ships from Africa, none of which visited New England at this time. Hence, it is doubtful that the disease at Plymouth Bay was YF.

Marr and Cathey (2010) realized that the reported symptoms of the disease among Native Americans were similar to those caused by leptospirosis and that black rats and mice were common on sailing ships and are reservoir hosts for leptospires. Marr and Cathey (2010) hypothesized that some of these rodents left the ships and established themselves in Native American villages and households. From there, these exotic rodents could have quickly spread the disease to native rodents and to humans.

Prior to 1960, large epidemics of leptospirosis in Europe involved thousands of people. Epidemics still occur today, but fewer people are infected during one. In the United Kingdom, there were 54 cases of leptospirosis during 2009; 16 patients acquired leptospirosis while traveling out of the country (mostly to Southeast Asia and Central America), and the other 38 acquired the disease locally. Two patients succumbed from their infection (Department for Environment, Food and Rural Affairs 2011). Germany had 317 reported cases of leptospirosis from 1998 to 2003; 84% of patients contracted the disease in Germany. From 1962 to 2002, leptospirosis caused 234 deaths in Germany, a case fatality rate of 9%. In both the United Kingdom and Germany, human leptospirosis is caused primarily by the serovar Icterohaemorrhagiae (Jansen et al. 2005).

SIDEBAR 12.2 OUTBREAK OF LEPTOSPIROSIS AMONG PARTICIPANTS OF AN ADVENTURE RACE IN FLORIDA (STERN ET AL. 2010)

The U.S. Adventure Racing Association held its 2005 national championship race in the Hillsborough River State Park near Tampa, Florida. The endurance-length swamp race covered a distance of more than 100 miles (160 km) and took about 24 hours to complete. The race involved padding, cycling, and hiking. A hurricane and heavy rains swept through the area prior to the race and flooded the park, exposing race participant for long periods to water in the area's creeks, swamps, and the Hillsborough River. Nearly all racers suffered cuts and scrapes during the race from submerged tree stumps, cypress roots, and plants. Among the 200 racers, 44 were diagnosed with leptospirosis, and three were hospitalized. No deaths occurred, and all patients recovered. Racers who consumed wet food or swallowed surface water were more likely to have contracted leptospirosis than other racers.

Worldwide, the incidence rate (number of new cases annually) per 100,000 people is 1 per 100,000 people in temperate climates and 10–100 per 100,000 in the humid tropics (Bovet et al. 1999), but incidence rates vary considerably among adjacent countries (Table 12.1). In India, about 12% of all patients admitted to hospitals due to high fevers had leptospirosis (Victoriano et al. 2009). In parts of Peru, 20%–30% of patients seeking medical attention for fever had leptospirosis; 9% of all men in the Seychelles had experienced a recent infection of leptospirosis,

TABLE 12.1
Annual Incidence Rate of Leptospirosis per 100,000 People in the Asia–Pacific Region

High (>10)	Moderate (1–10)	Low (<1)
Bangladesh	American Samoa	Australia
Cambodia	China	Hong Kong
Fiji	India	Japan
French Polynesia	Indonesia	South Korea
Laos	Malaysia	Taiwan
Nepal		Mongolia
New Caledonia		New Zealand
Sri Lanka		Philippines
Thailand		Marshall Islands
Vietnam		

Source: Victoriano, A.F.B. et al., *BMC Infect. Dis.*, 9, 147, 2009.

and 37% had an infection at some time in the past (Bharti et al. 2003). In the alpine regions of Italy, more than 10% of farmers and forest workers had antibodies against *Leptospira*, as did 14% of all people living in Yucatan, Mexico (Vijayachari et al. 2008). Worldwide, more than 500,000 people develop leptospirosis annually (WHO 1999).

Among Native Americans living along James Bay, Canada, 28% had antibodies for *Leptospira* antigens, as did 6% living in Nunavik, Canada, and 5% of fur trappers in Quebec (Lévesque et al. 1995, Messier et al. 2011, Sampasa-Kanyinga et al. 2012). The high prevalence among people in northern Canada probably results from frequent consumption of game species (e.g., caribou, moose, black bears and grizzly bears, and several species of marine mammals).

12.2 SYMPTOMS IN HUMANS

Within minutes of invading a host, leptospires migrate to the lymphatic and circulatory systems. Most infected people are asymptomatic or have only a mild illness. If symptoms develop, they occur between 2 and 28 days after exposure (10 days on average). Onset of symptoms is sudden with 90% of patients experiencing incapacitating headaches. Other common symptoms include fever, dry cough, muscle pain, nausea, vomiting, diarrhea, and chills. Leptospirosis should be suspected when conjunctivitis (i.e., "pinkeye") occurs in both eyes combined with some of the following symptoms: sensitivity to bright light, muscle aches in the calf or lower back, neck stiffness, and headaches. When these symptoms are combined with a possible leptospirosis exposure within the last month, there is sufficient evidence for a presumptive diagnosis of leptospirosis (Leptospirosis Information Center 2009). Another indication of leptospirosis is the presence of a red rash that has an irregular border and does not turn white (i.e., blanch) if the skin is pressed. The rash results from the leakage of blood from capillaries into the tissue beneath the skin. Medline Plus (2010) recommends seeking medical attention if a person exhibits any of these symptoms and may have been exposed to water contaminated with leptospires. Most patients recover after 3–7 days, but sometimes the infection enters a second phase that can include meningitis and severe bleeding; leptospirosis can be fatal without medical treatment. Patients with leptospirosis are subject to mood swings, depression, confusion, and aggression. Hallucinations may occur (Leptospirosis Information Center 2009).

Leptospirosis can be manifested in any of four broad types of illness: (1) a mild, flu-like illness, which is the most common form; (2) jaundice, myocarditis (inflammation of the heart muscle), heart arrhythmias (abnormal heartbeat), and renal failure (this manifestation is called Weil's disease); (3) meningitis (an infection of the lining of the brain or spinal cord) or meningoencephalitis (a medical condition that resembles both meningitis and encephalitis); and (4) pulmonary hemorrhaging and respiratory failure. Most patients fully recover from leptospirosis, but this can take months or years. For this reason, leptospirosis patients should be monitored for several years after recovery (WHO 2003, Puliyath and Singh 2012). Mortality rate for patients diagnosed with leptospirosis ranges from less than 5%–30% in different parts of the world (WHO 2003, Victoriano et al. 2009). Death from leptospirosis results from renal failure, cardiopulmonary failure, or widespread bleeding. No association has been discovered between any particular leptospire serovar and disease symptoms in humans (Bharti et al. 2003), but this is not true for many mammal species.

12.3 *LEPTOSPIRA* INFECTIONS IN ANIMALS

Different animal species have varied reactions when infected by the same serovar. Species that serve as a reservoir host for a serovar usually suffer little or no adverse effects from it (Table 12.2). Reservoir hosts shed leptospires in their urine and allow virulent leptospires to exist in the environment. In most areas, reservoir host species include mice, rats, voles, shrews, and hedgehogs. Other vertebrate species are accidental hosts and may become ill if infected by the same serovar that causes no illness in reservoir hosts. *Leptospira* populations are unable to maintain themselves in accidental hosts, but accidental hosts may still excrete leptospires for a few days or weeks (Faine 1994).

While serovars produce little or no signs of illness in their reservoir host species, other serovars produce illness in the same animal. For instance, red deer in New Zealand are the reservoir host for serovar Hardjo–Bovis; this serovar causes little or no disease in them. Yet, red deer infected with serovar Pomona can suffer kidney damage, jaundice, low weight gain, reproductive problems, abortions, and death (Ayanegui-Alcerreca et al. 2007).

Dogs are the reservoir host for serovar Canicola, which produces little illness in dogs, but two other serovars, Grippotyphosa and Pomona, cause canine leptospirosis (Merck 2010). Symptoms of leptospirosis in dogs include excessive drinking and

TABLE 12.2
Common Reservoir Hosts of Common *Leptospira* Serovars from across the World

Hosts	Serovars
Bats	Cynopteri, Wolffi
Cattle	Hardjo, Pomona
Dogs	Canicola
Horses	Bratislava
Marsupials	Grippotyphosa
Mice	Ballum, Arborea, Bim
Pigs	Pomona, Tarassovi
Raccoons	Grippotyphosa
Rats	Icterohaemorrhagiae, Copenhageni
Sheep	Hardjo

Sources: Bharti, A.R. et al., *Lancet Infect. Dis.*, 3, 757, 2003; *Merck Veterinary Manual*, 10th edn., Merck, Whitehouse Station, NJ, 2011.

TABLE 12.3
Number of Times that Different Serovars of *Leptospira interrogans* Were Confirmed in Dogs, Cattle, Pigs, and Horses in the United Kingdom

Test Results	Dogs	Cattle	Pigs	Horses
Number of animals tested	3802	4926	546	270
Positive for Canicola	1571	0	0	3
Positive for Icterohaemorrhagiae	681	0	0	2
Positive for Copenhageni	9	0	0	7
Positive for Zanoni	7	0	0	0
Positive for Bratislava	3	8	223	3
Positive for Hardjo	2	1210	0	0

Source: Department for Environment, Food, and Rural Affairs, Zoonoses report, UK 2009, London, U.K., 2011.

urinating. More serious symptoms include renal failure, pulmonary hemorrhaging, inflammation of lung tissue, and death. Dogs can be vaccinated against leptospirosis (serovars Icterohaemorrhagiae, Canicola, Grippotyphosa, and Pomona). Vaccinations protect dogs from illness and reduce the prevalence of leptospirosis in humans by preventing dogs from transmitting leptospires to people. Dogs with canine leptospirosis can be treated by providing oral doses of amoxicillin or doxycycline for 7–10 days (Merck 2010). Leptospires rarely sicken domestic cats.

Cattle are the reservoir host for Hardjo and Bovis, and these serovars do not make them ill (Merck 2010). Instead, disease in cattle is due primarily to serovars Hardjo, Pomona, and Grippotyphosa (Table 12.3). Illness can be severe in calves when infected by the Pomona serovar. Calves can experience anorexia (i.e., an unwillingness to eat), pulmonary congestion, anemia (a decrease in red blood cells), fever of 105°F–106°F (40°C–41°C), and death. Leptospire infections in cows cause abortions, a decrease in milk production, and discoloration of milk. Tetracycline and oxytetracycline have been used successfully to treat cows (Merck 2010). Twenty-two percent of cattle in Texas had antibodies against leptospires, as did 3% of cattle in France, 8% in Spain, 26% in the Netherlands, and 34% in Great Britain (Alonso-Andicoberry et al. 2001, Bharti et al. 2003).

Some host–serovar associations are the same across the globe. For example, rats almost always serve as the reservoir host for the Icterohaemorrhagiae serovar. Sometimes, the same mammal species serves as the reservoir host for different serovars in different parts of the world. This is especially true for exotic species such as the Indian mongoose (Figure 12.2), which was deliberately released in many sugarcane-producing countries in an attempt to control the rat population. The mongoose is the reservoir host for serovars Sejroe and Icterohaemorrhagiae in Hawaii, Jules and Icterohaemorrhagiae in Jamaica, Brasiliensis and Icterohaemorrhagiae in Grenada, and Canicola in Trinidad (Bharti et al. 2003).

FIGURE 12.2 Indian mongoose is a reservoir host for many different serovars of *L. interrogans*. (Courtesy of Bryan Harry and the U.S. National Park Service.)

Given the large number of different serovars, it is not surprising that almost all mammal species can be infected by some *Leptospira* serovars. Few carnivorous mammals are reservoir hosts to leptospires because their urine is too acidic; but many herbivores, however, are reservoir hosts because their urine is less acidic. Antibodies against leptospires have also been found in birds, reptiles, amphibians, arthropods, and mollusks (Torten and Marshall 1994).

Infection rates vary widely among wildlife species. In the Florida Panhandle and southwest Georgia, leptospires were isolated from less than 4% of rabbits, fox squirrels, and gray squirrels but from 38% of raccoons and 41% of opossums (Shotts et al. 1975). Infection rates also vary by area. Antibodies against leptospires were found in 0% of the white-tailed deer from northern Mexico, 8% from Mississippi, 21% from Tennessee, 30% from Virginia, and 60% from Minnesota (Shotts and Hayes 1970, Goyal et al. 1992, New et al. 1993, Cantu et al. 2008).

Leptospire infections occur in marine mammals, including northern fur seals, northern elephant seals, harbor seals, and California sea lions along the Pacific Coast from southern California to British Columbia and into the Bering Sea. These marine mammals are likely exposed either in breeding colonies or haul-out areas where individuals congregate. If infected by the serovar Pomona, marine mammals may develop leptospirosis that can result in renal failure and death; leptospirosis in pregnant females can result in abortions and stillbirths (Cameron et al. 2008). In some years, hundreds of California sea lions infected with leptospires are found stranded either sick or dead along the California coast (Gulland et al. 1996). Outbreaks of

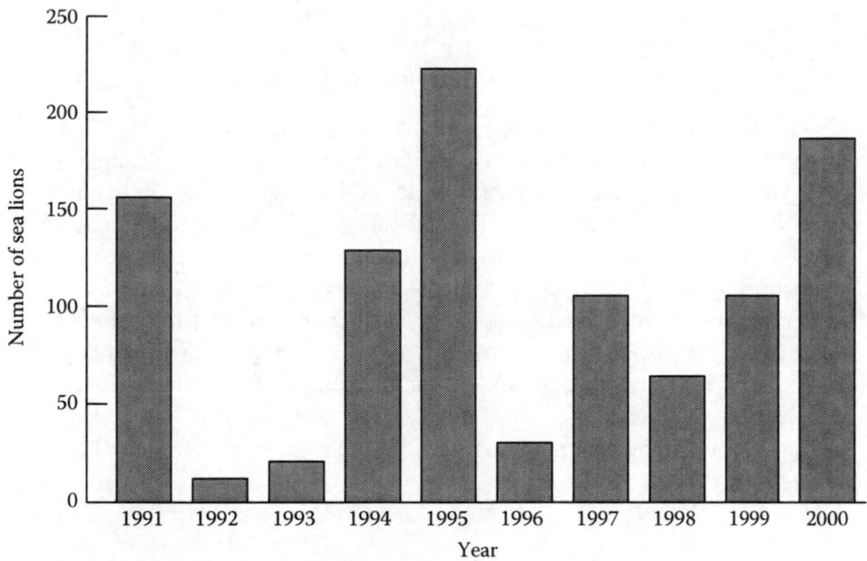

FIGURE 12.3 Number of sea lions stranded along the California coast that were infected with leptospires. (Data from Greig, D.J. et al., *Aquat. Mammals*, 31, 11, 2005.)

leptospirosis among marine mammals regularly occur every 4–5 years along the California coast (Figure 12.3; Lloyd-Smith et al. 2007). Among stranded California sea lions, 38% had antibodies against the Pomona serovar (Colagross-Schouten et al. 2002). Sea lions with leptospirosis are often stranded along California beaches that are frequented by people, pets, and rodents. These leptospires pose a potential risk for both marine mammals and people when viable leptospires occur in sand contaminated by the sea lions' urine and feces (Cameron et al. 2008).

12.4 HOW HUMANS CONTRACT LEPTOSPIROSIS

Human-to-human spread of leptospirosis is rare, although people can excrete leptospires in urine for a few weeks after being infected. Instead, most people contract leptospirosis when they (1) enter contaminated canals, ponds, rivers, caves, or drainage systems; (2) have physical contact with infected animals; (3) have contact with blood or urine from infected animals; and (4) spend time in areas with high densities of infected rodents. Incidence of leptospirosis has decreased substantially in recent decades in developed countries where most patients contract the disease from exposure to contaminated water during recreational activities. In contrast, the incidence of leptospirosis is increasing in developing countries (Faine 1994, Vijayachari et al. 2008).

Leptospires are unable to penetrate undamaged human skin but can enter the body through mucous membranes of the nose, mouth, or eye and through cut or abraded skin. Leptospires can also invade through the lungs when aerosol droplets containing leptospires are inhaled. Outbreaks of leptospirosis often occur after floods when people get wet entering flooded areas.

SIDEBAR 12.3 WHY RATS DO NOT MAKE GOOD PETS (JANSEN AND SCHNEIDER 2011)

If your children are begging to adopt a rat for a pet, you should find the experience of this pet owner interesting. He went to an emergency room complaining of fever, a severe headache, and muscle aches. He had jaundice and conjunctivitis in both eyes (Figure 12.4). The patient was transferred to the intensive care unit due to a rapid loss of kidney function. Medical doctors suspected leptospirosis and began an intravenous antibiotic treatment. After several weeks, the patient recovered completely. Testing showed that the patient had antibodies for *L. interrogans* serovars Icterohaemorrhagiae or Copenhageni. The patient was interviewed to determine the source of the infection, and he revealed that he owned four pet rats. The rats were located and tested; all had leptospires in their kidneys and were hosts for the same serovars that had infected their owner.

Professions with an elevated risk of contracting leptospirosis include livestock producers, rice farmers, mine workers, freshwater fishermen, veterinarians, wildlife biologists, military personnel, sewer workers, and sugarcane cutters. Livestock producers may be exposed when milking infected cows or when touching dead fetuses, amniotic fluid, or placentas. Rice farmers are at risk when they irrigate their crops using surface water that has been contaminated with animal urine. Mine workers are at risk if mines contain rats or mice. The general public is at risk when engaging in outdoor freshwater activities, such as swimming, fishing, or wading. Children may be exposed when playing in water puddles or mud that has been contaminated with urine from dogs, pigs, or rats. Entire villages are at risk if their source of drinking water is contaminated.

Leptospirosis is considered a rural disease, but outbreaks also occur in cities where rats and mice can serve as reservoir hosts for leptospires (Sidebar 12.3). Three people in Baltimore, Maryland, developed leptospirosis after walking barefoot and cutting themselves on glass in inner-city alleys. Twenty-two Norway rats were trapped in the same alleys, and 18 of them had antibodies against leptospires, leading health authorities to conclude that the patients contracted the disease from exposure to rat urine. In Germany, 12% of patients contracted the disease in urban areas (Vinetz et al. 1996, Jansen et al. 2005).

12.5 MEDICAL TREATMENT

Early antibiotic intervention is the most significant factor in recovery, but their effectiveness decreases rapidly after symptoms develop. For this reason, any patient suspected of having leptospirosis should be placed on antibiotics immediately. Several different antibiotics (e.g., amoxicillin, erythromycin, doxycycline, or ampicillin) are effective for mild infections and can be given orally. Severe infections are often managed with benzylpenicillin given intravenously (i.e., through the veins),

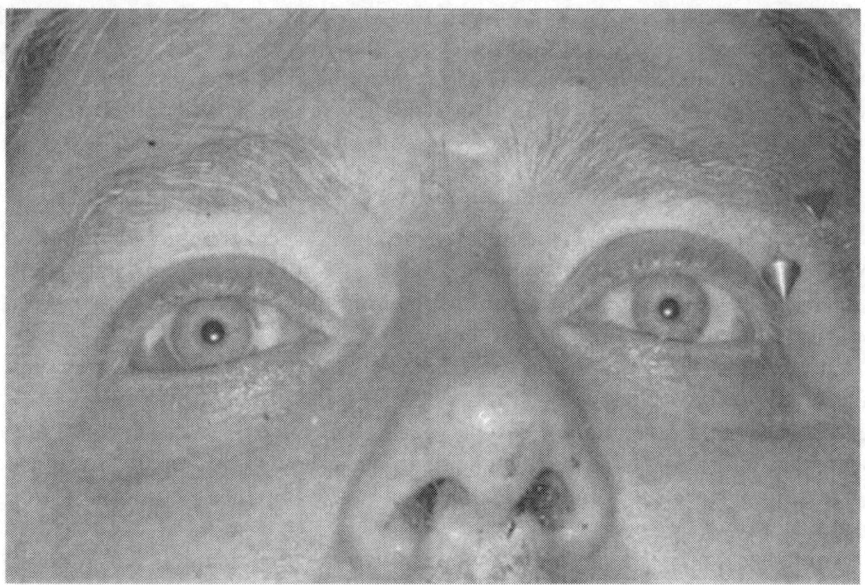

FIGURE 12.4 This person contracted leptospirosis from his pet rats. His symptoms include jaundice, conjunctivitis, and nose bleeds. (Courtesy of A. Jansen.)

but leptospires are usually resistant to vancomycin, chloramphenicol, rifampicin, and metronidazole (Leptospirosis Information Center 2009).

Leptospirosis patients should be monitored closely for serological, hepatic, urinary, and cardiac functions. With severe illness, ECG monitoring is important as heart arrhythmias are common. Fluid and electrolyte balances need to be monitored; kidney dialysis may be required if renal function is impaired. Patients with severe infection may experience psychological problems and may require sedation if they become aggressive or psychotic. These problems are temporary, but long-term depression or fatigue may result (Leptospirosis Information Center 2009).

If a woman becomes infected when pregnant, leptospires can cross the placenta and infect the fetus, even if the mother is asymptomatic. In most cases, a full recovery will occur, but leptospirosis can result in fetal death or developmental abnormalities for the baby. Fortunately, leptospirosis is a highly curable condition if treated early with antibiotics (Leptospirosis Information Center 2009, Puliyath and Singh 2012).

12.6 WHAT PEOPLE CAN DO TO REDUCE THEIR RISK OF CONTRACTING LEPTOSPIROSIS

Most people contract leptospirosis from direct contact with the urine or water contaminated with urine from infected animals. People should avoid exposure to animal urine. Leptospires can survive in freshwater or moist soils for weeks or months when conditions are suitable (WHO 2003) but can be easily killed by desiccation or disinfectants. Equipment and small areas, such as floors, can be disinfected but this is not possible for large areas or for lakes and streams. Hence, people should avoid swimming or wading in water that may be contaminated with animal urine. This is especially true with shallow ponds or slow-moving streams where animal urine can concentrate. People should also consider wearing waterproof boots when walking in mud or water that might be contaminated with animal urine.

Education programs should target people, who have an elevated risk of exposure to leptospires, and inform them of their potential risk, symptoms of leptospirosis, and the need to seek prompt medical attention if the disease is suspected. Education programs should contain information about preventative measures to reduce risk of infection.

The incidence of leptospirosis can be diminished by decreasing the population of animals serving as the reservoir hosts of leptospires. To be useful, this approach necessitates the identification of animal species that serve as reservoir hosts in the local area. Serovars of leptospires found in animals should be compared to serovars isolated from patients to identify which animal species are the reservoir hosts of the virulent serovars.

Animal host populations can be reduced through lethal methods (e.g., toxicants or traps), but the most effective method is to eliminate the shelter and food sources upon which the reservoir host species depends for survival. For example, miners have a high risk of contracting leptospirosis when mines contain rats. By discouraging miners from bringing food into the mine or requiring miners to dispose of leftover food in rodent-proof containers can reduce rat populations in mines.

Antibiotics can prevent leptospirosis in humans if provided prior to leptospire infections. Hence, they are sometimes taken by military personnel or travelers to areas where leptospirosis is common. The incidence of leptospirosis was reduced to 95% among American soldiers training in Panama who were given weekly doses of doxycycline (Takafuji et al. 1984). Antibiotics can also be used to protect residents of endemic areas during disease outbreaks. One such test was conducted on North Andaman Island in the Bay of Bengal where leptospirosis outbreaks are seasonal and last about 3 weeks. Residents who received doxycycline weekly were less likely to develop leptospirosis than residents who received a placebo (a pill that contains no active ingredients). Both groups actually had similar rates of infection, but the antibodies apparently stopped the invading leptospires from increasing to the point where they caused illness (Sehgal et al. 2000).

Other preventative steps can reduce the incidence of leptospirosis in humans. Prior to 1960, more than 200 Japanese died of leptospirosis each year. Most of the victims were people who worked in rice fields. Fifty years later, the annual number of human fatalities dropped to 20 due to the mechanization of Japan's agriculture, farmers' use of rubber boots when working in the fields, and the use of vaccines (Victoriano et al. 2009). Leptospirosis can also be reduced by using screens or fences to exclude host animals from buildings and water sources. Finally, the incidence of leptospirosis in humans can be reduced by vaccinating dogs and livestock (Torten and Marshall 1994).

People who have already contracted leptospirosis may acquire some immunity against reinfection by the same or similar serovars of leptospires, but not against other types. People can also gain immunity by being vaccinated, but the vaccines are only effective for specific serovars. Which vaccine should be administered is based upon which serovars occur locally. The protection provided by the vaccines is often of short duration, and the vaccinations can provide side effects, such as fever and pain at the site where the vaccine was injected (WHO 2003). For these reasons, vaccines are only available in countries where leptospirosis is common.

During an outbreak of leptospirosis, medical doctors may prescribe antibiotics to be taken as a preventative measure after balancing the risk of the disease against potential drug side effects. The preventative use of antibiotics may be useful for adventure travelers, ecotourists, or military personnel when in endemic areas.

The deliberate movement by humans of livestock and pets across the world, as well as the accidental movement of rodents and other exotic species, has allowed the inadvertent movement of different serovars of leptospires to new areas. A recently introduced serovar can cause a severe outbreak among local animals and humans soon after its arrival. Exotic serovars may also become endemic to the new area either by infecting native animals or the exotic animals that humans brought with them. Controlling the importation of livestock and domestic animals helps prevent infections by new serovars. Importation of animals should only be allowed if the animals are tested and found to be free of leptospires.

12.7 ERADICATING LEPTOSPIROSIS FROM A COUNTRY

Eradicating leptospirosis from an endemic area is difficult because so many wildlife species can serve as a reservoir host for one or more of the serovars. Instead, the focus should be on preventative measures to stop exotic serovars from entering the country and infecting livestock, other domestic animals, and wildlife.

LITERATURE CITED

Alonso-Andicoberry, C., F. J. Garćia-Peña, J. Pereira-Bueno, E. Costas, and L. M. Ortega-Mora. 2001. Herd-level risk factors associated with *Leptospira* spp. seroprevalence in dairy and beef cattle in Spain. *Preventative Veterinary Medicine* 52:109–117.

Ayanegui-Alcerreca, M. A., P. R. Wilson, C. G. Mackintosh, J. M. Collins-Emerson et al. 2007. Leptospirosis in farmed deer in New Zealand: A review. *New Zealand Veterinary Journal* 55:102–108.

Bharti, A. R., J. E. Nally, J. N. Ricaldi, M. A. Malthaias et al. 2003. Leptospirosis: A zoonotic disease of global importance. *Lancet Infectious Diseases* 3:757–771.

Bovet, P., C. Yersin, F. Merien, C. E. Davis, and P. Perolat. 1999. Factors associated with clinical leptospirosis: A population-based case-control study in the Seychelles (Indian Ocean). *International Journal of Epidemiology* 28:583–590.

Cameron, C. E., R. L. Zuerner, S. Raverty, K. M. Colegrove et al. 2008. Detection of pathogenic *Leptospira* bacteria in pinniped populations via PCR and identification of a source of transmission for zoonotic leptospirosis in the marine environment. *Journal of Clinical Microbiology* 46:1728–1733.

Cantu, A., J. A. Ortega-S, J. Mosqueda, Z. Garcia-Vazquez et al. 2008. Prevalence of infectious agents in free-ranging white-tailed deer in northeastern Mexico. *Journal of Wildlife Diseases* 44:1002–1007.

CDC. 2013. Summary of notifiable diseases—United States, 2011. *Morbidity and Mortality Weekly Report* 60(53):1–117.

Colagross-Schouten, A. M., J. A. K. Mazet, F. M. D. Guiland, M. A. Miller, and S. Hietala. 2002. Diagnosis and seroprevalence of leptospirosis in California sea lions from coastal California. *Journal of Wildlife Diseases* 38:7–17.

Department for Environment, Food, and Rural Affairs. 2011. Zoonoses report; UK 2009. London, U.K.

Ellis, T., A. Imrie, A. R. Katz, and P. V. Effler. 2008. Underrecognition of leptospirosis during a dengue fever outbreak in Hawaii, 2001–2002. *Vector-Borne and Zoonotic Diseases* 8:541–547.

Faine, S. 1994. *Leptospira and Leptospirosis*. CRC Press, Boca Raton, FL.

Goyal, S. M., L. D. Mech, and M. E. Nelson. 1992. Prevalence of antibody titers to *Leptospira* spp. in Minnesota white-tailed deer. *Journal of Wildlife Diseases* 28:445–448.

Greig, D. J., F. M. D. Gulland, and C. Kreuder. 2005. A decade of live California sea lion (*Zalophus californianus*) strandings along the central California coast: Causes and trends, 1991–2000. *Aquatic Mammals* 31:11–22.

Gulland, F. M. D, M. Koski, L. J. Lowenstine, A. Colagross et al. 1996. Leptospirosis in California sea lions (*Zalophus californianus*) stranded along the central California coast, 1981–1994. *Journal of Wildlife Diseases* 32:572–580.

Jansen, A. and T. Schneider. 2011. Weil's disease in a rat owner. *Lancet Infectious Diseases* 11:152.

Jansen, A., I. Schöneberg, C. Frank, K. Alpers et al. 2005. Leptospirosis in Germany, 1962–2003. *Emerging Infectious Diseases* 11:1048–1054.

Leptospirosis Information Center. 2009. Leptospirosis. http://www.leptospirosis.org (accessed November 15, 2012).

Lévesque, B., G. De Serres, R. Higgins, M.-A. D'Halewyn et al. 1995. Seroepidemiologic study of three zoonoses (leptospirosis, Q fever, and tularemia) among trappers in Quebec, Canada. *Clinical Diagnostic and Laboratory Immunology* 2:496–498.

Lloyd-Smith, J. O., D. J. Greig, S. Hietala, G. S. Ghneim et al. 2007. Cyclical changes in seroprevalence of leptospirosis in California sea lions: Endemic and epidemic disease in one host species? *BMC Infectious Diseases* 7(125):1–11.

Marr, J. S. and J. T. Cathey. 2010. New hypothesis for cause of epidemic among Native Americans, New England, 1616–1619. *Emerging Infectious Diseases* 16:281–286.

Medline Plus. 2010. Leptospirosis. http://www.nlm.nih.gov/medlineplus/ency/article/001376.htm (accessed June 1, 2012).

Meites, E., M. T. Jay, S. Deresinski, W. Shieh et al. 2004. Reemerging leptospirosis, California. *Emerging Infectious Diseases* 10:406–412.

Merck 2010. *Merck Veterinary Manual*, 10th edition. Merck, Whitehouse Station, NJ.

Messier, V., B. Lévesque, J. F. Proulx, L. Rochette et al. 2011. Seroprevalence of seven zoonotic infections in Nunavik, Quebec (Canada). *Zoonosis and Public Health* 59(2):107–117.

New, Jr., J. C., W. G. Wathen, and S. Dlutkowski. 1993. Prevalence of *Leptospira* antibodies in white-tailed deer, Cades Cove, Great Smoky Mountains National Park, Tennessee, USA. *Journal of Wildlife Diseases* 29:561–567.

Puliyath, G. and S. Singh. 2012. Leptospirosis in pregnancy. *European Journal of Clinical Microbiology and Infectious Diseases* 31:2491–2496.

Sampasa-Kanyinga, H., B. Lévesque, E. Anassour-Laouan-Sidi, S. Côté et al. 2012. Zoonotic infections in native communities of James Bay, Canada. *Vector-Borne and Zoonotic Diseases* 12:473–481.

Sehgal, S. C., A. P. Sugunan, M. V. Murhekar, S. Sharma, and P. Vijayachari. 2000. Randomized controlled trial of doxycycline prophylaxis against leptospirosis in an endemic area. *International Journal of Antimicrobial Agents* 13:249–255.

Shotts, Jr., E. B., C. L. Andrews, and T. W. Harvey. 1975. Leptospirosis in selected wild mammals of the Florida panhandle, and southwestern Georgia. *Journal of the American Veterinary Medical Association* 167:587–589.

Shotts, Jr., E. B. and F. A. Hayes. 1970. Leptospiral antibodies in white-tailed deer of the southeastern United States. *Journal of Wildlife Diseases* 6:295–298.

Stern, E. J., R. Galloway, S. V. Shadomy, K. Wannemuehler et al. 2010. Outbreak of leptospirosis among adventure race participants in Florida, 2005. *Clinical Infectious Diseases* 50:843–849.

Takafuji, E. T., J. W. Kirkpatrick, R. N. Miller, J. J. Karwaki et al. 1984. An efficacy trial of doxycycline chemoprophylaxis against leptospirosis. *New England Journal of Medicine* 310:497–500.

Torten, M. and R. B. Marshall. 1994. Leptospirosis. In G. W. Beran and J. H. Steele, editors. *Handbook of Zoonoses*, 2nd edition. CRC Press, Boca Raton, FL.

Victoriano, A. F. B., L. D. Smythe, N. Gloriani-Barzaga, L. L. Cavinta et al. 2009. Leptospirosis in the Asia Pacific region. *BMC Infectious Diseases* 9:147.

Vijayachari, P., A. P. Sugunan, and A. N. Shriram. 2008. Leptospirosis: An emerging global public health problem. *Journal of Bioscience* 33:557–569.

Vinetz, J. M., G. E. Glass, C. E. Flexner, P. Mueller, and D. C. Kaslow. 1996. Sporadic urban leptospirosis. *Annals of Internal Medicine* 125:794–798.

WHO. 1999. Leptospirosis worldwide 1999. *Weekly Epidemiological Record* 74:237–244.

WHO. 2003. *Human Leptospirosis: Guidance for Diagnosis, Surveillance and Control.* World Health Organization, Geneva, Switzerland.

13 Lyme Disease

Otzi the Iceman, the 5,300-year-old mummy that was found in the Italian Alps when a glacier retreated, was suffering from Lyme disease when he died, making him the oldest person known to have had Lyme disease.

Paraphrased from Hall (2011)

13.1 INTRODUCTION AND HISTORY

Lyme disease is caused by a Gram-negative, spirochetal bacterium from the genus *Borrelia* (Figure 13.1). There are more than 37 species in the genus, several of which cause Lyme disease in humans, including *B. burgdorferi* in the United States and parts of Europe and *B. afzelii* and *B. garinii* in Europe and Asia (Oliver et al. 2003). *Borrelia* species has been classified into two groups based on genetic differences: a Lyme disease group, which are discussed in this chapter, and the relapsing fever group, which are discussed in Chapter 14.

Lyme disease was first described by Allen Steere in 1977. He discovered an unusual concentration of children who had arthritis in Lyme, Connecticut, which accounts for the disease's name. However, the symptoms of Lyme disease had been described in Europe as early as 1909 when a Swedish dermatologist described an expanding bull's-eye-shaped rash that occurred following a tick bite. He described the rash as erythema migrans, a name still used today for this characteristic rash of Lyme disease. A tick, initially identified as *Ixodes dammini*, was identified as the main vector for *B. burgdorferi* in the eastern United States, but entomologists later realized that *I. dammini* was actually the blacklegged tick (*I. scapularis*), the name used for that tick thereafter.

Lyme disease occurs throughout the temperate regions of the Northern Hemisphere, including North America, Europe, Russia, China, and Japan. Lyme-like diseases also have been reported in Africa and South America. The number of reported cases of Lyme disease increased rapidly in the United States and Canada after its initial discovery, rising to 9,908 cases in 1992 and more than 30,000 cases by 2009 (Figure 13.2). However, CDC (2013a) estimated that the actual number of people in the United States diagnosed with Lyme disease each year is about 300,000—10 times the number reported to the CDC. In 2011, more than 33,000 confirmed and probable cases of Lyme disease were reported to the CDC (2013b).

There are three endemic areas for Lyme disease in the United States: the northeast, north–central region, and the Pacific Coast. Over 90% of cases occur in just 10 states: Connecticut, Delaware, Maryland, Massachusetts, Minnesota, New Jersey, New York, Pennsylvania, Rhode Island, and Wisconsin (Figure 13.3). The distribution of Lyme disease in the United States and Canada corresponds roughly with distribution of its main vectors, which are blacklegged ticks in the eastern United States and western blacklegged tick (*I. pacificus*) on the West Coast (Figure 13.4). The range of

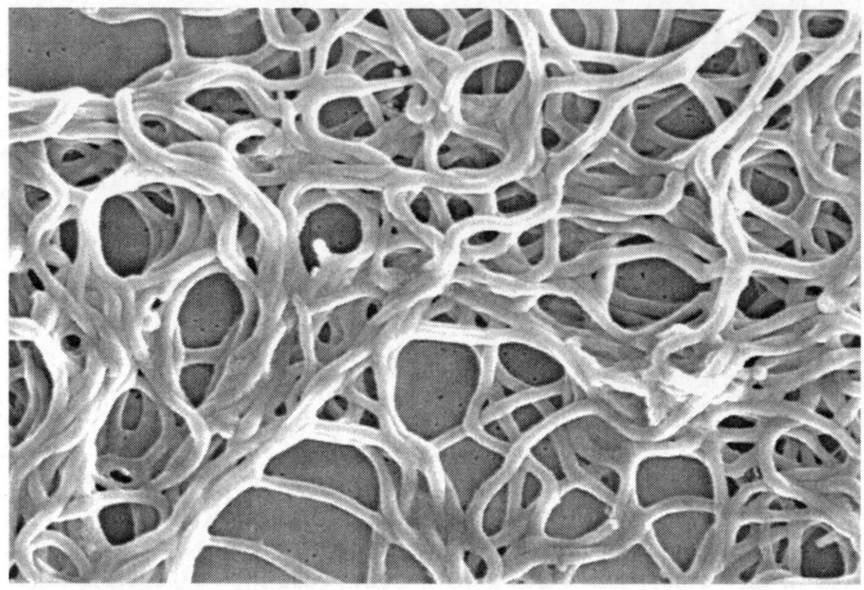

FIGURE 13.1 Photo of *B. burgdorferi* taken through a scanning electron microscope. (Courtesy of Claudia Molins and the CDC.)

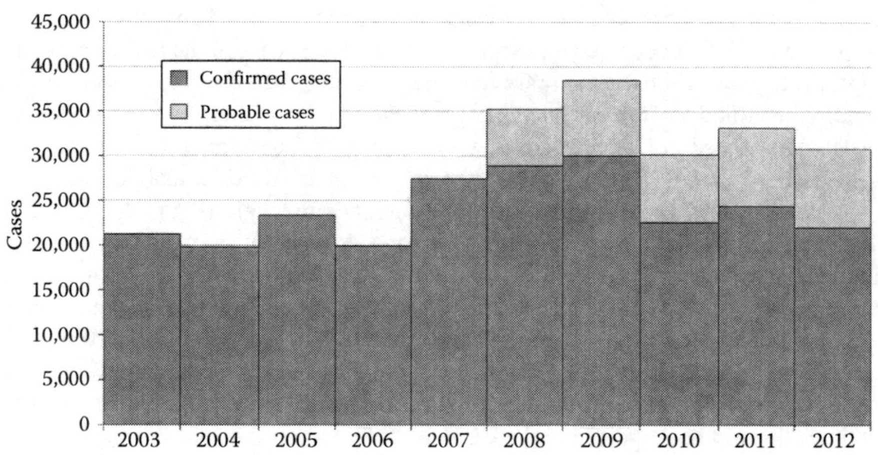

FIGURE 13.2 Number of reported Lyme disease cases in the United States from 2003 through 2012. (From CDC, Lyme disease data, http://www.cdc.gov/lyme/stats/index.html, 2013c.)

blacklegged tick has increased in recent years, and not surprisingly, Lyme disease has spread into the same new areas (Bacon et al. 2008).

Lyme disease occurs in almost every European country; thousands of Europeans contract the disease annually. The disease is most prevalent in central–eastern

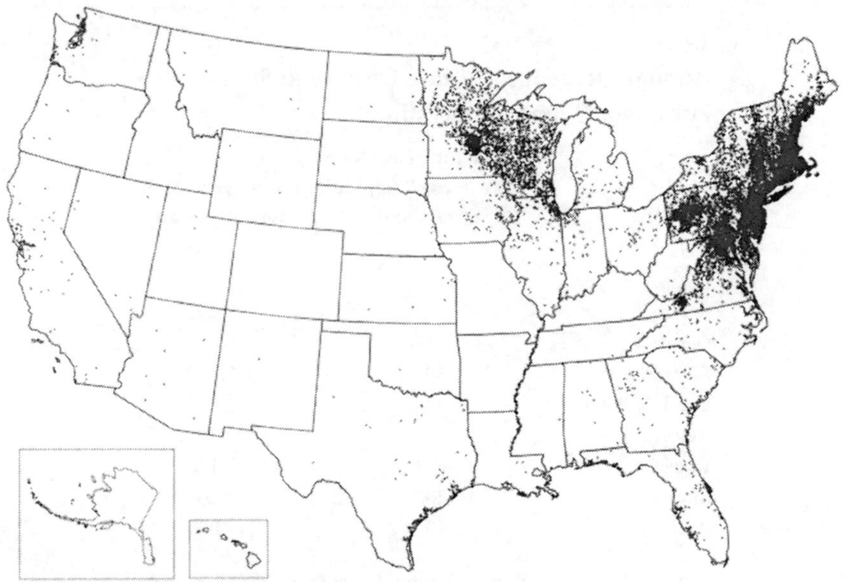

FIGURE 13.3 Distribution of cases of Lyme disease in the United States in 2012. One dot has been placed within the county of residence for each confirmed case. (From CDC, Lyme disease data, http://www.cdc.gov/lyme/stats/index.html, 2013c.)

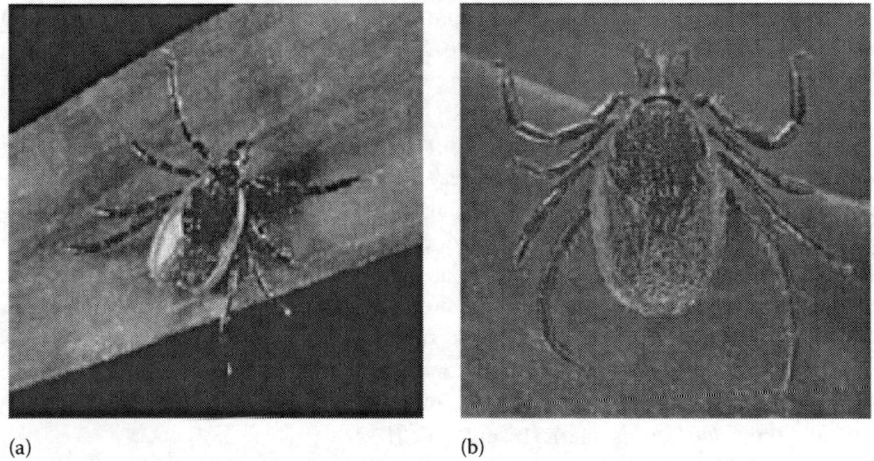

FIGURE 13.4 Photo of a (a) blacklegged tick (*I. scapularis*) and (b) western blacklegged tick (*I. pacificus*). (Photos courtesy of Chapman and the CDC.)

Europe, though pockets of high incidence occur throughout Europe. O'Connell et al. (1998) estimated that 45,000 people contracted Lyme disease annually in selected European countries, but the total number of cases in Europe is higher than this because several countries (e.g., Spain, France, Italy, Poland, Hungary) were not included in the estimate (Table 13.1). Recently, there has been an increase in

TABLE 13.1
Annual Incidence of Lyme Disease in Some European Countries as Estimated

Country	Annual Incidence Rate (New Cases/100,000 Population)	Number of New Cases Annually
United Kingdom	0.3	200
Ireland	0.6	30
France	16.0	7,200
Germany	25.0	20,000
Switzerland	30.4	2,000
Czech Republic	39.0	3,500
Bulgaria	55.0	3,500
Sweden (south)	69.0	7,120
Slovenia	120.0	2,000
Austria	130.0	14,000

Source: O'Connell, S. et al., *Zentralbl. Bakteriol.*, 287, 229, 1998.

patients diagnosed with Lyme disease making the estimates of O'Connell et al. (1998) conservative. For example, O'Connell et al. (1998) estimated that there were 200 cases annually in the United Kingdom, but then 1395 cases were confirmed in 2009 (Department for Environment, Food, and Rural Affairs 2011), including 973 in England and Wales, 420 in Scotland, and two in Northern Ireland.

The life cycle of *Ixodes* ticks typically involves four stages over the course of 2 years: egg, larva, nymph, and adult. Adult females lay a single clutch of eggs containing 800–3000 eggs during spring (year 1). The larvae hatch during the summer and begin to seek a host, usually a small mammal, lizard, or bird. After feeding for 3–5 days on the host and obtaining a blood meal, the larvae detach from the host, molt into a nymph, and overwinter. The next spring or summer, the nymph seeks a new host, usually a bird or small- to medium-sized mammal, although both immature stages will also feed on larger hosts, such as deer. After feeding for 4–5 days, the nymph disengages from the host and molts into an adult. The adult then seeks a blood meal during the fall, winter, or the following spring. Finally, adult females lay eggs, and the cycle begins again (Eisen et al. 2012).

Blacklegged ticks seek blood meals from a wide variety of hosts including birds, reptiles, and mammals. In the United States, 125 vertebrate species (54 mammals, 57 birds, and 14 lizards) are parasitized by this tick species (Keirans et al. 1996). Larval and nymphal blacklegged ticks usually feed on birds or small- or medium-sized mammals such as mice, voles, chipmunks, squirrels, shrews, skunks, and raccoons. Adult ticks feed primarily on large mammals, especially white-tailed deer. The most important reservoir host for *B. burgdorferi* is the white-footed mouse (Figure 13.5), which can remain infected with *B. burgdorferi* throughout its entire life. In some areas where Lyme disease is endemic, nearly all white-footed mice

FIGURE 13.5 Photo of a white-footed mouse. (From CDC, Lyme disease data, http://www.cdc.gov/lyme/stats/index.html, 2013c.)

were infected with *B. burgdorferi* by summer's end. Furthermore, a high proportion of *Ixodes* ticks feeding on an infected white-footed mouse became infected themselves. In one study, about 90% of the nymphs that fed on white-footed mice as larvae were infected with *B. burgdorferi* versus an infection rate of less than 10% for larvae that fed on raccoons or opossums. Infected ticks retained the pathogen throughout their lives. In contrast, white-tailed deer rarely serve as a reservoir for *B. burgdorferi*. Still, white-tailed deer have been important for the maintenance of blacklegged tick populations in an area and thus are important for the maintenance of *B. burgdorferi* in an area (LoGiudice et al. 2003, Bunikis et al. 2004, Eisen et al. 2012).

In California, 108 vertebrate species serve as hosts for at least one stage of western blacklegged tick, but some species are more important than others for maintaining populations of this tick. The western fence lizard and the alligator lizard are the primary hosts for larval and nymphal western blacklegged tick with up to 90% of these ticks feeding on lizards. Birds, especially passerines that forage on the ground, also are hosts for western blacklegged tick and transport both ticks and *B. burgdorferi* widely. Mule deer are the primary host for adult western blacklegged tick in California. Interestingly, neither the west fence lizard nor alligator lizard serves as reservoir host for *B. burgdorferi* because the lizard's immune system kills them. Instead, the western gray squirrel serves as an important reservoir host for *B. burgdorferi* in California, along with the dusky-footed woodrat, North American deer mouse, and California kangaroo rat. While mice are important reservoir hosts for *B. burgdorferi* in the eastern United States, mice are rarely parasitized by western blacklegged tick and probably play only a minor role in maintaining *B. burgdorferi* in the western United States (Castro and Wright 2007, Swei et al. 2011).

In the southeastern United States, there are two *Borrelia* species—*B. burgdorferi* and *B. bissettii*. Three mammal species serve as their reservoir hosts: cotton mice, hispid cotton rats, and eastern woodrats. All of these species occur in high densities, and a high percentage of them are infected with *B. burgdorferi*; Oliver et al. (2003) found that 17% of cotton mice, 15% of hispid cotton rats, and 39% of eastern woodrats were infected. Three tick species (*I. scapularis*, *I. affinis*, and *I. minor*) transmit *Borrelia* in the south. The last two tick species are important in spreading *Borrelia* among the three mammal species. The blacklegged tick is the primary tick responsible for spreading Lyme disease to humans.

Lyme disease also occurs in Europe and Asia, but in Europe, Lyme disease in humans is primarily caused by three species: *B. burgdorferi*, *B. garinii*, and *B. afzelii*. Four additional species, *B. valaisiana*, *B. spielmanii*, *B. bissettii*, and *B. lusitaniae*, occasionally infect humans. *B. garinii* and *B. afzelii* are widely distributed, extending across Europe and Russia and into Japan. In Europe and Asia, the ticks *I. ricinus* and *I. persulcatus* serve as the main vector for the different species of *Borrelia*. The bank vole and several species of mice (*Apodemus* spp.) are the main hosts for *I. persulcatus* in Europe. In Russia, China, and Japan, *I. persulcatus* serves as the vector for Lyme disease, and rodents serve as reservoir hosts for the pathogen. In Europe and Asia, the reservoir hosts for *Borrelia* include rodents (e.g., wood mice and bank voles) along with several avian species (e.g., Eurasian blackbird). Deer are important in maintaining *Ixodes* ticks and Lyme disease in North America, but not in Europe (Ishiguro et al. 2000, Richter et al. 2004, Gil et al. 2005, Miller and Mead 2011).

13.2 SYMPTOMS IN HUMANS

One of the first characteristic clinical signs of Lyme disease is an expanding red rash centered at the site of the tick bite. Either the rash may disappear at the center but remain at the margins (bull's-eye shape) or it may be a solid rash. The rash occurs as the body reacts to the outward movement of spirochetes through the skin from the point of infection. This characteristic rash occurs in 70% of Lyme disease patients and usually occurs 7–14 days after the tick bite, but the rash may occur as soon as 3 days or as long as 30 days (CDC 1999, 2010). A red rash less than 2 in. (5 cm) in diameter at the site of a tick bite is more likely caused by an allergic reaction to the tick bite rather than an indication of Lyme disease (Miller and Mead 2011).

Other early signs of Lyme disease include a mild fever, fatigue, pain in the muscles and joints, headache, swelling of lymph nodes, and conjunctivitis. Within a few days, these symptoms may disappear, but *B. burgdorferi* remains. Weeks or months later, the disease can reoccur with more serious consequences, such as arthritis, usually in a weight-bearing joint such as the knee. This condition results when spirochetes invade the joints and are experienced by 30% of patients (Hayes and Piesman 2003, CDC 2010).

Cardiac problems (i.e., Lyme carditis) can result if the spirochetes invade heart tissue. Among confirmed cases of Lyme disease, 0.8% experience this condition. During 2012 and 2013, three Lyme carditis patients in the United States suffered fatal heart attacks that resulted from a blockage of neural signals that synchronize the beating of the heart chambers. Signs of a heart block include chest pains, heart palpitations, shortness of breath, lightheadedness, and fainting. Lyme carditis can occur 4 days to 7 months after infection. Fortunately, Lyme carditis responds well to antibiotics (Fish et al. 2008, Ray et al. 2013).

As spirochetes invade nerves and the brain, neural problems can occur including Bell's palsy (temporary paralysis of facial muscles), meningitis (inflammation of the protective membranes surrounding the brain and spinal cord), and encephalitis (inflammation of the brain). Bell's palsy occurred in 8% of confirmed cases of Lyme disease; meningitis or encephalitis occurred in 2% of cases. While these conditions are serious and debilitating, Lyme disease is almost never fatal. Late-stage Lyme disease may require intravenous antibiotic therapy, which is usually successful. However, about 30% of patients with late-stage Lyme disease continue to have residual symptoms after antibiotic treatment. These residual symptoms may disappear over time. Up to 10% of patients with Lyme arthritis have persistent joint inflammation that does not respond to antibiotics (Hayes and Piesman 2003, CDC 2010).

Chronic Lyme disease can produce a wide variety of neurological problems. These include memory loss, hearing loss, headaches, mood changes, lethargy, and sleep disturbance. Depression is the most common psychiatric syndrome associated with late-stage Lyme disease; it is much more common in patients with Lyme disease than patients with other chronic illnesses. Suicidal thoughts are common among chronic Lyme disease patients who have encephalopathy (brain disorders); 33% of them have suicidal tendencies and 15% have homicidal tendencies. Several suicides and one combined homicide–suicide have occurred among these patients. In contrast, suicidal tendencies are lower among patients suffering from other chronic illnesses, such as cancer, heart disease, and diabetes (Logigian et al. 1990, Bransfield 2012).

The reasons why so many chronic Lyme disease patients suffer depression are unclear. Endocrine disorders such as hypothyroidism can result from Lyme disease, and these disorders can cause depression. Neural dysfunction resulting from Lyme disease can also cause depression. Lastly, Lyme disease patients can be psychologically overwhelmed by the multitude of symptoms caused by this disease, resulting in a vicious cycle of grief, stress, disappointment, and demoralization. These neurological problems often improve with antibiotic therapy and wane over time (Logigian et al. 1990, Bransfield 2012).

13.3 *BORRELIA BURGDORFERI* INFECTIONS IN ANIMALS

White-tailed deer populations have increased severalfold during the last century due to the implementation of game laws that restricted their killing by hunters. These deer have also moved into human-modified habitat, such as areas dominated by agriculture or housing (e.g., suburban areas). Concomitantly, blacklegged tick populations have increased severalfold, and their range has increased along with white-tailed deer, which are an important maintenance host for adult blacklegged ticks. Both deer and rodent hosts, nonetheless, must be abundant in an area to maintain populations of this tick.

B. burgdorferi have a minimal impact on their primary host, the white-footed mouse, which is why adult mice are such a competent host for *B. burgdorferi* (Moody et al. 1994, Schwanz et al. 2011). In a 2-year study, Hofmeister et al. (1999) found that infections of *B. burgdorferi* had no impact on the survival of white-footed mice.

13.4 HOW HUMANS CONTRACT LYME DISEASE

Lyme disease is transmitted to humans primarily through the bite of nymphal or adult female *Ixodes* ticks. These ticks are tiny: nymphs are the size of a poppy seed, and adults are about the size of a sesame seed. More people are infected from the bite of nymphs than adult ticks because ticks have to remain attached for at least 24 hours before they start to transmit *B. burgdorferi*; and the smaller nymph is easily overlooked. Most people in the United States contract Lyme disease during the summer when nymphal ticks are active and people spend time outdoors (Figure 13.6). Confirmed cases of Lyme disease are more common among boys than girls because boys typically spend more time outdoors. Reported cases of Lyme diseases peak among children 5–15 years old and among adults 35–55 years old (Figure 13.7).

Lyme disease is not spread by direct contact with infected animals. It is not spread by person-to-person contact, and there are few reports of pregnant women passing the infection through the placenta to their unborn child. The impact of infection on a fetus remains unclear. There are no reports of *B. burgdorferi* being transmitted to an infant through breast milk. *B. burgdorferi* can survive for several weeks in stored blood, but the risk of infection through blood transfusion is minimal (CDC 1999).

People at risk for Lyme disease are those who spend time outdoors in areas where they are exposed to ticks. This includes people who work outdoors such as

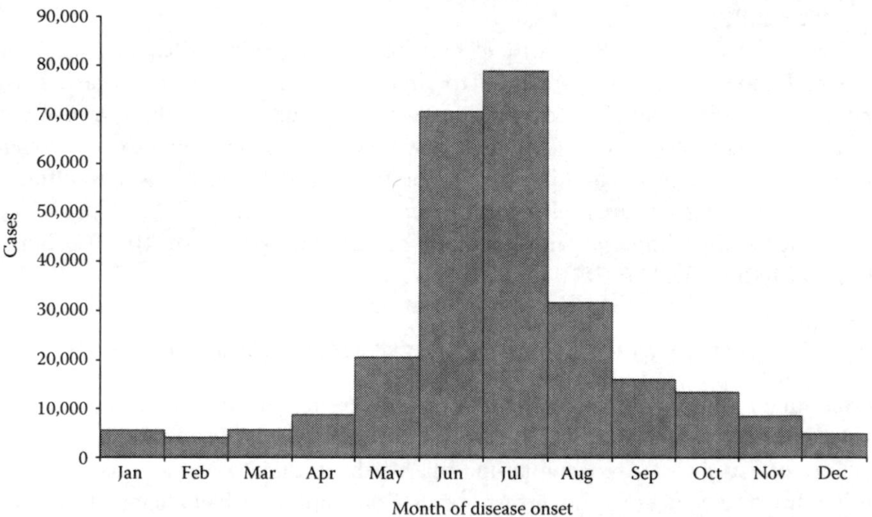

FIGURE 13.6 Confirmed cases of Lyme disease in the United States from 2001 to 2010 by month when symptoms began. (From CDC, Lyme disease, http://www.cdc.gov/diseaseconditions/, 2010.)

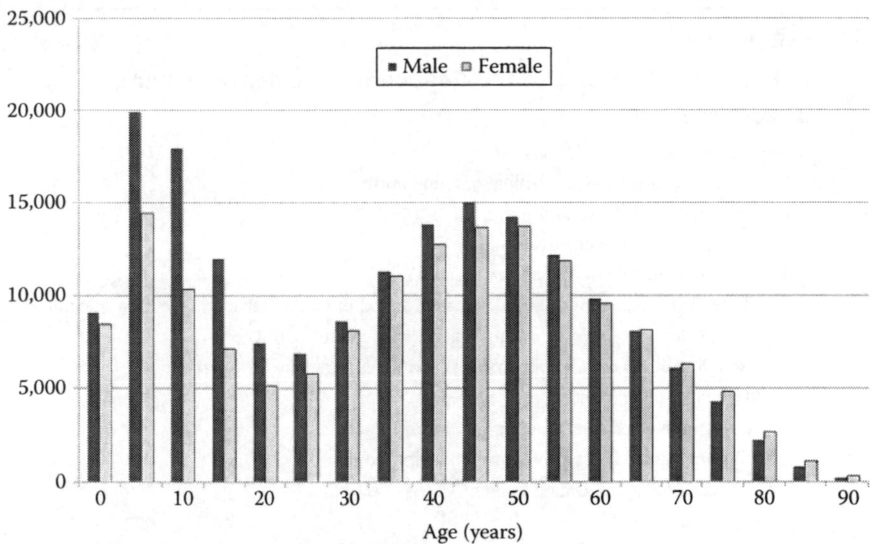

FIGURE 13.7 Number of reported cases of Lyme disease by age group. (From CDC, Lyme disease data, http://www.cdc.gov/lyme/stats/index.html, 2013c.)

wildlife biologists, foresters, landscapers, park employees, people who participate in outdoor recreation, and people who garden or play in their yards located within high-risk neighborhoods.

13.5 MEDICAL TREATMENT

During the initial stage of infection, Lyme disease is easy to treat with orally administered antibiotics. This is true even for cases involving Bell's palsy. If left untreated during this early stage, the disease becomes more difficult to treat. In some cases, repeated and prolonged intravenous antibiotic treatment may be required. For patients with later-stage symptoms of Lyme disease, a diagnosis of Lyme disease can be made with diagnostic tests. However, antibodies often persist for months or years after the elimination of *B. burgdorferi* from the body. Thus, the presence of antibodies does not mean the presence of an active infection (Wormser et al. 2006).

13.6 WHAT PEOPLE CAN DO TO REDUCE THEIR RISK OF CONTRACTING LYME DISEASE

There are several methods that can be used to reduce the incidence of Lyme disease (Table 13.2). The best preventative measure against the transmission of Lyme disease is personal protection from tick bites. Wearing long-sleeved shirts and tucking pants into socks or boots can reduce a tick's ability to reach a person's skin. Applying insect repellents containing DEET to clothing and skin and careful body inspection for ticks are both recommended. People can also reduce their risk of tick bites by avoiding areas where ticks are abundant or by avoiding specific behaviors that

TABLE 13.2
Methods that Can Be Used to Reduce the Incidence of Lyme Disease in Humans

- Reduce *B. burgdorferi* populations.
  - Treat mice and small rodents with a systemic antibiotic.
  - Treat deer with a systemic antibiotic.
- Reduce tick populations using acaricides.
  - Broadcast granular acaricides around homes.
  - Distribute tubes containing acaricide-treated cotton or bait boxes to kill ticks on mice.
  - Use four-poster deer feeders that apply acaricides to the fur of deer.
- Reduce small mammal populations that serve as reservoir hosts for *Borrelia*.
  - Modify landscape or habitat.
    - Reduce fragmented forest habitat and forest edges.
    - Modify habitat so it does not favor white-footed mice.
  - Modify yards and areas around homes.
    - Keep lawns mowed.
    - Remove woodpiles and other objects that mice can use for cover.
  - Use rodenticides to kill mice.
- Eradicate deer populations.
  - Use hunts or sharpshooters to eliminate all deer on islands or other isolated areas.
- Prevent ticks from biting people.
  - Apply DEET or other insecticides to clothing and exposed skin.
  - Wear long-sleeved shirts and pants, and tuck pants into socks and shoes.
  - Wear light-colored clothing so crawling ticks can be seen more easily.
  - Check body daily for any ticks.

might attract host-seeking ticks. For example, hikers are more likely to be bitten by ticks if they sit on logs than if they sit on the ground (Lane et al. 2004). Daily body checks will help prevent Lyme disease because it takes 1–2 days for a tick to attach itself and start feeding. For the same reason, wearing light-colored clothes will help people locate ticks, which are more conspicuous when crawling over light surfaces. However, the ticks are small and can be overlooked (Figure 13.8).

The second approach to reducing Lyme disease is to reduce tick density in the general area. Tick densities are high in areas that are shaded and moist, with leaf cover, and in low-lying vegetation. Williams et al. (2009) reduced both the density of blacklegged ticks and the proportion infected with *B. burgdorferi* by reducing the densities of the invasive shrub Japanese barberry. Schulze et al. (1995) used hand rakes and leaf blowers during spring and early summer to remove leave litter in a forested residential community to reduce nymphal tick densities by more than 73%. Well-maintained lawns also have lower tick densities than areas that are brushy or wooded (Maupin et al. 1991).

There are two tiny (0.1 in. or 2 mm in length) wasps, *Ixodiphagus hookeri* and *I. texanus* that prey upon ticks. These wasps were released in the United States and Russia to reduce tick populations. Populations of *I. hookeri* still exist on several

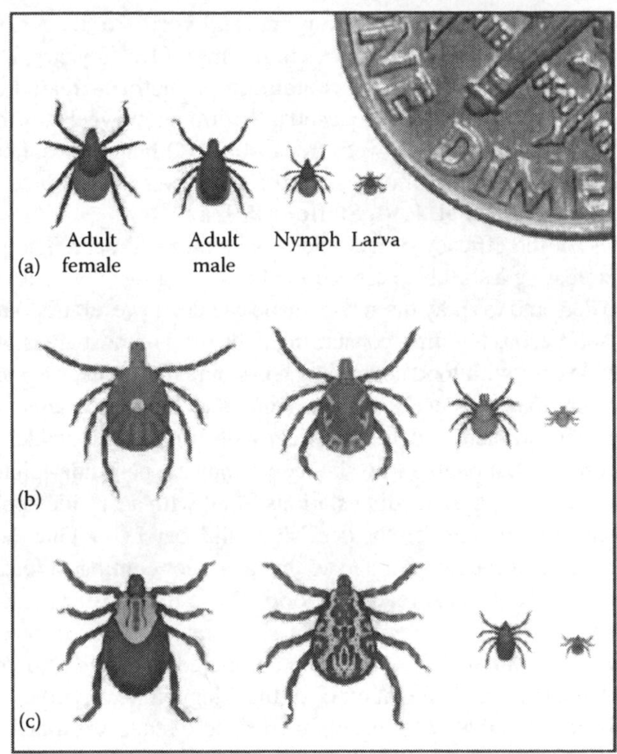

FIGURE 13.8 Actual size of (a) the blacklegged tick, (b) the lone star tick, and (c) the American dog tick relative to a dime. (Courtesy of Chapman and the CDC.)

islands in New England and New York where both deer and tick populations are extremely high; unfortunately, the proportion of ticks parasitized by the wasps is too low to limit tick populations (Stafford et al. 2003). Only 4% of the ticks on Fire Island, New York, were parasitized by the wasps (Ginsberg and Zhioua 1999).

Pesticide applications have been successful in reducing tick populations near homes. Schulze et al. (1994) suppressed the number of *Ixodes* ticks on mice by more than 70% after applying a granular acaricide (i.e., pesticide that kills ticks and mites) to the shrub layer and wooded buffers of a forested residential community. Another approach is to provide food for white-footed mice that have been treated with a systemic acaricide. The acaricide will get into the mouse's blood and kill ticks seeking a blood meal from it. It is also possible to apply acaricide to the fur of small mammals by distributing bait boxes that have a wick treated with a pesticide so that a mouse has to brush against the wick whenever it enters or leaves the bait box. After bait boxes containing fipronil were distributed on Mason's Island in Connecticut, nymphal and larval tick infestations on white-footed mice were reduced by more than 50%. There also was a reduction in proportion of ticks that were infected with *B. burgdorferi* by 67% (Dolan et al. 2004). Another approach to treat mice with acaricide is to distribute cotton balls impregnated with acaricide owing to the mouse's tendency to collect cotton balls for nesting material and take the cotton balls to its nest where the acaricide will kill any ticks in the nest. Over 3 years, Deblinger and Rimmer (1991) distributed 2000 tubes containing permethrin-treated cotton over a 17 acres (7 ha) site in Ipswich, Massachusetts. Within a few weeks of the first application, virtually all mice at the site were free of ticks. Other studies, however, failed to reduce host-seeking nymphal ticks when the tubes were distributed over smaller residential areas (Daniels et al. 1991, Stafford 1992).

Other studies on the efficacy of acaricides have targeted deer. The problem with this approach is treating a sizeable proportion of free-ranging deer. One technique has been to livetrap deer and to spray them with an acaricide; however, trapping deer is difficult, and this approach is too time-consuming to be used in most areas. Another technique is to provide deer with food containing a systemic pesticide, such as ivermectin. Following consumption by a deer, ivermectin enters its blood and kills ticks that later feed on the deer. The problem with treating deer with systemic acaricides, such as ivermectin, is the concern that people who shoot deer may eat pesticide-tainted venison.

To avoid this problem, deer-feeding stations fitted with acaricide applicators have been used to place an acaricide on the deer's fur (Sidebar 13.1). One model has four posts containing an acaricide positioned so that any deer coming to feed will have to brush against one of the posts to reach the food. These four-poster feeders were used to control ticks at seven sites (each more than 123 acres or 500 ha in size) scattered in five eastern states (Pound et al. 2009). When compared to paired untreated sites, the four-post deer feeders reduced the number of blacklegged tick nymphs in the treated sites by 71% (Brei et al. 2009). One problem with these feeders is that many different mammals like to eat corn or any other bait used to attract deer, and the method is not

SIDEBAR 13.1 USING FOUR-POSTER DEER FEEDERS TO KILL TICKS FEEDING ON FREE-RANGING DEER (POUND ET AL. 2000, 2012)

White-tailed deer are important to maintaining populations of *I. scapularis*. Therefore, one approach to preventing Lyme disease in humans is to reduce tick populations on white-tailed deer. But treating free-ranging deer with an acaricide is difficult. One approach is to use food to lure deer into rubbing against a post that will dispense a pesticide onto their fur. One such device is a four-poster feeder that functions by attracting deer to a small feeder containing whole corn. Each post is an acaricide-soaked paint roller. When deer extend their heads to get a bite of food, their head, neck, and ears brush against at least one of the posts and rollers, transferring some acaricide onto their fur (Figure 13.9). Later when the deer grooms itself, the chemical is distributed more widely to the deer's skin. Harmon et al. (2011) tested the ability of these feeders to reduce tick (*Amblyomma americanum*) abundance at a retirement community in Tennessee. Within 328 yards (300 m) of a feeder, numbers of larval, nymphal, and adult ticks were reduced by 91%, 68%, and 49%, respectively. Beyond this distance, the feeders had no impact on either nymphal or adult ticks.

FIGURE 13.9 Bucks feeding at a four-poster feeder and getting a dose of acaricide on their fur. (Courtesy of Mat Pound.)

effective if these nontarget species consume the bait instead of deer. To alleviate this problem, feeders can be surrounded with an electrified fence low enough to stop feral hogs, raccoons, and other medium-sized mammals from entering but which deer can easily jump.

A third approach is to reduce the abundance of those species that serve as host reservoirs for *Ixodes* ticks and *B. burgdorferi*, especially white-footed mice. In small areas, such as a yard or housing development, this has been achieved by removing woodpiles and debris that provide shelter to mice.

In the western United States, the western fence lizard hosts up to 90% of larval and nymphal western blacklegged tick, but the lizard's immune system kills any *B. burgdorferi* that are in the digestive system of ticks when they feed upon it;

this may explain why a lower proportion of western blacklegged ticks in California are infected with *B. burgdorferi* than blacklegged ticks in the east (Swei et al. 2011). Removal of western fence lizards may reduce tick densities but increase the proportion of them that carry *B. burgdorferi*. To test this, Swei et al. (2011) removed lizards from six plots, each 2.5 acres (1 ha) in size, and found that lizard removal reduced nymph tick densities but not the proportion of ticks that had *B. burgdorferi*.

Changes in plant communities can influence local populations of white-footed mice or white-tailed deer. The size of the area is important; vegetation changes within a small area (less than a couple of acres) can impact mice densities, while vegetation changes of a much larger area (more than 2500 acres or 1000 ha) are needed to impact deer densities. Deer and mice densities are also lower in areas where unpalatable or poisonous plants dominate than in areas where deer and mice can feed on a plethora of nutritional plants. Changes to the forest structure or the composition of the plant community also impact mammal densities. In closed-canopy forests, less light reaches the ground, causing a decrease in understory vegetation upon which both deer and mice depend. Hence, densities of white-footed mice and white-tailed deer are lower in closed-canopy forests and higher in forests with open canopies. Other studies indicate that forest fragmentation favors white-footed mice, which are abundant in small forest patches. In New York, small patches of forests (less than 5 acres or 2 ha) had higher densities of *Ixodes* nymphs, and a higher proportion of the nymphs carried *B. burgdorferi* than larger patches of forests (Allan et al. 2003).

Other suggested ways to reduce the density of white-footed mice in an area include increasing the number of foxes, mink, kestrels, or other raptors, hoping that these predators will suppress mouse populations. However, most scientific studies have not found a link between small mammal densities and predator numbers.

Another approach is to set a controlled fire to areas to reduce tick numbers. The rationale is that fire has an immediate effect by killing ticks and removing leaf litter and by reducing rodent populations by depriving them of food and shelter. Longer term, a fire will stimulate more abundant plant growth, which will both attract deer and allow mouse populations to rebound. Hence, fire may reduce tick densities for a few weeks or months until plants start to grow again, allowing an increase in tick densities (Stafford et al. 1998).

Reducing deer populations in an area would seem, intuitively, to reduce the incidence of Lyme disease among people in the area. Most studies, however, have not demonstrated such a relationship except on isolated islands and peninsulas where deer immigration is limited. White-tailed deer were introduced to Maine's Monhegan Island in 1955. With them came blacklegged ticks, and from 1990 to 1998, densities of host-seeking ticks during the fall ranged from 2.4 to 6.9 per acre (6 to 17/ha). By 1996, 13% of the island's population had contracted Lyme disease, and the island's residents decided to eliminate deer. Removing the deer required 3 years, but the last deer was removed from Monhegan Island in March 1999. That same fall, host-seeking tick densities sky-rocketed to 11 per acre (28/ha), presumably because many ticks have difficulty locating a deer. Thereafter, numbers of host-seeking ticks decreased until there were less than 2.4 per acre (1/ha) during the fall of 2003 (Rand et al. 2004). On one of Massachusetts' barrier islands, a reduction of deer from 72 to 18 per square mile (28/km^2 to 7/km^2) through a controlled hunt ultimately caused a 50% reduction in larval and nymphal tick densities on small mammal hosts, though host-seeking tick densities increased for the first few years after the deer reduction (Deblinger et al. 1993).

In 1984, deer were virtually eliminated from a 0.9 mile^2 (2.4 km^2) peninsula on Cape Cod, Massachusetts. This was followed during the next few years by a decline in larval and nymphal tick densities. Among the 220 people that lived on the island, 35 had contracted Lyme disease prior to the controlled deer hunt. Only two cases of Lyme disease were reported during the 5 years after the hunt (Wilson et al. 1988, Telford 2002, Rand et al. 2003).

Densities of white-tailed deer populations in two isolated tracts (Lake Success Business Park and Bluff Point) in Connecticut were reduced from more than five-fold. After the deer removal, there was a 10-fold decrease at Lake Success in the densities of host-seeking nymphal *Ixodes* and a threefold to 10-fold decrease in densities of host-seeking larval *Ixodes*, but the results at Bluff Point were not as encouraging. Lower deer densities there had no significant impact on densities of nymphal ticks, while densities of larval ticks were correlated with deer densities (Stafford et al. 2003).

Reducing deer densities can be achieved by erecting deer-proof fences, but doing this does not always reduce tick densities. On Fire Island, New York, numbers of host-seeking ticks within deer-proof exclosures were higher than outside them (Ginsberg and Zhioua 1999). The apparent reason for this was because a high proportion of adult ticks inside the exclosures were seeking a host, while most ticks outside the exclosure had already found one. In Westchester County, New York, Daniels et al. (1993) compared *Ixodes* densities within deer-proof exclosures and found that the exclosures had 83% fewer host-seeking nymphs and 90% fewer host-seeking larva than outside. Daniels and Fish (1995) checked small mammals for *Ixodes* inside two deer exclosures located again in Westchester County; mice inside the exclosures were infected with fewer ticks than mice outside the exclosures, but densities of adult ticks were similar inside and outside. Not surprisingly, the size of the deer exclosures is an important variable. Large deer exclosures reduce host-seeking tick numbers, but small ones (less than 6 acres or 2.5 ha) increase the densities of host-seeking ticks (Perkins et al. 2006).

A vaccine has been developed to protect people from Lyme disease. Although effective, the vaccine became controversial, and the manufacturer withdrew the vaccine from the United States in 2002 (Sidebar 13.2).

Perhaps the most effective approach in reducing the risk of Lyme disease is an integrated approach to tick management that combines a variety of methods to reduce tick densities and the proportion of ticks infected with *B. burgdorferi*. Such an approach requires an assessment of local disease risk, targeting those areas where the risk is highest. Knowledge of the local host species and tick ecology is also required. For example, spraying vegetation with acaricide will only be effective if timed to occur during those weeks when the larvae or nymphs are actively seeking hosts (Eisen et al. 2012). Likewise, efforts to reduce tick populations on host species will only be effective if the host species that are important locally have been identified.

SIDEBAR 13.2 DEATH BY RUMOR: WHAT HAPPENED TO THE VACCINE AGAINST LYME DISEASE? (ABBOTT 2006, NATURE 2006, SOHN 2011)

One of the mysteries in the history of Lyme disease prevention is what happened to the Lyme disease vaccine. During 1998, the pharmaceutical company SmithKline Beecham (now part of GlaxoSmithKline) started to market a vaccine in the United States to protect people from Lyme disease. The vaccine, called LYMErix, was made from a protein on the surface of *B. burgdorferi*; when injected into someone, that person's immune system produced antibodies against that protein. The vaccine was effective, preventing Lyme disease in about 80% of adults.

During the late 1990s, many people were worried about the adverse effects of vaccines, and LYMErix was not above suspicion. Soon, hundreds of people vaccinated with LYMErix claimed they were suffering from the vaccine's side effects. They worried that the vaccine might cause their immune systems not only to attack *B. burgdorferi* but also to attack human proteins, causing serious autoimmune diseases. Soon, SmithKline Beecham was facing a class action lawsuit and numerous individual suits. The U.S. Food and Drug Administration and Centers for Disease Control and Prevention investigated about 900 adverse case reports and found that there was no evidence of a link between the complaints and the vaccine. Still, the adverse publicity caused sales of LYMErix to plummet. By 2002, GlaxoSmithKline stopped making the vaccine.

Many doctors believe this is a public health fiasco. Failing to provide a vaccine when one is available, they say, is failing the many millions of people who live in areas where Lyme disease is endemic. Dr. Markus Simon of the Max Plank Institute for Immunobiology summed up the frustration of many when he said, "This just shows how irrational the world can be. There was no scientific justification for the [Lyme disease] vaccine being pulled."

Although considerable research has demonstrated that many techniques can be used to reduce reservoir host densities and tick numbers, few of the techniques are actually employed to protect people from Lyme disease. There are two reasons for this. First, the techniques are labor intensive and costly. Second, most of these studies report a decrease in mammal or tick densities rather than in the incidence of Lyme disease among people. Hence, it is difficult for public health agencies to determine if a method is cost-effective based on disease prevention (Eisen et al. 2012).

While public health agencies play a major role in disease prevention for most zoonotic diseases, such is not true for Lyme disease or other zoonotic diseases transmitted by ticks. For example, local governments take the lead in the prevention of mosquito-borne diseases such as equine encephalitis (Chapter 20) and West Nile virus (Chapter 21), often through mosquito control districts. Whether local governments should take the lead in preventing Lyme disease (perhaps by establishing tick control districts) is an interesting question. Consequently, prevention of Lyme disease primarily falls on individuals who are responsible for their own personal safety and landowners who are responsible for deciding if tick control should be conducted on their property. During 2013, the CDC called for local governments and communities to adopt a broader approach to reduce tick densities involving entire communities. The CDC (2013a) envisioned a community approach that would involve homeowners killing ticks in their own yards and communities addressing issues involving rodents and deer, which are reservoir hosts for the blacklegged tick and *B. burgdorferi*.

Before a person will employ a method of self-protection, such as the use of insect repellents, he or she must decide if the risk of infection and its consequences exceeds the cost of employing the method. The first part of this cost-to-benefit equation—costs—can be assessed by each individual. The cost of using an insect repellent includes the cost of the repellent, the inconvenience of having to find the repellent whenever it is needed, any perceived discomfort of having the repellent on one's skin, and the perceived risk to clothing and personal health of using the repellent. People are much less able to ascertain the benefit portion of the cost-to-benefit equation (i.e., a decrease in disease risk) because they probably know few people who have Lyme disease and have minimal personal knowledge of disease outcomes. Hence, they have to rely upon others in public health agencies to assess their risk of disease and its consequences. Unfortunately, identifying someone's actual risk of contracting Lyme disease is difficult because the risk will vary greatly from one neighborhood to another and differ from one person to another based on how much time a person spends in areas where ticks are common. Equally difficult is the effort to quantify how much of a reduction in risk is associated with specific actions. If someone applies DEET when camping or hiking in the woods, how much does this behavior reduce the person's risk of contracting Lyme disease? Answers to such questions are elusive because each person's circumstances, behavior, and disease risk differ from others. Yet, such information is needed before people can truly make informed decisions about how to avoid Lyme disease.

13.7 SOUTHERN TICK–ASSOCIATED RASH ILLNESS

Sometimes when people are bitten by the lone star tick (*A. americanum*; Figure 13.10), they develop a circular rash similar to the one caused by Lyme disease. Patients may also experience fatigue, fever, headache, and muscle pains. This illness is called the southern tick-associated rash illness (STARI). Despite its similarity to Lyme disease, STARI is not caused by *B. burgdorferi*, and the lone star tick does not transmit Lyme disease. The cause of the STARI rash has not been determined, but its symptoms are resolved following treatment with doxycycline, an oral antibiotic. The lone star tick occurs in the eastern United States from as far north as Maine, south into Mexico, and as far west as Texas and Oklahoma. Lone star ticks feed on many wildlife species including white-tailed deer and on livestock. All three life stages (larvae, nymphs, and adults) feed on dogs, cats, and humans. Pets can bring these ticks into homes where they pose a risk to people living in the house (CDC 2011).

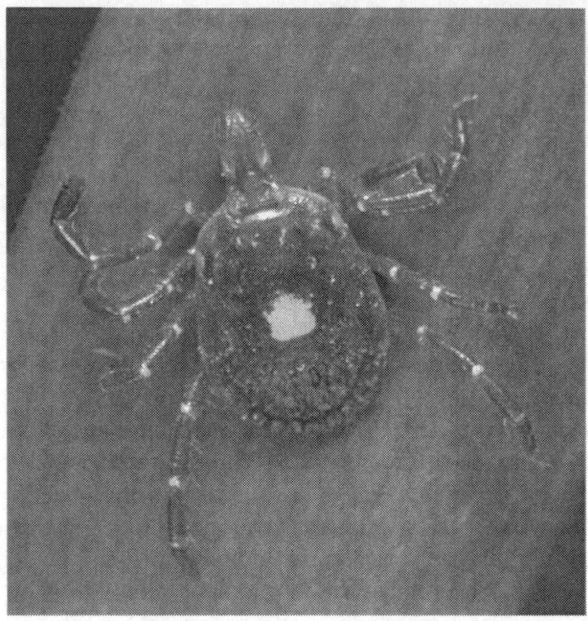

FIGURE 13.10 Photo of a lone star tick (*A. americanum*). (Courtesy of the CDC.)

13.8 ERADICATING LYME DISEASE FROM A COUNTRY

The difficulty in eradicating Lyme disease from a country where *B. burgdorferi* is endemic occurs because so many wildlife species can serve as a reservoir host. Another difficulty is that even if *B. burgdorferi* was eradicated from an area, the pathogen can be reintroduced by migrating birds.

LITERATURE CITED

Abbott, A. 2006. Lyme disease: Uphill struggle. *Nature* 429:509–510.

Allan, B. F., F. Keesing, and R. S. Ostfeld. 2003. Effect of forest fragmentation on Lyme disease risk. *Conservation Biology* 17:267–272.

Bacon, R. M., K. J. Kugeler, and P. S. Mead. 2008. Surveillance for Lyme disease—1992–2006. *Morbidity and Mortality Weekly Report* 57(SS10):1–9.

Bransfield, R. C. 2012. Lyme, depression, and suicide. http://www.mentalhealthandillness.com/Articles/LymeDepressionAndSuicide.htm (accessed June 29, 2013).

Brei, B., J. S. Brownstein, J. E. George, J. M. Pound et al. 2009. Evaluation of the United States Department of Agriculture northeast area-wide tick control project by meta-analysis. *Vector-Borne and Zoonotic Diseases* 9:423–430.

Bunikis, J., J. Tsao, C. J. Luke, M. G. Luna et al. 2004. *Borrelia burgdorferi* infection in a natural population of *Peromyscus leucopus* mice: A longitudinal study in an area where Lyme borreliosis is highly endemic. *Journal of Infectious Diseases* 189:1515–1523.

Castro, M. B. and S. A. Wright. 2007. Vertebrate hosts of *Ixodes pacificus* (Acari: Ixodidae) in California. *Journal of Vector Ecology* 32:140–149.

CDC. 1999. Recommendations for the use of Lyme disease vaccine: Recommendations of the Advisory Committee on Immunization Practices (ACIP). *Morbidity and Mortality Weekly Report* 48(RR7):1–17.

CDC. 2010. Lyme disease. http://www.cdc.gov/diseaseconditions/ (accessed December 16, 2012).

CDC. 2011. Lone star tick a concern, but not for Lyme disease. http://www.cdc.gov/stari/disease (accessed August 15, 2012).

CDC. 2013a. CDC provides estimate of Americans diagnosed with Lyme disease each year. http://www.cdc.gov/media/releases/2013/p0819-lyme-disease.html (accessed December 15, 2013).

CDC. 2013b. Summary of notifiable diseases—United States, 2011. *Morbidity and Mortality Weekly Report* 60(53):1–117.

CDC. 2013c. Lyme disease data. http://www.cdc.gov/lyme/stats/index.html (accessed December 3, 2013).

Daniels, T. J. and D. Fish. 1995. Effect of deer exclusion on the abundance of immature *Ixodes scapularis* (Acari: Ixodidae) parasitizing small and medium-sized mammals. *Journal of Medical Entomology* 32:5–11.

Daniels, T. J., D. Fish, and R. C. Falco. 1991. Evaluation of host-targeted acaricide for reducing risk of Lyme disease in southern New York state. *Journal of Medical Entomology* 28:537–543.

Daniels, T. J., D. Fish, and I. Schwartz. 1993. Reduced abundance of *Ixodes scapularis* (Acari: Ixodidae) and Lyme disease risk by deer exclusion. *Journal of Medical Entomology* 30:1043–1049.

Deblinger, R. D. and D. W. Rimmer. 1991. Efficacy of a permethrin-based acaricide to reduce the abundance of *Ixodes dammini* (Acari: Ixodidae). *Journal of Medical Entomology* 28:708–711.

Deblinger, R. D., M. L. Wilson, D. W. Rimmer, and A. Spielman. 1993. Reduced abundance of immature *Ixodes dammini* (Acari: Ixodidae) following incremental removal of deer. *Journal of Medical Entomology* 30:144–150.

Department for Environment, Food, and Rural Affairs. 2011. Zoonoses report UK 2009. Department for Environment, Food, and Rural Affairs, London, U.K.

Dolan, M. C., G. O. Maupin, B. S. Schneider, C. Denatale et al. 2004. Control of immature *Ixodes scapularis* (Acari: Ixodidae) on rodent reservoirs of *Borrelia burgdorferi* in a residential community of southeast Connecticut. *Journal of Medical Entomology* 41:1043–1054.

Eisen, R. J., J. Piesman, E. Zielinski-Gutierrez, and L. Eisen. 2012. What do we need to know about disease ecology to prevent Lyme disease in the northeastern United States? *Journal of Medical Entomology* 49:11–22.

Fish, A. E., Y. B. Pride, and D. S. Pinto. 2008. Lyme carditis. *Infectious Disease Clinics of North America* 22:275–288.

Gil, H., M. Barral, R. Escudero, A. L. García-Pérez, and P. Anda. 2005. Identification of a new *Borrelia* species among small mammals in areas of northern Spain where Lyme disease is endemic. *Applied and Environmental Microbiology* 71:1336–1345.

Ginsberg, H. S. and E. Zhioua. 1999. Influence of deer abundance on the abundance of questing adult *Ixodes scapularis* (Acari: Ixodidae). *Journal of Medical Entomology* 36:376–381.

Hall, S. A. 2011. Iceman autopsy. *National Geographic Magazine*. http://ngm.nationalgeographic.com/2011/11/iceman-autopsy/hall/text (accessed July 7, 2012).

Harmon, J. R., G. J. Hickling, M. C. Scott, and C. J. Jones. 2011. Evaluation of 4-poster acaricide applicators to manage tick populations associated with disease risk in a Tennessee retirement community. *Journal of Vector Ecology* 36:404–410.

Hayes, E. B. and J. Piesman. 2003. How can we prevent Lyme disease? *New England Journal of Medicine* 348:2424–2430.

Hofmeister, E. K., B. A. Ellis, G. E. Glass, and J. E. Childs. 1999. Longitudinal study of infection with *Borrelia burgdorferi* in a population of *Peromyscus leucopus* at a Lyme disease-enzootic site in Maryland. *American Journal of Tropical Medicine and Hygiene* 60:598–609.

Ishiguro, F., N. Takada, T. Masuzawa, and T. Fukui. 2000. Prevalence of Lyme disease *Borrelia* spp. in ticks from migratory birds on the Japanese mainland. *Applied and Environmental Microbiology* 66:982–986.

Keirans, J. E., H. J. Hutcheson, L. A. Durden, and J. S. H. Klompen. 1996. *Ixodes (Ixodes) scapularis* (Acari: Ixodidae); redescription of all active stages, distribution, hosts, geographical variation, and medical and veterinary importance. *Journal of Medical Entomology* 33:297–318.

Lane, R. S., D. B. Steinlein, and J. Mun. 2004. Human behavior elevating exposure to *Ixodes pacificus* (Acari: Ixodidae) nymphs and their associated bacterial zoonotic agents in a hardwood forest. *Journal of Medical Entomology* 41:239–248.

Logigian, E. L., R. F. Kaplan, and A. C. Steere. 1990. Chronic neurologic manifestations of Lyme disease. *New England Journal of Medicine* 323:1438–1444.

LoGiudice, K., R. S. Ostfeld, K. A. Schmidt, and F. Keesing. 2003. The ecology of infectious disease: Effects of host diversity and community composition on Lyme disease risk. *Proceedings of the National Academy of Sciences of the United States of America* 100:567–571.

Maupin, G. O., D. Fish, J. Zultowsky, E. G. Campos, and J. Piesman. 1991. Landscape ecology of Lyme disease in a residential area of Westchester County, New York. *American Journal of Epidemiology* 133:1105–1113.

Miller, J. R. and P. S. Mead. 2011. Chapter 3—Infectious diseases related to travel. *Travel, Yellowbook.* Centers for Disease Control and Prevention, Atlanta, GA.

Moody, K. D., G. A. Terwilliger, G. M. Hansen, and S. W. Barthold. 1994. Experimental *Borrelia burgdorferi* infection in *Peromyscus leucopus. Journal of Wildlife Diseases* 30:155–161.

Nature. 2006. When a vaccine is safe. *Nature* 439:509.

O'Connell, S., M. Granström, J. S. Gray, and G. Stanek. 1998. Epidemiology of European Lyme borreliosis. *Zentralblatt für Bakteriologie* 287:229–240.

Oliver Jr., J. H., T. Lin, L. Gao, K. L. Clark et al. 2003. An enzootic transmission cycle of Lyme borreliosis spirochetes in the southeastern United States. *Proceedings of the National Academy of Sciences of the United States of America* 100:11642–11645.

Perkins, S. E., I. M. Cattadori, V. Tagliapietra, A. P. Rizzoli, and P. J. Hudson. 2006. Localized deer absence leads to tick amplification. *Ecology* 87:1981–1986.

Pound, J. M., K. H. Lohmeyer, R. B. Davey, L. A. Soliz, and P. U. Olafson. 2012. Excluding feral swine, javelinas, and raccoons for deer bait stations. *Human–Wildlife Interactions* 6:169–177.

Pound, J. M., J. A. Miller, J. E. George, and D. Fish. 2009. The United States Department of Agriculture northeast area-wide tick control project: History and protocol. *Vector-Borne and Zoonotic Diseases* 9:365–370.

Pound, J. M., J. A. Miller, J. E. George, and C. A. LeMeilleur. 2000. The '4-poster' passive topical treatment device to apply acaricide for controlling ticks (Acari: Ixodidae) feeding of while-tailed deer. *Journal of Medical Entomology* 37:588–594.

Rand, P. W., C. Lubelczyk, M. S. Holman, E. H. Lacombe, and R. P. Smith Jr. 2004. Abundance of *Ixodes scapularis* (Acari: Ixodidae) after the complete removal of deer from an isolated offshore island, endemic for Lyme disease. *Journal of Medical Entomology* 41:779–784.

Rand, P. W., C. Lubelczyk, G. R. Lavigne, S. Elias et al. 2003. Deer density and the abundance of *Ixodes scapularis* (Acari: Ixodidae). *Journal of Medical Entomology* 40:179–184.

Ray, G., T. Schulz, W. Daniels, E. R. Daly et al. 2013. Three sudden cardiac deaths associated with Lyme carditis—United States, November 2012-July 2013. *Morbidity and Mortality Weekly Report* 62:993–996.

Richter, D., B. Klug, A. Spielman, and F.-R. Matuschka. 2004. Adaption of diverse Lyme disease spirochetes in a natural rodent reservoir host. *Infection and Immunity* 72:2442–2444.

Schulze, T. L., R. A. Jordan, and R. W. Hung. 1995. Suppression of subadult *Ixodes scapularis* (Acari: Ixodidae) following removal of leaf litter. *Journal of Medical Entomology* 32:730–733.

Schulze, T. L., R. A. Jordan, L. M. Vasvary, M. S. Chomsky et al. 1994. Suppression of *Ixodes scapularis* (Acari: Ixodidae) nymphs in a large residential community. *Journal of Medical Entomology* 31:206–211.

Schwanz, L. E., M. J. Voordouw, D. Brisson, and R. S. Ostfeld. 2011. *Borrelia burgdorferi* has minimal impact on the Lyme disease reservoir host *Peromyscus leucopus*. *Vector-Borne and Zoonotic Diseases* 11:117–124.

Sohn, E. 2011. Lyme disease: Where's the vaccine? http://discovery.com/human/lyme-disease-ticks-caccine-110617.html (accessed July 1, 2012).

Stafford III, K. C. 1992. Third-year evaluation of host-targeted permethrin for the control of *Ixodes dammini* in southeastern Connecticut. *Journal of Medical Entomology* 29:717–720.

Stafford III, K. C., A. J. DeNicola, and H. J. Kilpatrick. 2003. Reduced abundance of *Ixodes scapularis* (Acari: Ixodidae) and the tick parasitoid *Ixodiphagus hookeri* (Hymenoptera: Encyrtidae) with reduction of white-tailed deer. *Journal of Medical Entomology* 40:642–652.

Stafford III, K. C., J. S. Ward, and L. A. Magnarelli. 1998. Impact of controlled burns on *Ixodes scapularis* (Acari: Ixodidae). *Journal of Medical Entomology* 35:510–513.

Swei, A., R. S. Ostfeld, R. S. Lane, and C. J. Briggs. 2011. Impact of the experimental removal of lizards on Lyme disease risk. *Proceedings of the Royal Society B* 278:2970–2978.

Telford III, S. R. 2002. Deer tick-transmitted zoonoses in the eastern United States. In: A. A. Aguirre, R. S. Ostfeld, G. M. Tabor, C. House, and M. C. Pearl, editors. *Conservation Medicine: Ecological Health in Practice 2002*. Oxford University Press, New York, pp. 310–324.

Williams, S. C., J. S. Ward, T. E. Worthley, and K. C. Stafford III. 2009. Managing Japanese barberry (Ranunculales: Berberidaceae) infestations reduces blacklegged tick (Acari: Ixodidae) abundance and infection prevalence with *Borrelia burgdorferi* (Spirochaetales: Spirochaetaceae). *Environmental Entomology* 38:977–984.

Wilson, M. L., S. R. Telford III, J. Piesman, and A. Spielman. 1988. Reduced abundance of immature *Ixodes dammini* (Acari: Ixodidae) following elimination of deer. *Journal of Medical Entomology* 25:224–228.

Wormser, G. P., R. J. Dattwyler, E. D. Shapiro, J. J. Halperin et al. 2006. The clinical assessment, treatment, and prevention of Lyme disease, human granulocytic anaplasmosis, and babesiosis: Clinical practice guidelines by the Infectious Diseases Society of America. *Clinical Infectious Diseases* 43:1089–1134.

14 Tick-Borne Relapsing Fever

Louse-borne relapsing fever epidemics occurred frequently in Europe during the early 20th Century. Between 1919 and 1923, 13 million cases resulting in 5 million deaths occurred in the social upheaval that overtook Russia and eastern Europe.

CDC (2012a)

The incidence of TBRF [tick-borne relapsing fever] at the community level is the highest… for any bacterial disease [in Africa].

Vial et al. (2006)

14.1 INTRODUCTION AND HISTORY

There are two recognized types of relapsing fever based on the species that serve as vectors of the disease organism to humans: louse-borne relapsing fever (LBRF) and tick-borne relapsing fever (TBRF). LBRF is caused by the spirochete *Borrelia recurrentis* and is transmitted by human body lice (*Pediculus humanus*), which also serve as a host reservoir (Figure 14.1). After being consumed as part of a blood meal by lice (louse is the singular form of lice), the spirochetes pass into the coelomic cavity where they multiply. Lice remain infected their entire lives, which can last several weeks. People become infected not by the bite of an infected louse, but when the louse is crushed by scratching, which releases spirochetes inside the louse and contaminates human skin. The spirochete can gain entry through cut or abraded skin or through mucous membranes (Goubau 1984).

Not surprisingly, more than 15 million people during the twentieth century have had LBRF, and 5 million of them have died from the disease, with most deaths occurring in Russia, eastern Europe, and Africa (Dworkin et al. 1998). Epidemics occurred during both World War I and II (Figure 14.2). Fatality rates from LBRF vary widely among epidemics but can be as high as 40% without medical treatment and 5% with it (Goubau 1984). In recent decades, this disease has disappeared from most of the world, owing to improved standards of living and hygiene and the use of DDT and other insecticides to eradicate lice. In 2012, LBRF occurred mainly in Africa, particularly within the countries of Ethiopia, Sudan, Eritrea, and Somalia. It is common in refugee camps and areas affected by wars or extreme poverty (CDC 2012a). Humans and human body lice are the only known reservoir for *B. recurrentis*, so it is not a zoonotic disease and will not consider it further for this reason.

TBRF is a zoonotic disease first identified in the United States in 1915 (Spach et al. 1993). It occurs primarily in the western United States (Figures 14.3 and 14.4). TBRF is caused by several different species of *Borrelia*, but usually the species

FIGURE 14.1 Human body lice. (Courtesy of the CDC.)

FIGURE 14.2 American soldiers during World War I searching their clothing for lice. (Courtesy of the U.S. National Medical Library.)

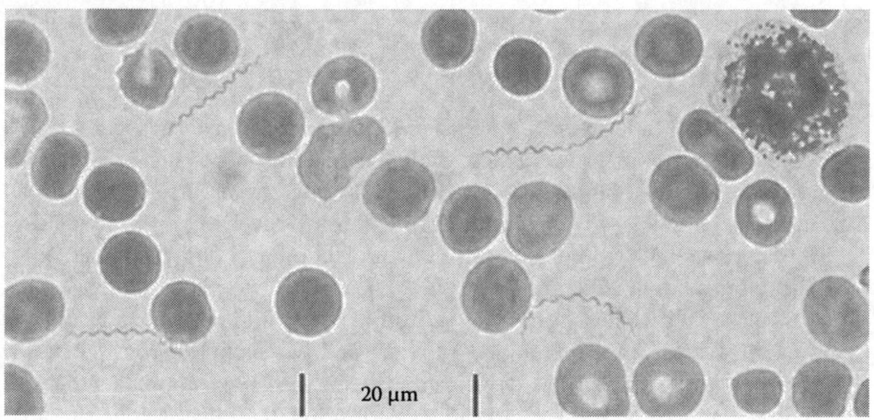

FIGURE 14.3 Photo of a blood smear showing a TBRF spirochete (wavy snake-like objects), red blood cells, and one white blood cell in the upper right corner. (From CDC, Tick-borne relapsing fever [TBRF], http://www.cdc.gov/relapsing-fever/, 2012b.)

FIGURE 14.4 Distribution of TBRF in the United States. Each dot was placed in the county where a patient contracted the disease from 1990 to 2011. (From CDC, Tick-borne relapsing fever [TBRF], http://www.cdc.gov/relapsing-fever/, 2012b.)

SIDEBAR 14.1 OUTBREAK OF TBRF IN THE MOUNTAINS OF NEW MEXICO (CDC 2003)

In late July 2002, approximately 40 individuals attended a family gathering at a cabin in northern New Mexico, half of whom spent the night. Located in a remote mountainous area (elevation approximately 8000 ft or 1524 m), the cabin had previously been uninhabited. After the gathering, 14 individuals (36%) demonstrated symptoms consistent with laboratory-confirmed borreliosis, including a combination of fever, chills, headache, body aches, and a type of skin rash. Spirochetes were observed on peripheral blood smear samples from nine of those 14 symptomatic patients, and two samples from patients not demonstrating spirochetes developed *B. hermsii*. Three family members visited the cabin several days early to clean it prior to the gathering, and all three became symptomatic. Patients received antibiotic treatment, and eight individuals who were asymptomatic received antibiotic prophylaxis.

A site inspection revealed gaps in the external areas and foundation of the cabin, which allowed rodents to enter and harbor nesting materials and droppings in cabin walls. A second site visit to trap rodents recovered four chipmunks, a woodrat, and two deer mice, all of which were negative for spirochetes. One live soft tick (*O. hermsi*) infected with *B. hermsii* was also recovered from the cabin site, supporting the transmission of TBRF.

responsible for the disease in North America are *B. hermsii* and *B. turicata*. In Central and South America, TBRF is caused by *B. venezuelensis*.

In North America, TBRF is transmitted by soft-bodied ticks (Argasidae). Two tick species are known to be vectors—*Ornithodoros hermsi* and *O. turicata*—and the range of TBRF in North America is closely linked with the range of these ticks. *O. hermsi* are found in the Pacific Northwest, California, and mountainous parts of the southwest (Sidebar 14.1). It is normally associated with conifer forests where its natural hosts are pine squirrels, other tree squirrels, and western chipmunks. People who stay in rustic cabins are often bitten by this tick. *O. turicata* has a sporadic distribution from Kansas to California and south into Mexico. Its natural hosts include rodents, snakes, tortoises, hogs, and cattle. It has been found in buildings, caves, and burrows occupied by ground squirrels, prairie dogs, burrowing owls, and snakes. Although people in Texas who explore caves have contracted TBRF, a new TBRF spirochete (named *B. johnsonii*) was isolated from the bat tick *Carios kelleyi* in Iowa (Schwan et al. 2009). It is unknown if this new spirochete can cause TBRF in humans, but the bat tick is known to feed on humans when bat colonies occur in homes.

Ornithodoros ticks differ from the ticks with which most people may be familiar. They feed mainly at night and bite people who are sleeping. These ticks feed quickly, often within an hour, and are no longer attached when most people awaken. Hence, most people bitten by the tick never see it and do not know they have been bitten.

In the United States, TBRF has a patchy distribution with most cases occurring in just 10 counties throughout the west. These locations may reflect the distribution

SIDEBAR 14.2 TBRF AMONG VACATIONERS AT A CABIN IN COLORADO (TREVEJO ET AL. 1998)

In 1995, 23 family members stayed overnight at a rental cabin in Estes Park, Colorado. Within 18 days, 11 of them had symptoms consistent with TBRF. Health authorities located 30 other individuals from five families that also stayed at the same cabin that year; five people (17%) also reported TBRF symptoms. *B. hermsii* was isolated from one patient, and three others had antibodies against the spirochete. TBRF patients were more likely to have slept on the floor or in a top bunk than others staying at the cabin who remained disease-free. An examination of the cabin found rodent nesting material, a dead chipmunk, and a dead mouse in the crawl space beneath the house as well as chipmunk feces in the attic. The nesting material contained an adult soft-bodied tick (*O. hermsi*). A trapping program within 109 yards (100 m) of the cabin caught seventeen Uinta chipmunks, ten golden-mantled ground squirrels, three deer mice, and one long-tailed vole; all were tested and *B. hermsii* were cultured from two of the chipmunks.

of *Ornithodoros* ticks or areas where ticks are infected with the pathogen. In the United States, more than 450 cases of TBRF were reported between 1977 and 2000 (Roscoe and Epperly 2005), but many cases are not diagnosed or reported. Often, the discovery of one TBRF patient leads to the discovery of additional ones who had previously been unknown. Many times a cluster of TBRF cases occurs among people who visited the same site, although sometimes months or years apart (Sidebar 14.2). For example, 27 employees and 35 overnight guests at the North Rim of Grand Canyon National Park in Arizona were diagnosed in 1973; most patients had slept in rustic log cabins at the park (Boyer et al. 1977).

Most people contract TBRF during summer (Figure 14.5), both because the ticks are more active and because it is a popular time for camping, with many people staying in rustic cabins. However, there is a difference in seasonality in TBRF between the northwest and southwestern United States, with the latter having more cases during late fall and early winter (Dworkin et al. 2002).

In the Middle East and central Asia (e.g., Cyprus, Egypt, Jordan, Syria, Iraq, Iran, Turkey, Uzbekistan, Tajikistan, and Russia), *B. persica* is responsible for most human cases of TBRF, but other spirochetes involved include *B. caucasica*, *B. latyschewii*, *B. microti*, and *B. baltazardi*. The most important vector of these pathogens in the Middle East and Asia is *O. tholozani*—a tick that lives in caves, old buildings, and cowsheds. Not surprisingly, people who spend the night in caves and old buildings in these regions are at risk of contracting TBRF. Reported cases average 8 per year in Israel (Assous and Wilamowski 2009) and 141 in Iran (Masoumi et al. 2009). Many patients reside in rural households located near animal shelters. Most Iranian patients are under 10 years old, and one-third are less than 5 years old. The reason young children are at risk is because most (92%) of the patients acquire the illness while sleeping at home.

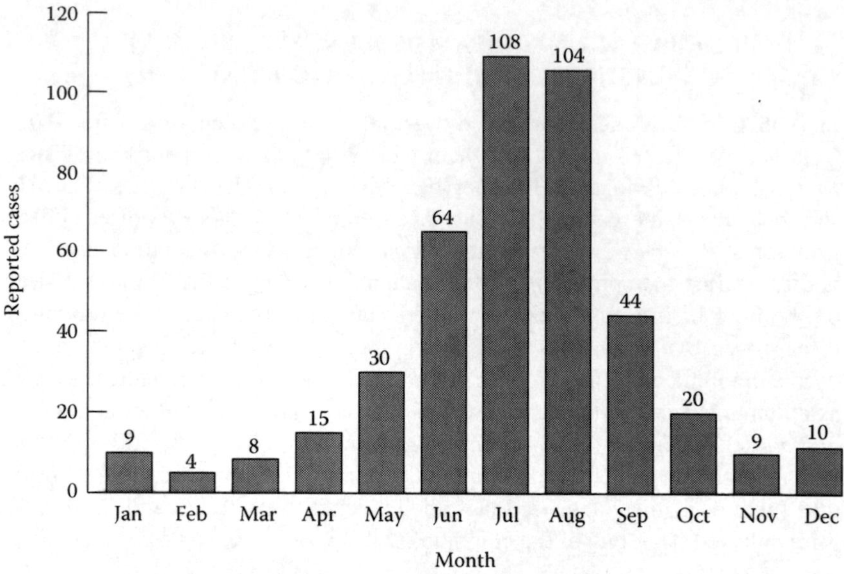

FIGURE 14.5 Distribution of cases of TBRF by month. (Data from Dworkin, M.S. et al., *Am. J. Trop. Med. Hyg.*, 66, 753, 2002.)

TBRF is much more common in Africa than other continents where the causative agents are *B. duttonii* and *B. crocidurae* and vectors include soft-bodied tick *O. porcinus* and *O. sonrai*, respectively (Cutler et al. 2009). In Senegal, the annual incidence rate is 5% (Lecompte and Trape 2003). Vial et al. (2006) examined TBRF in 30 villages located in Senegal, Mauritania, and Mali, reporting that the annual incidence rate was 11,000 per 100,000 people, making it the most common bacterial disease in these countries. In Mitwaba, Zaire, 4%–7% of all outpatients at the local hospital were seeking treatment for TBRF, and 6% of pregnant women in the maternity ward had TBRF (Dupont et al. 1997). In a rural hospital in Zaire, the mortality rate among women with TBRF was 1.5%; among pregnant women with the disease, the risk of premature delivery was 58%, resulting in a perinatal mortality (mortality occurring from 5 months prior to birth to 1 month after it) rate of 44%. This disease often caused maternal deaths and stillbirths (Dupont et al. 1997) (Figure 14.6).

14.2 SYMPTOMS IN HUMANS

The incubation period for TBRF in humans is normally 7 days (ranging from 4 to 18 days). TBRF produces a broad range of symptoms including abdominal pain, fever, chills, headaches, and altered senses. The spirochetes infect many body parts, accounting for a multitude of symptoms (Table 14.1). The pathogens that cause TBRF have the ability to alter the proteins on their surface. This ability allows the number of spirochetes inside a patient to resurge after the person's immune system has reduced the initial infection (Goubau 1984; Dworkin et al. 1998, 2002).

FIGURE 14.6 Scenes from the Dielmo village in Senegal. (From Parola, P. et al., *Emerg. Infect. Dis.*, 17, 883, 2011; used with permission from Didier Raoult.)

These resurging populations of spirochetes repeatedly stimulate the immune system and cause the relapse that gives the disease its common name. Without medical treatment, three to five relapses of fever normally occur, spaced about 8 days apart (Spach et al. 1993). In most cases, the severity of relapse wanes over time (Figure 14.7). Each fever relapse ends with a sequence of symptoms that collectively are called a fever crisis. It starts with a period of 10–30 minutes when fever suddenly rises, the patient's heart rate escalates, and breathing rate becomes rapid. After this period, the patient experiences a rapid decrease in temperature and excessive sweating (CDC 2012b).

TABLE 14.1
Common Symptoms of TBRF

Symptom	Frequency in Patients (%)
Headache	94
Muscle ache	92
Chills	88
Nausea	76
Joint ache	73
Vomiting	71
Fever over 104°F (40°C)	50
Abdominal pain	44
Confusion	38
Dry cough	27
Eye pain	26
Diarrhea	25
Dizziness	25
Sensitivity to light	25

Source: Dworkin, M.S. et al., *Clin. Infect. Dis.*, 26, 122, 1998.

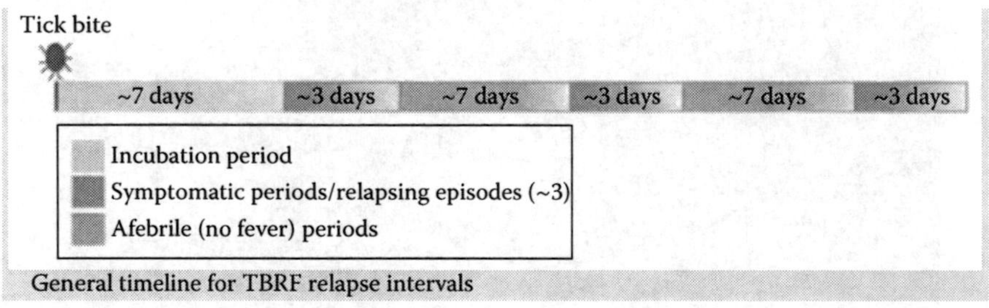

FIGURE 14.7 Progression of TBRF over time showing the pattern of 3 days of illness separated by a week of better health. It is this pattern that gives the disease its name of relapsing illness. (From CDC, Tick-borne relapsing fever [TBRF], http://www.cdc.gov/relapsing-fever/, 2012b.)

TBRF should be considered a possible diagnosis when a patient has recurring fevers of more than 102°F (39°C) and has past exposure to *Ornithodoros* ticks. Diagnosis can be confirmed through lab tests or when spirochetes are detected in blood. When a person is first infected, the fever often lasts 3–6 days and is accompanied by headaches, chills, muscle aches, vomiting, abdominal pain, and pain in the joints (Table 14.1). Confusion, pain in the eye, and sensitivity to light occur in more than 25% of patients, and an enlarged spleen or liver occurs in about 10% of patients. Inflammation of the eye can cause visual impairment. Neurological signs, such as confusion, delirium, and Bell's palsy, may occur, especially during subsequent fever episodes. TBRF can cause serious cases of meningitis and encephalitis that can result in paralysis on one side of the body and impairment of speech or hearing. Mortality rates for TBRF range from 0% to 8% in different countries (Goubau 1984). With proper medical attention, the mortality rate is less than 1%. Severe cases of TBRF are more likely to occur in pregnant women and infants. TBRF in pregnant women can cause miscarriages or stillbirths (Roscoe and Epperly 2005). Mothers with TBRF can infect their fetuses when spirochetes cross the placenta.

The clinical signs of TBRF in Africa differ from those in the United States, perhaps because the species of *Borrelia* that cause illness in North America and Africa are different. In Africa, TBRF causes much higher rates of miscarriages and stillbirths than in the United States (Jongen et al. 1997).

14.3 *BORRELIA* INFECTIONS IN ANIMALS

Little is known about which animals in the United States serve as reservoir hosts for the spirochete or are important to maintaining populations of *Ornithodoros* ticks that serve as vectors for *Borrelia*. Chipmunks, squirrels, and rodents are considered the reservoir hosts for TBRF spirochetes and also maintain populations of *Ornithodoros* ticks in many places. Deer may also play an important role; Nieto et al. (2012) reported that 7.7% of mule deer in Nevada were infected by *B. hermsii*, and *B. coriaceae* was isolated from one deer. TBRF spirochetes were also found in one mule deer from northern California and one from Arizona. In Africa, domestic chickens and pigs serve as hosts to both the spirochete and *Ornithodoros* ticks.

14.4 HOW HUMANS CONTRACT TBRF

People contract TBRF after being bitten by an infected tick. This can happen when people stay in a cabin that has been inhabited by rodents or visit a cave or other area where high densities of small mammals occur or have occurred in the past. *Ornithodoros* ticks may remain dormant in a cave or cabin waiting for a blood meal. They normally feed on rodents and can remain alive and infected with *Borrelia* for several years without feeding. *Borrelia* spirochetes occur in the salivary gland of these ticks, and some are transferred to its host species during feeding. Ticks usually do not feed on humans but may do so when rodent numbers have declined (Spach et al. 1993). When humans enter areas where the ticks reside, their heat and carbon dioxide stimulate ticks to search for a suitable host. In many cases, people are unaware that they have been bitten by the tick; they typically feed at night while people are sleeping, cause painless bites, and feed quickly, often obtaining an adequate blood meal in less than an hour (Figure 14.8).

Most (57%) TBRF patients contract the disease while visiting or living in a cabin or rural dwelling; other patients (17%) become infected while engaging in an outdoor activity (e.g., hiking and cave exploring) or while in their own home (16%). Most TBRF patients remembered that rodents were present at the site where they were

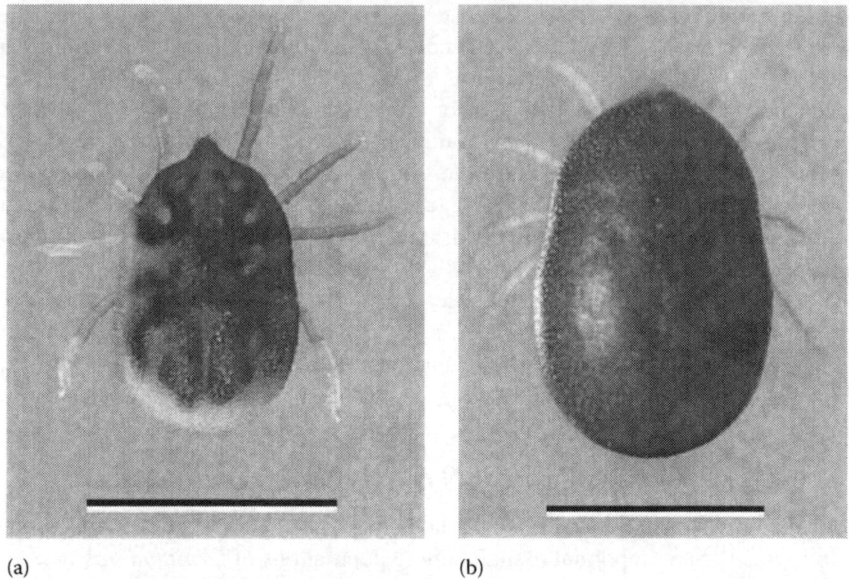

(a) (b)

FIGURE 14.8 Photo of the soft-bodied tick (*O. hermsi*). The photos show a nymphal tick before feeding (a) and after feeding (b). (Photo from Schwan, T.G. et al., *Emerg. Infect. Dis.*, 15, 1026, 2009; provided courtesy of the CDC and Tom Schwan.)

infected (Dworkin et al. 2002). Spirochetes have been isolated from the urine of patients with acute TBRF, so human-to-human infection is possible (Dworkin et al. 1998) but rarely happens.

14.5 MEDICAL TREATMENT

TBRF is usually treated by giving the patient an oral dose of tetracycline every 6 hours for 10 days. Daily injections of ceftriaxone for 10–14 days can be used when the central nervous system is affected (CDC 2012b). Antibiotic-resistant TBRF has not been detected (Roscoe and Epperly 2005). In about half of North American patients with TBRF, symptoms become worse within 2 hours of being treated with antibiotics, as toxins are released by the dying spirochetes. This is called a Jarisch–Herxheimer reaction and usually lasts for less than 4 hours, though it may continue for a day (CDC 2012b).

People can contract TBRF multiple times (CDC 2012b). However, patients who have had TBRF in the past develop milder symptoms of the relapsed disease than other patients (Goubau 1984).

14.6 WHAT PEOPLE CAN DO TO REDUCE THEIR RISK OF CONTRACTING TBRF

There is no vaccine against TBRF for use by either humans or animals, but the risk of contracting TBRF can be reduced by rodent-proofing homes and cabins.

Using acaricides has been shown to reduce tick numbers in buildings as well as the risk of contracting TBRF (Talbert et al. 1998). Rodent numbers inside buildings can also be reduced by using mouse or rat traps, but rodent numbers will quickly increase as long as they have access to food. People should keep all food—especially grains and foods made from them (e.g., bread, cookies, pasta)—stored in rodent-proof containers, which can also lead to reduced numbers of small mammals inside buildings. The presence of chipmunks and squirrels indoors should be discouraged. Over the short term, reducing rodent numbers inside a building may make the problem worse by increasing the willingness of *Ornithodoros* ticks to bite people. For this reason, efforts to reduce small mammal numbers in buildings infected with *Ornithodoros* ticks should be combined with acaricide applications. Although most patients who contracted TBRF had stayed in tick-infected cabins and caves, gloves should be worn when handling dead squirrels and chipmunks; one person in California contracted TBRF after getting blood from an infected animal on his hands. Because *Ornithodoros* ticks feed mainly at night, people are at minimal risk of being bitten during the day. It is not known if wearing long pants and long-sleeved shirts, tucking pants into socks, and spraying bare skin and clothes with an insect repellent reduce the risk of a sleeping person being bitten by an *Ornithodoros* tick, but these practices should help.

In Africa and Eurasia where livestock and chickens are reservoir hosts for *Ornithodoros* ticks and *Borrelia*, most people contract TBRF from sleeping at home (Figure 14.9). Often the patient lives in a house where livestock are kept nearby and

FIGURE 14.9 Part of the village of Mitwaba in southern Zaire where there is a high incidence rate of TBRF. (From Dupont, H.T. et al., *Clin. Infect. Dis.*, 25, 139, 1997; used with permission from Didier Raoult.)

the house is infected with *Ornithodoros* ticks. In such cases, the incidence of TBRF can be reduced by housing animals away from homes, taking measures to rodent-proof homes, and applying an acaricide—measures which unfortunately may be difficult to achieve.

14.7 ERADICATING TBRF FROM A COUNTRY

TBRF could be eradicated by either eliminating *Ornithodoros* ticks or eliminating *Borrelia* from these populations. Either might be possible given the clustered nature of TBRF. However, given the limited resources of many countries in Africa and Asia, such is unlikely in the immediate future.

LITERATURE CITED

Assous, M. V. and A. Wilamowski. 2009. Relapsing fever borreliosis in Eurasia—forgotten but certainly not gone! *Clinical Microbiology and Infection* 15:407–414.

Boyer, K. M., R. S. Munford, G. O. Maupin, C. P. Pattison et al. 1977. Tick-borne relapsing fever: An interstate outbreak originating at Grand Canyon National Park. *American Journal of Epidemiology* 105:469–479.

CDC. 2003. Tickborne relapsing fever outbreak after a family gathering—New Mexico, August 2002. *Morbidity and Mortality Weekly Report* 52(34):809–812.

CDC. 2012a. Louse-borne relapsing fever (LBRF). http://www.cdc.gov/relapsing-fever/resources/louse.html (accessed July 14, 2012).

CDC. 2012b. Tick-borne relapsing fever (TBRF). http://www.cdc.gov/relapsing-fever/ (accessed December 20, 2013).

Cutler, S. J., A. Abdissa, and J. F. Trape. 2009. New concepts for the old challenge of African relapsing fever borreliosis. *Clinical Microbiology and Infection* 15:400–406.

Dupont, H. T., B. La Scola, R. Williams, and D. Raoult. 1997. A focus of tick-borne relapsing fever in southern Zaire. *Clinical Infectious Diseases* 25:139–144.

Dworkin, M. S., D. E. Anderson Jr., T. G. Schwan, P. C. Shoemaker et al. 1998. Tick-borne relapsing fever in the northwestern United States and southwestern Canada. *Clinical Infectious Diseases* 26:122–131.

Dworkin, M. S., P. C. Shoemaker, C. L. Fritz, M. E. Dowell, and D. E. Anderson Jr. 2002. The epidemiology of tick-borne relapsing fever in the United States. *American Journal of Tropical Medicine and Hygiene* 66:753–758.

Goubau, P. F. 1984. Relapsing fevers. A review. *Annales de la Societe Belge de Medecine Tropicale* 64:335–364.

Jongen, V. H., J. van Roosmalen, J. Tiems, J. Van Holten, and J. C. Wetsteyn. 1997. Tick-borne relapsing fever and pregnancy outcome in rural Tanzania. *Acta Obstetricia et Gynecologica Scandinavica* 76:834–838.

Lecompte, Y. and J. F. Trape. 2003. West African tick-borne relapsing fever. *Annales de Biologie Clinique (Paris)* 61:541–548.

Masoumi Asl, H., M. M. Goya, H. Vatandoost, S. M. Zahraei et al. 2009. The epidemiology of tick-borne relapsing fever in Iran during 1997–2009. *Travel Medicine and Infectious Disease* 7:160–164.

Nieto, N. C., M. B. Teglas, K. M. Stewart, T. Wasley, and P. L. Wolff. 2012. Detection of relapsing fever spirochetes (*Borrelia hermsii* and *Borrelia coriaceae*) in free-ranging mule deer (*Odocoileus hemionus*) from Nevada, United States. *Vector-Borne and Zoonotic Diseases* 12:99–105.

Parola, P., G. Diatta, C. Socolovschi, O. Mediannikov et al. 2011. Tick-borne relapsing fever borreliosis, rural Senegal. *Emerging Infectious Diseases* 17:883–885.

Roscoe, C. and T. Epperly. 2005. Tick-borne relapsing fever. *American Family Physician* 72:2039–2044.

Schwan, T. G., S. J. Raffel, M. E. Schrumpf, J. S. Gill, and J. Piesman. 2009. Characterization of a novel relapsing fever spirochete in the midgut, coxal fluid, and salivary glands of the bat tick *Carios kelleyi*. *Vector-Borne and Zoonotic Diseases* 9:643–647.

Schwan, T. G., S. J. Raffel, M. E. Schrumpf, L. S. Webster et al. 2009. Tick-borne relapsing fever and *Borrelia hermsii*, Los Angeles County, California, U.S. *Emerging Infectious Diseases* 15:1026–1031.

Spach, D. H., W. C. Liles, G. L. Campbell, R. E. Quick et al. 1993. Tick-borne diseases in the United States. *New England Journal of Medicine* 329:936–947.

Talbert, A., A. Nyange, and F. Molteni. 1998. Spraying tick-infected houses with lambda-cyhalothrin reduces the incidence of tick-borne relapsing fever in children under five years old. *Transactions of the Royal Society of Tropical Medicine and Hygiene* 92:251–253.

Trevejo, R. T., M. E. Schriefer, K. L. Gage, T. J. Safranek et al. 1998. An interstate outbreak of tick-borne relapsing fever among vacationers at a Rocky Mountain cabin. *American Journal of Tropical Medicine and Hygiene* 58:743–747.

Vial, L., G. Diatta, A. Tall, E. H. Ba et al. 2006. Incidence of tick-borne relapsing fever in west Africa: Longitudinal study. *Lancet* 368:37–43.

Section III

Rickettsial Diseases

15 Rocky Mountain Spotted Fever and Other Spotted Fevers

The first case of 'spotted fever,' 'black fever,' or 'blue fever' in the Bitter Root Valley in Montana occurred in 1873. At the time, there were but few white men in the valley... Since its first appearance probably 200 cases of the severe type have occurred, 70 to 80 percent of which have been fatal.

Wilson and Chowning (1904)

[Rocky Mountain Spotted Fever] is among the most virulent infections identified in human beings and its diagnosis often presents a dilemma for clinicians.

Dantas-Torres (2007)

15.1 INTRODUCTION AND HISTORY

Rocky Mountain spotted fever (RMSF) was first identified in Idaho during the 1800s by Edward Maxey, initially referred to as the spotted fever of Idaho. In 1904, Louis Wilson and William Chowning concluded that ticks of the genus *Dermacentor* were responsible for spreading the infection to humans. By 1910, Howard Ricketts first identified the pathogen and isolated it in both ticks and mammals. Later this pathogen was named *Rickettsia rickettsii* in honor of him. *Rickettsia* are small (0.3–2.0 μm), polymorphic coccobacilli (Figure 15.1), Gram negative, and die quickly outside of a cell. *Rickettsia* populations can double every 10–13 hours, relatively long for a disease that can spread throughout the body so rapidly. The speed of infection is accelerated because *Rickettsia* can pass through a cell membrane without killing the cell. This allows the pathogen to spread quickly from one cell to another (Dumler 1994). Other species of *Rickettsia* that occur in North America include *R. montanensis*, *R. rhipicephali*, and *R. parkeri*, but only the last species causes human illness (Parola et al. 2005).

Several species of the hard tick (Ixodidae) are hosts for *R. rickettsii*. Ticks can become infected in two ways: an infected female tick can produce infected eggs that develop into infected nymphs, larva, and adults or a tick can become infected by feeding on an infected mammal. The tick most responsible for transmitting *R. rickettsii* to people in the eastern United States and eastern Canada is the American dog tick (*D. variabilis*; Figure 15.2), while the Rocky Mountain wood tick (*D. andersoni*; Figure 15.2) is the primary vector in the western United States and western Canada. The brown dog tick (*Rhipicephalus sanguineus*) transmits the pathogen to

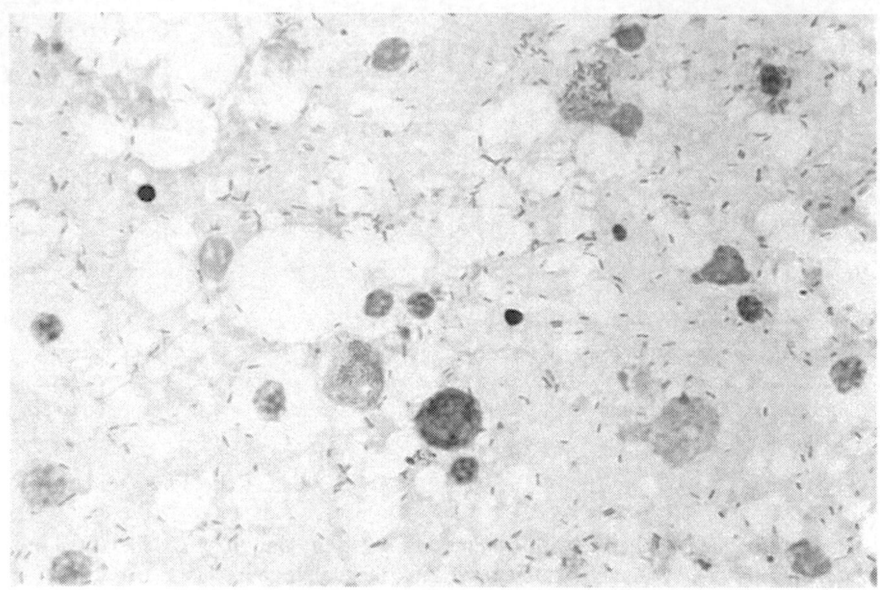

FIGURE 15.1 Photo showing a smear from a yolk sac that is infected with *R. rickettsii*, which can be seen in the photo as small rods. (Courtesy of the CDC.)

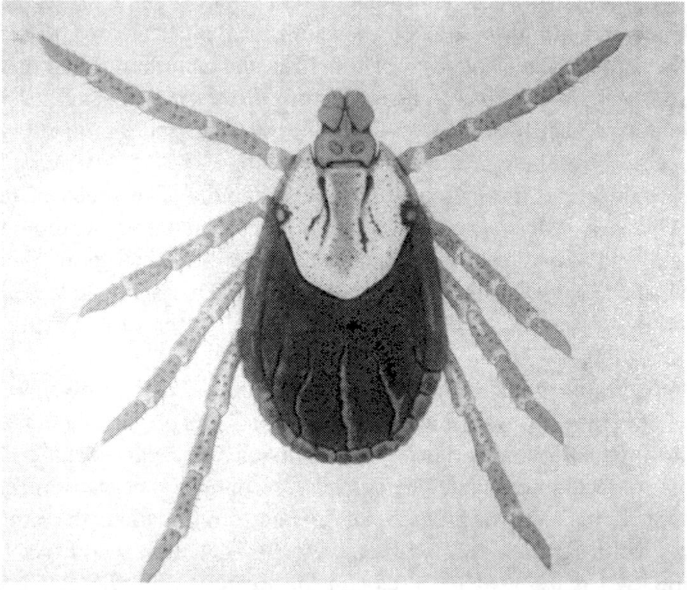

FIGURE 15.2 Photo of the American dog tick that is a primary vector of RMSF in North America. (Courtesy of Tom Schwan and the CDC.)

people in Mexico, and the cayenne tick (*Amblyomma cajennense*) is a common vector in Central America, South America, and parts of Texas. Several North American ticks are vectors for *R. rickettsii* including *D. parumapertus*, *D. occidentalis*, *Haemaphysalis leporispalustris*, *A. americanus*, *Ixodes scapularis*, *I. pacificus*, *I. dentatus*, *I. brunneus*, *I. texanus*, and *I. cookei* (McDade and Newhouse 1986).

RMSF is confined to the Western Hemisphere and occurs in Canada, United States, Mexico, Panama, Costa Rica, Columbia, Brazil, and Argentina. The disease has been reported from all of the lower 48 states except Vermont and Maine. Despite its name, most cases of RMSF do not occur in the Rocky Mountains. Instead most cases (56%) occur in just five states in the middle of the United States: Arkansas, Oklahoma, Tennessee, North Carolina, and South Carolina (Figure 15.3). Historically, there were between 250 and 1200 reported cases of RMSF annually in the United States, but RMSF and other spotted fevers have become more common in recent years. During 2011, there were 2800 reported cases—the first year the number exceeded 2000 in the United States. The annual incidence rate in the United States was two cases per one million people from 1997 through 2002 but climbed to nine cases per million people during 2011, the highest incidence rate of RMSF and other spotted fevers since records began during the 1920s (CDC 2010, 2013). Annually, there were 5–39 reported fatalities from RMSF in the United States from 1983 to 1998 and perhaps an additional 400 deaths during this same 15-year period that were not reported to the CDC (Dantas-Torres 2007). However, the mortality rate has been declining since the 1920s, and the case fatality rate was less than 0.5%

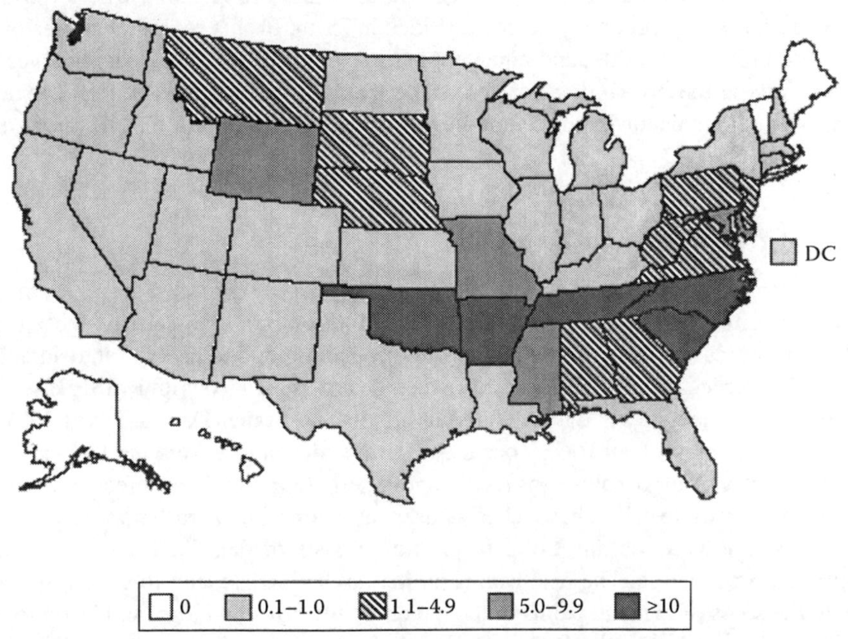

FIGURE 15.3 Annual incidence rate of RMSF from 1997 to 2002. (Courtesy of Alice Chapman and the CDC.)

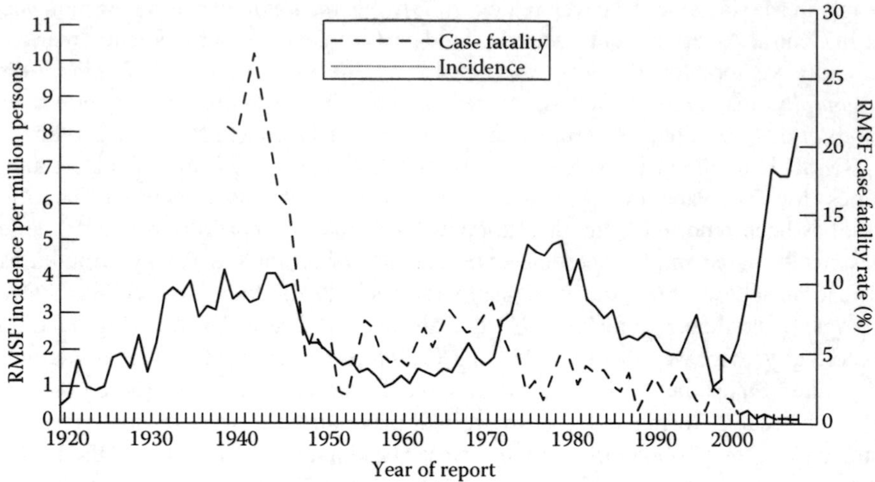

FIGURE 15.4 Incidence of RMSF in the United States from 1920 to 2008. (From CDC, Rocky Mountain spotted fever [RMSF], http://www.cdc.gov/rmsf/index.html, 2010.)

during 2008 (Figure 15.4). In Central America and South America, the disease is called Brazilian spotted fever, Minas Gerais exanthematic typhus, Sao Paulo exanthematic typhus, or febre maculosa.

Recently, RMSF has been reported in parts of eastern Arizona where the disease had not been previously detected (Sidebar 15.1). In this region, 90 cases have been identified as of 2009, and nine people have died (CDC 2010). The tick vector in this area is the brown dog tick, the same tick that transmits RMSF in Mexico. Almost all cases occurred in communities with a large population of free-roaming dogs (CDC 2010).

15.2 SYMPTOMS IN HUMANS

R. rickettsii bacteria gain entry into the human body through tick bites. In humans, *R. rickettsii* live and multiply within the cells that line blood vessels. As these cells die, small holes develop in the blood vessels, causing blood to leak out into adjacent tissue. This process causes the characteristic red rash associated with RMSF. If severe enough, this bleeding will damage internal organs and tissue (Dantas-Torres 2007).

The first symptoms of RMSF occur 5–10 days after a bite from an infected tick. Early symptoms are usually nonspecific and resemble many other infections or diseases: fever, muscle aches, headaches, nausea, and vomiting. A rash normally occurs 2–4 days following the onset of a fever and consists of flat, red freckles that turn white when pushed in. The rash begins on the ankles, wrists, and forearms, spreading to the soles of feet and palms of the hands, to the arms and legs, and lastly to the abdomen by the end of the first week. By then, the red spots develop small bumps on their surface (Figure 15.5). The rash occurs in 80% of adult patients and 90% of children (Spach et al. 1993, Chapman 2006) (Table 15.1).

SIDEBAR 15.1 AN UNEXPECTED CLUSTER OF PATIENTS WITH RMSF IN ARIZONA (DEMMA ET AL. 2005, CDC 2010)

RMSF is transmitted by the American dog tick and the Rocky Mountain wood tick, neither of which is common in Arizona due to its hot, dry weather. RMSF is rare in the state, with only three cases reported there from 1981 to 2001. Hence, medical physicians were surprised when 16 patients were identified from June 2002 through October 2004 with RMSF. All patients lived in the same rural area of eastern Arizona. In fact, seven patients came from one small community of 21,000 people, and nine came from another small community located 50 miles (80 km) away with a population of 10,000. Fifteen patients experienced a fever of 100.5°F (38°C) or higher, and the same number of patients had the characteristic RMSF rash (red spots). Fifteen patients required hospitalization and six spent time in the intensive care unit. Unfortunately, RMSF was not initially suspected for two patients, leading to a delay in antibiotic treatment. These two patients did not survive.

All patients had come into contact with tick-infected dogs, and four patients reported recent tick bites. A tick was found attached to one patient, but it was a brown dog tick and neither of the other two tick species that are the usual vectors RMSF in the United States. Over a thousand ticks were collected from the homes, yards, and neighborhoods of the patients, and all were brown dog ticks. *R. rickettsii* was cultured from a tick found attached to a dog owned by one of the patients, and *R. rickettsii* DNA was detected from three other ticks. Four dogs belonging to three different patients showed high titer levels to *R. rickettsii*.

In Mexico, the brown dog tick is a vector for *R. rickettsii*. However, this clustering of patients in Arizona is the first demonstration that the brown dog tick serves as a vector for RMSF in the United States. Furthermore, the annual incidence rate of RMSF among children and teenagers in this part of Arizona (1800 cases per million) is much higher than the national average (5.6 cases per million) in the United States. The results of Demma et al. (2005) are a cause for concern about the introduction of RMSF into areas where the disease has not previously been reported.

RMSF is a severe illness, and patients commonly require hospitalization. Up to 20% of untreated patients die from the disease, but mortality rates decrease to between 5% and 10% with proper medical care (Chapman 2006). Death occurs when RMSF results in kidney failure, a severe infection of the central nervous system, or respiratory collapse (Dumler 1994). RMSF is more likely to be fatal if antibiotics are not provided within the first 5 days of illness (Dantas-Torres 2007). For this reason, people who suspect they may have contracted RMSF should seek medical attention and inform their physician if they have been bitten by a tick. In severe cases, actual renal failure, multiple organ failure, and respiratory distress are often observed.

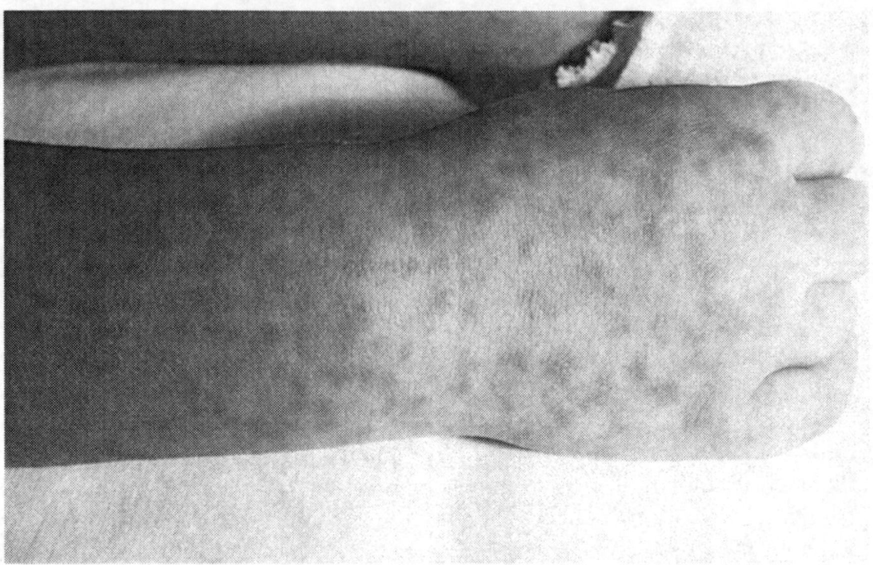

FIGURE 15.5 The typical rash caused by RMSF. This photo is of a child's hand and wrist. (From CDC, Rocky Mountain spotted fever [RMSF], http://www.cdc.gov/rmsf/index.html, 2010.)

TABLE 15.1
Common Symptoms of RMSF

Symptom	Frequency in Patients (%)
Abrupt fever	100
Malaise	95
Severe frontal headache	90
Rash	80
Muscle ache	80
Vomiting	60
Loss of appetite	60

Source: Data from Spach, D.H. et al., *N. Engl. J. Med.*, 329, 936, 1993.

Neurological problems are observed in about 40% of patients including lethargy, sensitivity to bright lights, amnesia, and bizarre behavior (Dantas-Torres 2007). RMSF patients are more likely to develop a severe case if they are over 60 years old, male, African-American, chronic abusers of alcohol, or having a genetic mutation that makes them deficient in glucose-6-phosphate-dehydrogenase. In patients with this deficiency, RMSF develops rapidly and can be fatal within 5 days of the onset of fever (Chapman 2006). Some patients with severe cases of RMSF experience long-term or permanent problems, including partial paralysis of the legs and feet, movement disorders, loss of bowel or bladder control, blindness, hearing loss, and speech disorders. Gangrene can develop in the fingers, arms, toes, and legs that is severe enough to require amputation.

15.3 *RICKETTSIA RICKETTSII* INFECTIONS IN ANIMALS

After being ingested by a tick, *R. rickettsii* infects the epithelial cells (i.e., cells that line the interior of hollow organs or the skin) of the tick's midgut. The pathogen multiplies there without any apparent damage to the tick, then migrates to other parts of the tick, including the ovaries and salivary glands. This results in infected ticks that produce infected eggs and also infect the mammals upon which they feed. Mammal hosts are vital for the life cycle of *R. rickettsii* because they provide the only mechanism by which an uninfected tick can acquire the pathogen. If not for mammal hosts, *R. rickettsii*-infected ticks would be eliminated over time because infected ticks have lower viability or produce fewer eggs than uninfected ticks (McDade and Newhouse 1986, Parola et al. 2005).

Antibodies against *R. rickettsii* have been identified in several birds (Table 15.2), and ticks infected with *R. rickettsii* have been found on birds (McDade and Newhouse 1986). Hence, birds have the potential to spread the pathogen into new areas.

Several mammal species serve as hosts for *R. rickettsii* or the larval and nymphal ticks that transmit it and provide a mechanism for amplification of the pathogen. *R. rickettsii* have been isolated from the cotton rat, white-footed mouse, meadow vole, woodland vole, golden-mantled ground squirrel, opossum, eastern cottontail rabbit, snowshoe hare, and yellow-pine chipmunk (McDade and Newhouse 1986). For most of these species, being infected by *R. rickettsii* produces either mild or no illness.

Domestic dogs are the main host of adult American dog ticks in eastern Canada and United States, but adult ticks will also feed on livestock (e.g., horses, cattle, sheep, and hogs) and several wildlife species (e.g., bears, raccoons, opossums, porcupines, coyotes, and foxes). Several species of mice, voles, and rabbits serve as hosts for larval and nymphal American dog ticks. The Rocky Mountain wood tick will feed on virtually any warm blooded animal. The primary tick vectors for *R. rickettsii* in Mexico—the brown dog tick and the cayenne tick—feed primarily on domestic dogs (McDade and Newhouse 1986).

R. rickettsii causes serious disease in domestic dogs and can be fatal. Dogs infected with the pathogen may have purple or red spots on the skin, swollen lymph nodes, swollen limbs from edema, conjunctivitis, and blood in their urine or feces. Infected dogs can be treated with antibiotics. They are not a reservoir host for *R. rickettsii*, and few ticks will acquire the pathogen by feeding on them. Still, dogs play a major role in the disease cycle of RMSF by serving as a host of the ticks and bringing infected ticks into homes. Dogs and their owners often contract RMSF at the same time (McDade and Newhouse 1986).

15.4 HOW HUMANS CONTRACT RMSF

Most people in North America contract RMSF from the bite of an infected tick, usually from the American dog tick or Rocky Mountain wood tick. Human-to-human

TABLE 15.2
Avian Species that Tested Positive for Antibodies against *R. rickettsii*

Symptom	Number of Individuals Tested	Number of Individuals with Antibodies
Snowy egret	13	1
Downy woodpecker	26	1
Eastern wood peewee	7	1
Horned lark	89	1
Blue jay	33	1
American crow	1	1
Catbird	55	1
Brown thrasher	11	1
American robin	23	1
Myrtle warbler	3	1
American redstart	2	1
House sparrow	50	1
Eastern meadowlark	1	1
Redwing blackbird	15	2
Common grackle	95	21
Brown-headed cowbird	54	9
Northern cardinal	18	2
Rufous-sided towhee	13	1
White-throated sparrow	3	1

Source: Data from McDade, J.E. and Newhouse, V.F., *Ann. Rev. Microbiol.*, 40, 287, 1986.

spread of RMSF is rare but can occur through blood transfusions. Humans can also unwittingly self-inoculate themselves with RMSF by removing an infected tick, usually from a dog, and rupturing the tick in the process. *R. rickettsii* from the ruptured tick may then gain entry into the human body when fingers carry the pathogen to moist membranes (e.g., by rubbing one's eyes) or through cut or abraded skin. For this reason, it is advisable to wear gloves when removing ticks and to wash one's hands promptly after tick removal (Dumler 1994).

15.5 MEDICAL TREATMENT

RMSF is often treated with doxycycline for at least 3 days after the fever has subsided but may be extended for severe or complicated cases. Doxycycline or other tetracyclines are usually not given to pregnant women because the drug may deform teeth or bone in the fetus. Instead, chloramphenicol is often administered to pregnant women with RMSF (Chapman 2006). If there are clinical signs involving the central nervous system, intravenous injections of chloramphenicol are sometimes administered (Spach et al. 1993).

RMSF is difficult to diagnose because the symptoms are similar to numerous other diseases. It is rare enough that many physicians would not initially consider it as a potential diagnosis. However, the decision to begin treatment should not be delayed until laboratory confirmation. Any patients with a fever or rash should be monitored carefully and considered for hospital admission and treatment. Patients should notify their physician if they have recently been bitten by a tick (Dantas-Torres 2007).

15.6 WHAT PEOPLE CAN DO TO REDUCE THEIR RISK OF CONTRACTING RMSF

Currently, there are no vaccines against RMSF that are approved by the U.S. Food and Drug Administration. Antibiotics are not recommended for people who have been bitten by a tick as a means to prevent RMSF because most ticks are not infected with *R. rickettsii*. Hence, the risk of infection from any single tick bite is minimal. Using antibiotics as a preventative treatment delays but does not prevent RMSF (Dumler 1994, Chapman 2006).

Although cases of RMSF have been reported in large cities, most people contract the disease in rural settings. People who live in wooded areas and who own a dog have an elevated risk of contracting the disease. Most RMSF patients (90%) contract the disease between April and September when ticks are active (Figure 15.6). In the United States and Canada, most cases of RMSF are widely dispersed and seldom are clustered together. This dispersed nature of RMSF makes it difficult for people or public health agencies to take steps to reduce the spread of the disease because it is hard to predict where the next case will occur. In 95% of cases, only one member

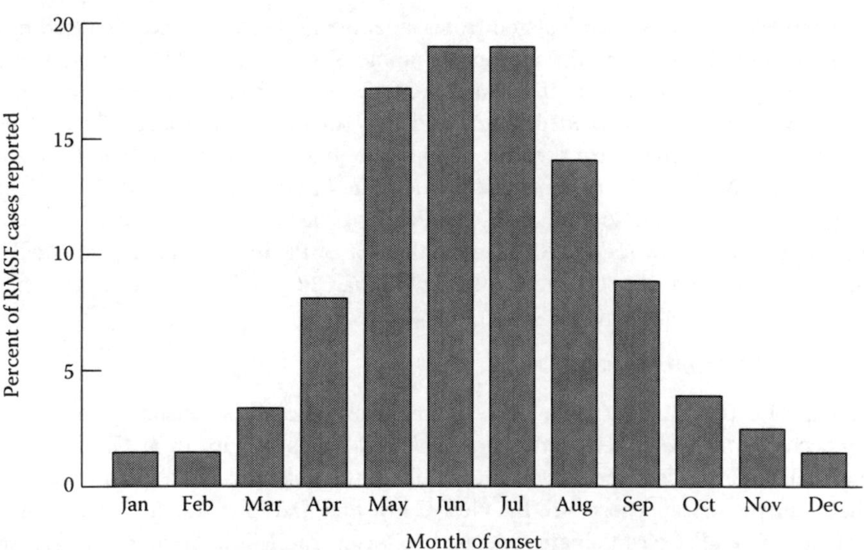

FIGURE 15.6 Percent of cases of RMSF in the United States by month. (From CDC, Rocky Mountain spotted fever [RMSF], http://www.cdc.gov/rmsf/index.html, 2010.)

of a family became infected, although domestic dogs and people living in the same house tend to contract RMSF at the same time. There is no vaccine to protect dogs from *R. rickettsii*, but the risk of a dog becoming infected can be reduced by keeping domestic dogs free of ticks. This often requires the use of tick collars, checking dogs daily, and dusting or spraying dogs with an acaricide on a regular basis. When RMSF is linked to free-ranging dogs, communities may be able to reduce the incidence of RMSF by reducing the number of wild dogs (Dantas-Torres 2007).

People can reduce their risk of contracting RMSF by checking themselves frequently for ticks when visiting tick-infested areas where RMSF is endemic. However, 40% of RMSF patients did not know they had been bitten by a tick (Dantas-Torres 2007).

15.7 OTHER SPOTTED FEVER DISEASES CAUSED BY *RICKETTSIA* IN THE UNITED STATES OR CANADA

There are many species of *Rickettsia* throughout the world, some of which are known to cause human diseases, while others do not. Human diseases caused by different species of *Rickettsia* in North America are discussed in the following based on information from the CDC (2012) and other sources. Many years may elapse between the discovery of a new *Rickettsia* and the recognition that it causes disease in humans. For example, *R. canadensis* has recently been isolated from *H. leporispalustris* ticks feeding on rabbits in Ontario, Canada. It is not known if it poses a risk to people. Thus, the list of *Rickettsia* species known to cause human disease may increase in the future.

15.7.1 364D Rickettsiosis

This spotted fever has been isolated from patients in California and is caused by a newly identified pathogen with a proposed name of *R. philippi*. The disease is currently referred to by the unusual name of 364D rickettsiosis. Symptoms are mild compared to other *Rickettsia* diseases and may include fever, headache, fatigue, myalgia, and an eschar (i.e., a scab or dead tissue that covers a wound) at the site of the tick bite. None of the patients had a red spotted rash (Shapiro et al. 2010). This pathogen is transmitted to humans by the Pacific Coast tick, which occurs throughout California. Wikswo et al. (2008) found that 8% of Pacific Coast ticks collected in southern California were infected with *R. philippi*.

15.7.2 Maculatum Infection

This spotted fever, also known as *R. parkeri* rickettsiosis, is caused by *R. parkeri* and occurs along the eastern and southern United States (Figure 15.7). Clinical signs may include fever, headache, eschar at the site of the tick bite, and a spotted red rash. The main vectors and hosts are the Gulf Coast tick and lone star tick. The animal hosts for the Gulf Coast tick are rodents. In Florida and Mississippi, *R. parkeri* was identified in 10%–40% of adult Gulf Coast ticks. *R. parkeri* has also been isolated in several South American countries (Paddock et al. 2004, 2010).

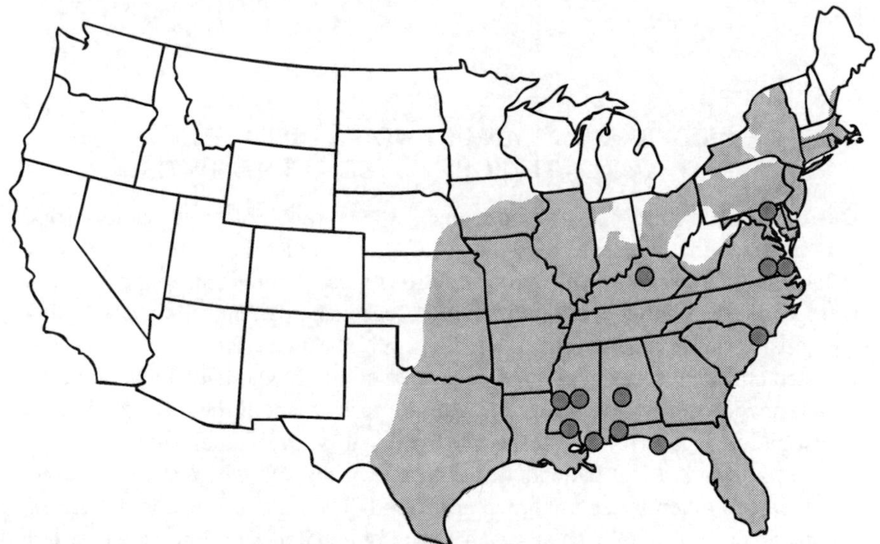

FIGURE 15.7 Range of lone star ticks, which completely overlaps the range of the Gulf Coast tick. *R. parkeri* has been isolated in both tick species. Dots mark the locations of confirmed and suspected cases of *R. parkeri* infection in humans. (From Cohen et al. 2009; used courtesy of the authors, the CDC, and *Emerging Infectious Diseases*.)

15.7.3 *Rickettsia felis*

During the 1990s, a new rickettsial species, *R. felis*, was identified and the first human case was reported in the United States. This pathogen causes an illness in humans that is similar to typhus and has been identified from patients in the United States, Mexico, Brazil, Spain, Germany, Tunisia, Thailand, Laos, and South Korea (Pérez-Osorio et al. 2008). In the United States, it has been isolated from patients in southern California and Texas (Civen and Ngo 2008). But from 1994 through 2008, there was only one confirmed case in the United States (Campbell et al. 2009).

The cat flea is the primary vector and reservoir host of *R. felis*, and cat fleas infected with this pathogen have been recovered from dogs, cat, rodents, opossums, and hedgehogs. In other parts of the world, *R. felis* has also been isolated from ticks and other flea species (Pérez-Osorio et al. 2008).

15.7.4 *Rickettsia amblyommii*

This *Rickettsia species* has been isolated from lone star ticks throughout the tick's range in the southern and midwestern parts of the United States. The pathogen has also been detected in ticks collected from forests in Brazil's western Amazon region. It is unclear if this species of *Rickettsia* causes human illness, but antibodies against it have been detected in a few patients that have had mild symptoms similar to other spotted fevers.

SIDEBAR 15.2 AN UNKNOWN DISEASE IN NEW YORK CITY (GREENBERG ET AL. 1947)

During the hot summer of 1946, doctors in the Borough of Queens, New York City, realized that what was thought to be a local epidemic of chickenpox was in fact another disease—with some odd differences. Unlike chickenpox, adults were as likely as children to contract the disease, and patients all lived in a single housing development; other people living in the area were unaffected. Also, infected children were not contagious unlike children who had chickenpox.

Alarmed, local doctors alerted health authorities about the disease, and an investigation began. About 2000 people lived in this development, and 125 (6%) had become sick. Most patients had a papule at the site where they appeared to have been bitten by an unknown arthropod. The papule was about 1 cm in width and developed into a black eschar. The lymph nodes draining the infected area were enlarged. A week after the papule first appeared, the patients experienced fever, chill, sweats, headache, and backache. This was followed a few days later by a rash of red bumps similar to other spotted fevers. The illness and rash lasted 7–10 days, after which all of the patients recovered, with no complications or deaths. The same strain of *Rickettsia* was isolated from the blood of two patients, and 80% of patients had antibodies specific to this *Rickettsia.* The disease was named rickettsialpox, and the pathogen named *R. akari*.

Health authorities began a search for the source of the *R. akari*. The housing development did not allow pets, and the few dogs located in the area were free of ticks and fleas. However, mice and their feces were seen in the basements and courtyards, and all of the interviewed tenants reported that rodents were abundant in the buildings. Blood-sucking mites, *L. sanguineus*, were discovered in the basement and on apartment walls. House mice were trapped in the buildings, and the same mites were found on them. No other rodent species were trapped, and this mite was the only ectoparasite (i.e., parasites found out the skin on an animal) found on house mice. *R. akari* was isolated from these mites and house mice. During the investigation, the disease was also discovered in an eleven-story building located in the Bronx, New York City, where 10 people had the same symptoms a few years earlier. Blood was drawn from two of the patients that had antibodies against *R. akari* antigens. Once again house mice and the blood-sucking mite (*L. sanguineus*) were found in the building. Three mice were trapped, and one had antibodies against *R. akari*. After the completion of the investigation, building landlords were ordered to clean up the basements, to eliminate food sources for the mice, and to take measures to kill the mice.

15.7.5 Rickettsialpox

This disease is caused by *R. akari*, a member of the spotted fever group of rickettsiae, but it differs from the other diseases because it is not spread by a tick, but rather by a mite, *Liponyssoides sanguineus*. House mice, brown rats, and black rats are hosts for both *R. akari* and the mite. All of these rodents are native to the Old World, and rickettsialpox has been reported in Korea, Ukraine, Croatia, Turkey, Egypt, and South Africa (Ozturk et al. 2003). These rodents are exotic species in North America and occur in close association with humans. Hence, it is not surprising that rickettsialpox occurs mainly in U.S. eastern cities, such as Boston, Cleveland, Pittsburg, West Hartford, Baltimore, and New York City (Sidebar 15.2). About half of the confirmed cases in the United States came from New York City (Krusell et al. 2002). Intravenous drug users living in inner cities have an increased risk for rickettsialpox. One study of 631 intravenous drug users found that 16% had antibodies against *R. akari* (Comer et al. 1999).

Rickettsialpox is mild compared to other spotted fevers; most patients experience fever, chills, headaches, muscle aches, swollen lymph glands, and a spotted red rash. Symptoms usually last 7–10 days, and most patients make a full recovery. Complications and fatalities are rare.

15.8 ERADICATING RMSF FROM A COUNTRY

The ubiquitous nature of dog ticks and other *Ixodes* ticks, as well as their mammal hosts, makes it almost impossible to eradicate *R. rickettsii* or their tick vectors from an area.

LITERATURE CITED

Campbell, J., M. E. Eremeeva, W. L. Nicholson, J. McQuiston et al. 2009. Outbreak of *Rickettsia typhi* infection—Austin, Texas, 2008. *Morbidity and Mortality Weekly Report* 58:1267–1270.

CDC. 2010. Rocky Mountain spotted fever (RMSF). http://www.cdc.gov/rmsf/index.html (accessed December 14, 2013).

CDC. 2012. Other tick-borne spotted fever rickettsial infection. http://www.cdc.gov/otherspottedfever/ (accessed August 20, 2012).

CDC. 2013. Summary of notifiable diseases—United States, 2011. *Morbidity and Mortality Weekly Report* 60(53):1–117.

Chapman, A. S. 2006. Diagnosis and management of tickborne rickettsial diseases: Rocky Mountain spotted fever, ehrlichioses, and anaplasmosis—United States. *MMWR Recommendations and Reports* 55(RR04):1–27.

Civen, R. and V. Ngo. 2008. Murine typhus: An unrecognized suburban vectorborne disease. *Clinical Infectious Diseases* 46:913–918.

Cohen, S. B., M. J. Yabsley, L. E. Garrison, J. D. Freye, et al. 2009. *Rickettsia parkeri* in *Amblyomma americanum* ticks, Tennessee and Georgia, USA. *Emerging Infectious Diseases* 15:1471–1473.

Comer, J. A., T. Tzianabos, C. Flynn, D. Vlahov, and J. E. Childs. 1999. Serologic evidence of rickettsialpox (*Rickettsia akari*) infection among intravenous drug users in inner-city Baltimore, Maryland. *American Journal of Tropical Medicine and Hygiene* 60:894–898.

Dantas-Torres, F. 2007. Rocky Mountain spotted fever. *Lancet Infectious Diseases* 7:724–732.

Demma, L. J., M. S. Traeger, W. L. Nicholson, C. D. Paddock et al. 2005. Rocky Mountain spotted fever from an unexpected tick vector in Arizona. *New England Journal of Medicine* 353:587–594.

Dumler, J. S. 1994. Rocky Mountain spotted fever. In: G. W. Beran, editor. *Handbook of Zoonoses*, 2nd edition. CRC Press, Boca Raton, FL, pp. 417–427.

Greenberg, M., O. J. Pellitteri, and W. L. Jellison. 1947. Rickettsialpox—A newly recognized rickettsial disease. *American Journal of Public Health* 37:860–868.

Krusell, A., J. A. Comer, and J. Sexton. 2002. Rickettsialpox in North Carolina: A case report. *Emerging Infectious Diseases* 8:727–728.

McDade, J. E. and V. F. Newhouse. 1986. Natural history of *Rickettsia rickettsii. Annual Review of Microbiology* 40:287–309.

Ozturk, M. K., T. Gunes, M. Kose, C. Coker, and S. Radulovic. 2003. Rickettsialpox in Turkey. *Emerging Infectious Diseases* 9:1498–1499.

Paddock, C. D., P.-E. Fournier, J. W. Sumner, J. Goddard et al. 2010. Isolation of *Rickettsia parkeri* and identification of a novel spotted fever group *Rickettsia* sp. from Gulf Coast ticks (*Amblyomma maculatum*) in the United States. *Applied and Environmental Microbiology* 76:2689–2696.

Paddock, C. D., J. W. Sumner, J. A. Comer, S. R. Zaki et al. 2004. *Rickettsia parkeri*: A newly recognized cause of spotted fever rickettsiosis in the United States. *Clinical Infectious Diseases* 38:805–811.

Parola, P., C. D. Paddock, and D. Raoult. 2005. Tick-borne rickettsioses around the world: Emerging diseases challenging old concepts. *Clinical Microbiology Reviews* 18:719–756.

Pérez-Osorio, C. E., J. E. Zavala-Velázquez, J. J. A. León, and J. E. Zavala-Castro. 2008. *Rickettsia felis* as emergent global threat for humans. *Emerging Infectious Diseases* 14:1019–1023.

Shapiro, M. R., C. L. Fritz, K. Tait, C. D. Paddock et al. 2010. *Rickettsia* 364D: A newly recognized cause of eschar-associated illness in California. *Clinical Infectious Diseases* 50:541–548.

Spach, D. H., W. C. Liles, G. L. Campbell, R. E. Quick et al. 1993. Tick-borne diseases in the United States. *New England Journal of Medicine* 329:936–947.

Wikswo, M. E., R. Hu, G. A. Dasch, L. Krueger et al. 2008. Detection and identification of spotted fever group rickettsiae in *Dermacentor* species from southern California. *Journal of Medical Entomology* 45:509–516.

Wilson, L. B. and W. M. Chowning. 1904. Studies in pyroplasmosis hominis. ("spotted fever" or "tick fever" of the Rocky Mountains). *Reviews of Infectious Diseases* 1:31–57.

16 Q Fever and Coxiellosis

The public health and military medical significance of Q fever unfolded during World War II after military physicians documented that Q fever had caused significant morbidity during military operations. Epidemics occurred among Allied and German troops operating in the Mediterranean countries, the Balkans, and southern Europe.

Williams and Sanchez (1994)

Wind in November, Q fever in December.

Tissot-Dupont et al. (2004)

16.1 INTRODUCTION AND HISTORY

During the early 1930s, meat workers in Queensland, Australia, complained of an illness that produced symptoms unlike any of the known diseases of the area. The unknown nature of this new disease led to the illness initially being called query fever or Q fever, until more was known about the disease and a more suitable name was found. However, the name stuck and is still used today. The bacterium responsible for the disease was first identified during 1935 in blood and urine collected from Australian patients. Two years later, a new rickettsial bacterium was isolated in Australia from the bandicoot tick (also called the slender opossum tick). At the same time but a world away, an unknown pathogen was isolated from *Dermacentor andersoni* ticks in Montana and named Nine Mile agent. These disparate events were linked when an unidentified doctor accidently pricked himself with a needle containing the pathogen from Montana and developed Q fever. This accident led to the discovery that the pathogen that caused Q fever in Australia, the rickettsial bacterium isolated from ticks in Australia, and the pathogen recovered in Montana were all the same; it was named *Coxiella burnetii* (McDade 1990).

C. burnetii is a small (0.2–1.0 μm) Gram-negative bacterium (Figure 16.1). It has two distinct morphological forms: large cells that are found inside hosts and small cells that are dormant and can survive for long periods of time outside of a host. The small cells are often referred to as spores because of their resilience, but they are not true spores. The spore-like cells can survive for weeks outside of a host and account for the pathogen's ability to spread via the air from one host to another as well as to humans. When these spore-like cells enter a host cell, they change into the large-cell form. These large cells can double in number every 8–12 hours. *C. burnetii* infections are concentrated in female reproductive tissue (mammary glands and uterus). Infected animals shed large quantities of the pathogen during partition through milk, vaginal fluids, placentas, and dead fetuses. For this reason, people often contract Q fever after infected animals give birth. Infected animals also shed *C. burnetii* in milk, nasal secretions, urine, and feces (Oyston and Davies 2011).

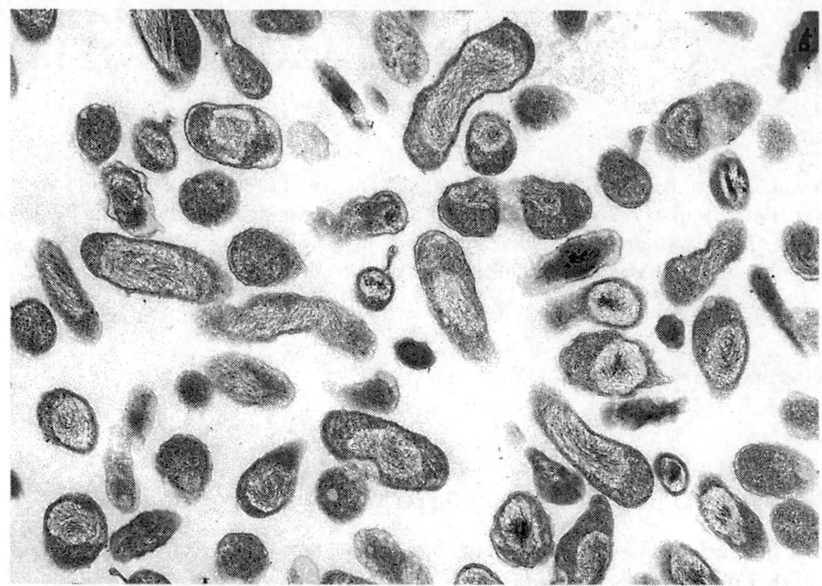

FIGURE 16.1 Microphotograph of *C. burnetii*. (Courtesy of Rocky Mountain Laboratories, National Institute of Allergy and Infectious Diseases, National Institutes of Health.)

Q fever occurs in humans around the world with the exception of New Zealand. There is an average of 27–100 confirmed cases annually in Germany, 67–170 in England and Wales, 30–90 in Switzerland, 50 in Spain, 202–860 in Australia, 7–46 in Japan, and 25 in Canada's Nova Scotia (Maurin and Raoult 1999). From 2000 through 2004, 436 cases were reported in the United States (Figure 16.2). These reported cases were concentrated in California (181), Colorado (67), and Oregon (23). Numbers of Q fever cases have increased in the United States since 1999 when Q fever became a notifiable disease (Figure 16.3). During 2011, there were 134 reported cases of Q fever in the United States (110 acute cases and 24 chronic cases). These numbers included patients from two outbreaks of Q fever during 2011; five people were infected in Michigan from drinking unpasteurized milk, and 20 people in Montana and Washington were infected after exposure to goats at a farm. Case fatality rate for Q fever is less than 2% in the United States. In the United States, people contract Q fever year round, but the risk of infection increases during spring and early summer with the birthing season for livestock and an increase in the amount of time people spend outdoors (Figure 16.4; McQuiston et al. 2006, CDC 2011, 2013). The annual incidence rate of Q fever is two cases per one million people in England and Wales, 38 in Australia, and 500 in France (Houwers and Richardus 1987, Gardon et al. 2001, McQuiston et al. 2006).

In the United States, 3% of the U.S. population and 10%–20% of people in high-risk occupations (e.g., farmers and veterinarians) have antibodies against *C. burnetii* indicating past exposure to the pathogen (CDC 2011). These percentages are much higher than expected given that there were less than 200 cases reported annually in the United States. Perhaps most people infected by *C. burnetii*

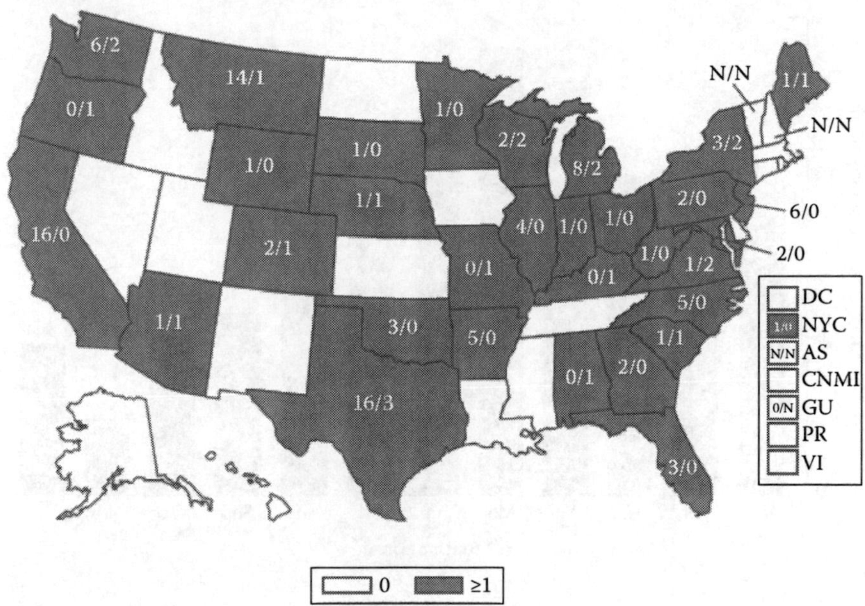

FIGURE 16.2 The darkened states are those where someone was diagnosed with Q fever during 2011. The first number in these states gives the number of acute cases and the second number shows the number of chronic cases. (From CDC, *Morbid. Mortal. Wkly. Rep.*, 60(53), 1, 2013.)

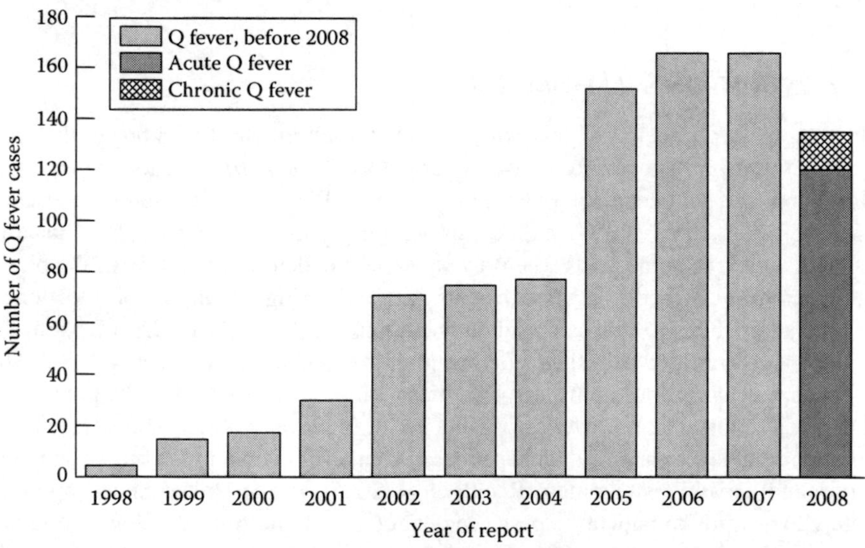

FIGURE 16.3 Number of Q fever cases in the United States reported annually to the CDC from 1998 to 2008. (From CDC, *Morbid. Mortal. Wkly. Rep.*, 60(53), 1, 2011.)

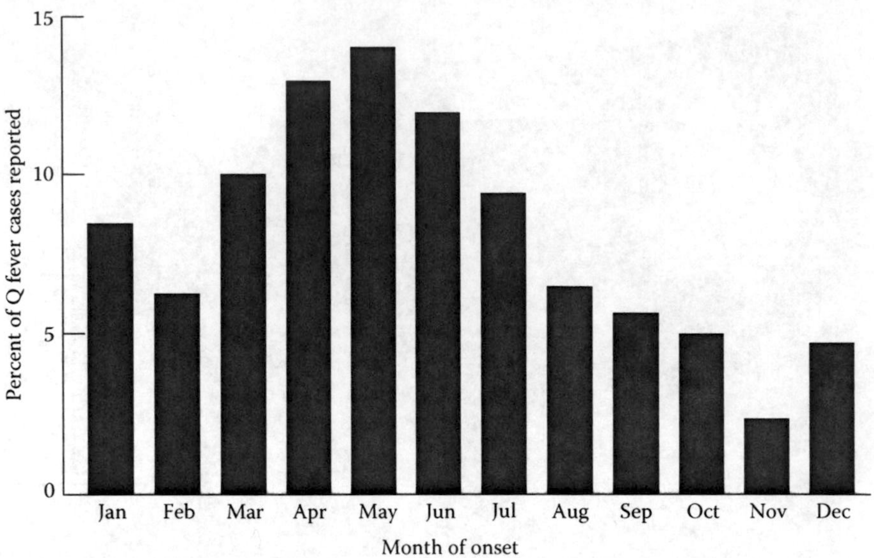

FIGURE 16.4 Percent of Q fever cases in the United States that occurred during each month of the year based on when the symptoms began. (From CDC, Q fever, http://www.cdc.gov/qfever/index.html, 2011.)

in the United States are asymptomatic, do not seek medical attention, or are not diagnosed as having Q fever. Q fever is much more common in other countries than in the United States. In Crete, 49% of the population is seropositive for *C. burnetii* (Vranakis et al. 2012).

16.2 SYMPTOMS IN HUMANS

Most people are infected with *C. burnetii* when they inhale the pathogen that was shed by livestock or a pet. Response of people to *C. burnetii* infections is variable. Most people do not become ill or have only mild flu-like illness, but some experience a fever of up to 104°F (40°C), chills, muscle pains, and malaise. Symptoms usually last 1–2 weeks. In some areas where *C. burnetii* is endemic, up to 11% of the population can have antibodies against the pathogen indicating a high rate of infection.

Inhalation of *C. burnetii* can result in pneumonia (Table 16.1). In Great Britain, 1% of all patients admitted to hospitals for pneumonia had Q fever (Mann et al. 1986). The infection may spread to the liver, spleen, heart, and other organs, but the prevalence of different clinical signs differ across the world. In Nova Scotia, pneumonia is more common in Q fever patients than hepatitis; in southern Spain and France, pneumonia is rare and hepatitis is common (Raoult et al. 2005). In the United States, approximately two-thirds of patients experience hepatitis, and one-third develop respiratory problems (McQuiston et al. 2006). Rare but serious complications of Q fever include infections of the heart and central nervous system. The mortality rate ranges from 0.5% to 1.5%, with death resulting from heart problems or respiratory distress.

TABLE 16.1
Occurrence of Different Symptoms of Q Fever among 1009 Patients during the 2007–2009 Epidemic in the Netherlands and Worldwide

	Occurrence in Patients (%)	
Symptom	**Netherlands**	**Worldwide**
Fever	92	88–100
Fatigue	78	97–100
Headache	69	68–98
Night sweats	67	31–98
Shortness of breath	61	n.a.
Muscle pain	57	47–69
Cough	48	24–90
Nausea or vomiting	32	22–49
Neurological problems	20	n.a.
Skin rash	12	5–21

Sources: Maurin, M. and Raoult, D., *Clin. Microbiol. Rev.*, 12, 518, 1999; Dijkstra, F. et al., *FEMS Immunol. Med. Microbiol.*, 64, 3, 2012.

Note: n.a., not available.

Months or years after the initial (i.e., acute) infection, Q fever can develop into a chronic illness with endocarditis, vascular infection, hepatitis, bone infection, or arthritis. Heart problems caused by Q fever are rare in the United States but more common elsewhere, especially in Nova Scotia. Six percent of patients in Spain developed heart complications as did 9% of patients in the United Kingdom and 3% in the Netherlands. People, who already have valvular heart disease, have an elevated risk of developing endocarditis (Williams and Sanchez 1994, CDC 2011).

Approximately 15% of Q fever patients experience chronic fatigue, night sweats, muscle aches, breathlessness, blurred vision, and trembling, which may continue for more than a decade after the initial infection (Marmion et al. 1996, Ayres et al. 1998). This condition is called post-Q fever fatigue syndrome (QFS). There are two hypotheses to explain what causes QFS. First, QFS may result because antibodies produced by the immune system to fight off an acute infection of Q fever may impair the body's ability to regulate the production of cytokine (Oyston and Davies 2011). Second, QFS may result from a persistent *C. burnetii* infection (Harris et al. 2000).

16.3 *COXIELLA BURNETII* INFECTIONS IN ANIMALS

Almost all mammals are susceptible to infection by *C. burnetii*, including marine mammals, such as Steller sea lions, Pacific harbor seals, and harbor porpoises. Rabbits and hare are more likely to be infected than other wild mammals. Animals can become infected by consuming placentas or milk from infected livestock, being bitten by an infected tick, or inhaling the pathogen. The pathogen has been detected in golden-crowned sparrows, house sparrows, coots, blackbirds, crows, pigeons, American robins, and mallards (Duncan et al. 2012, Kersh et al. 2012).

In Mendocino County, California, antibodies against *C. burnetii* were detected in 17 of 21 wild mammal species, and the pathogen was isolated from 9 of 20 species of wild mammals (Enright et al. 1971). The highest seropositive rates were found in coyotes (78%), foxes (55%), and brush rabbits (53%); mule deer had a rate of 22%. In the same area, seropositive rates among wild birds (38%) collected around a dairy farm were more than twice as high as birds collected elsewhere (13%). Seropositive rates were over 50% among carrion-eating birds (e.g., crows, raven, and vultures) and 20% among birds that consumed plants or insects (Enright et al. 1971).

Among wildlife species, *C. burnetii* infections are generally not harmful except in species that are accidental hosts. In the latter species, the disease caused by *C. burnetii* is called coxiellosis. It may result in an increase in the rate of abortions, lower birth weights, and reduced fertility. Coxiellosis in dogs and cats can cause fever, lethargy, loss of appetite, and lack of coordination. Infection may lead to reduced fertility or early death of kittens or puppies (Maurin and Raoult 1999).

Cattle, sheep, and goats serve as reservoir hosts for *C. burnetii.* In livestock, there is no specific symptom that is characteristic of Q fever. When a herd of large mammals is initially infected with *C. burnetii*, an increase in abortions and stillbirths may result, but once the herd has acquired some immunity, there is little reduction in herd fertility (Williams and Sanchez 1994). In infected herds of cattle, sheep, and goats, economic losses from coxiellosis range from 5% to 15%, owing to higher abortion rates and a loss in fertility (Williams and Sanchez 1994).

Most livestock become infected after inhaling dust contaminated with *C. burnetii.* Once inhaled, cows can shed the pathogen for months in their milk. Months after sheep have been infected, the pathogen can be isolated from their placentas, vaginal mucus, and feces (Marrie 2011).

Many arthropods, including hard ticks, become infected with *C. burnetii* by feeding on an infected host. Once infected, ticks shed the pathogen through their feces. Arthropods are not essential for the maintenance of *C. burnetii* within a herd of livestock. However, ticks play an important role in transmitting *C. burnetii* among wildlife species that do not live in large social groups.

Although most people contract Q fever primarily from *C. burnetii* that have been shed by cattle, sheep, and goats, wildlife species serve as important host reservoirs and sometimes are responsible for infecting livestock herds. For example, black rats, brown rat, and European roe deer are believed to be the reservoir hosts of *C. burnetii* in the Netherlands. They can spread the pathogen to sheep and goats, which in turn amplified the pathogen and spread it to people (Sidebar 16.1). In the Netherlands, 23% of roe deer and 50% of rats collected at goat farms were seropositive for *C. burnetii* antibodies (Reusken et al. 2011, Rijks et al. 2011, Dijkstra et al. 2012).

16.4 HOW HUMANS CONTRACT Q FEVER

Most people acquire Q fever by inhaling *C. burnetii* in aerosols and contaminated dust, which had been released by infected pets or livestock, especially cattle, sheep,

SIDEBAR 16.1 THE 2007–2010 Q FEVER EPIDEMIC IN THE NETHERLANDS (ROEST ET AL. 2011, DIJKSTRA ET AL. 2012)

The Netherlands is a densely settled country of 1272 people/mile^2 (491/km^2) and the province of Noord-Brabant also has high densities of cattle (324/mile^2 or 125/km^2), goats (60/mile^2 or 23/km^2), and sheep (52/mile^2 or 20/km^2). Prior to 2007, only 10–20 patients were diagnosed annually with Q fever in the Netherlands, but a large epidemic of Q fever broke out from 2007 through 2010, primarily in Noord-Brabant where 77% of the patients lived (Figure 16.5). There were 193 cases during 2007, 929 during 2008, and 2142 during 2009. Most patients (61%) were diagnosed with pneumonia, 3% with endocarditis, and 0.4% with hepatitis.

Health authorities were caught off guard by the size of this epidemic. However, they soon learned that 28 dairy goat farms and 2 dairy sheep farms had experienced a wave of abortions caused by *C. burnetii* and that most were in same area as the epidemic. Air-borne dispersal of the pathogen from these goat and sheep farms was identified as the source of the epidemic. In response, control measures were implemented at these farms that included vaccinating herds, prohibiting the movement of animals from infected farms, banning visitors from infected farms, and culling pregnant sheep and goats. A mandatory monitoring program of bulk tank milk on dairy goat and dairy farms was implemented, and all pregnant animals on infected farms were culled (Table 16.2). These measures, combined with the increase in the proportion of people that had antibodies against *C. burnetii* (as a result of infection in the preceding years), resulted in a sharp decline in the number of Q fever cases during 2010 (Figure 16.5).

and goats. Some people become infected by drinking water or unpasteurized milk contaminated with *C. burnetii* or by consuming cheese or other milk products made from the contaminated milk. Ranchers, sheep herders, dairy farmers, veterinarians, and slaughterhouse workers are 10 times more likely to be infected with *C. burnetii* than the general population (McQuiston and Childs 2002). In Switzerland, 415 people (21% of people living in the Val de Bagnes) developed Q fever after infected sheep were herded along the road where the people lived (Dupuis et al. 1987). There was an outbreak at a French kindergarten when nearby pastures were fertilized using cattle manure infected with *C. burnetii.* In southern France, the incidence of Q fever is correlated with sheep densities. The incidence of Q fever also increases following a period of windy weather presumably because high wind increases the amount of the pathogen in the air (Tissot-Dupont et al. 2004). Over 3000 people contracted Q fever during an epidemic in the Netherlands that involved dairy goats and dairy sheep (Figure 16.6). In French Guiana where wildlife are the host reservoir for the pathogen, the incidence of Q fever increased after a period of rainy weather (Gardon et al. 2001).

Ticks, fleas, lice, mites, flies, and cockroaches can transmit *C. burnetii*, but few people become infected through the bites of arthropods. Instead, the feces of infected

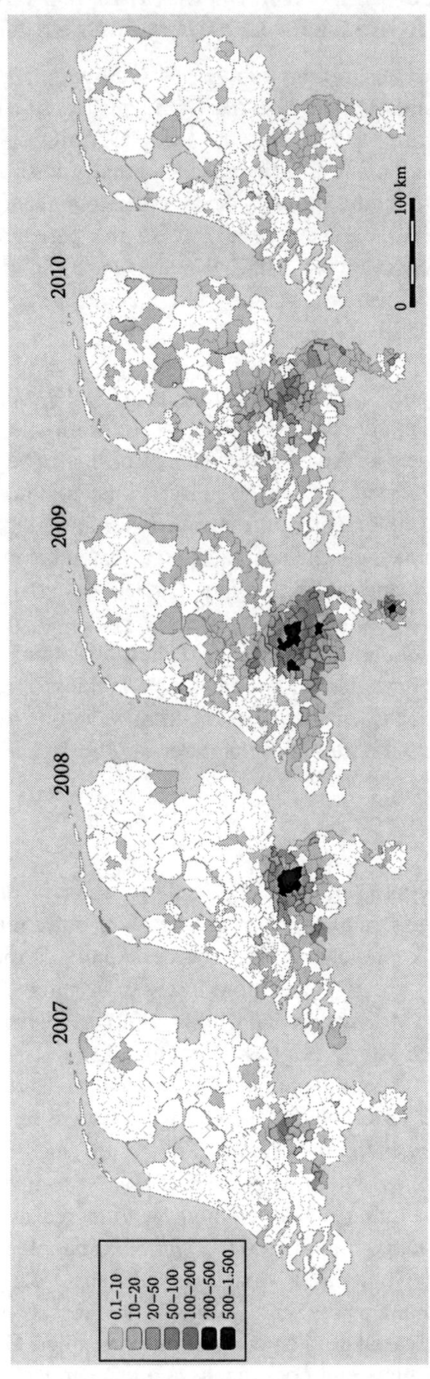

FIGURE 16.5 Map of the Netherlands showing the reported incidence per 100,000 people of acute Q fever cases during the 2007–2010 epidemic by municipality. (Courtesy of W. van der Hoek and the National Institute for Public Health and the Environment, the Netherlands.)

TABLE 16.2
Animal Husbandry Legislation Passed in the Netherlands in Response to the Q Fever Epidemic

Date	Legislation
June 2008	Dairy goat and sheep farmers must report abortion outbreaks (+5% abortion rates).
June 2008	Prohibition against visitors and removing manure from infected farms.
October 2008	Farmers compensated for purchasing Q fever vaccine.
February 2009	Farmers with more than 50 dairy goats or sheep must implement hygienic measures.
April 2009	Mandatory vaccination of dairy sheep and goats on farms with more than 50 head.
October 2009	Mandatory milk monitoring for Q fever every 2 months.
October 2009	Ban on moving dairy sheep or goats from infected farm.
December 2009	Ban on reproduction of goats.
	Ban on increasing the number of dairy sheep or goats on a farm.
	Mandatory milk monitoring for Q fever every 2 weeks.
	Ban on removing manure from a stable within 30 days of lambing season.
	Mandatory killing of all pregnant goats and sheep at infected farms.
January 2010	Mandatory vaccination of all dairy sheep and goats.

Source: Roest, H.I.J. et al., *Epidemiol. Infect.*, 139, 1, 2011.

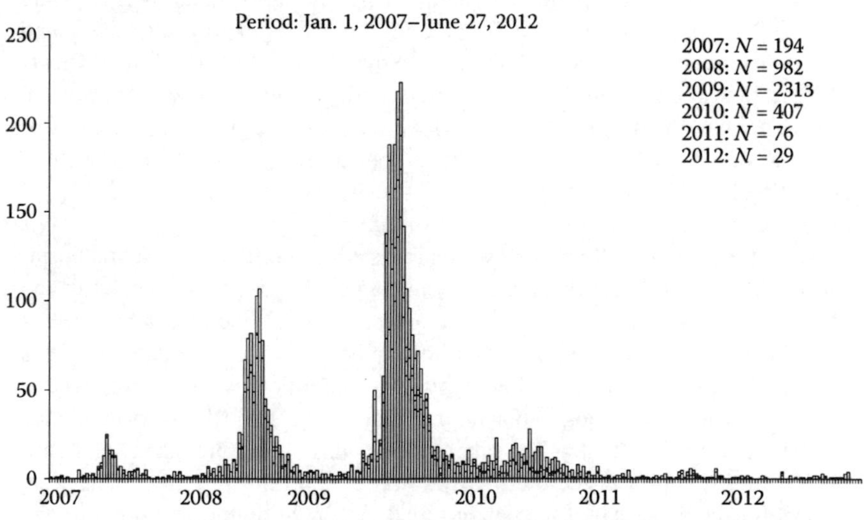

FIGURE 16.6 Number of reported acute Q fever patients in the Netherlands during the 2007–2010 epidemic and during the years following it. (Courtesy of W. van der Hoek and the National Institute for Public Health and the Environment, the Netherlands.)

arthropods contain high densities of *C. burnetii*, and these are dispersed by the wind where they are inhaled by people. People have also become infected from exposure to pigeons, although it is not clear if *C. burnetii* was transmitted to humans through ticks or inhalation of contaminated pigeon feces.

Most Q fever patients in Europe became infected from *C. burnetii*, which had been shed by sheep and goats. In the United States, cattle are an important source of *C. burnetii*, as well as sheep and goats. The situation was very different in Nova Scotia and Quebec where only 1% of Q fever patients had been exposed to an infected cow. Instead, most Q fever patients had been exposed to infected dogs, cats, white-tailed deer, or snowshoe hares (Sidebar 16.2). In Quebec, 15% of fur trappers were seropositive for *C. burnetii*, suggesting that fur-bearing mammals can serve as a host or vector (Lévesque et al. 1995). Wildlife (perhaps the rodent *Proechimys* spp. or the opossum *Philander opossum*) are believed to be the primary source in French Guiana (Gardon et al. 2001) while domestic cats are a significant source of *C. burnetii* responsible for Q fever outbreaks in humans in Japan (Porter et al. 2011).

Among livestock, coxiellosis can be treated by giving an oral dose of tetracycline for 2–4 weeks. Its spread within an infected herd can be reduced by isolating

SIDEBAR 16.2 Q FEVER IN NOVA SCOTIA, CANADA (MARRIE ET AL. 1988, BUHARIWALLA ET AL. 1996, MAURIN AND RAOULT 1999)

Q fever was first detected in Nova Scotia during 1981 when a pneumonia patient was diagnosed with Q fever. From 1980 to 1987, there were 174 confirmed cases with male patients outnumbering female patients by 2:1. While most Q fever patients in Europe became infected after exposure to sheep and goats, the situation was very different in Nova Scotia. Many Q fever patients in Nova Scotia had been exposed to infected dogs and cats and often present when a pet gave birth. Other risk factors were exposure to wild snowshoe hares or white-tailed deer. At least one person contracted Q fever after skinning a snowshoe hare. Tests revealed that 50% of the snowshoe hares and 24% of domestic cats are seropositive for *C. burnetii.*

As one example, a 40-year-old woman was admitted to a Nova Scotia hospital complaining of headache, nausea, vomiting and myalgia. A chest x-ray and other tests revealed pneumonia and the presence of antibodies against *C. burnetii.* Within the next week, both her husband and son had similar symptoms and also were diagnosed with pneumonia. An epidemiology study revealed that the family had a female dog (rabbit hound) that had given birth 8 days before the woman became ill. The dog had caught rabbits during her pregnancy and gave birth to four puppies inside the family's house, and all family members were present for the birth. All of the puppies died within 24 hours. The son also had antibodies against *C. burnetii* as did the dog. The husband was not tested for antibodies. All three patients were treated with clarithromycin and recovered.

pregnant animals, burning or burying placentas and other tissues associated with birth, and providing uninfected animals with tetracycline daily (Merck 2010).

16.5 MEDICAL TREATMENT

Acute cases of Q fever are treated by giving patients doxycycline for 2 or 3 weeks. Antibiotics are most successful if given within the first 3 days of illness. For this reason, antibiotic therapy is recommended for patients who are suspected of having Q fever or may have been exposed to *C. burnetii*; treatment should not be delayed while waiting for laboratory tests to confirm the diagnosis (Jay-Russell et al. 2002, CDC 2011). Chronic cases of Q fever require long periods of antibiotic treatment; Q fever can reappear in patients after cessation of the antibiotics treatment. For chronic cases, a combination of doxycycline and hydroxyl chloroquine or doxycycline and ofloxacin may be recommended (Maurin and Raoult 1999, Oyston and Davies 2011). Heart valves damaged by *C. burnetii* may need to be replaced. Pregnant women diagnosed with acute Q fever can be treated with co-trimoxazole to reduce the risk to the fetus (CDC 2011).

16.6 WHAT PEOPLE CAN DO TO REDUCE THEIR RISK OF CONTRACTING Q FEVER

Despite much research, there has been only partial success in developing a human vaccine that is effective against *C. burnetii.* Russia developed a vaccine using a live attenuated strain during the 1960s and has used it extensively since then. This vaccine has not been adopted in Europe, Canada, or the United States due to safety concerns over the use of live bacteria in a vaccine. Australia developed a vaccine called Q-Vax using formalin-inactivated *C. burnetii.* This vaccine has been effective in protecting people from Q fever but common side effects include swelling and tenderness at the inoculation site, headaches, and flu-like symptoms. Giving a Q fever vaccine to someone who already has antibodies against *C. burnetii* can cause severe reactions. Fortunately, this risk can be reduced by testing people for these antibodies prior to administering the vaccine (Oyston and Davies 2011). During the Q fever epidemic in the Netherlands, the country's health council recommended vaccinating people who had cardiac valve problems or specific cardiac or vascular disorders, owing to their risk of developing serious complications if they contract Q fever. During 2011, Q-Vax was given to 1366 people in the Netherlands (Dijkstra et al. 2012). This vaccine is not commercially available in the United States.

Livestock vaccines have been developed that provide some level of protection against *C. burnetii.* Vaccinating disease-free animals usually prevents them from suffering a loss of fertility or shedding *C. burnetii* if they become infected. Vaccinations have no effect on animals that are already infected and will not stop them from continuing to shed the pathogen (Astobiza et al. 2010, Hogerwerf et al. 2011). A widespread program to vaccinate cattle in Slovakia reduced the local incidence of Q fever but did not eliminate it (Kovacova and Kazar 2002). A livestock vaccine against Q fever is not available in the United States.

C. burnetii spore-like cells are resistant to desiccation or temperature extremes. They also are immune to many disinfectants that are lethal to other bacteria. UK health authorities report that it is difficult to decontaminate surfaces of *C. burnetii* and suggest using a solution containing 2% formaldehyde, 1% Lysol, 5% hydrogen peroxide, or 70% ethanol (Oyston and Davies 2011).

Most people acquire Q fever from infected pets and livestock. Pet owners are at a high risk when exposed to newborn or stillborn animals. High-risk groups also include people who spend time around livestock. In Germany, 38% of veterinarians were seropositive for *C. burnetii* antigens (Bernard et al. 2012). Men are twice as likely as women to contract Q fever even when there is no difference in exposure rates between men and women (Raoult et al. 2005). Children are less likely to contract Q fever than adults, and adults are more likely to contract the disease as they become older (Figure 16.7).

The best approach for ranchers or dairy farmers with infected herds is an integrated one that combines vaccinating disease-free animals and restricting the movement of livestock from infected farms. Tick control and good animal hygiene are also important. Aborted fetuses, placentas, and contaminated bedding material should be buried with lime or incinerated. The manure from infected herds should be managed carefully and, in some cases, treated with lime or calcium cyanide. The Netherlands was able to stop a fever epidemic in people by following these procedures (Table 16.2). Any integrated approach should determine if rats are a reservoir host for *C. burnetii* in the local area. If so, it may be necessary to reduce rodent populations at livestock facilities.

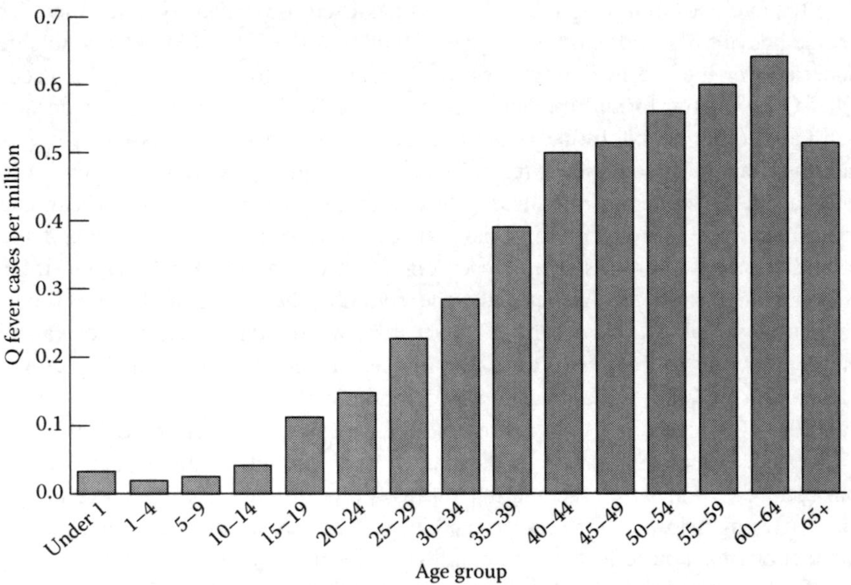

FIGURE 16.7 Incidence of Q fever in the United States (number of cases/1,000,000 people) among people of different age groups. (From CDC, *Morbid. Mortal. Wkly. Rep.*, 60(53), 1, 2011.)

16.7 ERADICATING Q FEVER FROM A COUNTRY

There are no practical means of eradicating these diseases from areas where *C. burnetii* is endemic due to the ubiquitous distribution of *C. burnetii* and the multitude of wild animals that can serve as its reservoir hosts.

LITERATURE CITED

Astobiza, I., M. Barral, F. Ruiz-Fons, J. F. Barandika et al. 2010. Molecular investigation of the occurrence of *C. burnetii* in wildlife and ticks in an endemic area. *Veterinary Microbiology* 147:190–194.

Ayres, J. G., N. Flint, E. G. Smith, W. S. Tunnicliffe et al. 1998. Post-infection fatigue syndrome following Q fever. *Quarterly Journal of Medicine* 91:105–123.

Bernard, H., S. O. Brockmann, N. Kleinkauf, C. Kline et al. 2012. High seroprevalence of *Coxiella burnetii* antibodies in veterinarians associated with cattle obstetrics, Bavaria, 2009. *Vector-Borne and Zoonotic Diseases* 12:552–557.

Buhariwalla, F., B. Cann, and T. J. Marrie. 1996. A dog-related outbreak of Q fever. *Clinical Infectious Diseases* 23:753–755.

CDC. 2011. Q fever. http://www.cdc.gov/qfever/index.html (accessed October 1, 2012).

CDC. 2013. Summary of notifiable diseases—United States, 2011. *Morbidity and Mortality Weekly Report* 60(53):1–117.

Dijkstra, F., W. van der Hoek, N. Wijers, B. Schimmer et al. 2012. The 2007–2010 Q fever epidemic in the Netherlands: Characteristics of notified acute Q fever patients and the association with dairy goat farming. *FEMS Immunology and Medical Microbiology* 64:3–12.

Duncan, C., G. J. Kersh, K. A. T. Spraker, K. A. Patyk et al. 2012. *Coxiella burnetii* in northern fur seal (*Callorhinus ursinus*) placentas from St. Paul Island, Alaska. *Vector Borne and Zoonotic Diseases* 12:192–195.

Dupuis, G., J. Petite, O. Péter, and M. Vouilloz. 1987. An important outbreak of human Q fever in a Swiss alpine valley. *International Journal of Epidemiology* 16:282–287.

Enright, J. B., C. E. Franti, D. E. Behymer, W. M. Longhurst et al. 1971. *Coxiella burnetii* in a wildlife-livestock environment. Distribution of Q fever in wild mammals. *American Journal of Epidemiology* 94:79–90.

Enright, J. B., W. M. Longhurst, M. E. Wright, V. J. Dutson et al. 1971. Q-fever antibodies in birds. *Journal of Wildlife Diseases* 7:14–21.

Gardon, J., J.-M. Héraud, S. Laventure, A. Ladam et al. 2001. Suburban transmission of Q fever in French Guiana: Evidence of a wild reservoir. *Journal of Infectious Diseases* 184:278–284.

Harris, R. J., P. A. Storm, A. Lloyd, M. Arens, and B. P. Marmion. 2000. Long-term persistence of *Coxiella burnetii* in the host after primary Q fever. *Journal of Epidemiology and Infection* 124:543–549.

Hogerwerf, L., R. van den Brom, H. I. J. Roest, A. Bouma et al. 2011. Reduction of *Coxiella burnetii* prevalence by vaccination of goats and sheep, the Netherlands. *Emerging Infectious Diseases* 17:379–386.

Houwers, D. J. and J. H. Richardus. 1987. Infections with *Coxiella burnetii* in man and animals in the Netherlands. Zentralblatt für Bakteriologie, Mikrobiologie, und Hygiene. *Series A: Medical Microbiology, Infectious Diseases, Virology, Parasitology* 267:30–36.

Jay-Russell, M., J. Douglas, C. Drenzek, J. Stone et al. 2002. Q fever—California, Georgia, Pennsylvania, and Tennessee, 2001–2001. *Morbidity and Mortality Weekly Report* 51:924–927.

Kersh, G. J., D. M. Lambourn, S. A. Raverty, K. A. Fitzpatrick et al. 2012. *Coxiella burnetii* infection of marine mammals in the Pacific Northwest, 1997–2010. *Journal of Wildlife Diseases* 48:201–206.

Kováčová, E. and J. Kazár. 2002. Q fever—Still a query and underestimated infectious disease. *Acta Virologica* 46:193–210.

Lévesque, B., G. De Serres, R. Higgins, M. A. D'Halewyn et al. 1995. Seroepidemiologic study of three zoonoses (leptospirosis, Q fever, and tularemia) among trappers in Quebec, Canada. *Clinical and Diagnostic Laboratory Immunology* 2:496–498.

Mann, J. S., J. G. Douglas, J. M. Inglis, and A. G. Leitch. 1986. Q fever: Person to person transmission within a family. *Thorax* 41:974–975.

Marmion, B. P., M. Shannon, I. Maddocks, P. Storm, and I. A. Penttila. 1996. Protracted debility and fatigue after acute Q fever. *Lancet* 347:977–978.

Marrie, T. J., H. Durant, J. C. Williams, E. Mintz, and D. M. Waag. 1988. Exposure to parturient cats: A risk factor for acquisition of Q fever in maritime Canada. *Journal of Infectious Diseases* 158:101–108.

Marrie, T. J. 2011. Q fever. In: S. R. Palmer, L. Soulsby, P. R. Torgerson, and D. W. G. Brown, editors. *Oxford Textbook of Zoonoses*. Oxford University Press, New York City, NY, pp. 158–173.

Maurin, M. and D. Raoult. 1999. Q fever. *Clinical Microbiology Reviews* 12:518–553.

McDade, J. E. 1990. Historical aspects of Q fever. 1990. In: T. L. Marrie, editor. *Q Fever. Volume 1: The Disease*. CRC Press, Boca Raton, FL, pp. 5–21.

McQuiston, J. H., and J. E. Childs. 2002. Q fever in humans and animals in the United States. *Vector Borne and Zoonotic Diseases* 2:179–191.

McQuiston, J. H., R. C. Holman, C. L. McCall, J. E. Childs et al. 2006. National surveillance and the epidemiology of human Q fever in the United States 1978–2004. *American Journal of Tropical Medicine and Hygiene* 75:36–40.

Merck. 2010. Q fever. *Merck Veterinary Manual*, 10th edition. Merck, Whitehouse Station, NJ.

Oyston, P. C. F. and C. Davies. 2011. Q fever: The neglected biothreat agent. *Journal of Medical Microbiology* 60:9–21.

Porter, S. R., G. Czaplicki, J. Mainil, Y. Horii, et al. 2011. Q fever in Japan: An updated review. *Veterinary Microbiology* 149:298–306.

Raoult, D., T. J. Marrie, and J. L. Mege. 2005. Natural history and pathophysiology of Q fever. *Lancet Infectious Diseases* 5:219–226.

Reusken, C., R. van der Plaats, M. Opsteegh, A. de Bruin, and A. Swart. 2011. *Coxiella burnetii* (Q fever) in *Rattus norvegicus* and *Rattus rattus* at livestock farms and urban locations in the Netherlands; could *Rattus* spp. represents reservoirs for (re)introduction? *Preventative Veterinary Medicine* 101:124–130.

Rijks, J. M., H. I. J. Roest, P. W. van Tulden, M. J. L. Kik et al. 2011. *Coxiella burnetii* infection in roe deer during Q fever epidemic, the Netherlands. *Emerging Infectious Diseases* 17:2369–2371.

Roest, H. I. J., J. J. H. C. Tilburg, W. van der Hoek, P. Vellema et al. 2011. The Q fever epidemic in the Netherlands: History, onset, response and reflection. *Epidemiology and Infection* 139:1–12.

Tissot-Dupont, H., M.-A. Amadei, M. Nezri, and D. Raoult. 2004. Wind in November, Q fever in December. *Emerging Infectious Diseases* 10:1264–1269.

Vranakis, I., S. Kokkini, D. Chocklakis, V. Sandalakis et al. 2012. Serological survey of Q fever in Crete, southern Greece. *Comparative Immunology, Microbiology, and Infectious Diseases* 354:123–127.

Williams, J. C. and V. Sanchez. 1994. Q fever and coxiellosis. In: G. W. Beran, editor. *Handbook of Zoonoses*, 2nd edition. CRC Press, Boca Raton, FL, pp. 429–446.

17 Epidemic Typhus and Murine Typhus

After World War I, 20–30 million people died in Eastern Europe from this disease [epidemic typhus], and an additional several million died during and after World War II. Crowding, the scarcity of clean clothes, and dirt were the principal factors enabling the spread of typhus.

Gross (1996)

17.1 INTRODUCTION AND HISTORY

The typhus group of rickettsiae includes two diseases—epidemic typhus and murine typhus—which are produced by *Rickettsia prowazekii* and *R. typhi*, respectively. Both pathogens are small (0.3–2.0 μm), Gram-negative coccobacilli (Figure 17.1). Neither disease should be confused with typhoid fever, which is a disease caused by *Salmonella* bacteria.

Epidemic typhus is also known as louse-borne typhus, camp fever, hospital fever, and ship fever. It is a contagious disease and is spread among humans by the human body louse (*Pediculus humanus corporis*; Figure 17.2). Dr. Charles Nicolle discovered in 1909 that lice spread epidemic typhus from one person to another; he later admitted that his discovery was made quite by accident. While a physician in Tunis (Tunisia's largest city) during an epidemic, he observed that people waiting to be admitted to the hospital often spread typhus to others, including workers in the hospital's laundry room, but that once patients were admitted to the hospital, received a hot bath, and dressed in hospital clothing, the patients ceased to be infectious. Nicolle realized that the infection was spread by the patients' clothing and hypothesized that lice in the clothing were the vector responsible for transmitting the pathogen from one person to another. His discovery earned him the Nobel Prize in 1928 (Gross 1996).

The origins of epidemic typhus are unclear. Some believe that it originated in the Old World and was an ancient scourge of humans. Others believe that it is a relatively new human disease that began as a wildlife disease in North America and jumped to humans during the 1700s or 1800s. Regardless of its origin, *R. prowazekii* spread throughout the world during the 1800s and became a significant cause of death among humans (Sidebar 17.1). During and after World War I, epidemic typhus killed millions of people in Europe. During World War II, it killed thousands of people being held in Nazi concentration camps. Among its many victims was a 15-year-old Jewish girl from Holland—Anne Frank.

The widespread use of DDT and other insecticides during and following World War II greatly diminished the threat of epidemic typhus by reducing the numbers of

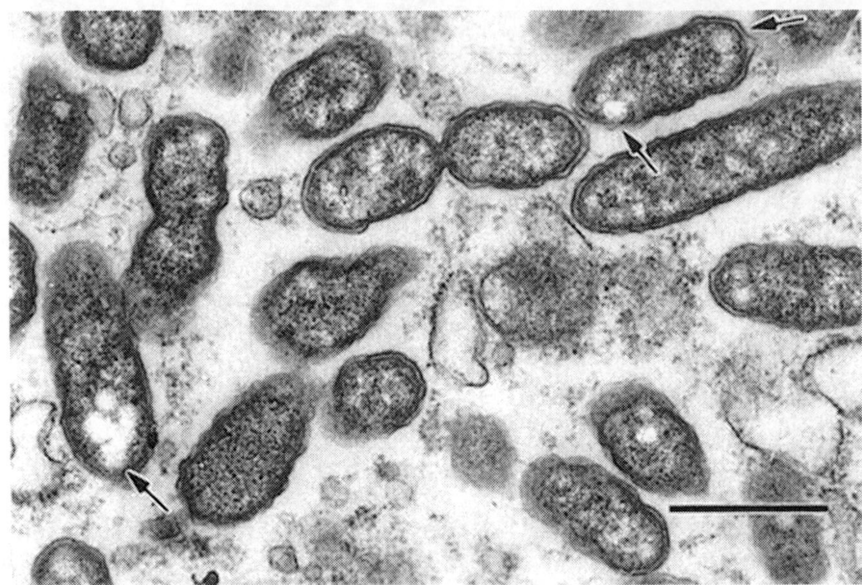

FIGURE 17.1 The dark cells in this electron photomicrograph are *R. typhi*. (Courtesy of David Walker and Macmillan Publishers Ltd.: *Laboratory Investigations* 80(9), Copyright 2000.)

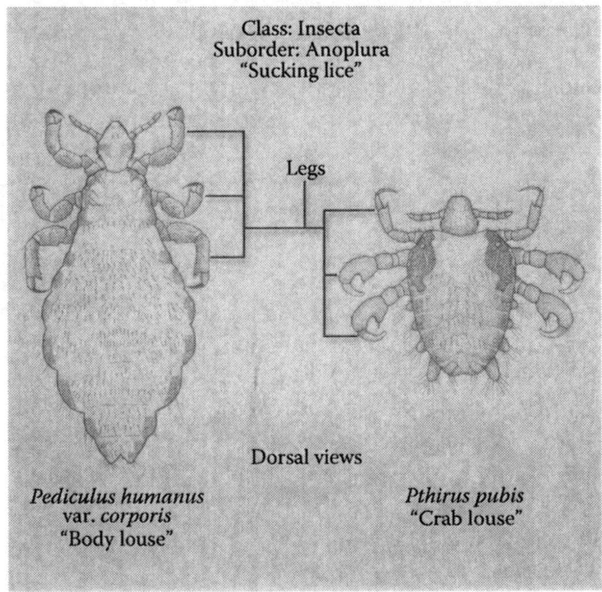

FIGURE 17.2 Human body lice (*P. humanus*) and the pubic or crab louse (*P. pubis*). (Courtesy of the CDC.)

SIDEBAR 17.1 DID SOLDIERS IN NAPOLEON'S GRAND ARMY SUFFER FROM EPIDEMIC TYPHUS? (RAOULT ET AL. 2006, DAVID 2012)

Many historical accounts indicate the presence of louse-borne infectious diseases among soldiers during times of war in the nineteenth and early twentieth centuries. Raoult et al. (2006) examined human remains discovered in Vilnius, Lithuania. Local archives indicated it was the mass grave site of French troops garrisoned in Vilnius during the retreat of Napoleon's Grand Army in 1812. DNA of *R. prowazekii* were detected in the dental pulp of three soldiers. In totality, nearly a third of Napoleon's army buried at Vilnius had some type of louse-borne disease, including epidemic typhus. Although the cause of death for the three soldiers cannot be ascertained by the presence of *R. prowazekii* in their dental pulp, it indicates that they were infected when they died. Discovery of the pathogen supports earlier speculation that an epidemic of typhus struck Napoleon's soldiers during their retreat from Russia that destroyed the French army. Napoleon's retreat was truly catastrophic. His Grand Army numbered over 600,000 when it invaded Russia but contained only 10,000 soldiers capable of fighting by the time the retreat ended. Typhus along with other diseases, battle deaths, desertions, starvation, and the Russian winter had destroyed Napoleon's army and the French Empire. For every 20 soldiers that invaded Russia, only one lived to see his home again (Figure 17.3).

FIGURE 17.3 Lithograph by Victor Adam entitled the "Retreat of Napoleon from Russia, 3 November 1816." (Courtesy of the McGill University Libraries.)

human lice. Since then, epidemics of typhus have been reported in refugee camps, prisons, and other areas where clothing is limited, human lice are common, antibiotics are scarce, and insecticides are lacking. By 1995, epidemic typhus looked like it might be eradicated; Ethiopia appeared to be the last country where cases of epidemic typhus were being spread by human body lice. Unfortunately, epidemic typhus has made a remarkable comeback since then with 100,000 people infected in Burundi during its civil war and more people infected during minor epidemics in Russia, Peru, and northern Africa (Houhamdi and Raoult 2007). Epidemic typhus is still found in the mountainous regions of central and eastern Africa, Central America, South America, and Asia and among homeless people in Europe and North America (WHO 2011).

In the United States, epidemic typhus is not a notifiable disease, so the incidence rate is unknown. Some U.S. cases of epidemic typhus involved travelers who recently visited foreign countries and contracted the disease overseas. Epidemic typhus, nonetheless, has been diagnosed in patients who never left the United States and have had no known exposure to human lice or typhus patients. Medical authorities were baffled by how these people became infected, but during the 1970s, scientists discovered that *R. prowazekii* can maintain itself in North America by infecting flying squirrels and their ectoparasites, making flying squirrels the only known mammal host of *R. prowazekii* other than humans. About a third of the epidemic typhus patients in the United States had recent contact with flying squirrels (Sidebar 17.2).

Murine typhus is also called endemic typhus, sylvatic typhus, Mexican typhus, or flea-borne typhus. Murine typhus differs from epidemic typhus in having fleas as the vector rather than human body lice. Murine typhus is widespread, occurring on every continent except Antarctica. The reservoir hosts are brown rats, black rats, and their fleas. The disease occurs in seaports and urban areas with large populations of rats, including cities in southern Europe, Africa, Southeast Asia, Australia, and the United States. Within the United States, murine typhus is endemic in southern Texas, southern California, and Hawaii, with most cases occurring in Texas. There were 20–60 confirmed cases of murine typhus in Texas between 1998 and 2003, but the number increased to 140 and 160 between 2006 and 2008 (Figure 17.4). During 2008, an outbreak of murine typhus occurred in and around Austin, Texas—a part of the state where only two cases were reported from 1997 to 2007. During the 2008 outbreak, there were 53 reported cases (33 were confirmed cases) around Austin (Campbell et al. 2009).

Fleas serve as both vectors and reservoirs for *R. typhi*. Fleas become infected when they ingest blood from an infected mammal. Upon ingestion by a flea, the pathogens penetrate the epithelial cells of the flea's midgut and multiply. From there, the pathogens are excreted in the flea's feces. In some fleas, such as the oriental rat flea (*Xenopsylla cheopis*), *R. typhi* infects the flea's reproductive system, and the pathogen is transmitted from females to their offspring. In some endemic areas, more than half the fleas are infected (Rawlings and Clark 1994). Another disease cycle involving *R. typhi* is the cat flea (*Ctenocephalides felis*) as a vector and opossums, dogs, and cats as reservoir hosts.

A new rickettsial species, *R. felis*, was identified in 1990. It causes an illness in humans that is similar to typhus and has been identified in patients from California

SIDEBAR 17.2 TWO CASES OF EPIDEMIC TYPHUS IN THE EASTERN UNITED STATES (REYNOLDS ET AL. 2003)

Epidemic typhus has been reported sporadically in the United States, mostly in the eastern United States. Here are two case studies of patients who acquired *R. prowazekii* from flying squirrels.

West Virginia—A 44-year-old man arrived at an emergency room in West Virginia during February 2002. His symptoms included headache, abdominal pain, fever, chills, vomiting, and blood in his urine. The patient was admitted to the hospital and placed on two antibiotics: levofloxacin and metronidazole. His condition worsened, and by the fourth day, he had bloodshot eyes, muscle aches, malaise, and a temperature of 100°F (38°C). In response, the two antibiotics were replaced by doxycycline; his condition improved thereafter, and he was released from the hospital 3 days later. During a follow-up visit, the patient still had muscle aches and bloodshot eyes, but his fever and abdominal pain had ended. At this time, he was tested for typhus, and a serum sample showed that he had antibodies against *R. prowazekii.* The patient reported that he had spent several nights in a hunting cabin located in West Virginia that was occupied by flying squirrels every winter. He removed rodent nesting material and debris from the cabin's wall space 10–15 days prior to becoming ill.

Georgia—During March 2002, a 57-year-old man sought medical attention in Georgia with symptoms including fever, headache, muscle aches, malaise, vomiting, confusion, and lack of coordination. His doctors made a presumptive diagnosis of bacterial meningitis, and the patient was placed on the following antibiotics: cefepime, ampicillin, and gentamicin. The patient's wife told the doctors that her husband had removed a flying squirrel carcass from the air-intake chamber of his furnace and had taken the air filter outside to brush it clean of dust and animal hair. Suspecting typhus, his doctors then prescribed doxycycline. Their diagnosis of typhus was confirmed when blood samples showed that the patient had antibodies against *R. prowazekii*. The patient fully recovered and was discharged from the hospital on the 10th day.

and Texas (Azad et al. 1997, Civen and Ngo 2008, Pérez-Osorio et al. 2008). Based on its DNA, *R. felis* belongs to the spotted fever group of *Rickettsia* species, though its life cycle is similar to the typhus group. *R. felis* is discussed in more detail in Chapter 15.

17.2 SYMPTOMS IN HUMANS

Following an incubation period of 10–14 days, symptoms of epidemic typhus begin abruptly and include a frequent cough, high fever of up to 104°F (40°C), chills, joint pain, headache, muscle pain, and low blood pressure. There is no eschar at the site of the louse bite (Houhamdi and Raoult 2007). Patients may have a rash that starts as a light rose color and later becomes dullish red; the rash does not blanch when

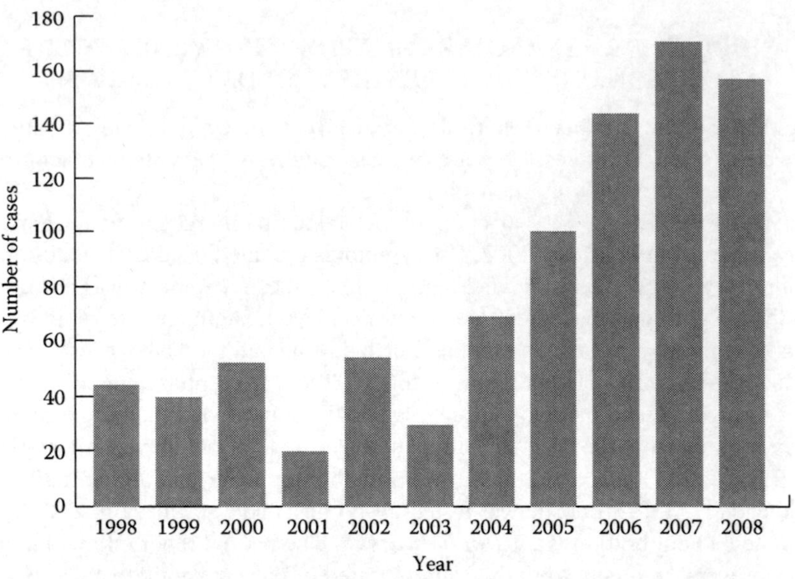

FIGURE 17.4 Number of confirmed cases of murine typhus in Texas between 1998 and 2008. (Data from Campbell, J. et al., *Morbid. Mortal. Wkly. Rep.*, 58, 1267, 2009.)

compressed (MedlinePlus 2012). Patients may also develop small areas of bleeding beneath the skin. Many patients experience problems related to the central nervous system including hearing problems, seizures, confusion, drowsiness, and coma (Houhamdi and Raoult 2007). Dangerous complications are renal failure and pneumonia. During typhus epidemics in Africa, 10%–25% of patients perish without medical treatment (Massung et al. 2001), but the mortality rate drops to 4% with the proper antibiotics (Bechah et al. 2008). Mortality rates increase when patients are over 60 years old or are malnourished. In the United States, cases of epidemic typhus are rare and sporadic, and medical doctors may not initially consider typhus as a possible diagnosis for a patient's symptoms. Recovery can be delayed if no antibiotics are prescribed or the wrong ones are employed (Reynolds et al. 2003).

People who had epidemic typhus in the past may experience a relapse of symptoms called Brill–Zinsser disease. The relapse may occur months or years after the initial infection, often following a weakening of the patient's immune system. Clinical signs of the Brill–Zinsser disease are similar to an acute case of epidemic typhus. Without treatment, the fatality rate for Brill–Zinsser patients is 1.5% (Houhamdi and Raoult 2007).

Typically, murine typhus is a mild disease produced when *R. typhi* infects the cells lining the walls of blood vessels, but it can become severe. It usually begins 7–14 days after the initial infection and lasts 3–7 days; symptoms are flu-like and include headache, fever, nausea, malaise, and muscle aches (Table 17.1). Several days after the illness begins, most patients (60%) have a rash that initially occurs on the middle of the body and spreads to the arms and legs but not to the palms of the

TABLE 17.1
Occurrence of Different Symptoms of Murine Typhus in Humans

Symptom	Occurrence in Patients (%)
Fever	98–100
Headache	41–90
Pain in the joints	40–77
Swollen liver	24–29
Rash	20–80
Cough	15–40
Abdominal pain	11–60
Diarrhea	5–40
Swollen spleen	5–24
Nausea or vomiting	3–48
Confusion	3–12

Source: Civen, R. and Ngo, V., *Clin. Infect. Dis.*, 46, 913, 2008.

hands or soles of the feet. The infection can cause abdominal pains, the result of an enlarged liver or spleen; the lungs can become inflamed. In rare cases, delirium, stupor, and coma can result. Serious complications include meningitis and a rupturing of the spleen. Historically, about 5% of murine typhus cases were fatal, but this has dropped to less than 1% with the use of antibiotics (Rawlings and Clark 1994, Civen and Ngo 2008). During the 2008 outbreak around Austin, Texas, 73% of murine typhus patients required hospitalization. Delayed diagnosis and delays in administering appropriate antibiotics may have contributed to the severity of the illness experienced by some patients (Campbell et al. 2009).

17.3 *RICKETTSIA PROWAZEKII* AND *RICKETTSIA TYPHI* INFECTIONS IN ANIMALS

R. prowazekii, the pathogen that causes epidemic typhus, can maintain itself either by infecting humans and human body lice or by infecting flying squirrels in North America and their fleas or lice (Figure 17.5). Up to 57% of flying squirrels caught in the immediate vicinity of where an epidemic typhus patient lived had antibodies against *R. prowazekii* (Chapman et al. 2009). When laboratory mice were experimentally infected with *R. prowazekii*, some became sick with the infection spreading to their lungs, liver, or brain; their clinical signs were similar to those observed in human patients (Bechah et al. 2008). Whether other mammalian species can be infected or become ill remains unknown.

Murine typhus is named for the Murinae: a group of rodents that evolved in the Old World. It is these Old World rodents, especially the brown rat and black rat, that are the reservoir hosts for *R. typhi* and the fleas that serve as vectors.

FIGURE 17.5 Photo of a flying squirrel. (Courtesy of Ken Thomas.)

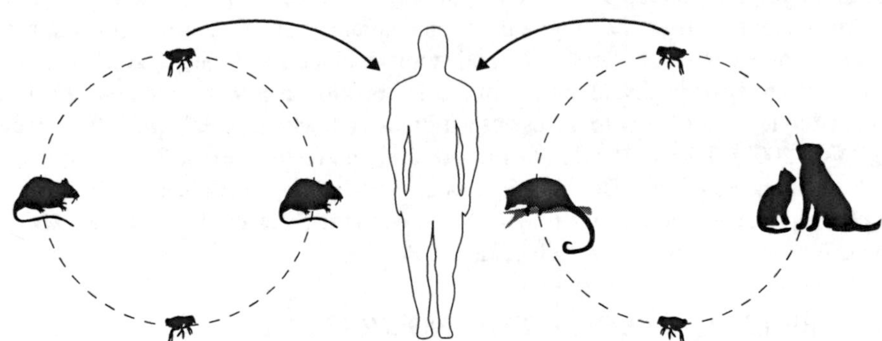

FIGURE 17.6 The two transmission cycles of *R. typhi*, the causative agent of murine typhus in the United States. One cycle, involving opossums, cats, and dogs, occurs mostly in suburban areas. The other cycle involves brown rats and black rats and occurs primarily in urban areas. (Adapted from Azad, A.F. et al., *Emerg. Infect. Dis.*, 3, 319, 1997.)

These commensal rodents live in close association with humans and are abundant in urban areas, buildings, and ships. These rodents now occur worldwide due to their affinity for living on ships. *R. typhi* causes either no or only mild infection in these two species.

In most of the world, *R. typhi* is maintained through this flea–rat–flea cycle and is an urban disease where black and brown rats are abundant (Figure 17.6). But in the United States, murine typhus also occurs in rural and suburban areas

FIGURE 17.7 A photo of an opossum. (Courtesy of the USFWS.)

where there are few Old World rodents. In these areas, the reservoir hosts for *R. typhi* are opossums and domestic cats, and the cat flea serves as the principal vector (Boostrom et al. 2002). In the United States, opossums live not only in rural areas but also in urban and suburban environments where they exhibit little fear of humans or cats (Figure 17.7). In the suburbs of Los Angeles, California, Sorvillo et al. (1993) documented that 90% of domestic cats and 42% of opossums living in neighborhoods of murine typhus patients had antibodies against *R. typhi*, but only 3% of rats were seropositive. Cat fleas were common on both opossums and domestic cats, but no rat fleas were found. During the 2008 outbreak of murine typhus in Austin, Texas, health authorities collected cats, dogs, opossums, raccoons, and rats from around the homes of patients. Antibodies against *R. typhi* were found in 3 of 17 cats tested, 4 of 9 dogs, and 12 of 18 opossums. All tests conducted on rats were negative (Campbell et al. 2009).

17.4 HOW HUMANS CONTRACT TYPHUS

There are three different types of human lice. Human head lice (*P. humanus capitis*) are common worldwide, including in the United States and Canada and are especially common among children. They infect the head and neck and attach their eggs to hair. They do not serve as a vector for *R. prowazekii* and play no role in the spread of typhus. Pubic lice (*Pthirus pubis*), also called crabs, are found attached to hair in the pubic area but are sometimes found on coarse hair elsewhere on the body, such as eyebrows, beards, and armpits. They spread from one person to another primarily through sexual contact and are mostly found on adults. They also do not serve as a vector for *R. prowazekii* and play no role in spreading typhus. The third type is human body lice (*P. humanus corporis*), which live and lay eggs within the folds of clothing; they are found on human skin only when feeding. They occur worldwide among people who lack access to clean clothing. Human body lice are the only human lice that can spread *R. prowazekii*. Lice acquire *R. prowazekii* either from their mother via her eggs or by feeding on an infected person or flying squirrel.

Human-care workers from the United States and Canada are at risk of contracting epidemic typhus when working in countries where human body lice and typhus epidemics still occur. The risk is minimal for travelers to these countries because most epidemic typhus occurs in very remote and impoverished areas that are typically not visited by travelers.

In North America, *R. prowazekii* is endemic in populations of flying squirrels. It is unclear whether flying squirrels initially contracted *R. prowazekii* from humans or if humans contracted it from flying squirrels. From 1976 to 2001, 39 typhus patients in the United States acquired *R. prowazekii* from flying squirrels; these patients had no known contact with body lice or typhus patients (Reynolds et al. 2003). For example, four counselors of a Pennsylvania wilderness camp for troubled youth were diagnosed with epidemic typhus over a period of three winters. All had slept in the same cabin and three in the same bed (Figure 17.8). All patients reported seeing flying squirrels or hearing them in the cabin wall adjacent to the bed. Other staff members who slept elsewhere in the camp did not develop typhus. Fourteen flying squirrels were later captured at the camp, and 10 had been infected by *R. prowazekii* (Chapman et al. 2009).

One hypothesis on how *R. prowazekii* is transmitted from flying squirrels to humans involves people inhaling *R. prowazekii* that have been shed in the feces of infected lice or fleas or passage of the pathogen through a mucous membrane or skin abrasion or wound. Another hypothesis involves people being bitten by lice or fleas from flying squirrels; at least one flea (*Orchopeas howardii*) of flying squirrels is known to bite humans and could transmit *R. prowazekii* from infected squirrels to people (Reynolds et al. 2003). People are more likely to get bitten by this flea or lice from flying squirrels after the squirrel has died, and its lice and fleas are seeking a new host.

Usually, *R. prowazekii* infections are limited to a single person in a household or building, indicating that causal exposure to flying squirrels rarely results in infections. Instead, most typhus patients in the United States had close physical contact with flying squirrels or their nests. A person who contracts epidemic typhus from a flying squirrel is capable of infecting other people and starting an epidemic, but this is unlikely to happen due to the absence of human body lice.

Humans contract murine typhus when they are bitten by fleas infected with *R. typhi*. Humans are an accidental host for *R. typhi*. These bacteria are usually transmitted by the fleas, such as the oriental rat flea (*X. cheopis*), which feed on the Old World rodents. These fleas do not normally seek humans as hosts but will do so if they cannot locate a suitable rodent. Humans cannot be infected by the bite of the oriental rat flea, but flea bites often itch, and scratching can rub flea feces into

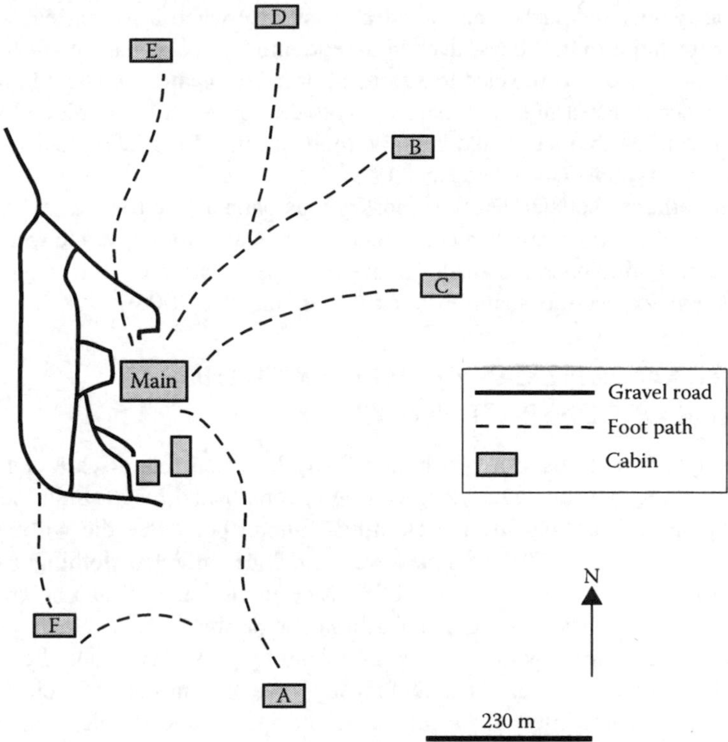

FIGURE 17.8 Map of a Pennsylvania wilderness camp showing the six cabins (A through F on the map) and the buildings in the main part of the camp. Four counselors were diagnosed with epidemic typhus, and all had slept in cabin A. (Redrawn from Chapman, A.S. et al., *Emerg. Infect. Dis.*, 15, 1005, 2009; courtesy of the CDC and *Emerging Infectious Diseases*.)

abraded skin. *R. typhi* can also gain entry into the human body by being rubbed into the eye and inhaled via dust contaminated with flea feces. Humans can be infected if bitten by an infected cat flea (Eremeeva and Dasch 2013).

In most of the world, human cases of murine typhus peak in late summer and early fall when rat fleas are most abundant. But this pattern is not manifested in the United States; instead, most cases occur during the spring in Texas and during the summer in California. In Hawaii, there is no seasonal pattern (Civen and Ngo 2008).

17.5 MEDICAL TREATMENT

A challenge in treating typhus is that the disease is difficult to diagnose and rare enough in Canada and the United States that most medical doctors may not consider it as a possible diagnosis. Epidemic typhus should be considered a possibility when a patient has a persistent fever for more than 3 days and has had direct contact with a flying squirrel or its nesting material or has recently traveled to another country where typhus is endemic.

Murine typhus should be considered a possibility when a patient has a persistent fever for more than 3 days, lives in an endemic area and has been exposed to opossums or cats, or has traveled to a tropical or subtropical area where high densities of rats occur. Most murine typhus patients do not recall being bitten by a flea. Treatment with antibiotics should begin without waiting for a laboratory confirmation of murine typhus (Civen and Ngo 2008).

As with other rickettsial diseases, most typhus patients are treated with the antibiotic doxycycline. Chloramphenicol is also effective and often given to women who are pregnant. Quinolones are also effective, but the duration of illness typically is shorter when doxycycline is administered (Civen and Ngo 2008).

17.6 WHAT PEOPLE CAN DO TO REDUCE THEIR RISK OF CONTRACTING TYPHUS

Epidemic typhus is transmitted by human body lice, and this disease can be prevented by eradicating lice. Delousing can be accomplished by washing the patient and changing and boiling infected clothing. Human body lice die within 5 days when deprived of blood, so the simplest way to delouse infected clothing is to leave it unworn for a week (Bechah et al. 2008). Lice in clothes can also be eradicated with an insecticide, such as DDT, malathion, or permethrin. In North America, most people are infected with *R. prowazekii* from flying squirrels and their fleas or lice. Flying squirrels are nocturnal, and most people are unaware of their presence. Patients often become infected during winter and spring when flying squirrels enter homes and cabins, seeking shelter from the cold.

At least four people contracted epidemic typhus during 2004 and 2005 at a Pennsylvania wilderness camp where most flying squirrels were infected with *R. prowazekii*. In response to the epidemic, all cabin holes large enough for a flying squirrel to gain entry were closed with wire hardware cloth, the inside wall boards of the cabins were removed, and insulation was replaced. The cabins were treated with an insecticide to kill ectoparasites, and camp staff and students were educated about epidemic typhus. Since then, no additional cases of typhus were detected (Chapman et al. 2009). These same procedures should be followed whenever a building is linked with typhus and *R. prowazekii*. Traps can be set to remove any flying squirrels or rodents that are living inside the building. However, flying squirrels should not be removed without the concomitant use of insecticides, otherwise the fleas and lice, lacking flying squirrel host, will be more willing to bite humans.

Any dying or dead flying squirrels found outside a building should be left alone; they should not be handled or examined closely. Any dead flying squirrel found inside a building should be carefully removed; the carcass should be placed in an airtight plastic bag and disposed. The person handling the carcass should wear waterproof gloves and a mask capable of filtering out dust particles that may be contaminated with the pathogen. Both the mask and gloves should also be discarded in an airtight plastic bag. Lice and fleas will die soon after a flying squirrel has died, but if the carcass is fresh, the area around the carcass should be sprayed with an insecticide. A professional can be hired to remove the carcass. Some people catch flying squirrels

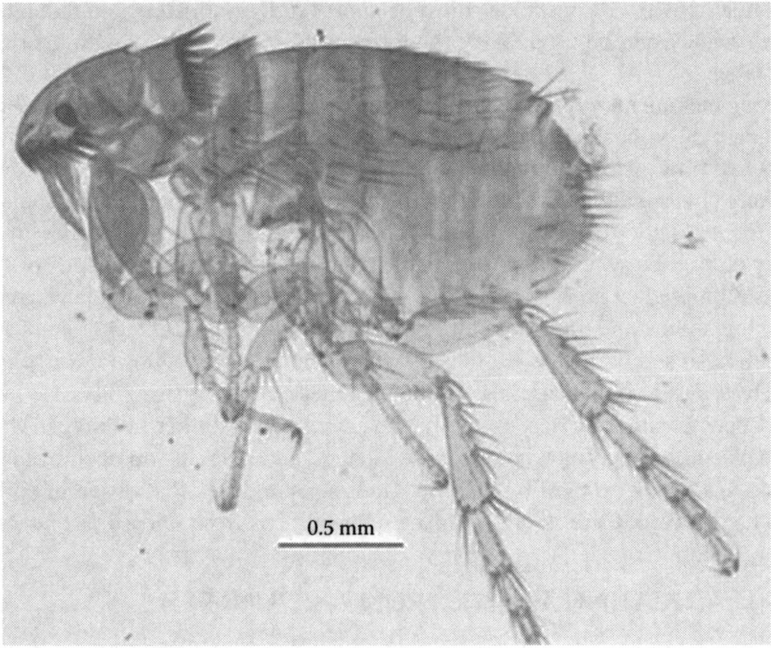

FIGURE 17.9 Photo of a cat flea. (Courtesy of the DPDx Parasite Image Library.)

and keep them as pets even though this is illegal. This practice increases their risk of contracting typhus.

When someone has been diagnosed with epidemic typhus, neighbors should contact their local health authorities for guidance, especially if they find a dead or dying flying squirrel either in their home or yard. Discouraging flying squirrels from staying in your neighborhood is unnecessary unless someone locally has contracted epidemic typhus. When the latter happens, it is wise to discourage flying squirrels from spending time near one's house. For example, feeding birds should be suspended because flying squirrels are attracted to the food at night when they are foraging.

The cat flea is the most common flea found on both cats and dogs in the United States (Figure 17.9) and can transmit *R. typhi* to humans. Hence, one way to diminish the risk of contracting murine typhus is to reduce flea densities on cats and dogs. Pets can be treated for fleas using collars, pills, sprays, dusts, and shampoos that contain insecticides. Carbaryl (Sevin®), fipronil (Frontline®), imidacloprid (Advantage®), malathion, propoxur, pyrethrins, and permethrin are insecticides that have been approved for use on cats and dogs by the federal government (Williams and Bennett 2010).

Dog kennels and homes with high densities of fleas can be treated with insecticides including carbaryl (Sevin®), bendiocarb (Ficam®), propoxur (Baygon®), pyriproxyfen (Nylar® and Archer®), methoprene (Precor®), and pyrethrins (Williams and Bennett 2010). All pesticides must be approved by both state and federal agencies before they can be used. They should be used in accordance with instructions on the label.

During the murine typhus outbreak in Austin, Texas, 79% of patients with a confirmed case of typhus owned a dog or cat, and most patients did not regularly treat their pet with an insecticide. Almost all patients (95%) lived in a household where there were obvious signs of wildlife on the premises and/or wildlife attractants, such as pet food, outside the house. Not surprisingly, health officials were able to trap a number of feral cats, feral dogs, opossums, and rats from the home sites of typhus patients (Campbell et al. 2009). The risk of contracting murine typhus can be reduced by making homes and yards less attractive to opossums and rats. Dog and cat food and open trash should not be left outside where it may attract wildlife. Garbage cans should be secured so that animals cannot open them. Buildings should be rat-proofed to deny opossum and rat access into homes, garages, and other buildings. This can involve placing screens on windows, crawl spaces, and blocking all openings into the building that might be used by wildlife. Opossums and rats that reside in buildings should be removed. Care should be taken to avoid fleas from opossums and rats.

17.7 ERADICATING TYPHUS FROM A COUNTRY

For most countries, epidemic typhus can be eradicated by eliminating human body lice and treating patients so they will not be contagious. Eradicating epidemic typhus from the United States will be more difficult because *R. prowazekii* is endemic in flying squirrels and their ectoparasites. Moreover, eliminating murine typhus from the United States, Mexico, and other countries remains elusive because *R. typhi* is endemic in house mice, brown rats, black rats, and opossums. After a pathogen has established itself within a wildlife population, there are few if any methods that can be used to treat all of the infected individuals that serve as reservoir hosts. Unless all infected animals are removed or treated, the few remaining individuals that still harbor the pathogen can reinfect the rest.

LITERATURE CITED

Azad, A. F., S. Radulovic, J. A. Higgins, B. H. Noden, and J. M. Troyer. 1997. Flea-borne rickettsioses: Ecologic considerations. *Emerging Infectious Diseases* 3:319–327.

Bechah, Y., C. Capo, J.-L. Mege, and D. Raoult. 2008. Epidemic typhus. *Lancet Infectious Diseases* 8:417–426.

Boostrom, A., M. S. Beier, J. A. Macaluso, K. R. Macaluso et al. 2002. Geographic association of *Rickettsia felis*-infected opossums with human murine typhus, Texas. *Emerging Infectious Diseases* 8:549–554.

Campbell, J., M. E. Eremeeva, W. L. Nicholson, J. McQuiston et al. 2009. Outbreak of *Rickettsia typhi* infection—Austin, Texas, 2008. *Morbidity and Mortality Weekly Report* 58:1267–1270.

Chapman, A. S., D. L. Swerdlow, V. M. Dato, A. D. Anderson et al. 2009. Cluster of sylvatic epidemic typhus cases associated with flying squirrels, 2004–2006. *Emerging Infectious Diseases* 15:1005–1011.

Civen, R. and V. Ngo. 2008. Murine typhus: An unrecognized suburban vectorborne disease. *Clinical Infectious Diseases* 46:913–918.

David, S. 2012. Napoleon's failure: For the want of a winter horseshoe. *BBC News Magazine* (February 8, 2012).

Eremeeva, M. E. and G. A. Dasch. 2013. Rickettsial (spotted and typhus fevers) and related infections (anaplasmosis and ehrlichiosis). Chapter 3, Infectious diseases related to travel. http://www.cdc.gov/travel/yellowbook/2014/chapter-3-infectious-diseases-related-to-travel/rickettsial-spotted-and-typhus-fevers-and-related-infections-anaplasmosis-and-ehrlichiosis (accessed December 14, 2013).

Gross, L. 1996. How Charles Nicolle of the Pasteur Institute discovered that epidemic typhus is transmitted by lice: Reminiscences from my years at the Pasteur Institute in Paris. *Proceedings of the National Academy of Science, USA* 93:10539–10540.

Houhamdi, L. and D. Raoult. 2007. Chapter 5, louse-borne epidemic typhus. In: D. Raoult and P. Parola, editors. *Rickettsial Diseases*. Informa Healthcare, New York, pp. 51–61.

Massung, R. F., L. E. Davis, K. Slater, D. B. McKechnie, and M. Puerzer. 2001. Epidemic typhus meningitis in the southwestern United States. *Clinical Infectious Diseases* 32:979–982.

MedlinePlus. 2012. Typhus. U.S. National Library of Medicine, National Institutes of Health. http://www.nlm.nih.gov/medlineplus/ency/article/001363.htm (accessed June 16, 2013).

Pérez-Osorio, C. E., J. E. Zavala-Velázquez, J. J. Arias León, and J. E. Zavala-Castro. 2008. *Rickettsia felis* as emergent global threat for humans. *Emerging Infectious Diseases* 14:1019–1023.

Raoult, D., O. Dutour, L. Houhamdi, R. Jankauskas et al. 2006. Evidence for louse-transmitted diseases in soldiers of Napoleon's Grand Army in Vilnius. *Journal of Infectious Diseases* 193:112–120.

Rawlings, J. A. and K. A. Clark. 1994. Murine typhus. In: G. W. Beran, editor. *Handbook of Zoonoses*, 2nd edition. CRC Press, Boca Raton, FL, pp. 457–461.

Reynolds, M. G., J. W. Krebs, J. A. Comer, J. W. Sumner et al. 2003. Flying squirrel-associated typhus, United States. *Emerging Infectious Diseases* 9:1341–1343.

Sorvillo, F. J., B. Gondo, R. Emmons, P. Ryan et al. 1993. A suburban focus of endemic typhus in Los Angeles County: Association with seropositive domestic cats and opossums. *American Journal of Tropical Medicine and Hygiene* 48:269–273.

WHO. 2011. Typhus fever (epidemic louse-borne typhus). *International Travel and Health*. World Health Organization, Geneva, Switzerland.

Williams, R. E. and G. W. Bennett. 2010. Household and structural fleas. Purdue Extension E-8-W, Purdue University, West Lafayette, IN.

18 Ehrlichiosis and Anaplasmosis

Only recently have ehrlichiosis and anaplasmosis been recognized as causing human disease. Ehrlichiosis was first described during 1987 based on a critically ill patient who had been bitten by a tick in Arkansas. Anaplasmosis was first recognized during 1990 when a person died in Wisconsin.

Paraphrased from Thomas et al. (2009)

Each spring, migrating birds bring into Sweden hundreds of thousands of ticks infected with the pathogen that causes anaplasmosis in humans.

Paraphrased from Bjöersdorff et al. (2001)

If you play golf in areas where ehrlichiosis is endemic and want to reduce your risk of becoming ill, you need to keep yourself and your golf ball on the fairway.

Paraphrased from Standaert et al. (1995)

18.1 INTRODUCTION AND HISTORY

Anaplasmosis and ehrlichiosis are similar in that both diseases are caused by Gram-negative rickettsial bacteria, infect blood cells or blood platelets, and are transmitted by ticks. Anaplasmosis, which is also called human granulocytic anaplasmosis, is produced by *Anaplasma phagocytophilum*. This pathogen infects blood neutrophils; these blood cells are 10–12 μm in diameter and have a multilobed nucleus (Figure 18.1). Neutrophils function in phagocytosis and engulf bacteria and fungi inside the body; their deaths in great numbers form pus.

Ehrlichiosis is caused by several rickettsial species within the genera *Ehrlichia*. In North America, these species include *E. chaffeensis*, *E. ewingii*, and a third species currently referred to as "*E. muris*-like pathogen." Most human infections are caused by *E. chaffeensis*, which infects blood monocytes and is the causative agent for human monocytic ehrlichiosis, cited hereafter as ehrlichiosis. Blood monocytes are 8–10 μm in diameter, have a kidney-shaped nucleus, and serve in phagocytosis by engulfing foreign bacteria in the blood (Figure 18.1). Another species, *E. ewingii*, infects neutrophils and causes *E. ewingii* ehrlichiosis in humans.

The number of reported cases of ehrlichiosis in the United States has increased steadily from 2000 to 2011, rising from 200 to 850 cases (Figure 18.2). There were 10 reported cases of *E. ewingii* ehrlichiosis in the United States during 2010. A new disease was first detected in the United States during 2009; it was produced by a bacterium named *E. muris*-like pathogen. By 2011, six patients residing in Wisconsin

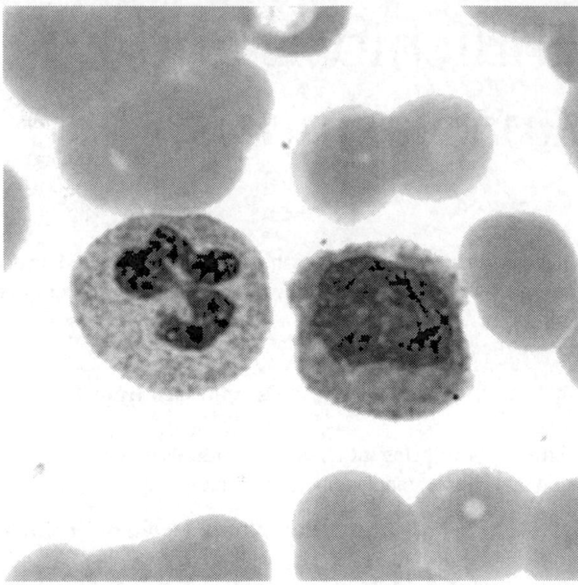

FIGURE 18.1 Photo of a neutrophil (left) and a monocyte (right). (Courtesy of the CDC, DPDx Parasite Image Library.)

and Minnesota had been diagnosed with this disease. This low incidence rate is due, in part, to the difficulty of distinguishing between infections caused by this pathogen and those by *E. chaffeensis* (CDC 2012, 2013).

During the 1990s, there were 100–200 reported cases of anaplasmosis annually in the United States, jumping to over 2575 in 2005. However, the prevalence of anaplasmosis may be much higher than these data suggest. In some parts of the United States, up to 10% of healthy people are seropositive to *A. phagocytophilum*, indicating past exposure to this or a similar pathogen (Adams et al. 2012, CDC 2012, 2013).

Anaplasmosis was first reported in Europe during 1995. Between then and 2005, 66 cases were reported in Europe, yet 6% of the human population in some parts of Europe have antibodies against *A. phagocytophilum*. This could be accounted for if there are strains of *A. phagocytophilum* that can infect humans without causing illness (Dumler et al. 2005, Thomas et al. 2009). *A. phagocytophilum* is primarily transmitted to humans in the United States by the blacklegged tick and the western blacklegged tick—the same two ticks that are the primary vectors for Lyme disease in the United States. Many anaplasmosis patients are located in Minnesota, Wisconsin, or the New England states (Figure 18.3) where blacklegged ticks are abundant. Anaplasmosis also occurs in northern California where the western blacklegged tick is the vector.

In North America, *E. chaffeensis* and *E. ewingii* are transmitted to humans by the lone star tick, and the U.S. distribution of ehrlichiosis and *E. ewingii* ehrlichiosis diseases reflects the tick's range. Ehrlichiosis occurs primarily among people living in

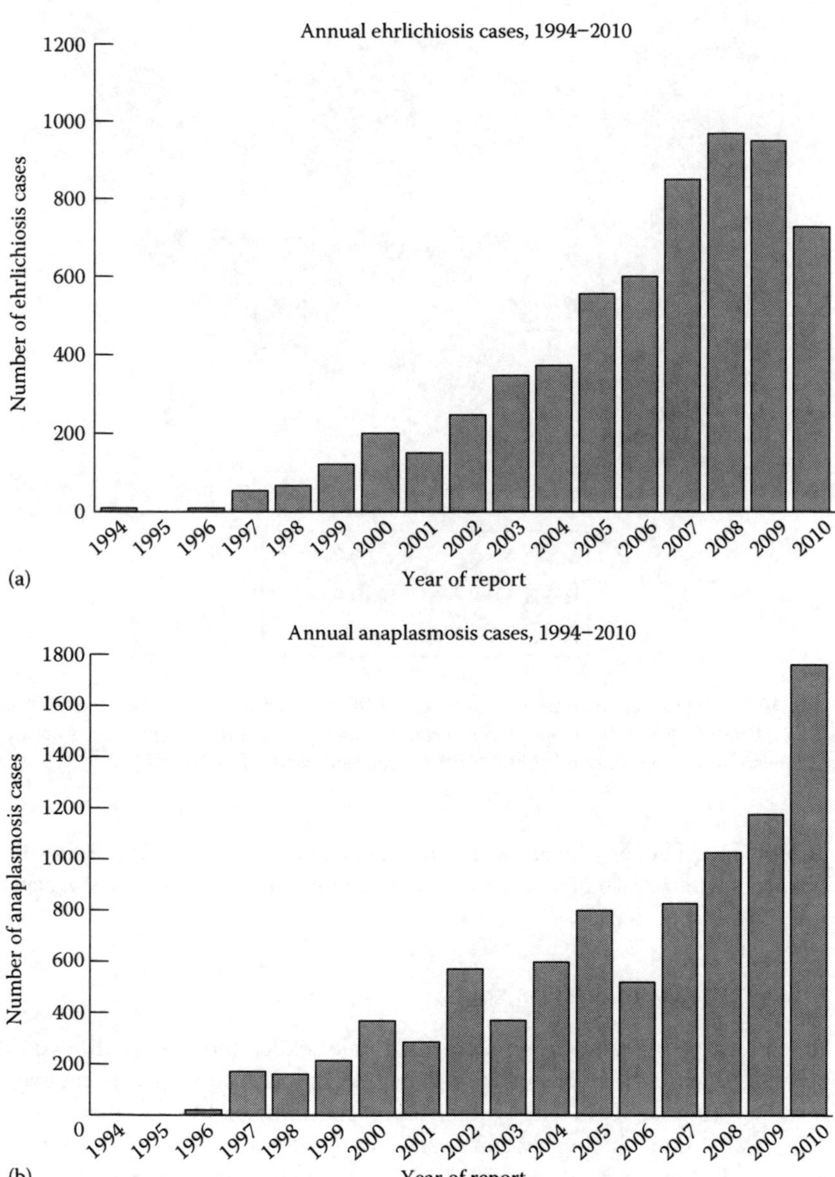

FIGURE 18.2 Annual changes in the number of reported cases of (a) ehrlichiosis and (b) anaplasmosis in the United States. (From CDC, Anaplasmosis: Symptoms, diagnosis, and treatment, http://www.cdc.gov/Anaplasmosis/symptoms/index.html, 2012, accessed November 1, 2012.)

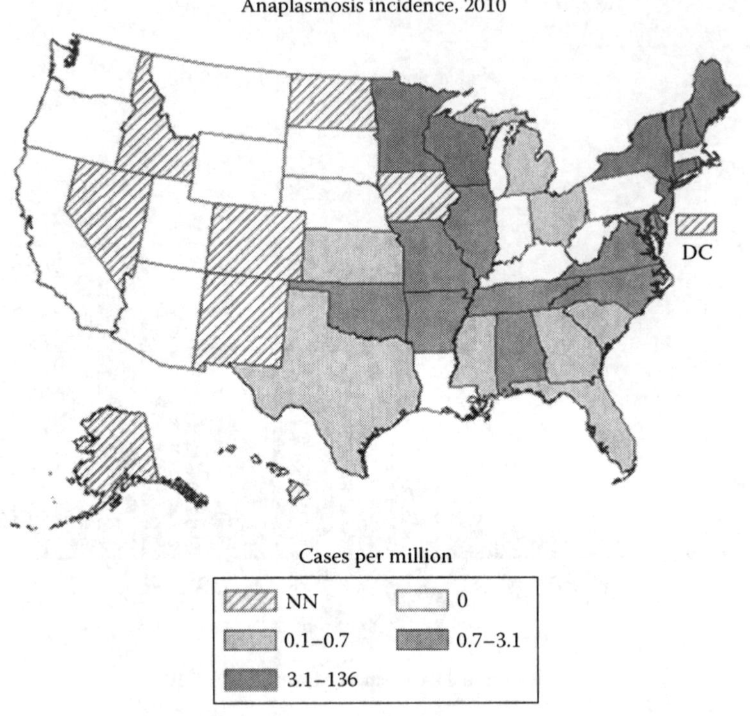

FIGURE 18.3 Incidence rate (number of cases/1,000,000 people) of anaplasmosis in different states. (From CDC, Anaplasmosis: Symptoms, diagnosis, and treatment, http://www.cdc.gov/Anaplasmosis/symptoms/index.html, 2012 (accessed November 1, 2012).

the eastern United States who are within the range of the lone star tick (Figure 18.4); most patients with *E. ewingii* ehrlichiosis live in Missouri, Tennessee, and Delaware (Adams et al. 2012).

18.2 SYMPTOMS IN HUMANS

Within a blood cell, *Ehrlichia* forms microcolonies called morulae (Latin word for mulberries) because the micro-colonies have the appearance of purple mulberries when stained (Figure 18.5). The type of blood cell infected by the micro-colonies provides insight into which pathogen is causing the illness; morulae in neutrophils suggest an infection by *A. phagocytophilum* or *E. ewingii* while morulae in monocytes indicate an infection by *E. chaffeensis* (Thomas et al. 2009, CDC 2011).

The first symptoms of ehrlichiosis begin after an incubation period of 1–3 weeks. Symptoms are similar to other rickettsial diseases (e.g., Rocky Mountain spotted fever, typhus, and Q fever) and include malaise, muscle aches, headaches, fever, chills, nausea, vomiting, and diarrhea. Patients may exhibit conjunctivitis and confusion.

People infected with *E. chaffeensis* can develop a non-itchy rash of small red or purple spots where blood has leaked from capillaries in the skin. The rash may

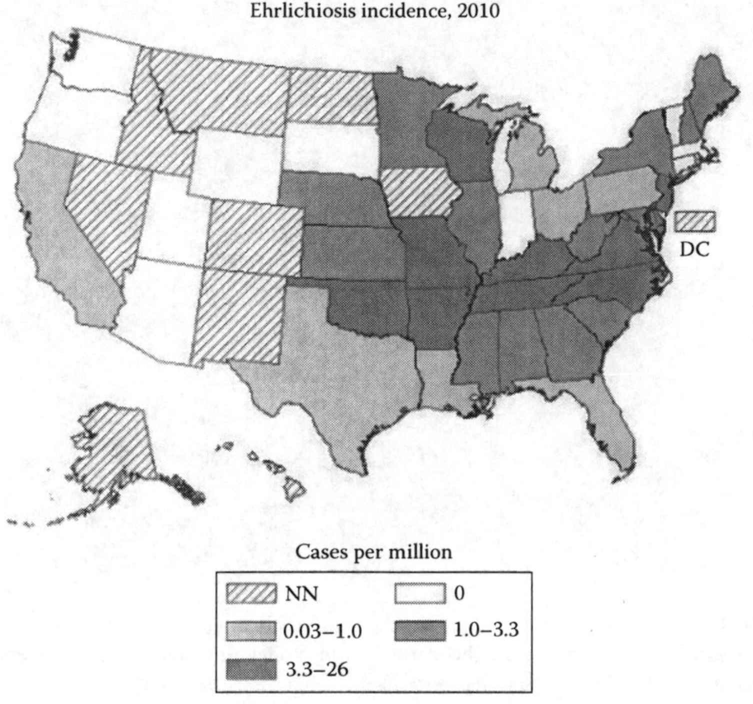

FIGURE 18.4 Incidence rate (number of cases/1,000,000 people) of ehrlichiosis caused by *E. chaffeensis* in different states. (From CDC, Ehrlichiosis: Symptoms, diagnosis, and treatment, http://www.cdc.gov/Ehrlichiosis/symptoms/index.html, accessed October 1, 2012, 2011.)

spread to the palms of the hand and soles of the feet but usually not the face. Some patients develop a rash that resembles a sun burn (e.g., a widespread reddening of the skin), and the skin may later peel where the rash was located. More children (60%) develop rashes than adults (30%); rashes are more common with patients infected with *E. chaffeensis* than patients infected with *E. ewingii* or *E. muris*-like pathogen (CDC 2011).

Most ehrlichiosis patients have a mild illness that lasts less than 30 days even without antibiotic treatment. However, ehrlichiosis can develop into a serious disease; some patients have difficulty breathing, and others have excessive bleeding. The case fatality rate is 1.8% (CDC 2011).

A. phagocytophilum infects white blood cells, resulting in anaplasmosis: a disease that can range from a mild fever to a fatal illness. Initial symptoms of anaplasmosis occur 7–10 days after a tick bite and include fever, chills, headache, muscle aches, and nausea (Table 18.1). Some patients develop a cough, have mild liver damage, or exhibit confusion. Few anaplasmosis patients develop a rash; consequently, a rash may indicate Lyme disease or Rocky Mountain spotted fever. Half of anaplasmosis patients require hospitalization; 6% require intensive care. At least seven deaths have resulted from anaplasmosis; the risk of mortality rises when the

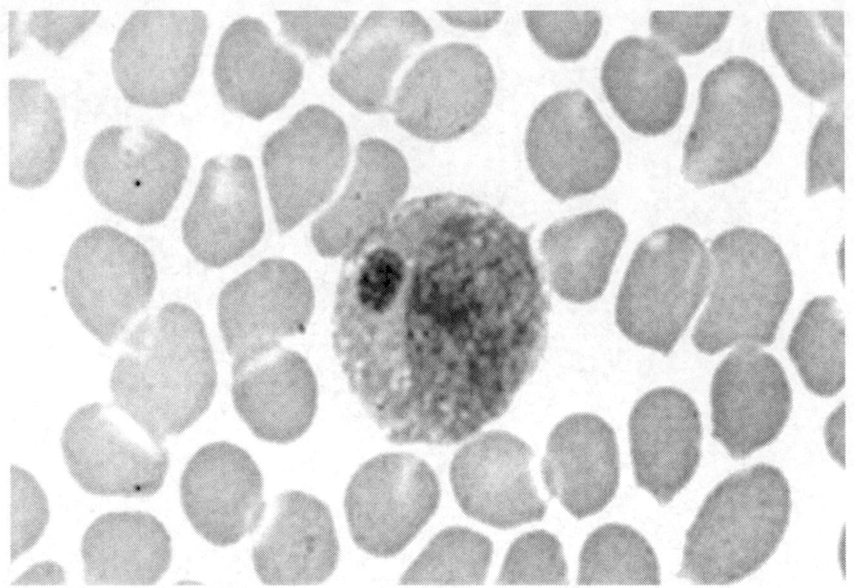

FIGURE 18.5 A photo of a blood smear showing a monocyte cell that has been infected with *E. chaffeensis*. (From CDC, Ehrlichiosis: Symptoms, diagnosis, and treatment, http://www.cdc.gov/Ehrlichiosis/symptoms/index.html, 2011, accessed October 1, 2012. With permission.)

TABLE 18.1
Characteristics of Ehrlichiosis, *E. ewingii* Ehrlichiosis, and Anaplasmosis in the United States

Characteristic	Ehrlichiosis	*E. ewingii* Ehrlichiosis	Anaplasmosis
Alternate name	Human monocytic ehrlichiosis	*ewingii* ehrlichiosis	Human granulocytic anaplasmosis
Pathogen	*E. chaffeensis*	*E. ewingii*	*A. phagocytophilum*
Geographic area	Southeast, central United States	Southeast, central United States	Northeast, Upper Midwest, mid-Atlantic United States
Vector	Lone star tick	Lone star tick	Blacklegged tick, western blacklegged tick
Primary reservoir host	White-tailed deer	White-tailed deer, dogs	White-footed mouse, dusky-footed woodrat
Target white-blood cell	Monocyte	Neutrophil	Neutrophil

Source: Thomas, R.J. et al., *Expert Rev. Anti Infect. Ther.*, 7, 709, 2009.

disease is misdiagnosed or antibiotic treatment is delayed. By destroying white blood cells, anaplasmosis enhances the risk that the patient will develop some other type of infection. Consequently, the risk of serious complications or death is higher for patients who already have another illness or suppressed immune system (Dumler et al. 2005, CDC 2011).

18.3 *EHRLICHIA* AND *ANAPLASMA* INFECTIONS IN ANIMALS

Little is known about the worldwide distribution of *E. chaffeensis*, but it has been found in Brazil where it infects the marsh deer and in Korea where it infects the sika deer (Figure 18.6). In the United States, white-tailed deer and domestic dogs are reservoir hosts for *E. chaffeensis* and *E. ewingii* as well as for the lone star tick, which is

FIGURE 18.6 In Brazil, marsh deer (a) are a reservoir host for *E. chaffeensis* and the ticks that serve as vectors for the pathogen in Brazil while the sika deer (b) serve those same functions in Asia. (Marsh deer: photo courtesy of Fabio Paschoal and Caiman Ecological Refuge, Mato Grosso do Sul, Brazil; sika deer: photo courtesy of John White and the USFWS.)

a vector for these pathogens (Arens et al. 2003). In the United States, 5%–15% of lone star ticks are infected with *E. chaffeensis*, and 1%–5% are infected with *E. ewingii* (Thomas et al. 2009). In domestic dogs, *E. ewingii* produces a disease called canine granulocytic ehrlichiosis, which has symptoms similar to ehrlichiosis in humans.

A. phagocytophilum is transmitted to humans primarily by the blacklegged tick in the eastern United States, the western blacklegged tick in northern California, *Ixodes ricinus* in Europe, and *I. persulcatus* in Asia. *A. phagocytophilum* has been isolated from *I. spinipalpis* in the western United States, but this tick species rarely bite humans and transmits *A. phagocytophilum* to humans. Nonetheless, this tick plays a major role in spreading the pathogen among different mammal species and thus contributes to the maintenance of *A. phagocytophilum* among the pathogen's reservoir host species (Thomas et al. 2009).

The white-footed mouse is the main reservoir host for *A. phagocytophilum* in the United States, but other hosts include raccoons, opossums, tree squirrels, chipmunks, and dusky-footed woodrats (Foley et al. 2008, Thomas et al. 2009). In Europe, red deer, roe deer, and rodents serve as reservoir hosts for *A. phagocytophilum*, and the pathogen has been detected in common shrews, feral hogs, European bison, moose, and red foxes (Petrovec et al. 2002, Stuen 2007, Rosef et al. 2009, Thomas et al. 2009). Long-distance transport of *A. phagocytophilum* is provided by birds when they carry infected ticks and quill mites (Bjöersdorff et al. 2001, Skoracki et al. 2006) and by humans, who move infected livestock.

A. phagocytophilum causes disease in sheep, cattle, and goats in Europe where the livestock disease is called tick-borne fever by farmers who realize that their livestock became ill soon after being bitten by *Ixodes* ticks. In other parts of Europe, the disease is referred to as pasture fever because it often occurs after cattle are released into pastures during the spring. A similar disease occurs in India and South Africa. In animals, *A. phagocytophilum* attacks white blood cells. The clinical signs in sheep include fever of 105°F–108°F (40°C–42°C), weight loss, increase in respiratory and pulse rate, cough, and dull appearance. In cattle, coughing, respiratory distress, loss of appetite, and a sudden reduction in milk production are common clinical signs. Tick-borne fever in sheep, goats, and cattle produces abortions and stillbirths in infected females and a decrease in the quality of semen in infected males. Tick-borne fever can impair the immune system of infected livestock, causing other infections to worsen (Merck 2010).

Horses and burros in the United States, Mexico, and parts of Europe contract a disease called equine granulocytic ehrlichiosis or tick-borne fever, which is caused by *A. phagocytophilum* and transmitted by *Ixodes* ticks. In horses, *A. phagocytophilum* infects neutrophils and weakens their immune system, making them vulnerable to other infections. Clinical signs in horses include fever, anorexia, swelling in the limbs, lack of muscle coordination, and a fast pulse rate. Horses can be successfully treated with oxytetracycline (Merck 2010).

Several species of *Anaplasma* infect the blood cells of different mammalian species but not humans. *A. ovis* infects red blood cells of bighorn sheep and also domestic sheep and goats (de la Fuente et al. 2006). *A. bovis* infects cattle primarily in Africa but has a worldwide distribution in cattle. The pathogen has been detected in sika deer in Asia (Kawahara et al. 2006, Ooshiro et al. 2008, Dergousoff and Chilton 2011).

Goethert and Telford (2003) reported that antibodies against *A. bovis* have been found in 18% of eastern cottontail rabbits on Nantucket Island, Massachusetts. Infectious cyclic thrombocytopenia in dogs is caused by *A. platys* in the United States, especially in the Southeast and Midwest, and is transmitted by ticks. Further, this pathogen has been reported in several countries in Europe, Asia, and South America where it occasionally has been detected in cats, sheep, and impalas (Alleman and Wamsley 2008). *A. marginale* occurs worldwide in cattle, producing bovine anaplasmosis, which is also known as gall sickness. Although *A. marginale* does not infect people, it is economically important because of its impact on livestock. Additionally, the pathogen exists in bison and mule deer (Aubry and Geale 2011). *A. centrale* infects cattle but produces a milder illness than *A. marginale.* This pathogen has not been reported in North America (Aubry and Geale 2011).

18.4 HOW HUMANS CONTRACT EHRLICHIOSIS OR ANAPLASMOSIS

Most people in the United States contract ehrlichiosis or *E. ewingii* ehrlichiosis after being bitten by an infected lone star tick. Anaplasmosis is contracted from being bitten by blacklegged ticks in the eastern United States or western blacklegged ticks in California. The disease can be contracted through a blood transfusion or organ transplant. The incidence of anaplasmosis increases with human age and is highest in people over 65 years old (Figure 18.7). The risk of anaplasmosis is higher for people who live in rural areas, work with wildlife or domestic animals, or spend time in areas where ticks are abundant.

18.5 MEDICAL TREATMENT

Ehrlichiosis and anaplasmosis are difficult to diagnose because their symptoms vary so much among patients and are similar to other diseases. Furthermore, diagnostic tests based on antibodies will produce negative results for the first week or two after the onset of disease; time is required for a person to produce a sufficient number of specific antibodies against these pathogens to be detectable in diagnostic tests. Antibiotic treatments are more effective if started early, so medical physicians usually start patients on antibiotics before a diagnosis of ehrlichiosis can be confirmed. A presumptive diagnosis of ehrlichiosis or anaplasmosis is easier to reach if the patient remembers being recently bitten by a tick while in an endemic area or if family members or pets have been diagnosed with the same disease (CDC 2012). Patients with ehrlichiosis are often clustered together in both time and space (Sidebar 18.1).

Doxycycline is the preferred antibiotic for use when ehrlichiosis or anaplasmosis is suspected. This antibiotic, if started early, can minimize the risk of complications. Adults are usually given doxycycline until at least 3 days after the fever subsides. The Center for Diseases Control and Prevention (CDC) also recommends doxycycline to treat suspected cases of ehrlichiosis or anaplasmosis in children. *Ehrlichia* has not developed resistance to this antibiotic; thus, relapses of disease have not been observed after completion of the recommended antibiotic treatment. Still,

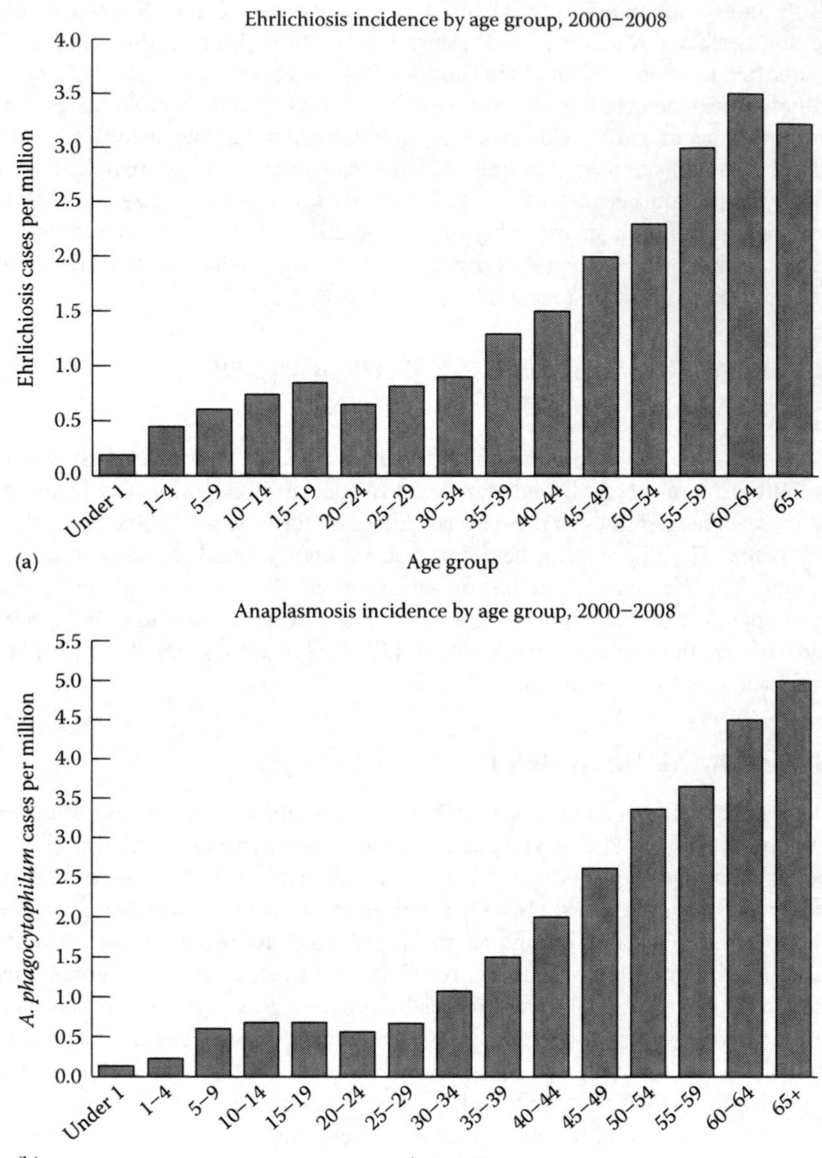

FIGURE 18.7 Annual incidence rate of (a) ehrlichiosis and (b) anaplasmosis among people of different ages. (From CDC, Anaplasmosis: Symptoms, diagnosis, and treatment, http://www.cdc.gov/Anaplasmosis/symptoms/index.html, 2012, accessed November 1, 2012. With permission.)

SIDEBAR 18.1 OUTBREAK OF EHRLICHIOSIS IN A TENNESSEE RETIREMENT COMMUNITY (STANDAERT ET AL. 1995)

During 1 week in June 1993, four men were hospitalized in the same local hospital; each complained of similar problems: fever, abdominal pain, vomiting, and a severe headache. Blood tests revealed that all had low counts of platelets and white blood cells. One patient developed acute renal and respiratory failure, followed by a coma. All four patients were diagnosed with acute ehrlichiosis and later recovered from their illnesses.

All four patients lived at the same retirement community in Cumberland County, Tennessee, which was adjacent to an 80,000 acre (32,389 ha) wildlife management area (Figure 18.8). The community was located on 15,000 acres (6,073 ha) of wooded land and contained four golf courses. White-tailed deer and other wildlife were abundant in the community. All four patients remembered being bitten by ticks while at the retirement community.

An investigation identified seven more people from the county who had confirmed cases of acute ehrlichiosis during the prior 6 months. Six of the seven new patients also lived in the same retirement community as the initial four victims. Tests revealed that 12% of residents from that endemic community (Figure 18.8) were seropositive for *E. chaffeensis*. In another golf-oriented retirement community from the same area, only 3% of residents were seropositive. One difference between the two communities was that thousands of lone star ticks were recovered from the endemic community but only three ticks from the reference community.

A disease risk assessment was conducted among residents from the endemic community. Golfers were more likely to be seropositive for ehrlichiosis than non-golfers. Among golfers, those who retrieved golf balls that were hit off the fairway and into the tall grass or woods were 3.7 times more likely to be infected than those who just used a new ball. This was likely due to higher tick densities in these areas. Interestingly, poorer golfers had a higher risk of infection than good golfers. Apparently in this golfing community, keeping one's golf ball on the fairway is healthy.

some patients may experience headaches, malaise, and weakness for several weeks after the completion of an antibiotic regime. Other antibiotics, including broad spectrum ones, are less effective for treating ehrlichiosis or anaplasmosis; use of sulfa drugs during an acute infection of ehrlichiosis or anaplasmosis may make the illness worse (Wormser et al. 2006, CDC 2011, 2012).

18.6 WHAT PEOPLE CAN DO TO REDUCE THEIR RISK OF CONTRACTING EHRLICHIOSIS OR ANAPLASMOSIS

Antibiotic treatment after a tick bite is ineffective in preventing ehrlichiosis or anaplasmosis in humans. At best, it may simply delay the onset of symptoms.

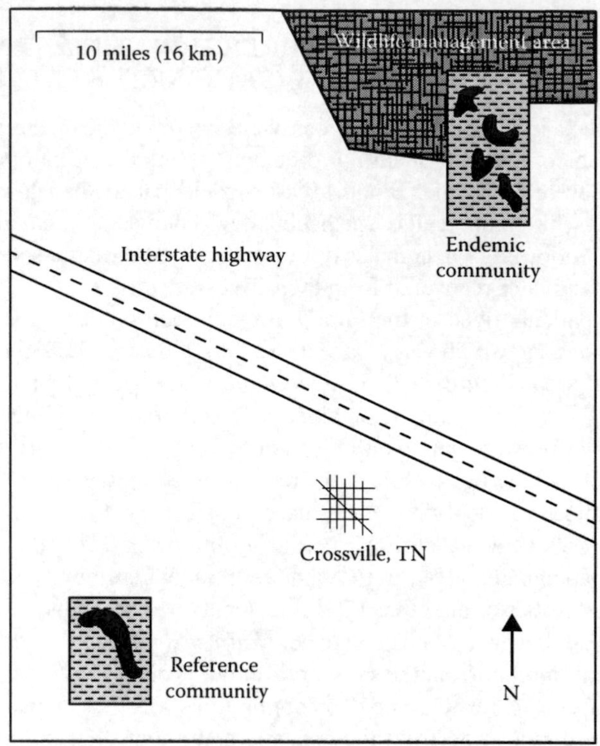

FIGURE 18.8 A map showing the retirement community located in Cumberland County, Tennessee, where ehrlichiosis was endemic and another retirement community that served as a reference community for a study by Standaert et al. (1995).

Instead, people who have been bitten by a tick in areas where *E. chaffeensis* or *A. phagocytophilum* are endemic should watch for symptoms of ehrlichiosis or anaplasmosis and seek medical attention if such symptoms develop (CDC 2011, 2012). Likewise, the rapid removal of a tick reduces the risk of infection but does not eliminate the risk entirely; a tick can transmit these pathogens within a few hours after attaching to a person.

To reduce the risk of contracting ehrlichiosis and anaplasmosis, the best approach is to limit exposure to ticks. One example is illustrative. In communities where ehrlichiosis is endemic, golfers have a higher risk of infection than non-golfers (Sidebar 18.1). Of course, most golfers are unwilling to give up the sport, but minor changes can decrease the risk of a tick bite. Wearing long sleeve shirts and long pants, inspecting oneself for ticks, and using an insect repellent will reduce the risk of being bitten by an infected tick. In a retirement community where ehrlichiosis was prevalent, golfers who did not use insect repellent had twice the risk of contracting ehrlichiosis than golfers who used insect repellents (Standaert et al. 1995). White-tailed deer preferred to bed down in areas where an exotic understory plant, Amur honeysuckle, provides abundant cover. By removing honeysuckle, Allan et al. (2010) was able to decrease the number of lone star ticks that were infected with *Ehrlichia.*

Cases of ehrlichiosis and anaplasmosis increase from less than 30 per month during January, February, and March to over 600 per month during June and July (Figure 18.9). This is not surprising because the ticks that transmit these diseases are more common and active during the summer than the winter, and people spend more time outside during the summer.

Several steps can be taken to reduce the risk of livestock becoming infected with *A. phagocytophilum* and developing tick-borne fever or pasture fever. One approach is to protect grazing livestock, especially pregnant females, from tick bites by keeping livestock in tick-free pastures or by treating them with an acaricide (i.e., a chemical used to kill mites or ticks). A second approach involves giving livestock tetracycline as a prophylactic measure against *A. phagocytophilum*. A third approach, based on this acquired immunity, is to infect susceptible animals with the pathogen to stimulate their immune systems and then treat them with oxytetracyclines before or immediately after the onset of fever. There is no vaccine against *A. phagocytophilum*, but sheep and cattle that have already been infected by *A. phagocytophilum* develop some immunity to the pathogen that can last for several months; during this period, subsequent infections by *A. phagocytophilum* are usually milder than the first one. There are many strains of *A. phagocytophilum*, and an acquired immunity to one strain only provides limited immunity to another. Hence, strains that are specific to the local area must be used for effectiveness (Merck 2010).

Humans can contract anaplasmosis from physical contact with infected deer. In the upper Midwest, three anaplasmosis patients contracted the disease after butchering an infected white-tailed deer. Each patient had numerous cuts on their hands and fingers when butchering, and all three used an electric saw that sent blood flying, so the patients may have either contracted the pathogen from a skin wound or by inhalation. None of the patients wore protective gloves or masks (Bakken et al. 1996). People can avoid the risk of infection when cleaning deer or other animals by wearing protective gloves and masks and by preventing blood from splattering. There is no evidence that reducing populations of wild ungulates can decrease the risk of humans, livestock, or companion animals being infected with *A. phagocytophilum*.

18.7 ERADICATING EHRLICHIOSIS OR ANAPLASMOSIS FROM A COUNTRY

Anaplasmosis is a serious disease in cattle and is often fatal. Its eradication is difficult in areas where the pathogen occurs in free-ranging deer. In such areas, there have been calls to eliminate *Anaplasma* from deer by reducing deer populations. Eliminating *Anaplasma* in deer is currently impossible given that infected animals cannot be easily identified and reducing deer populations is ineffective and controversial because many people are protective of deer. Bjöersdorff et al. (2001) estimated that during each spring, migratory birds bring into Sweden over 500,000 ticks infected with *A. phagocytophilum*. Their findings raise the concern that any eradication of the pathogen from an area would only be temporary because migratory birds might quickly reintroduce it.

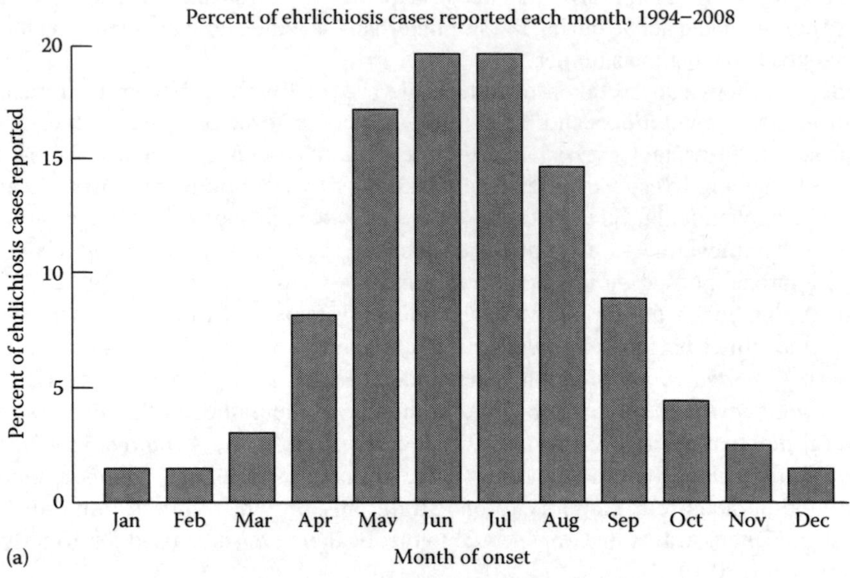

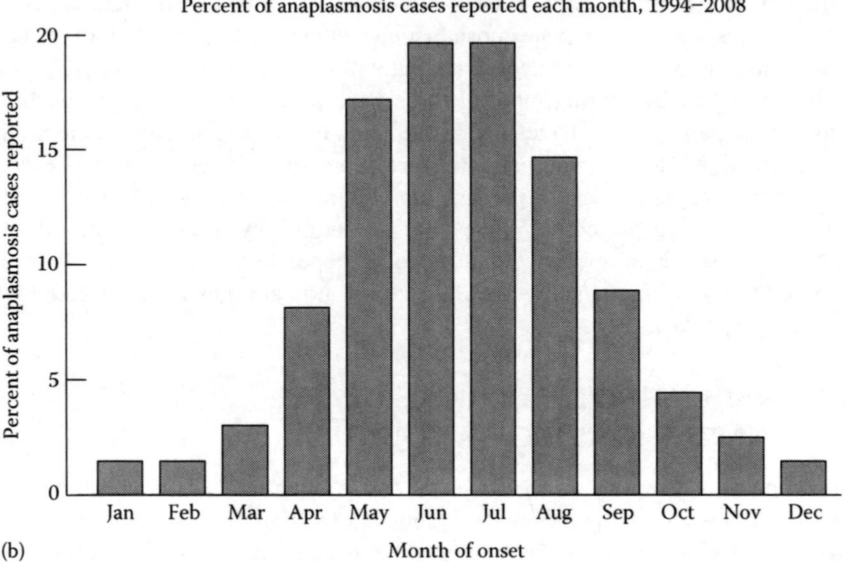

FIGURE 18.9 Seasonal changes in the number of reported cases of (a) ehrlichiosis and (b) anaplasmosis in the United States. (From CDC, Anaplasmosis: Symptoms, diagnosis, and treatment, http://www.cdc.gov/Anaplasmosis/symptoms/index.html, 2012, accessed November 1, 2012.)

LITERATURE CITED

Adams, D. A., K. M. Gallagher, R. A. Jajosky, J. Ward et al. 2012. Summary of notifiable disease—United States 2010. *Morbidity and Mortality Weekly Report* 59(53):1–111.

Allan, B. F., H. P. Dutra, L. S. Goessling, K. Barnett et al. 2010. Invasive honeysuckle eradication reduces tick-borne disease risk by altering host dynamics. *Proceedings of the National Academy of Sciences of the United States of America* 107:18523–18527.

Alleman, R. and H. L. Wamsley. 2008. An update on anaplasmosis in dogs. http://www.veterinarymedicine.com/dvm360 (accessed November 1, 2012).

Arens, M. Q., A. M. Liddell, G. Buening, M. Gaudreault-Keener et al. 2003. Detection of *Ehrlichia* spp. in the blood of wild white-tailed deer in Missouri by PCR assay and serologic analysis. *Journal of Clinical Microbiology* 4:1283–1285.

Aubry, P. and D. W. Geale. 2011. A review of bovine anaplasmosis. *Transboundary and Emerging Diseases* 58:1–30.

Bakken, J. S., J. K. Krueth, T. Lund, D. Malkovitch et al. 1996. Exposure to deer blood may be a cause of human granulocytic ehrlichiosis. *Clinical Infectious Diseases* 23:198.

Bjöersdorff, A., S. Bergstrom, R. F. Massung, P. D. Haemig, and B. Olsen. 2001. *Ehrlichia*-infected ticks on migrating birds. *Emerging Infectious Diseases* 7:877–879.

CDC. 2011. Ehrlichiosis: Symptoms, diagnosis, and treatment. http://www.cdc.gov/Ehrlichiosis/symptoms/index.html (accessed October 1, 2012).

CDC. 2012. Anaplasmosis: Symptoms, diagnosis, and treatment. http://www.cdc.gov/Anaplasmosis/symptoms/index.html (accessed November 1, 2012).

CDC. 2013. Summary of notifiable diseases—United States, 2011. *Morbidity and Mortality Weekly Report* 60(53):1–117.

de la Fuenta, J., M. W. Atkinson, J. T. Hogg, D. S. Miller et al. 2006. Genetic characterization of *Anaplasma ovis* strains from bighorn sheep in Montana. *Journal of Wildlife Diseases* 42:381–385.

Dergousoff, S. J. and N. B. Chilton. 2011. Novel genotypes of *Anaplasma bovis*, "*Candidatus* Midichloria" sp. and *Ignatzschineria* sp. in the Rocky Mountain wood tick, *Dermacentor andersoni*. *Veterinary Microbiology* 150:100–106.

Dumler, J. S., K.-S. Choi, J. C. Garcia-Garcia, N. S. Barat et al. 2005. Human granulocytic anaplasmosis and *Anaplasma phagocytophilum*. *Emerging Infectious Diseases* 11:1828–1834.

Foley, J. E., N. C. Nieto, J. Adjemian, H. Dabritz, and R. N. Brown. 2008. *Anaplasma phagocytophilum* infection in small mammal hosts of *Ixodes* ticks, western United States. *Emerging Infectious Diseases* 14:1147–1150.

Goethert, H. K. and S. R. Telford III. 2003. Enzootic transmission of *Anaplasma bovis* in Nantucket cottontail rabbits. *Journal of Clinical Microbiology* 41:3744–3747.

Kawahara, M., Y. Rikihisa, Q. Lin, E. Isogal et al. 2006. Novel genetic variants of *Anaplasma phagocytophilum, Anaplasma bovis, Anaplasma centrale*, and a novel *Ehrlichia* sp. in wild deer and ticks on two major island in Japan. *Applied and Environmental Microbiology* 72:1102–1109.

Merck. 2010. *Merck Veterinary Manual*, 10th edition. Merck, Whitehouse Station, NJ.

Ooshiro, M., S. Zakimi, Y. Matsukawa, Y. Katagiri, and H. Inokuma. 2008. Detection of *Anaplasma bovis* and *Anaplasma phagocytophilum* from cattle on Yonaguni Island, Okinawa, Japan. *Veterinary Parasitology* 154:360–364.

Petrovec, M., A. Bidovec, J. W. Sumner, W. L. Nicholson et al. 2002. Infection with *Anaplasma phagocytophila* in cervids from Slovenia: Evidence of two genotypic lineages. *Wiener Klinische Wochenschrift* 114:641–647.

Rosef, O., A. Paulauskas, and J. Radzijevskaja. 2009. Prevalence of *Borrelia burgdorferi* sensu lato and *Anaplasma phagocytophilum* in questing *Ixodes ricinus* ticks in relation to the density of wild cervids. *Acta Veterinaria Scandinavica* 51:47.

Skoracki, M., J. Michalik, B. Skotarczak, A. Rymaszewska et al. 2006. First detection of *Anaplasma phagocytophilum* in quill mites (Acari: Syringophilidae) parasitizing passerine birds. *Microbes and Infection* 8:303–307.

Standaert, S. M., J. E. Dawson, W. Schaffner, J. E. Childs et al. 1995. Ehrlichiosis in a golf-oriented retirement community. *New England Journal of Medicine* 333:420–425.

Stuen, S. 2007. *Anaplasma phagocytophilum*—The most widespread tick-borne infection in animals in Europe. *Veterinary Research Communications* 31(Supplement 1):79–84.

Thomas, R. J., J. S. Dumler, and J. A. Carlyon. 2009. Current management of human granulocytic anaplasmosis, human monocytic ehrlichiosis and *Ehrlichia ewingii* ehrlichiosis. *Expert Review of Anti-Infective Therapy* 7:709–722.

Wormser, G. P., R. J. Dattwyler, E. D. Shapiro, J. J. Halperin et al. 2006. The clinical assessment, treatment, and prevention of Lyme disease, human granulocytic anaplasmosis, and babesiosis: Clinical practice guidelines by the Infectious Diseases Society of America. *Clinical Infectious Diseases* 43:1089–1134.

Section IV

Viral Diseases

19 Rabies

In the exaggerated reflexes so characteristic of rabies, the author has watched rabid dogs snarl and attack driven by an encompassing internal threat, rabid foxes with mouths wide open and with paralyzed rear legs alternately attempting to strike or flee from any creature that moved nearby… and human rabies patients express unrelented fear as even the stimulus of a slight movement of air across their skin set off painful spasms. Yet this is not all, as some animal patients die quietly and some human patients are calm and at peace; characteristically both are alert or lucid until the final coma from which there is only the rarest return.

Beran (1994)

19.1 INTRODUCTION AND HISTORY

Rabies is one of the most terrifying diseases, owing both to its disturbing symptoms and its fatal outcome. It is an ancient disease described in some of man's earliest writings. The name rabies comes from the Latin word rabere, meaning to rave. Descriptions of rabies appear in the writings of ancient Mesopotamians (2300 BC) and Chinese (782 BC), indicating that it had already spread throughout Europe and Asia in antiquity (Rupprecht et al. 2001). In 1885 BC, the Middle Eastern city of Eshnunna required that "if a dog is mad and the authorities have informed the owner and he does not cage it, if it then bites a man and causes his death, the owner of the dog shall pay [40 shekels] of silver" (Beran 1994). The Greek philosophers Democritus, Aristotle, and Hippocrates described the disease around 400 BC. Rabies epidemics swept Europe throughout the Middle Ages (Beran 1994, Rupprecht et al. 2001, Conover 2002).

The rabies virus may have reached the Americas during the Ice Age when Arctic foxes crossed the Bering Land Bridge. Alternatively, Spanish conquistadors or other Europeans may have brought it with them (Conover 2002). Early Eskimo folklore describes a rabies-like disease in sled dogs and humans after being bitten by a fox. During the 1700s and 1800s, this fox strain of rabies had spread into southeastern Canada and the northeastern United States. The first recognized case in Canada occurred in 1819 when the governor general contracted rabies from a pet fox. Early explorers and pioneers of the Great Plains and California feared rabid animals. While the origin of rabies in bats, foxes, and skunks in North America is shrouded by the mists of time, it is much clearer that as Europeans moved across the world, they inadvertently transported the canine strain of rabies with them from Europe.

Canine-strain rabies was first recognized in Virginia in 1753 and in North Carolina in 1772. By 1785, rabies outbreaks were occurring throughout the 13 original states, and by 1860, canine rabies had spread across the entire continent. Elsewhere, canine rabies was introduced into South Africa during the late 1700s; into Peru, Argentina, Uruguay, and Columbia in the early 1800s; into Hong Kong in

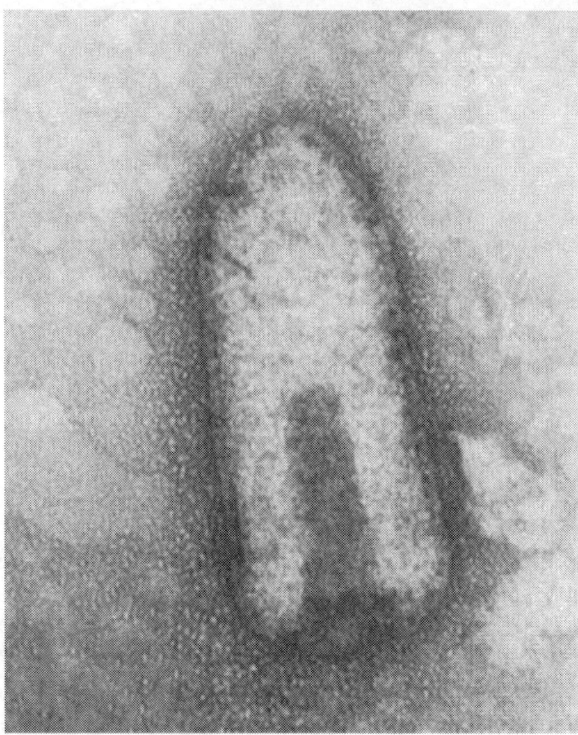

FIGURE 19.1 Electron micrograph of the rabies virus. (Courtesy of the Institute of Epidemiology, Friedrick-Loeffler Institute.)

the 1850s; and into China in the 1860s (Beran 1994). During the nineteenth century, Europeans also inadvertently brought mongoose-strain rabies to several Caribbean Islands including Cuba, Hispaniola, Grenada, and Puerto Rico when they imported the small Asian mongoose in a failed attempt to control rats in sugar cane plantations (Rupprecht et al. 2001).

Rabies is caused by a virus that belongs to the genus *Lyssavirus* (lyssa is the ancient Greek word for rabies); this genus is in the family Rhabdoviridae. All species of *Lyssavirus* are single-stranded RNA viruses with a distinctive bullet shape (Figure 19.1). The outer membrane is composed of lipids and is covered with spike-like projections. All species of *Lyssavirus* have evolved to infect and replicate in mammals; they cannot survive for long outside of their hosts. Rabies probably evolved in Africa, and bats are the reservoir hosts for most species of *Lyssavirus* (Rupprecht et al. 2001, Schatz et al. 2013).

At present, rabies is endemic throughout most of the world including the Americas, Europe, Africa, and Asia. During 2004, an estimated 55,000 people died of rabies with 56% of deaths in Asia and 44% in Africa. Most victims (84%) lived in rural areas. Children are most at risk because they are more likely to be bitten by dogs and suffer multiple bites to the head and neck. Such bites are more likely to result in rabies than a single bite to a different body part. Rabies costs African and

SIDEBAR 19.1 BAT-STRAIN RABIES CAUSES TWO HUMAN FATALITIES IN THE UNITED STATES DURING 2011 (BLANTON ET AL. 2012)

A 46-year-old woman sought medical attention at a South Carolina hospital during December 2011. Her symptoms included dizziness, shortness of breath, sweating, and numbness in the hands. Family members knew of no bites sustained by the patient but reported multiple occasions when bats were seen in the home, including one when the patient awoke to find a bat in her bedroom. Tissue sample sent to the CDC revealed that the woman had bat-strain rabies associated with the Mexican free-tailed bat (Figure 19.2). The patient's condition did not improve, and she died 16 days after being hospitalized.

That same month, a 63-year-old man visited an emergency room in Massachusetts complaining of elbow pain and a loss of appetite. Soon after being admitted to the hospital, his condition worsened, including the development of hydrophobia. Rabies was suspected, and samples sent to the CDC; testing confirmed bat-strain rabies associated with the bat genus *Myotis*. The patient reported waking to find a bat in his bedroom 2–3 months earlier. The patient was started on an experimental treatment for rabies, but his condition did not improve. He died 28 days after being hospitalized.

Asian countries more than $500 million annually with most money spent on postexposure treatment of bite victims (WHO 2005, CDC 2011). There were nine human cases of rabies in Europe during 2012. Rabies victims were in Belarus, Romania, Russia, Switzerland, Lichtenstein, Turkey, and the United Kingdom.

In the United States, 32 people contracted rabies from 2002 through 2011; of these, eight were infected in a foreign country, and 24 were infected while in the United States (Blanton et al. 2012). Of the 24 rabies patients infected in the United States, 21 were infected with bat-strain rabies (Sidebar 19.1).

After the rabies virus gains access into a mammal's body, usually through the bite of a rabid animal, it enters an eclipse phase that may last days or months. During this phase, the virus is hard to detect in the host and is vulnerable to cell-mediated immune responses, especially in animals that have been previously vaccinated for rabies. The viruses that do survive can infect either nerve cells near the wound or other cells. The infection then spreads along the neural pathways to the central nervous systems and the brain where the virus can replicate rapidly (Rupprecht et al. 2001, CDC 2011).

19.2 SYMPTOMS IN HUMANS

Most rabies patients become infected when the saliva of a rabid animal gains entry into the human body, usually after a bite. People can also become infected when the virus is inhaled or crosses mucosal surfaces, but the virus is unable to cross intact human skin (WHO 2005). The rabies incubation period in humans is usually several weeks but can be as short as 10 days and as long as several months or years.

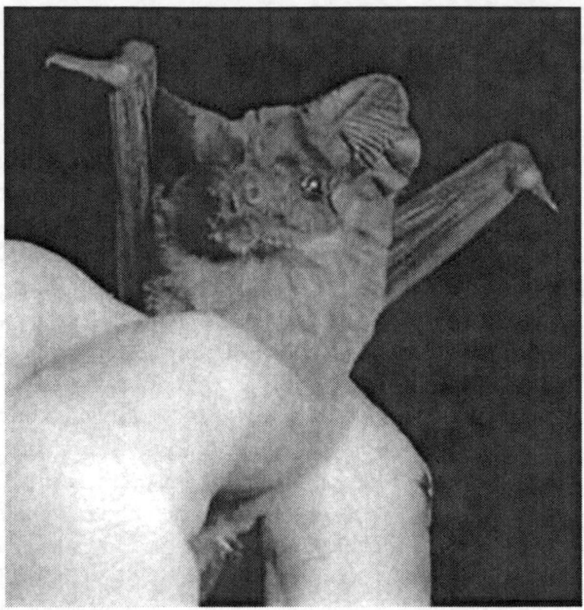

FIGURE 19.2 Photo of a Mexican free-tailed bat. (Courtesy of the U.S. National Park Service.)

Short incubation periods are associated with severe bites to the head, face, neck, or other parts of the body that contain high concentrations of nerves (Rupprecht et al. 2001, CDC 2011).

The first symptoms of rabies may be similar to influenza and can include loss of appetite, vomiting, diarrhea, and restlessness. There may also be a tingling, prickly, or itchy sensation around the bite wound; these localized sensations are believed to result from a viral infection of nerves. These sensations may progressively ascend toward the central nervous system. Weakness and trembling may develop in the affected limb. Patients progress into a hyperactive phase with enhanced sensitivity to visual, auditory, and tactile stimuli, which also elicit spasms that alternate with periods of agitation, confusion, and lucidity. Some patients exhibit unprovoked agitation, rage, or aggression directed at objects, animals, or people. Excessive salivation, perspiration, and watering from the eyes typically occur. When patients attempt to swallow, fluids are expelled forcefully through the mouth and nasal passages due to an exaggerated respiratory reflex caused by enhanced sensitivity of the oropharynx to the contact of fluid. This reflex is both terrifying and painful to the patients, who often develop a fear of swallowing liquids; this symptom provides one of the common names of rabies: hydrophobia, which means a fear of water (Beran 1994). In India, almost all (96%) rabies patients developed hydrophobia, 66% aerophobia (fear of the pain caused by wind or a draft), and 33% photophobia (Sudarshan et al. 2007).

This hyperactive phase is followed by a paralytic phase with the patient gradually developing stupor, paralysis, delirium, and coma. Once a rabies infection has progressed to the point where symptoms of rabies appear, it is almost always fatal to humans. Death may result from cardiac arrest or respiratory failure (Beran 1994).

19.3 RABIES VIRUS INFECTIONS IN ANIMALS

Clinical signs of rabies in wildlife involve changes in normal behavior that progressively become more abnormal. These clinical signs may include decreased wariness toward humans or domestic animals, a change in activity cycles (e.g., nocturnal animals being active in the day), extreme aggressive behavior, unusual vocalizations, and odd behavior (e.g., social animals seeking solitude). Animals may exhibit anger, extreme aggression, or furious behavior at the slightest provocation. This form of the disease is referred to as furious rabies or the mad-dog syndrome, although it can occur in any species. A dumb or paralytic form of rabies also exists in which animals are not vicious and rarely bite. Early clinical signs of rabies often include paralysis of the throat and lower jaw, resulting in drooling and the inability to swallow. Rabid dogs and livestock often have their mouths partially open and can appear to have something stuck in their mouth or throat. In response, owners frequently examine the mouths of rabid animals, exposing themselves to the rabies virus. Animals may have difficulty walking, become lethargic or prostrate, and exhibit paralysis that may start in the legs and progressively spread. This phase ends with coma and death, usually through heart, lung, or multiple-organ failure (Merck 2010).

One clinical sign of rabies in cattle is an unusual bellow that occurs intermittently. The normal, placid expression of cattle is replaced with alertness: rabid cattle follow movements with their eyes and ears. Horses and mules become distressed and agitated. Rabid livestock frequently have self-inflicted wounds. Rabid cattle and other livestock are dangerous because they often pursue and attack people and other animals. Such attacks can result in human fatalities.

In addition to infecting nerves, the rabies virus is able to infect the salivary glands, which accounts for the high density of the virus in saliva. Rabies is transmitted from one animal to another primarily through biting or exposure to the saliva of infected animals. There are more than 4000 species of mammals worldwide, and all are susceptible to an infection of rabies. Yet, very few species serve as reservoir hosts because in most species, the rabies virus kills its host so fast that there is not enough time for it to infect a new host. Strains of the rabies virus (hereafter called variants) have evolved that can maintain themselves by infecting specific mammalian species.

In North America, several variants of rabies exist and are named after their reservoir host. Domestic dogs and coyotes are the reservoir hosts for the canine variant of rabies found mostly in Mexico and southern Texas. The red fox and Arctic fox are the reservoir hosts of one fox variant in Alaska, Canada, and northeastern states, while the gray fox is the reservoir host for different fox variants in Texas and still another variant in Arizona and New Mexico. The striped skunk is the reservoir host for two different skunk variants: one infecting skunks in the southern Great Plains and another in the northern Great Plains. A third skunk variant infects skunks in California. There are several different bat variants that use different bat species as reservoir hosts (Rupprecht et al. 2001, Merck 2010).

Rabies variants associated with one reservoir host can spill over to another species when one rabid animal bites an individual of another species. For example, a dog can become infected by a raccoon bite and die from raccoon-variant rabies.

However, the raccoon variant cannot maintain itself through dog-to-dog transmission. The current rabies situation in different parts of the world is provided here.

19.3.1 Rabies in India

More than 20,500 people contract rabies in India each year. Most of the patients were males (71%) and from rural areas (76%). Most patients (96%) were infected by a rabid dog bite and 2% by a rabid cat bite. In most cases, the rabid dog was a stray (75%), a pet (11%), feral or free-ranging (4%), or unknown (10%). Relatively few patients were infected after being bitten by wildlife (Sudarshan et al. 2007).

19.3.2 Rabies in Africa

Rabies occurs across Africa, but reliable information on its presence in wildlife exists only for South Africa where two variants of rabies currently exist. The mongoose variant is associated with species of the Herpestidae (mongooses and suricates) and Viverridae (genets and civets). The yellow mongoose is its main reservoir host for the mongoose variant in South Africa, and the slender mongoose is the reservoir host in Zimbabwe (WHO 2005, Van Zyl et al. 2010). The other strain is the canine variant, which did not reach South Africa until 1950. Since then, canine-strain rabies has spread to several wildlife populations and is endemic in black-backed jackals, side-striped jackals, and bat-eared fox. There is concern that canine rabies threatens the survival of African wild dogs and Ethiopian wolves (WHO 2005).

19.3.3 Rabies in Europe

Canine-strain rabies has been widespread across Europe for centuries, but it started to disappear at the beginning of the 1900s for unknown reasons. During World War II, an outbreak of rabies began in eastern Europe and spread west in a wave-like fashion. It reached Germany during 1940, France during 1968, and Italy during 1980. Red foxes were the reservoir host. Rabies has been reported in several other species, but these are accidental hosts, and rabies cannot maintain itself by infecting them. Included in these accidental hosts are roe deer, cattle, and other livestock. The progress of the rabies wave stopped in areas where an oral vaccination program was successful in immunizing a large proportion of the fox population. The raccoon dog is susceptible to rabies; whether it is a reservoir host for rabies is unclear. Wolves are an accidental host for the rabies virus. Once a pack member is infected, the virus spreads quickly throughout the pack; wolves are territorial, however, and opportunities for the virus to spread from one pack to another are too limited to allow the virus to maintain itself by just infecting wolves (WHO 2005).

During 2012 in Europe, rabies was confirmed in 2462 domestic animals, 2382 wildlife species, and 40 bats. Wildlife infected with rabies included 2012 foxes, 188 raccoon dogs, 50 European marten, 21 wolves, 16 badgers, 8 wild boars, 6 roe deer, 6 fallow deer, and 2 raccoons (WHO 2012). Four different lyssavirus species have been identified among European bats. There are two different European bat lyssaviruses: EBLV 1, which is associated with the serotine bats

SIDEBAR 19.2 WILDLIFE CONSERVATIONIST IN SCOTLAND DIES OF BAT RABIES (FOOKS ET AL. 2003)

On November 8, 2002, a 55-year-old wildlife biologist in Scotland visited his physician and complained of a feeling of pins and needles in his left hand, pain in his left arm, and a stiff neck. He was prescribed pain killers, which provided little relief. After vomiting blood on November 11, he was referred to the Ninewells Hospital (Dundee, Scotland). The bloody vomit was attributed to his consumption of the pain killers. He complained of a persistent pain in his left arm, weakness in both arms, and difficulty swallowing. A computer tomography scan of this head found nothing unusual, cerebrospinal fluid did not reveal any pathogens, and a magnetic resonance image of the brain suggested increased density of the left cerebellum but otherwise was normal. On November 13, the patient was confused, aggressive, and agitated. He was transferred to the intensive care unit the next day for mechanical ventilation. He exhibited excessive salivation, and flaccid paralysis affected all of his limbs. Rabies was suspected and tissue samples were collected. On November 22, rabies was confirmed, and the rabies virus was identified as European bat lyssavirus 2.

Prior to the illness, the patient had been working with bats for the Scottish Natural Heritage. He had handled bats without wearing gloves on several occasions during 2002 and had been bitten in the past. The last known bite occurred on September 29 when a Daubenton's bat (Figure 19.3) bit him on the ring finger of his left hand. The patient had not been vaccinated against rabies. The patient died on November 24, 2002.

and isabelline bats, and EBLV 2 associated with pond bats and Daubenton's bats. There also is a Bokeloh bat lyssavirus (BBLV) associated with Natterer's bat and a west Caucasian bat virus (WCBV) associated with Schreiber's long-fingered bat (Schatz et al. 2013). The presence of EBLV in bats poses a public health problem in European countries. From 1977 to 2008, three people died of confirmed cases of EBLV rabies (Sidebar 19.2), and two other deaths were suspected but not confirmed (Dacheux et al. 2008).

19.3.4 Rabies in North America

During the 1800s and first part of the 1900s, the canine variant was endemic in the United States, but a massive vaccination program of dogs eradicated this variant. Unfortunately, canine-variant rabies moved north from Mexico to Texas and became endemic in coyotes. During 1994, the same variant jumped to parts of Alabama and Florida, where it appeared in domestic dogs. Apparently, these new foci were established when rabid animals were translocated into the new areas. During 1988, a new canine variant was found in gray fox in west-central Texas (Velasco-Villa et al. 2008).

FIGURE 19.3 Photo of a Daubenton's bat. (Courtesy of Gilles San Martin.)

In North America, rabies is not limited to dogs, foxes, and coyotes. Approximately 120,000 animals are tested for rabies each year in the United States, and about 6% are found to be rabid (Figure 19.4). More than 90% of rabid animals in the United States are wildlife species. Rabies was confirmed in 6031 animals in the United States during 2011 including 1981 raccoons (38% of all rabid animals), 1627 skunks (27%), 1380 bats (23%), 427 foxes (7%), 303 cats (5%), 70 dogs (1%), and 65 cattle (1%; Blanton et al. 2012). Reservoir hosts for the rabies virus in the United States include raccoons, skunks, foxes, and bats. Raccoon-strain rabies is endemic in the eastern states; skunk-strain in the Great Plains and California; and fox-strain in Arizona, New Mexico, and Texas (Figure 19.5). Additionally, mongoose-strain rabies is endemic in Puerto Rico where mongooses are the reservoir host. The Arctic fox and domestic dogs are the reservoir hosts for rabies in the Arctic (CDC 2011, Blanton et al. 2012).

Many bats in the United States have had past exposure to the rabies virus, especially the hoary bat and Mexican free-tailed bat (Table 19.1). In general, bat species that are solitary or live in small groups are less likely to test positive for bat virus antibodies than species that roost in large groups. Rabies in bats is caused by several variants of the rabies virus that have adapted to different bat species. Bats collected during the fall were more likely to have antibodies against the rabies virus (12%) than bats collected during the winter (3%), spring (8%), or summer (5%). Bats from the southwestern United States were more likely to have antibodies against the rabies virus (19%) than bats in the Northwest (7%), Southeast (7%), and Northeast (4%). Most (71%) people infected in the United States by rabid bats from 1993 to 2000 were infected by the rabies variant associated with the silver-haired bat and the eastern pipistrelle; 20% of people were

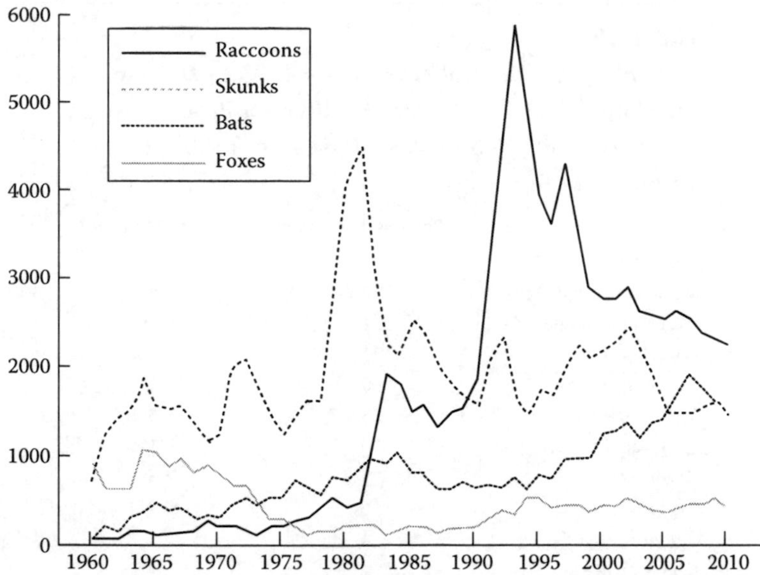

FIGURE 19.4 Numbers of rabid raccoons, skunks, bats, and foxes confirmed in the United States from 1960 to 2011. (From CDC, Rabies surveillance data in the United States. http://www.cdc.gov/rabies/location/usa/surveillance/wild_animals.html, 2012.)

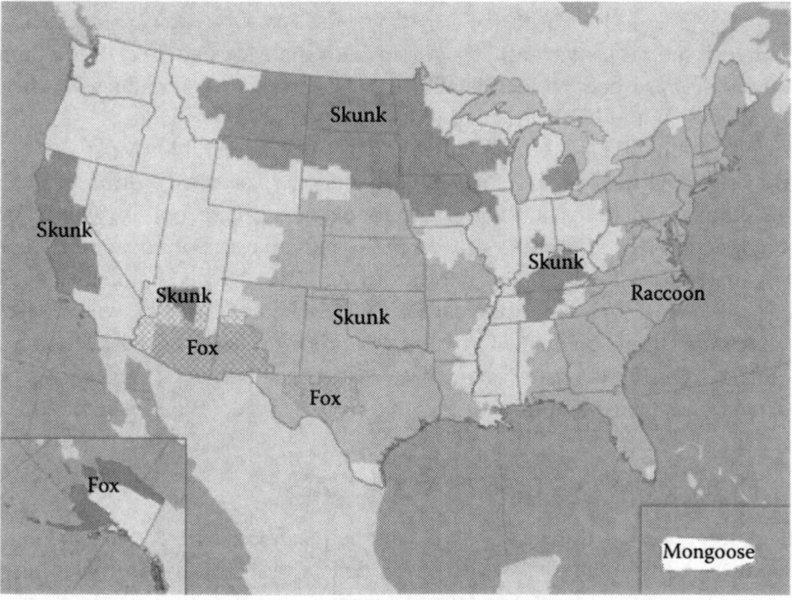

FIGURE 19.5 Distribution of raccoon, skunk, fox, and mongoose strains of rabies in the United States. (From CDC, Rabies surveillance data in the United States. http://www.cdc.gov/rabies/location/usa/surveillance/wild_animals.html, 2012.)

TABLE 19.1
Percentage of Bats Collected from 1993 to 2000 in the United States that Were Positive for Past Exposure to the Rabies Virus Based on Antibody Testing of Brain Issue

Bat Species	% Positive for Rabies	No. of Bats Tested
Hoary bat	38	254
Mexican free-tailed bat	32	673
Western pipistrelle	21	193
Pallid bat	21	100
Eastern pipistrelle	20	117
Silver-haired bat	13	566
Long-eared myotis	10	196
Eastern red bat	9	520
Big brown bat	6	20,911
California myotis	4	388
Keen's myotis	2	572
Little brown myotis	2	5,721
Yuma myotis	2	241

Source: Adapted from Mondul, A.M. et al., *J. Am. Vet. Med. Assoc.*, 222, 633, 2003.

infected with the rabies variant associated with the Mexican free-tailed bat. One or two people have been infected with rabies variants associated with the little brown myotis bat and big brown bat (Mondul et al. 2003).

Canada identified 115 animals during 2011 that were rabid. Most (92%) were wildlife, 3% were livestock, and 5% were cats or dogs. The rabid wildlife included 47 bats, 42 skunks, and 16 foxes. Rabies was not detected in raccoons or wolves during 2011 but was detected in a wolf a year later. No human cases of rabies were reported among Canadians who contracted rabies during 2011. Mexico confirmed rabies in 148 animals during 2011; 82% were cattle, 2% were horses, and 14% were dogs. Only four wild animals were confirmed as rabid, but this is probably because few were sent in for testing. During the same year, three Mexicans contracted rabies; two people contracted rabies from vampire bats and one from a skunk (Blanton et al. 2012).

19.4 HOW HUMANS CONTRACT RABIES

Rabid mammals shed the rabies virus in their saliva; both people and animals become infected from contact with virus-laden saliva. Usually, animals only shed the virus when they exhibit the signs of rabies, but some domestic dogs, cats, and ferrets can shed the virus for several days prior to the onset of the disease; skunks can shed the virus for up to 8 days prior to the onset. The rabies virus cannot penetrate intact human skin, but people can become infected when the virus is introduced into a bite or scratch wounds, open cuts, or skin abrasions. If inhaled, the virus can also gain entry through the lungs or through mucous membranes of the eyes, mouth, or nose. Most rabies patients, however, are infected after being bitten by a rabid animal. The risk of infection is greatest when people are bitten by animals that have long, sharp teeth (e.g., coyotes and dogs) because their bites can insert the virus deep into muscles (Merck 2010).

Being bitten by a rabid animal does not always result in human rabies, but the risk increases with the number and severity of the bites. The risk also depends on which part of the body is bitten. Knobel et al. (2005) calculated that in Africa the probability of developing rabies after a bite by a rabid dog was 45% if bitten on the head, 28% if bitten on arms or hands, and 5% if bitten on the feet or legs. A few people contracted rabies after visiting caves containing roosting bats. Rabid bats shed the virus in their saliva, and tiny droplets can be inhaled by people in the cave.

Person-to-person transmission of rabies is possible but rarely happens because a person is not contagious until symptoms of rabies have developed. People can acquire rabies from organ transplants if the donor was infected, but this is very rare. Nonetheless, four of eight rabies patients in the United States during 2004 were infected from organ transplants (Table 19.2). During 2013, a patient in Maryland died of raccoon-variant rabies after receiving an organ transplant (CDC 2013).

19.5 MEDICAL TREATMENT

All bite wounds should be washed immediately for a minimum of 15 minutes with copious amounts of soap, water, and an antiviral antiseptic to remove as much saliva as possible. Anyone who has been bitten by an animal should seek medical attention so that a physician can ascertain what medical steps are appropriate after consulting with local public health authorities. Often, these medical steps include administration of a complete series of rabies vaccines, beginning as soon as possible. Since the 1980s when new cell culture vaccines were developed, there have been no documented cases of rabies in the United States among patients who completed the rabies postexposure prophylaxis (rabies PEP). In the United States, rabies PEP includes human rabies immunoglobulin given on day 0 of the vaccine regime and a series of four doses of rabies vaccine (Table 19.3), which are given on days 0, 3, 7, and 14. For immune-compromised patients, five doses of rabies vaccine and HRIG may be provided. The vaccine is given in an upper-arm muscle. A tetanus vaccine may also be warranted (WHO 2005, CDC 2011).

Factors that physicians will consider in deciding whether to initiate rabies PEP include the species of animal, availability of the animal for observation or testing, nature of the contact or exposure, presence of rabies in the geographic area, and vaccination history of the animal. A history of rabies vaccination in an animal is not always a guarantee that the biting animal is not rabid.

Many U.S. rabies patients acquired the rabies virus from a bat and did not seek medical attention either because the patients dismissed the seriousness of the bites or were unaware that they had been bitten. The CDC (2011) recommends that the rabies PEP be initiated whenever someone has been bitten, scratched, or had mucous membrane exposure to a bat unless the bat has been tested and is negative. Rabies

TABLE 19.2
Human Cases of Rabies in the United States from 2002 to 2011

Date of Onset	Patient's Sex	Patient's Age	Patient's Location	Mode of Infection	Place of Infection	Rabies Strain
Dec. 2011	M	63	MA	Contact	United States	Bat (*Myotis*)
Dec. 2011	F	46	SC	Unknown	United States	Bat (Mexican free-tailed)
Sep. 2011	M	40	MA	Contact	Brazil	Dog
Aug. 2011	M	25	NY	Contact	Afghanistan	Dog
Jun. 2011	F	73	NJ	Bite	Haiti	Dog
Apr. 2011	F	8	CA	Unknown	United States	Unknown
Dec. 2010	M	70	WI	Unknown	United States	Bat (eastern pipistrelle)
Aug. 2010	M	10	LA	Bite	Mexico	Bat (vampire)
Oct. 2009	M	55	MI	Contact	United States	Bat (unknown)
Oct. 2009	M	42	VA	Contact	India	Dog
Oct. 2009	M	43	IN	Unknown	United States	Bat (eastern pipistrelle)
Feb. 2009	F	17	TX	Contact	United States	Bat (unknown)
Nov. 2008	M	55	MO	Bite	United States	Bat (silver-haired)
Mar. 2008	M	16	CA	Bite	Mexico	Fox
Sep. 2007	M	46	MN	Bite	United States	Bat (unknown)
Nov. 2006	M	11	CA	Bite	Philippines	Dog
Sep. 2006	F	10	IN	Contact	United States	Bat (silver-haired)
May 2006	M	16	TX	Contact	United States	Bat (Mexican free-tailed)
Sep. 2005	M	10	MS	Contact	United States	Bat (unknown)
Oct. 2004	M	22	CA	Unknown	El Salvador	Dog
Oct. 2004	F	15	WI	Bite	United States	Bat (unknown)
Jun. 2004	F	55	TX	Organ transplant	United States	Bat (Mexican free-tailed)
May 2004	M	18	TX	Organ transplant	United States	Bat (Mexican free-tailed)
May 2004	F	55	TX	Organ transplant	United States	Bat (Mexican free-tailed)
May 2004	F	50	TX	Organ transplant	United States	Bat (Mexican free-tailed)
Apr. 2004	M	20	TX (organ donor)	Bite	United States	Bat (Mexican free-tailed)
Feb. 2004	M	41	FL	Bite	Haiti	Dog
Aug. 2003	M	66	CA	Bite	United States	Bat (silver-haired)
May 2003	M	64	Puerto Rico	Bite	Puerto Rico	Mongoose
Feb. 2003	M	25	VA	Unknown	United States	Raccoon
Sep. 2002	M	20	IA	Unknown	United States	Bat (unknown)
Aug. 2002	M	13	TN	Contact	United States	Bat (eastern pipistrelle)
Mar. 2002	M	28	CA	Unknown	United States	Bat (Mexican free-tailed)

Source: Adapted from Blanton, J.D. et al., *J. Am. Vet. Med. Assoc.*, 241, 712, 2012.

TABLE 19.3
Rabies Vaccines and Immunoglobulins Used in the United States as Part of a Postexposure Vaccination Program

Drug Type	Drug Name	Mode of Administration
Purified chick embryo cell vaccine	RabAvert®	Intramuscular
Human diploid cell vaccine	Imovax® Rabies	Intramuscular
Human rabies immunoglobulin	HyperRAB® S/D	Local infusion around the wound and vaccination at distant site
Human rabies immunoglobulin	Imogam® Rabies-HT	Local infusion around the wound and vaccination at distant site

Source: CDC, Rabies, http://www.cdc.gov/rabies, 2011 (accessed May 1, 2013).

PEP should be considered if a bat was found in a room where someone was sleeping, and the bat is unavailable for testing.

Coyotes, foxes, raccoons, and skunks are the terrestrial mammals most likely to be rabid in the United States, but a bite by any wild carnivore can transmit the rabies virus to humans. The rabies PEP should be initiated as soon as possible after such exposure. The animal should be euthanized and tested for the rabies virus. The rabies PEP can be stopped if the animal tests negative for rabies (CDC 2011). Small mammalian pets (e.g., hamsters, domesticated guinea pigs, and gerbils), wild rodents (squirrels, chipmunks, rats, and mice), and wild rabbits and jackrabbits are almost never infected with the rabies virus and are not known to transmit rabies to humans. However, in areas where raccoon-variant rabies occurs, beaver and woodchucks (i.e., groundhogs) may be infected. Whenever someone is bitten by one of these animals, local and state health departments should be consulted (CDC 2011).

Dogs, cats, and livestock that are currently vaccinated against rabies should be immediately revaccinated if exposed to a rabid animal, and they should be observed for 45 days while under the owner's control. Any illness in the animal should be reported immediately to the attending veterinarian and the local health authorities. If symptoms suggest that the animal is rabid, it should be euthanized and its head shipped for testing. Unvaccinated dogs, cats, and livestock should be euthanized immediately. If the owner is unwilling to do this, the animal should be kept in isolation for 6 months. The proper course of action for dogs, ferrets, cats, and livestock with expired vaccinations should be evaluated on an individual basis by a public health veterinarian. Wild mammals or domesticated animals other than dogs, cats, and livestock should be euthanized if exposed to a rabid animal. When a wildlife species, such as a bat or wild mammal, bites someone, the animal should be captured and euthanized in a manner that leaves the brain intact so that it can be tested for rabies. State public health authorities should be contacted for guidance for other domestic animal exposures. No PEP regimen has been established for either domestic animals or wildlife (CDC 2011).

SIDEBAR 19.3 SURVIVING RABIES (WILLOUGHBY ET AL. 2005, HU ET AL. 2007)

A 15-year-old girl in Wisconsin was bitten on the finger by a bat, which she had rescued after it collided with a window. She did not seek medical attention, and no postexposure rabies vaccines were administered. A month later, she exhibited clinical signs of rabies including trouble with balance, difficulty with muscle coordination, blurred vision, weakness in the left leg, and difficulty walking. These clinical signs progressed to involuntary eye movement, tremors in arms, and slurred speech. She was transferred to the Medical College of Wisconsin where samples of serum and spinal fluid were collected and sent to the CDC. Tests of these samples confirmed the diagnosis of rabies. A decision was made by the parents to institute a program of aggressive care that combined an induced coma with antiviral and anti-excitatory drugs and supportive intensive care. Recovery was slow. On the 8th day in the hospital, she had an increase in antibodies against the rabies virus in her blood and spinal fluid. On the 12th day, she blinked when eye drops were administered and opened her mouth in response to pressure on the sternum. On the 19th day, she could wiggle her toes, squeeze hands in response to commands, and looked preferentially at her mother. On the 23rd day, she could sit up and hold her head erect. On the 76th day, she was discharged from the hospital. The patient continued her remarkable recovery. Twenty-seven months later, she had no difficulty interacting with her peers, completing college-level classes, or driving a car. She did, however, have an unsteady gait and some speech difficulties.

Once symptoms develop, rabies is almost always fatal in humans. Only a few people have recovered from rabies without receiving any rabies vaccines prior to the onset of symptoms (Sidebar 19.3). Of the 33 rabies cases in the United States from 2002 through 2011, all but three died. The survivors were 8-year-old, 15-year-old, and 17-year-old girls. Rabies patients remain conscious, and most are aware of the nature of their illness and are agitated. Patients with confirmed rabies should receive adequate sedation and emotional and physical support. Repeated intravenous morphine is effective in relieving the patient's agitation and anxiety. Patients should be cared for in private, quiet, and draft-free rooms (Blanton et al. 2012).

19.6 WHAT PEOPLE CAN DO TO REDUCE THEIR RISK OF CONTRACTING RABIES

Rabies should be suspected in any wild mammal that is behaving abnormally, exhibits no fear of humans, is active during the day, or is paralyzed. Rabid foxes, coyotes, and raccoons often invade yards and venture inside buildings where they may attack people or pets. Bats should be avoided if they are active in the daytime, fighting each other, attacking other animals, unable to fly, or resting on the ground. No bat should ever be handled with bare hands even if it is behaving normally. People who may have been exposed to rabid animals should seek medical attention.

Worldwide, 55,000 people die of rabies each year; yet rabies in humans is completely preventable with prompt medical care. The incidence of human rabies could be reduced with better education and better access to the rabies PEP in Asia and Africa where most fatalities occur. In India, for example, only 21% of rabies victims received rabies vaccines during 2002, while 60% of rabies victims resorted to indigenous treatment after being bitten. Indigenous treatments included faith healing and witchcraft (used by 17% of victims) and herbal therapy (6%). Only half of rabies victims were hospitalized; most others died at home (Sudarshan et al. 2007).

Two rabies vaccines are available for use in humans to prevent rabies prior to exposure: Imovax® Rabies uses human diploid cells, and RabAvert® uses purified chick embryo cells. Preexposure vaccines against rabies are not recommended for the general public, but they are recommended for people with an elevated risk of infection. These people include wildlife biologists, veterinarians, veterinarian assistants, wildlife rehabilitators, animal control officers, animal handlers, mammalogists, and spelunkers who visit caves containing bats. Preexposure vaccination involves three injections that are administered intramuscularly on days 0 and 7 and then on either day 21 or 28. The vaccine is given in the upper arm of adults and outer thigh of young children. People should be tested for rabies antibody titers every 2 years; a booster infection is recommended for people who have low levels of antibody titers. These preexposure vaccines do not eliminate the need for postexposure vaccines if a person is bitten or exposed to a rabid animal, but it does simplify the rabies PEP regimen. Another benefit of preexposure vaccines is that it helps protect people from unrecognized exposure to the rabies virus, such as a wildlife biologist who gets scratched while handling a cage or trap (CDC 2011).

Public health programs that require all domestic dogs to be vaccinated against rabies have been successful in eradicating canine-strain rabies in the United States, Canada, western Europe, Japan, and many parts of South America. Yet, canine-strain rabies remains widespread in the developing world and has been documented in more than 90 countries that contain half of the world's human population. Worldwide, 99% of rabies patients are infected from a dog bite. Recommended vaccination programs for dogs usually involve two vaccinations a year apart, followed by a booster shot every third year. The main challenge in eradicating canine-strain rabies is to ensure adequate coverage of the dog population. In several areas, vaccination coverage of 70% of dogs has been sufficient to control canine rabies (WHO 2005). Stray and feral dogs can be treated through the use of oral rabies vaccines (ORVs).

Effective vaccination programs to eradicate canine-strain rabies in dogs require a sound educational campaign, interagency cooperation, commitment of the local community, and an adequate supply of vaccines. An assessment of dog numbers is required before initiating the program to determine how many vaccines are required. The main difficulty with programs to vaccinate all dogs is the high cost. For instance, the annual expenditure for rabies prevention in the United States is more than $300 million, most of which is spent vaccinating dogs. Dogs have high reproductive rates, requiring vaccination programs to continue year after year (WHO 2005, CDC 2011).

Stopping the spread of rabies in animals is the best means of protecting humans from the disease. Historical methods to eliminate rabies in wildlife involved attempting to kill enough host animals that their densities drop below the level necessary to sustain the rabies virus. The methods employed in these efforts include hunting, trapping, using toxicants, and killing canids in dens through the use of toxic gasses. These methods have rarely been successful in preventing rabies. Mathematical models suggested that rabies would be eliminated in Europe if fox densities dropped to below 1.2–2 animals per mile^2 (0.3–0.5 animals per km^2) (Rupprecht et al. 2001). However, intensified hunting was insufficient to reduce fox densities below the required level. In South Africa, teams were sent out to control mongooses in areas where rabies was endemic. Efforts primarily involved using toxic gasses inside mongoose dens, but mongooses quickly repopulated areas where their numbers had been reduced. Hence, the program was soon abandoned as a failure. Likewise, efforts in South Africa to use toxins to reduce jackal densities failed (Zumpt and Hassel 1982).

More success was achieved trying to reduce populations of vampire bats in cattle-raising areas of Central America and South America. The focus of this program has been to destroy entire bat colonies through the use of toxins. Vampire bats within a single colony often share blood meals so that if one bat in a colony is infected, the disease has probably spread to others. The usual toxins are anticoagulants, such as warfarin, which can be injected into cattle so that bats feeding on the animal's blood will receive a toxic dose. Vampire bats regurgitate blood for other vampire bats in a colony so that a single blood meal can kill several vampire bats. Warfarin can also be applied to the fur of a captured vampire bat and released to contaminate other bats in the same colony. Bats ingest a toxic dose when licking their fur during grooming. Both methods are species specific because vampire bats are the only bats that drink blood and do not roost with other bat species (Acha and Malaga-Alba 1988).

Vampire bats do not occur in the United States, but many other bat species are found throughout the United States. In fact, rabid bats have been reported from every state except Hawaii, and at least 43 people in the United States have contracted rabies from bats from 1950 to 2011 (National Association of State Public Health Veterinarians 2011). The risk posed by bats in the United States cannot be reduced by killing them; nor is such an approach desirable. It is more effective to exclude them from homes and occupied buildings.

A more successful approach to controlling rabies in wildlife populations has been the use of oral vaccines to immunize wild animals. Switzerland tried this during 1978 by placing the SAD Bern virus vaccine in sealed blister packets and then inserting the packets inside pieces of chicken. The baits were distributed in the Rhone Valley and successfully stopped an impending outbreak of rabies (Rupprecht et al. 2001). This success led to an expansion of the ORV program throughout western Europe. During the 1990s, 8–17 million vaccine baits were distributed annually in Europe (Figure 19.6) at a bait density of 32–50 baits/mile^2 (13–20 baits/km^2). This program has been successful in reducing the number of animals infected with rabies (Figure 19.7). Excluding bat-strain rabies, seven European countries have become free of rabies as a result of the oral vaccination program; rabies have been eliminated

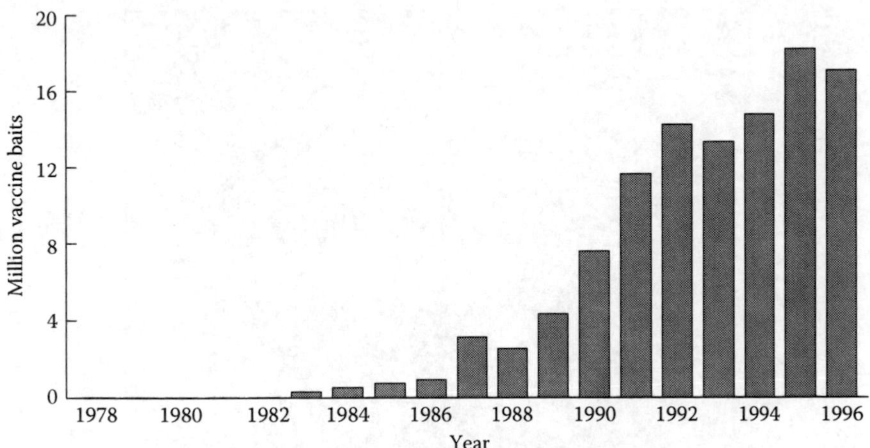

FIGURE 19.6 Annual number of rabies vaccine baits distributed in Europe from 1978 through 1996. (From Rupprecht, C.E. et al., Rabies, in E. S. Williams and I. K. Barker, eds., *Infectious Diseases of Wild Mammals*, 3rd edn., Blackwell, Ames, IA, pp. 3–36, 2001.)

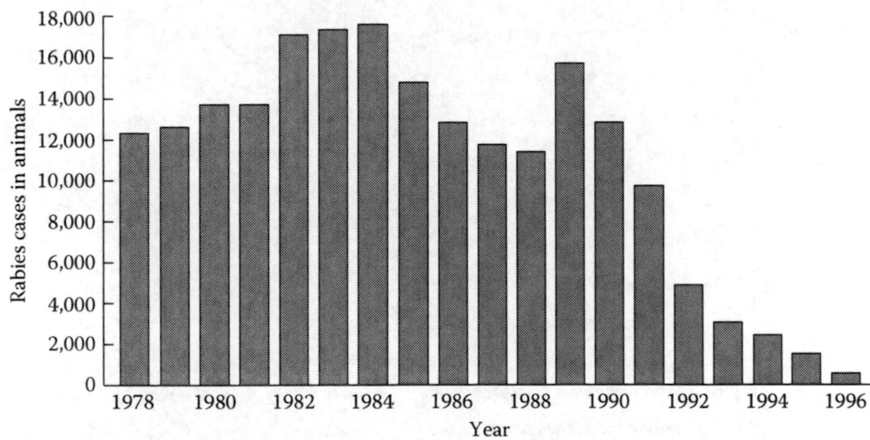

FIGURE 19.7 Annual number of animals that contracted rabies in European countries that participated in the rabies vaccine bait program. (From Rupprecht, C.E. et al., Rabies, in E.S. Williams and I.K. Barker, eds., *Infectious Diseases of Wild Mammals*, 3rd edn., Blackwell, Ames, IA, pp. 3–36, 2001.)

in Finland and the Netherlands since 1991, Italy since 1997, Switzerland since 1998, France since 2000, and Belgium and Luxembourg since 2001 (WHO 2005).

Canada started an ORV program during 1989 to eliminate the arctic fox variant of rabies from red foxes. For 7 years, ORV baits were dropped from airplanes at a rate of 50 baits/mile^2 (20 baits/km^2). This program was successful, and this variant has nearly been eliminated from most of Ontario although this variant still exists in

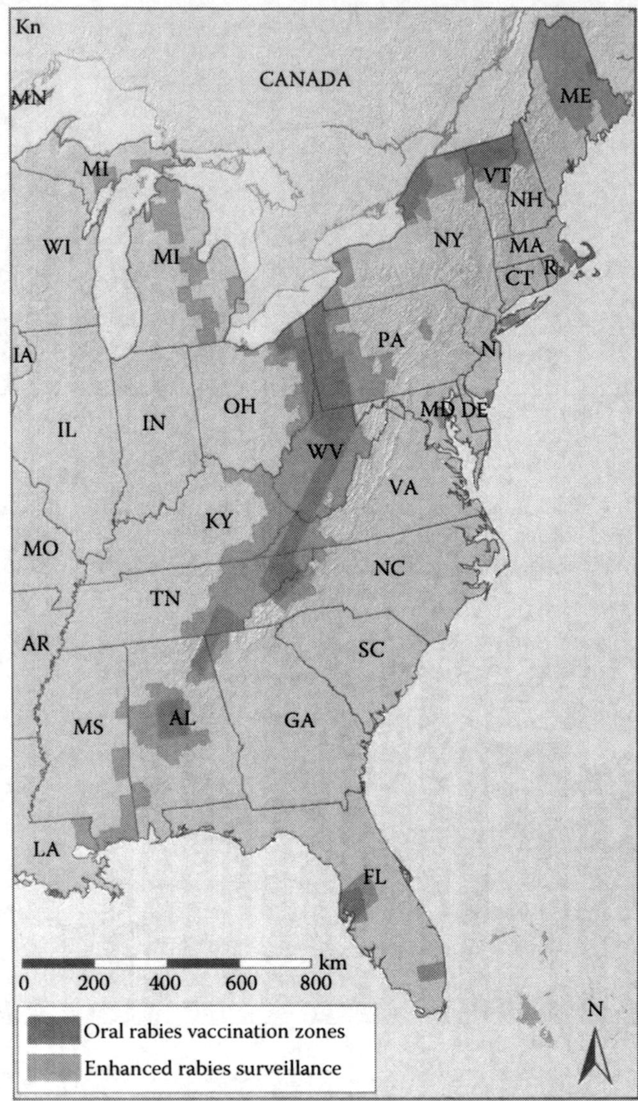

FIGURE 19.8 Oral rabies vaccination zones in the United States during 2004. (From Slate, D. et al., *Virus Res.*, 111, 68, 2005.)

skunks located in southwest Ontario (MacInnes and LeBer 2001, Slate et al. 2005, MacInnes et al. 2001).

The United States initiated an ORV program during the mid-1990s. In Texas, 700,000 ORV baits were distributed to eliminate the canine strain in coyotes and 1.8 million to eliminate a gray fox variant (Figure 19.8). The Texas ORV program was successful in eliminating the canine variant of rabies from coyotes by 2000. Since then, a 40 mile (65 km) buffer zone was created along the Rio Grande River

SIDEBAR 19.4 DANGERS OF TRANSLOCATING WILDLIFE (NETTLES ET AL. 1979, SLATE ET AL. 2005)

Many private hunting clubs purchase and release wild raccoons on their property so that hunting dogs can chase them. From 1976 to 1978, a raccoon club in Haywood County, North Carolina, purchased 137 raccoons in one shipment and 86 raccoons in a second shipment from Avon Park, Florida. Twenty of the raccoons died before arrival in North Carolina, and the rest were released at the gun club. Although unknown at the time, two of the dead raccoons had rabies. Prior to shipment, the raccoons were held together in a large pen and shipped with eight raccoons in a crate. Under these conditions, the two rabid raccoons had ample opportunity to infect some of the raccoons that were released. It is unclear if a new epizootic of the rabies virus became established from this particular shipment, but during the next 7 years, 1608 raccoons tested positive for rabies in Washington, DC, West Virginia, Virginia, Maryland, and Pennsylvania. Tests revealed that the rabies virus from this new area was identical to the rabies virus in Florida. These findings serve as a reminder that translocating wildlife can also result in the translocation of zoonotic diseases.

to prevent the reintroduction of the canine variant from Mexico where it is still endemic. Consequently, the United States was declared free of the canine variant of rabies in 2007. The ORV program is continuing in Texas to prevent the return of the canine variant from Mexico and to eliminate the gray fox variant from western Texas (Slate et al. 2009). During 2001, 19 skunks near Flagstaff were found to be infected with the big brown bat variant of rabies, indicating a viral host shift from bats to skunks. A trap–vaccinate–release program was initiated for skunks in the endemic area of Arizona. This program resulted in a quiescent period followed by reappearance of the variant in skunks during 2004–2005 and again during 2008. The same variant has also been confirmed in 17 gray foxes in Arizona (Slate et al. 2005, 2009).

Raccoon rabies was first detected in the United States during the 1940s and was confined to the south until the 1970s. Unfortunately, raccoons were translocated from the endemic area to Virginia and West Virginia, creating a new raccoon epizootic in the Mid-Atlantic states (Sidebar 19.4). This new epizootic spread to Maine, Vermont, and west to Ohio. In response, an ORV program was initiated in the eastern United States during 1995 to stop the spread of the raccoon variant. In the states where the ORV program was conducted, 33% of raccoons exhibited antibodies to rabies (Slate et al. 2005, 2009).

19.7 ERADICATING RABIES FROM A COUNTRY

The development of rabies PEP vaccines and the eradication of canine-strain rabies from many countries are some of the great achievements of modern medicine and epidemiology. Yet, thousands of people are killed by rabies each year in rural parts of Africa and Asia. In these areas, greater access to rabies vaccines and effective programs to eliminate canine-strain rabies are desperately needed.

Because of the slow dispersal rate of some wildlife strains of rabies, it is possible to stop its spread into new areas by vaccinating susceptible wildlife immediately in front of the spreading wave front. The most effective way to accomplish this is through the use of an ORV placed in baits that will be consumed by the reservoir hosts. The USDA Wildlife Services has been conducting a program of dropping baits for many years to stop the spread of rabies in the United States.

LITERATURE CITED

Acha, P. N. and A. Malaga-Alba. 1988. Economic losses due to *Desmodus rotundus*. In: A. M. Greenhall and U. Schmidt, editors. *Natural History of Vampire Bats*. CRC Press, Boca Raton, FL, p. 246.

Beran, G. W. 1994. Rabies and infections by rabies-related viruses. In: G. W. Beran, editor. *Handbook of Zoonoses*. CRC Press, Boca Raton, FL, pp. 307–357.

Blanton, J. D., J. Dyer, J. McBrayer, and C. E. Rupprecht. 2012. Rabies surveillance in the United States during 2011. *Journal of the American Veterinary Medical Association* 241:712–722.

CDC. 2011. Rabies. http://www.cdc.gov/rabies (accessed May 1, 2013).

CDC. 2012. Rabies surveillance data in the United States. http://www.cdc.gov/rabies/location/usa/surveillance/wild_animals.html (accessed December 4, 2013).

CDC. 2013. Human rabies due to organ transplantation, 2013. http://www.cdc.gov/rabies/resources/news/2013-03-15.html (accessed November 22, 2013).

Conover, M. R. 2002. *Resolving Human–Wildlife Conflicts: The Science of Wildlife Damage Management*. CRC Press, Boca Raton, FL.

Dacheux, L., F. Larrous, A. Mailles, D. Boisseleau et al. 2008. First description of European bat lyssavirus spill-overs to domestic cats in France and in Europe. *Rabies Bulletin Europe* 32:5–7.

Fooks, A. R., L. M. McElhinney, D. J. Pounder, C. J. Finnegan et al. 2003. Case report: Isolation of a European bat lyssavirus type 2a from a fatal human case of rabies encephalitis. *Journal of Medical Virology* 71:281–289.

Hu, W. T., R. E. Willoughby Jr., H. Dhonau, and K. J. Mack. 2007. Long-term follow-up after treatment of rabies by induction of coma. *New England Journal of Medicine* 357:945–946.

Knobel, D. L., S. Cleaveland, P. G. Coleman, E. M. Fèvre et al. 2005. Re-evaluating the burden of rabies in Africa and Asia. *Bulletin of the World Health Organization* 83:360–368.

MacInnes, C. D. and C. A. LeBer. 2001. Wildlife management agencies should participate in rabies control. *Wildlife Society Bulletin* 28:1156–1167.

MacInnes, C. D., S. M. Smith, R. R. Tinline, N. R. Ayers et al. 2001. Elimination of rabies from red foxes in eastern Ontario. *Journal of Wildlife Diseases* 37:119–132.

Merck. 2010. *Merck Veterinary Manual*, 10th edition. Merck, Whitehouse Station, NJ.

Mondul, A. M., J. W. Krebs, and J. E. Childs. 2003. Trends in national surveillance for rabies among bats in the United States (1993–2000). *Journal of the American Veterinary Medical Association* 222:633–639.

National Association of State Public Health Veterinarians. 2011. Compendium of animal rabies prevention and control, 2011. *Morbidity and Mortality Weekly Report* 60(RR06):1–14.

Nettles, V. F., J. H. Shaddock, R. K. Sikes, and C. R. Reyes. 1979. Rabies in translocated raccoons. *American Journal of Public Health* 69:601–602.

Rupprecht, C. E., K. Stöhr, and C. Meredith. 2001. Rabies. In: E. S. Williams and I. K. Barker, editors. *Infectious Diseases of Wild Mammals*, 3rd edition. Blackwell, Ames, IA, pp. 3–36.

Schatz, J., A. R. Fooks, L. McElhinney, D. Horton et al. 2013. Bat rabies surveillance in Europe. *Zoonoses and Public Health* 60:22–34.

Slate, D., T. P. Algeo, K. M. Nelson, R. B. Chipman et al. 2009. Oral rabies vaccination in North America: Opportunities, complexities, and challenges. *PLOS Neglected Tropical Diseases* 3(12):1–8.

Slate, D., C. E. Rupprecht, J. A. Rooney, D. Donovan et al. 2005. Status of oral rabies vaccination in wild carnivores in the United States. *Virus Research* 111:68–76.

Sudarshan, M. K., S. N. Madhusudana, B. J. Mahendra, N. S. Rao et al. 2007. Assessing the burden of human rabies in India: Results of a national multi-center epidemiological survey. *International Journal of Infectious Diseases* 11:29–35.

Van Zyl, N., W. Markotter, and L. H. Nel. 2010. Evolutionary history of African mongoose rabies. *Virus Research* 150:93–102.

Velasco-Villa, A., S. A. Reeder, L. A. Orciari, P. A. Yager et al. 2008. Enzootic rabies elimination from dogs and reemergence in wild terrestrial carnivores, United States. *Emerging Infectious Diseases* 14:1849–1854.

WHO. 2005. *WHO Expert Consultation on Rabies: First Report*. World Health Organization, Geneva, Switzerland.

WHO. 2012. The journal. Rabies—Bulletin—Europe. Rabies Information System of the WHO Collaboration Centre for Rabies Surveillance and Research. *Rabies Bulletin Europe* 36(3). http://www.who-rabies-bulletin.org (accessed April 30, 2013).

Willoughby Jr., R. E., K. S. Tieves, G. M. Hoffman, N. S. Ghanayem et al. 2005. Survival after treatment of rabies with induction of coma. *New England Journal of Medicine* 352:2508–2514.

Zumpt, I. F. and R. H. Hassel. 1982. The yellow mongoose as a rabies vector on the central plateau of South Africa. *South African Journal of Science* 78:417–418.

20 Equine Encephalitis *Eastern, Western, and Venezuelan*

Western equine encephalitis, forage poisoning, cerebrospinal meningitis, corn-stalk disease, harvest disease, sleeping sickness, and blind staggers were all terms used to describe the disease in horse, which was very likely WEE [Western equine encephalitis] in the western states of the United States and provinces of Canada prior to 1935.

Iversen (1994)

The survival and recovery of the endangered whooping crane was threatened during 1984 when seven of the 39 captive whooping cranes died after being infected with the eastern equine encephalitis virus.

Paraphrased from Dein et al. (1986)

Although the host range of epizootic VEEV [Venezuelan equine encephalitis virus] strains is wide and include humans, sheep, dogs, bats, rodents, and some birds, major epidemics in the absence of equine cases have never occurred.

Weaver et al. (2004)

20.1 INTRODUCTION AND HISTORY

Arboviruses are viruses that are transmitted by arthropods (usually ticks, mosquitoes, or other blood-feeding insects). In fact, "arbo" is derived from the first two letters of "*ar*thropod" and "*bo*rne," so that arbovirus is a shortened form of *ar*thropod-*bo*rne *virus*. Arboviruses are members of four different families of viruses: Togaviridae, Flaviviridae, Reoviridae, and Bunyaviridae (Table 20.1). More than 500 arboviruses have been recognized worldwide, and at least 5 of them were first isolated in Canada and 58 in the United States (Calisher 1994). In this chapter, we will discuss several arboviruses of the Togaviridae family that cause human encephalitis in the United States and Canada. These diseases include eastern equine encephalitis (EEE), western equine encephalitis (WEE), and Venezuelan equine encephalitis (VEE). In animals, diseases caused by these same viruses are often referred to as "encephalomyelitis" (inflammation of the brain and spinal cord) rather than "encephalitis" (CDC 2005).

The EEE, WEE, and VEE viruses are members of the *Alphavirus* genus. These viruses are spherical and have of a single-strand RNA (Figure 20.1). EEE virus was discovered during the 1930s, but deadly epidemics of encephalitis in horses have

TABLE 20.1
Classification of Arboviruses that Historically Caused Human Illnesses in the United States or Canada and the Chapter Where Each Virus Is Discussed

Family	Species	Chapter
Togaviridae	Eastern equine encephalitis virus	20
	Western equine encephalitis virus	20
	Venezuelan equine encephalitis virus	20
Flaviviridae	West Nile virus	21
	Yellow fever virus	21
	Dengue virus	21
	St. Louis encephalitis virus	21
	Powassan encephalitis virus	21
	Deer tick virus	21
Reoviridae	Colorado tick fever virus	22
Bunyaviridae	California encephalitis virus	22
	La Crosse encephalitis virus	22

occurred in the United States since the early 1800s and were likely the result of this virus. EEE occurs primarily in the eastern United States, particularly along the Eastern seaboard and Gulf Coast; its range, however, extends north into eastern Canada and south into Central America and South America (Figure 20.2). Ten human cases of EEE were reported in the United States during 2010, 4 reported during 2011, and 15 during 2012. Although few people develop EEE, it is a deadly disease. Three of the 4 EEE patients in the United States during 2011 succumbed, as did 5 of the 15 patients during 2012 (CDC 2012, 2013a,b).

WEE virus was first isolated in 1930. During 1941, a major epidemic of WEE swept through the north-central United States, Manitoba, and Saskatchewan and sickened more than 2800 people. The range of the WEE virus extends from Argentina, through the western United States (Figure 20.3), and north into western Canada. The disease is more prevalent in rural than urban areas. In the United States, there were 639 confirmed human cases of WEE from 1964 to 2005, but none were reported from 2006 through 2011 (Iversen 1994, CDC 2005, 2013a).

VEE was first identified during 1935 as a disease of horses, mules, and donkeys in Columbia when the virus was isolated from the brains of horses that died of the disease. Apparently, the VEE virus only evolved a few years earlier for there is no indication of the disease prior to 1935. It was not until the 1950s that the VEE virus was recognized as causing human disease (Weaver et al. 2004). VEE occurs mainly in South America and Central America where large outbreaks of the disease can occur in both horses and humans (Figure 20.4). One VEE outbreak in Venezuela between 1962 and 1964 produced more than 23,000 human cases and killed 156. An outbreak in Columbia during 1967 affected 220,000 humans. Another large outbreak began in South America during 1969 and reached as far north as Texas 2 years later, killing

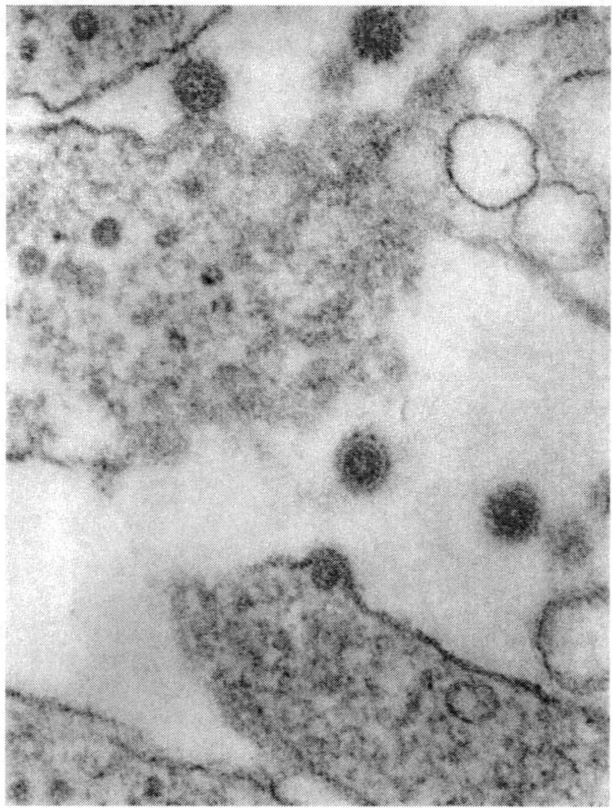

FIGURE 20.1 The dark spherical objects in this electron micrograph are the *Alphaviruses* that causes EEE. (Courtesy of the CDC, Atlanta, GA.)

tens of thousands of horses and infecting thousands of people. Interestingly, there was a gap of almost two decades with no known VEE outbreaks until one began during 1992. Soon thereafter (1995), a VEE outbreak in Venezuela and Colombia infected an estimated 90,000 people (Osorio and Yuill 1994, Weaver et al. 2004).

There are several different strains of the VEE virus, some of which cause epidemics in horses, and others which do not. Enzootic strains circulate constantly in tropical and subtropical swamps and wet forests of Central America and South America. Epizootic strains evolve from enzootic strains. One enzootic strain of the VEE virus is endemic in Florida (Figure 20.5) and is called the Everglades virus (Sidebar 20.1).

20.2 SYMPTOMS IN HUMANS

Symptoms of EEE in humans begin 4–12 days after the bite of an infected mosquito. Most infected people do not become ill or only have mild flu-like symptoms that include fever, headache, muscle aches, and malaise. In most people, the disease is self-limiting, but in some, the EEE virus becomes neuroinvasive (i.e., an infection of

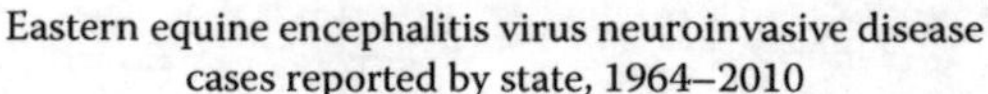

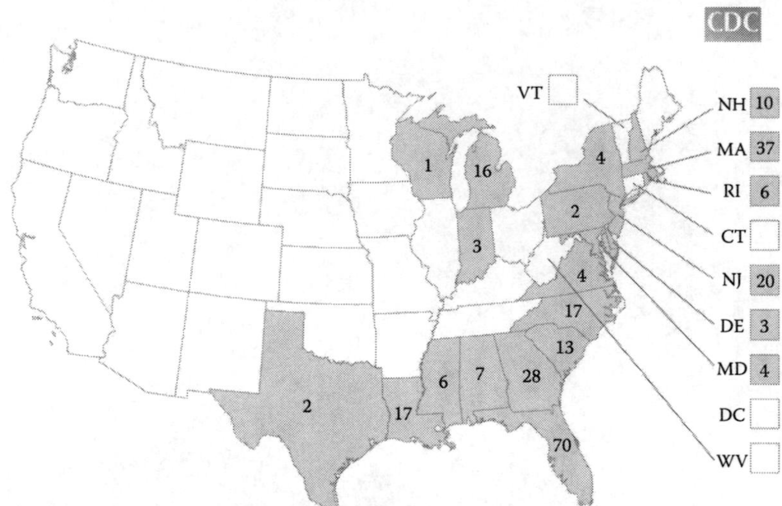

FIGURE 20.2 U.S. distribution of reported human cases from 1964 through 2010 of EEE that became neuroinvasive. (Courtesy of the CDC, Atlanta, GA.)

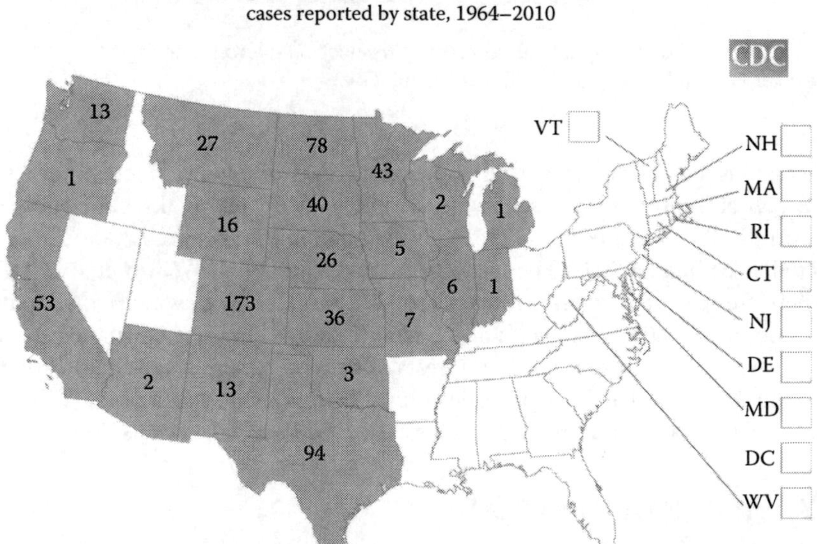

FIGURE 20.3 U.S. distribution of reported human cases from 1964 through 2010 of WEE that became neuroinvasive. (Courtesy of the CDC, Atlanta, GA.)

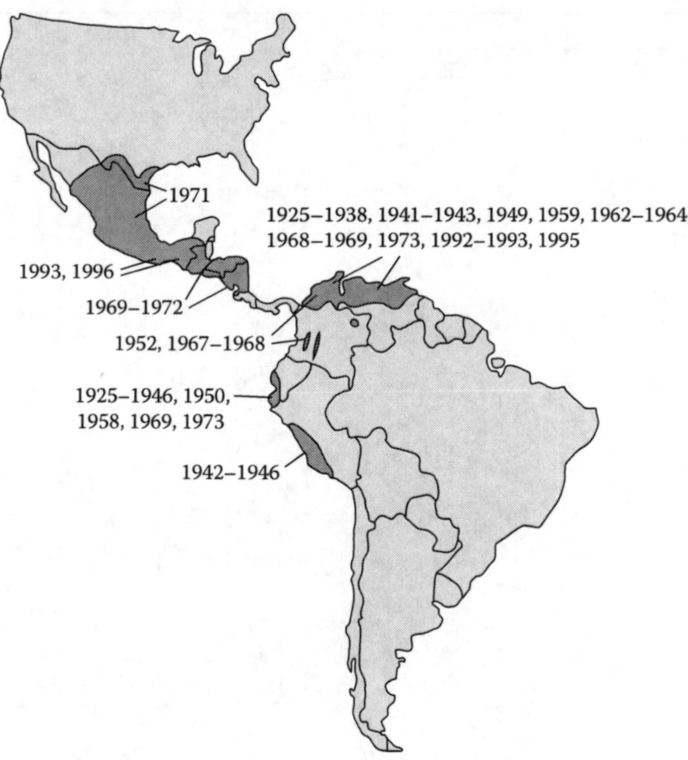

FIGURE 20.4 Map showing where different epidemics of VEE have occurred from 1938 to 2000 and the years when the outbreaks occurred. (From Weaver, S.C. et al., *Annu. Rev. Entomol.*, 49, 141, 2004.)

the nervous system) and produces encephalitis and/or meningitis. For these people, headaches and sensitivity to light can progress to weakness and lack of coordination and finally to seizures, paralysis, and coma. The mood of patients can alternate between periods of calm and agitation. EEE is the most severe of the arboviral diseases in humans and can result in death or permanent neurological problems. Its human case fatality rate is about 33% but increases to 40% among children and 70% among the elderly. A third of surviving EEE patients suffer from long-term neurological problems (Gibbs and Tsai 1994, CDC 2005, 2012).

Most people infected by WEE virus do not become ill or develop only mild symptoms. The first signs of the disease are abrupt and begin after an incubation period of 5–10 days. Symptoms of WEE include fever, chills, muscle aches in the legs and lower back, nausea, and vomiting. In some patients, the virus infects the central nervous system, causing encephalitis that often starts with sensitivity to light and headaches and can progress to weakness and altered mental states. The WEE case fatality rate is only 3% although recovery from WEE can be slow with fatigue, headaches, tremors, and irritability lasting over a year. Most adults make a complete recovery, but permanent neurological problems are possible. When infected with the

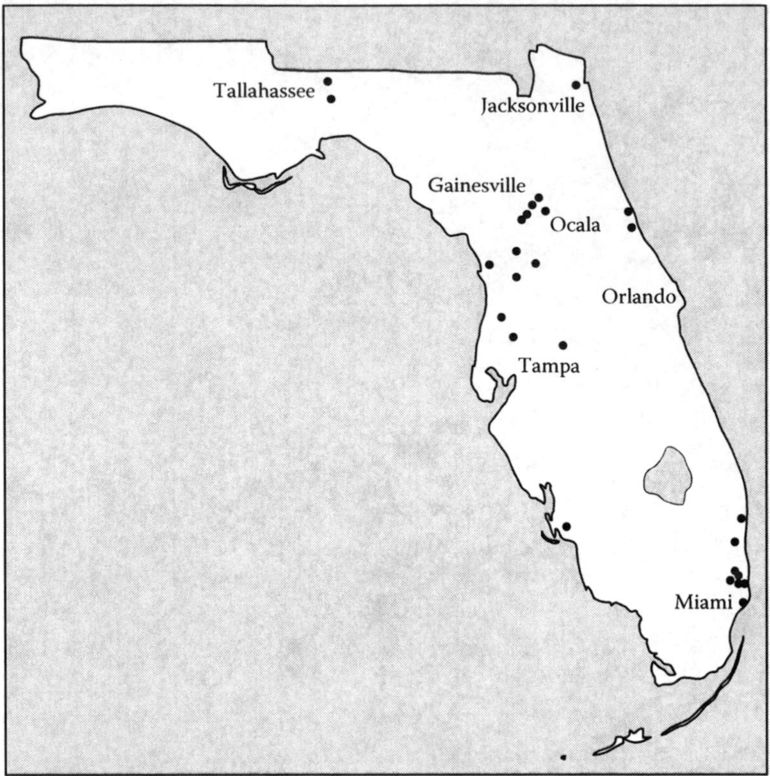

FIGURE 20.5 Range of the Everglades virus in Florida, United States. The dark circles indicate where Coffey et al. (2006) found domestic dogs that tested positive for antibodies against the virus.

WEE virus, babies and young children are more likely to develop encephalitis than adults. Thirty percent of surviving children who developed WEE encephalitis suffered permanent brain injuries (CDC 2005, Office International des Epizooties 2009).

Fever, chills, malaise, nausea, vomiting, and severe headaches are the initial symptoms of VEE; they begin after an incubation period of 1–6 days. Muscle aches are centered in the back and hips. This acute phase lasts 1–3 days and is followed by a 1 to 3-week period of lethargy and loss of appetite. Most patients then recover, but 1% of adult patients and 4% of children develop encephalitis. VEE encephalitis is usually self-limiting; fatalities and permanent neurological problems are rare (Weaver et al. 2004, Center for Food Security and Public Health 2008).

20.3 INFECTIONS OF EEE, WEE, OR VEE VIRUS IN ANIMALS

EEE, WEE, and VEE are able to replicate in some mosquito species that serve as their vectors. After a mosquito ingests a blood meal from an infected host, the epithelial cells in the mosquito's midgut become infected. The infection then spreads to

SIDEBAR 20.1 EVERGLADES VIRUS IN FLORIDA (WORK 1964, COFFEY ET AL. 2004, 2006)

There is one type of VEE virus that is endemic to Florida. It is called the Everglades virus because the first patients were Seminole Indians living in the Everglades (Figure 20.6). More recently, Coffey et al. (2006) found that 4% of domestic dogs in Florida had antibodies against the Everglades virus, suggesting that the virus is widespread in the state including Miami, Gainesville, and Tallahassee. The mosquito *Culex cedecei* is its primary vector, and the cotton mouse and the Hispid cotton rat serve as amplifying hosts.

People living in south Florida have frequently been exposed to the virus. Most of the Seminole Indians living north of Everglades National Park have antibodies against the virus as do 9% of people living in rural communities surrounding the park. Most people who tested positive cannot recall having symptoms similar to VEE. Hence, it appears that most people infected with the Everglades virus are either asymptomatic or have only a mild illness. In humans, the Everglades virus normally results in a flu-like illness featuring fever, muscle aches, enlarged lymph nodes, and diarrhea. Occasionally, infections progress to a severe neurological disease. How long this unique type of VEE virus has been in Florida in unknown.

FIGURE 20.6 Seminole family of Cypress Tiger at their camp in the Everglades near Kendall, Florida, 1916. (Photo taken by John Kunkel Small.)

the mosquito's hemocoel (blood), salivary glands, and other organs where the virus replicates. Once the salivary glands are infected, the mosquito can transmit the arbovirus to other hosts (Weaver et al. 2004).

Horses are particularly susceptible to the EEE, WEE, and VEE virus; the number of horses stricken by the EEE virus during an outbreak greatly exceeds the number of humans. The fatality rate for unvaccinated horses that develop EEE can approach 90%. An EEE outbreak in Louisiana and Texas during 1947 sickened 14,000 horses and mules and killed 12,000 of them. Some horses infected with the EEE virus experience only a mild illness, involving fever and loss of appetite that last 1–2 days as the virus infects muscle, connective tissue, and myeloid tissue. The fever can return a few days later as the virus spreads to the central nervous system; clinical signs of neurological infections include a hanging head, flaccid or paralyzed lips, drooping ears, and legs held in a wide stance due to problems maintaining balance. Most horses die within 4 days of the onset of clinical signs (Gibbs and Tsai 1994, Center for Food Security and Public Health 2008).

EEE virus occasionally causes encephalitis in sheep, cattle, deer, llamas, alpacas, and pigs. Some infected sheep have a fever and lack leg coordination. Signs of an EEE virus infection in deer include shortness of breath, emaciation, excessive salivation, circling, confusion, and fearlessness. EEE infections have been observed in bats, rodents, reptiles, and amphibians. Horses and all other mammals are dead-end hosts for the EEE virus (Center for Food Security and Public Health 2008).

Birds serve as amplifying hosts for the EEE virus, and many avian species native to North America do not become ill when infected, probably because the EEE virus and these birds have coexisted for eons; the virus, nonetheless, produces illness and death in exotic species, such as chukar, ring-necked pheasants, house sparrows, feral pigeons, and glossy ibis (Figure 20.7). In these exotic birds, disease signs are marked by diarrhea, lethargy, drowsiness, tremors, leg or neck paralysis, and loss of muscle coordination. Exotic birds can die within a day of symptom initiation (Gibbs and Tsai 1994).

The EEE virus has been isolated from more than 25 mosquito species. In North America, the virus is able to maintain itself through a disease cycle involving the mosquito *Culiseta melanura* and passerine birds. *C. melanura* lives in freshwater and forested swamps; its larvae are found in small pools of water at the base of trees. The mosquito feeds almost exclusively on songbirds and is active at night, especially the first 2 hours after sunset. *C. melanura* rarely bites humans or horses; instead, other mosquitoes (*Coquillettidia perturbans* and several species of *Aedes* mosquitoes) that feed on both mammals and birds serve as bridge species and play an important role in transmitting the EEE virus to humans and horses. In Central and South America, EEE virus maintains itself through a disease cycle involving the mosquito *Culex taeniopus* as the primary vector and native birds as amplifying hosts. Among captive birds, EEE infections can also spread among members of the same flock from feather picking or cannibalism; these behaviors are rare among free-ranging animals (Gibbs and Tsai 1994, Friend and Franson 1999, Center for Food Security and Public Health 2008).

The WEE virus maintains itself through a transmission cycle involving the mosquito *Culex tarsalis* as its primary vector and passerine birds, especially nestlings, as amplifying hosts. Throughout most of its range in the United States and Canada,

FIGURE 20.7 (a) Glossy ibis, (b) ring-necked pheasant, and (c) chukar are exotic species in North America and did not evolve there. Because of this, infections of the VEE virus make them ill and can result in their deaths. (Ibis photo is in the public domain; chukar photo courtesy of the National Biological Information Infrastructure Project, U.S. Geological Survey, Reston, VA; pheasant photo courtesy of the U.S. Forest Service, Washington, DC.)

house sparrows and house finches are the primary amplifying hosts. The WEE virus infects several different mammalian species; most of these species are dead-end hosts with the exception of the black-tailed jackrabbit. WEE virus can also infect snakes, frogs, and tortoises. Most horses infected with the WEE virus are asymptomatic, but the proportion that become ill varies among outbreaks. During the deadly outbreak of 1937–1938, over 350,000 horses and mules in the United States and Canada became ill, and 20%–30% of them succumbed. Neither the EEE nor WEE virus can survive for long outside of a host (Center for Food Security and Public Health 2008).

Unlike EEE and WEE viruses, the VEE virus can maintain itself by infecting rodents, which do not become ill when infected. The VEE virus has been isolated from cotton rats, common opossum, Derby's woolly opossum, gray fox, several bat species, and a variety of wild birds. Infected horses, cattle, and hogs can produce sufficient concentrations of the VEE virus in their blood to infect mosquitoes. In fact, horses are an important amplifying host for the VEE virus during outbreaks (Young et al. 1969).

Several mosquito species serve as vectors for the VEE virus. The virus can also infect ticks; however, ticks are less important as vectors than mosquitoes. During outbreaks, black flies and mites can serve as mechanical vectors if the virus contaminates their mouth parts, but the pathogen is unable to replicate in black flies (Linthicum et al. 1991).

VEE virus causes severe illnesses in horses, mules, burros, and donkeys. Although less common, deaths attributed to VEE have been reported in hogs, goats, sheep, domestic rabbits, and dogs (Osori and Yuill 1994). The first signs of VEE in horses include fever, muscle weakness, anorexia, colic, and diarrhea. If the VEE virus infects the central nervous system, horses may exhibit muscle spasms, convulsions, head pressing, incessant chewing, aimless walking, circling, lack of coordination, and hyperexcitability. Death can occur within a few hours after the onset of neurological signs in horses or after a long illness during which horses become dehydrated and lose weight. Mortality rates among infected horses are 40%–90% and vary from one epidemic to another (Osorio and Yuill 1994, Center for Food Security and Public Health 2008, Office International des Epizooties 2009).

20.4 HOW HUMANS CONTRACT EEE, WEE, OR VEE

Most EEE, WEE, and VEE patients become ill after being bitten by an infected mosquito. Unlike the YF virus, densities of EEE, WEE, or VEE viruses in human blood are not high enough to infect mosquitoes, and human-to-human transmission of these viruses do not occur except between pregnant women and their fetuses (Center for Food Security and Public Health 2008). EEE, WEE, and VEE outbreaks usually occur during summer or when mosquitoes are abundant. In many areas, one mosquito species is involved in maintaining the virus by transmitting the virus among amplifying hosts; another mosquito species then serves as a bridge vector that bites and infects humans. *Aedes* and other mosquito species that feed on both mammals and birds are bridge species and play an important role in transmitting the EEE virus to humans and horses (Friend and Franson 1999, Center for Food Security and Public Health 2008).

Most WEE patients in the western United States become infected after being bitten by the mosquito *C. tarsalis*. Other important mosquito vectors include *Aedes melanimon* in California, *Aedes campestris* in New Mexico, and *Aedes dorsalis* in Utah. The habitat of these mosquitoes includes irrigated farmland and associated drainage ditches. The expansion of irrigated farmland along the North Platte River has allowed WEE virus to expand its range into Nebraska and Wyoming (Iversen 1994, Center for Food Security and Public Health 2008).

Mosquitoes are the main vector responsible for the spread of VEE to humans. Unlike EEE and WEE, the VEE virus can survive outside its host and is found in dried blood or sputum of horses. Occasionally, other horses or humans are infected with VEE by direct contact of body fluids or inhalation of aerosolized viruses. Common disinfectants are effective in destroying any VEE viruses that may be on surfaces (Weaver et al. 2004, Center for Food Security and Public Health 2008).

20.5 MEDICAL TREATMENT

Antibiotics are not effective against EEE, WEE, or VEE; moreover, medical care is limited to supportive practices aimed at making patients comfortable and reducing the adverse affects of the disease, such as providing breathing assistance if needed. There are no human vaccines against EEE, WEE, or VEE, but there is a vaccine for horses that can be used in the United States that provides immunity against EEE, WEE, and VEE viruses. The risk to horses can be reduced by housing them in screened barns at night.

20.6 WHAT PEOPLE CAN DO TO REDUCE THE RISK OF CONTRACTING EEE, WEE, OR VEE

EEE, WEE, and VEE are seasonal diseases and occur mainly during the spring, summer, and fall when mosquito populations are high. In the United States and Canada, the risk of EEE and WEE declines in the fall when temperatures drop low enough to kill mosquitoes (Figure 20.8). During most years, the EEE, WEE, and VEE viruses are confined to freshwater marshes where their mosquito vectors are abundant. Some years, the viruses spread from their normal wet habitat into drier areas inhabited by horses and humans. When this happens, outbreaks of EEE, WEE, and VEE can result. The severity of outbreaks can be reduced if detected early. Horses are particularly susceptible to the EEE, WEE, and VEE viruses and usually become ill before humans during an outbreak. For this reason, public health authorities use horses as a sentinel species for EEE, WEE, or VEE outbreaks (Gibbs and Tsai 1994, Osorio and Yuill 1994).

The best method to reduce the risk of contracting an arbovirus is to reduce the frequency of mosquito bites. On an individual level, this often involves wearing long-sleeved shirts, long pants, socks, and shoes when outside and applying an insect repellent to bare skin. Replacing old screens and making sure that no gaps exist between the screens and the window sills can keep mosquitoes out of buildings. Reducing mosquito populations around homes can be accomplished by leaving no containers outside that

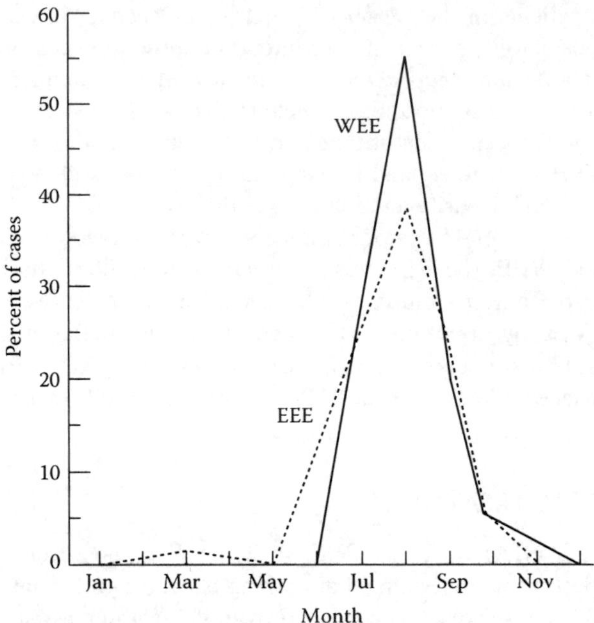

FIGURE 20.8 Seasonal distribution of human cases of EEE and WEE in the United States, 1964 through 1994. (Data from Gibbs, E.P.J. and Tsai, T.F., Eastern encephalitis, in: G.W. Beran, editor-in-chief, *Handbook of Zoonoses, Section B: Viral*, CRC Press, Boca Raton, FL, pp. 11–24, 1994.)

can hold water, drilling holes in the bottom of tire swings so they do not hold water, and emptying wading pools for children and bird baths at least weekly.

VEE epidemics usually begin in horses or donkeys, which serve as amplifying hosts for VEE virus. During outbreaks, they are infected weeks before humans develop VEE. Hence, an outbreak of VEE can be contained or reduced if horses in the endemic area are vaccinated against the virus. In Central and South America, interest in vaccinating horses often wanes after an epidemic has ended, increasing over time the proportion of horses susceptible to the VEE virus and the probability of a future epidemic. Once a VEE outbreak begins, the endemic area generally expands, owing to the movement of infected horses. Efforts to restrict the movement of infected horses are often unsuccessful because infected horses are initially asymptomatic. The VEE virus can infect several species of bats and birds; infected bats and birds may provide one mechanism for the long-distance transport of the VEE virus (Weaver et al. 2004, Office International des Epizooties 2009).

20.7 ERADICATING EEE, WEE, OR VEE FROM A COUNTRY

The viruses responsible for EEE, WEE, and VEE circulate among several species of mosquitoes and wild birds or mammals. This makes it difficult to eradicate these diseases in countries where the viruses are endemic.

LITERATURE CITED

Calisher, C. H. 1994. Medically important arboviruses of the United States and Canada. *Clinical Microbiology Reviews* 7:89–116.

CDC. 2005. Information on arboviral encephalitides. http://www.cdc.gov/ncidod/dvbid/arbor/arbdet.htm (accessed June 2, 2013).

CDC. 2012. Summary of notifiable diseases—United States, 2010. *Morbidity and Mortality Weekly Report* 59(53):1–111.

CDC. 2013a. Summary of notifiable diseases—United States, 2011. *Morbidity and Mortality Weekly Report* 60(53):1–117.

CDC. 2013b. West Nile virus and other arboviral diseases—United States, 2012. *Morbidity and Mortality Weekly Report* 62:513–517.

Center for Food Security and Public Health. 2008. Eastern equine encephalomyelitis, western equine encephalomyelitis, and Venezuelan equine encephalomyelitis. Center for Food Security and Public Health, Iowa State University, Ames, IA.

Coffey, L. L., A.-S. Carrara, S. Paessler, M. L. Haynie et al. 2004. Experimental Everglades virus infection of cotton rats (*Sigmodon hispidus*). *Emerging Infectious Diseases* 10:2182–2188.

Coffey, L. L., C. Crawford, J. Dee, R. Miller et al. 2006. Serologic evidence of widespread Everglades virus activity in dogs, Florida. *Emerging Infectious Diseases* 12:1873–1879.

Dein, F. J., J. W. Carpenter, G.G. Clark, R. J. Montali et al. 1986. Mortality of captive whooping cranes caused by eastern equine encephalitis virus. *Journal of the American Veterinary Medical Association* 189:1006–1010.

Friend, M. and J. C. Franson. 1999. Field manual of wildlife diseases. U.S. Geological Survey, Biological Resources Division, Information and Technology Report 1999-001.

Gibbs, E. P. J. and T. F. Tsai. 1994. Eastern encephalitis. In: G. W. Beran, editor-in-chief. *Handbook of Zoonoses, Section B: Viral.* CRC Press, Boca Raton, FL, pp. 11–24.

Iversen, J. O. 1994. Western equine encephalomyelitis. In: G. W. Beran, editor-in-chief. *Handbook of Zoonoses, Section B: Viral.* CRC Press, Boca Raton, FL, pp. 25–31.

Linthicum, K. J., T. M. Logan, C. L. Bailey, S. W. Gordon et al. 1991. Venezuelan equine encephalomyelitis virus infection in and transmission by the tick *Amblyomma cajennense* (Arachnide: Ixodidae). *Journal of Medical Entomology* 28:405–409.

Office International des Epizooties. 2009. *Venezuelan Equine Encephalitis.* Office International des Epizooties, Paris, France.

Osorio, J. E. and T. M. Yuill. 1994. Venezuelan equine encephalitis. In: G. W. Beran, editor-in-chief. *Handbook of Zoonoses. Section B: Viral.* CRC Press, Boca Raton, FL, pp. 33–46.

Weaver, S. C., C. Ferro, R. Barrera, J. Boshell, and J.-C. Navarro. 2004. Venezuelan equine encephalitis. *Annual Review of Entomology* 49:141–147.

Work, T. H. 1964. Serological evidence of arbovirus infection in the Seminole Indians of southern Florida. *Science* 145:270–272.

Young, N. A., K. M. Johnson, and L. W. Gauld. 1969. Viruses of the Venezuelan equine encephalomyelitis complex: Experimental infection of Panamanian rodents. *American Journal of Tropical Medicine and Hygiene* 18:290–296.

21 West Nile Virus and Other Diseases Caused by Flaviviruses

Yellow Fever, Dengue, St. Louis Encephalitis, Japanese Encephalitis, Powassan Encephalitis, and Deer Tick Virus

The 2002 WNV [West Nile virus] epidemic was the largest recognized arboviral meningoencephalitis epidemic in the Western Hemisphere and the largest WN meningoencephalitis epidemic ever recorded.

CDC (2003)

Victims of dengue often have contortions due to the intense joint and muscle pain. Hence [its common] name: breakbone fever.

MedicineNet.com (2013)

21.1 INTRODUCTION AND HISTORY

West Nile Virus (WNV) consists of a linear single-stranded RNA (Figure 21.1) and is one species of the *Flavivirus* genus and *Flavivirus* family. This genus also includes the pathogens for several human diseases, which will be discussed at the end of this chapter. WNV was not discovered until 1937 when it was isolated from a sick woman living in the West Nile district of Uganda, Africa (the disease and virus were named for the location where they were first identified). In this chapter, West Nile fever (WNF) is defined as the human disease caused by WNV, including neuroinvasive diseases (i.e., infections of the central nervous system that result in meningitis and encephalitis). The first recognized epidemics of WNF struck Israel in 1951 and Europe in 1962. Prior to the 1990s, WNV was not known to cause illness in birds, and the symptoms of WNF in humans were mild and complications rare. In Egypt, it was a common childhood illness, and over 80% of adults had antibodies to WNV (Taylor et al. 1956,

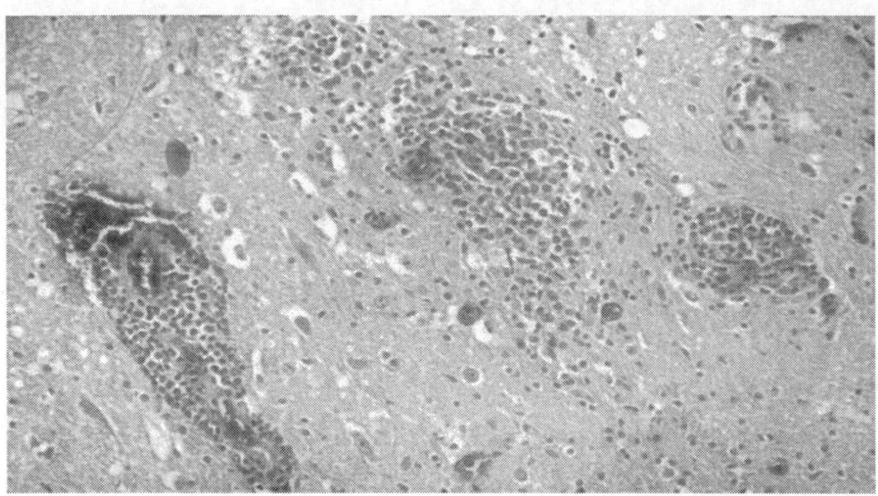

FIGURE 21.1 An electron micrograph showing WNV (small dark spheres) inside the brain tissue of a horse. (Courtesy of the CDC.)

Zeller and Schuffenecker 2004). However, the virus morphed in the 1990s when outbreaks of WNF in Europe, northern Africa, and the Middle East were unusually severe and lethal (Rossi et al. 2010). In Romania alone, more than 500 people were diagnosed with WNF in 1996–1997, and 10% of them died. In 2000, 417 WNF patients required hospitalization during a lethal outbreak in Israel; 35 perished. WNF outbreaks still occur sporadically in Europe, including Austria in 2008, Italy in 2009, and Greece in 2010 (Sidebar 21.1). Kunjin virus is a unique type of WNV that occurs in Australia (Hubálek and Halouzka 1999, Zeller and Schuffenecker 2004, Papa et al. 2010).

The first documented cases of WNF in the United States occurred in New York City in 1999 when birds, horses, and people became infected with WNV (Sidebar 21.2). How the virus crossed the Atlantic Ocean is unknown, but the strain was identical to a strain of WNV isolated from geese in Israel 1 year earlier. After reaching New York, WNV spread rapidly (Figure 21.3) and had reached 44 states by 2002, where it caused 4156 human cases of WNF, including 2942 patients with meningoencephalitis and 284 deaths (Table 21.1). WNV also spread rapidly among horses in the United States where there were 738 infected horses in 2001 and 14,571 horses a year later when the virus reached the western states (CDC 2003). The WNV is now endemic in all U.S. states except Hawaii and Alaska. From the United States, WNV quickly spread to other countries in the Western Hemisphere. It was first detected in Canada in 2001; just 2 years later, there were 1400 confirmed cases in nine provinces, with the epicenter in Saskatchewan. WNV reached Mexico, Jamaica, and Dominican Republic in 2002; Cuba, Bahamas, and Puerto Rico in 2003; and Columbia, Venezuela, and Argentina by 2004 (Kilpatrick 2011). Surprisingly, the Israeli strain of WNV that was introduced into the United States in 1999 had been replaced across the entire United States only 6 years later by a new strain that reputedly was more efficient in being transmitted by North American mosquitoes (Kilpatrick et al. 2008, Kilpatrick 2011).

SIDEBAR 21.1 THE 2010 OUTBREAK OF WNF IN GREECE (PAPA ET AL. 2010, GOMES ET AL. 2013)

On August 4, 2010, physicians from the Infectious Disease Hospital in Thessaloniki, Greece, notified public health authorities of an increase in patients with encephalitis during the prior month. Most of the patients were over 65 years old and resided in central Macedonia. WNF was not initially suspected because WNV had never been detected in humans from Greece. By the next day, however, 10 of the 11 patients were diagnosed with infections of WNV after blood samples showed antibodies against the virus. Before the outbreak ended, there were at least 262 people diagnosed with an infection of WNV; 197 with encephalitis, meningitis, or flaccid paralysis; and 35 fatalities. The outbreak was limited to the northern part of Greece, and most human cases occurred in central Macedonia (Figure 21.2). This region has lowland humid areas, which provide ideal conditions for birds. Many avian species migrating between Africa and Europe pass through the land corridor between the Mediterranean Sea and the Black Sea, and numerous birds use central Macedonia as a stopover. The hypothesis that migrating birds introduced WNV into Greece was supported by the finding that the WNV isolated during the outbreak was similar to a strain found in both Hungary and South Africa.

SIDEBAR 21.2 WNV REACHES THE NEW WORLD (ASNIS ET AL. 2000, MURRAY ET AL. 2010A, CDC 2013)

On August 12, 1999, a 60-year-old man, who had experienced several days of fever, weakness, and nausea, was admitted to the Flushing Hospital in New York City. An x-ray revealed a lung infection, and the patient was given antibiotics. Four days later, the patient was confused, exhibited muscle weakness, and had difficulty breathing and urinating. He was provided ventilator assistance. A lumbar puncture and a CT scan of his head were conducted, and he was treated for a possible Guillain–Barré syndrome. During the next month, seven more patients were admitted to the same hospital with fever, confusion, and weakness. Six of the patients were diagnosed with encephalitis and two with meningitis. Ultimately, four of these patients also developed flaccid paralysis and required ventilator support; three died. They were among the first people to contract WNF in North America, but they would not be the last. By the end of the year, almost 10,000 people in New York and the surrounding states were infected with WNV. Of these, 1700 developed WNF. By 2013, there were over 16,000 human cases of neuroinvasive WNF and 1,400 deaths in the United States, along with over 20,000 cases of WNF that did not become neuroinvasive.

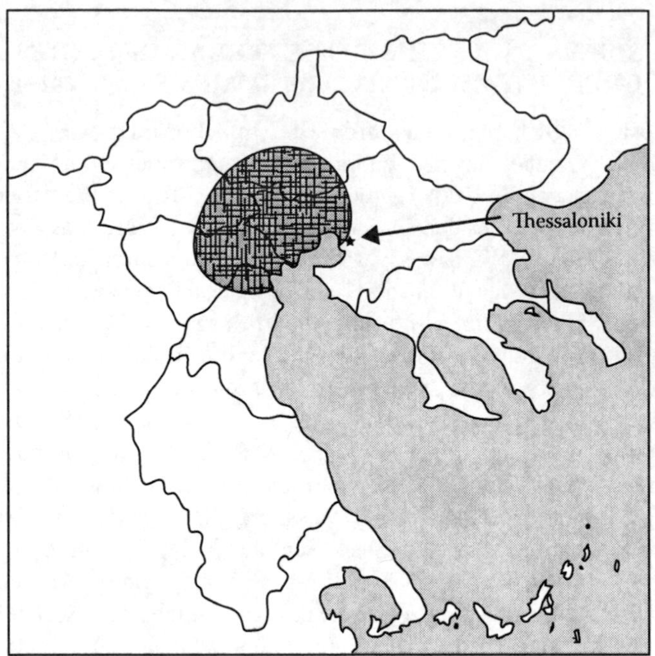

FIGURE 21.2 The 2010 outbreak of WNV was located in central Macedonia and west of the city of Thessaloniki. The endemic region is the shaded region on this map of Greece.

Epidemics of WNV in the Western Hemisphere have produced unusually high mortality rates in avian species, as well as high rates of encephalitis or meningitis among human patients (Murray et al. 2010a,b).

21.2 SYMPTOMS IN HUMANS

In 2012, there were 5774 reported cases of WNF in the United States; 2873 of these cases were neuroinvasive, and 2801 were not. The CDC (2013e) estimates that there are 30–70 nonneuroinvasive cases of WNF for each reported case of neuroinvasive WNF. By CDC estimates, between 86,000 and 200,000 people in the United States suffered from nonneuroinvasive cases of WNF in 2012.

WNF is manifested by fever, headaches, muscle aches, nausea, vomiting, swollen lymph nodes, or a skin rash on the chest and back. These symptoms usually occur 2–6 days after the bite of an infected mosquito, but the incubation period can extend for more than 2 weeks. Symptoms can last for a few days or a few weeks and typically resolve on their own. Elderly patients are more likely to develop neuroinvasive WNF (Figure 21.4). In 2012, 9% of patients with neuroinvasive WNF died as a result of their infection (CDC 2003, 2013e, Zeller and Schuffenecker 2004).

Symptoms of encephalitis or meningitis caused by WNV include a stiff neck, back pain, severe headache, stupor, disorientation, confusion, tremors, numbness, convulsions, and coma. Some patients experience flaccid paralysis that can involve

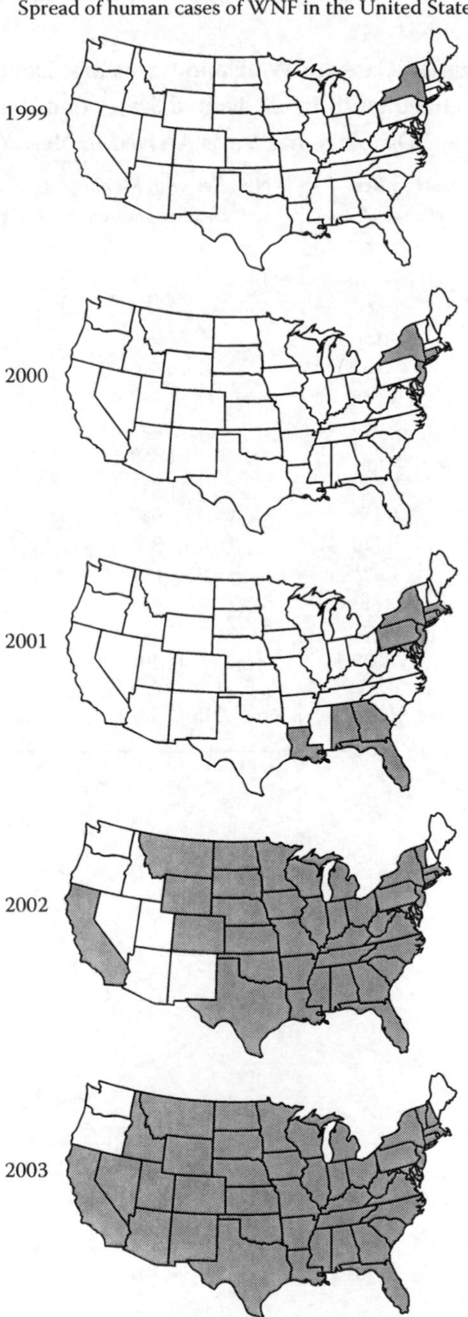

FIGURE 21.3 Maps of the United States showing the spread of WNF across the United States. States where humans have been diagnosed with WNF are shown in gray. (From CDC, West Nile Virus annual maps and data. http://www.cdc.gov/westnile/statsMaps/finalMaps Data/index.html, 2013f.)

TABLE 21.1
Number of Human Cases of WNF and Cases that Developed into Meningitis or Encephalitis in the United States that Were Reported to the CDC Since the Virus Arrived in New York in 1999

Year	Number of Human Cases	Number with Meningitis or Encephalitis	Number of Human Deaths
1999	62	59	7
2000	21	19	2
2001	66	64	10
2002	4,156	2,946	284
2003	9,862	2,866	264
2004	2,539	1,142	100
2005	3,000	1,294	119
2006	4,269	1,459	177
2007	3,630	1,217	124
2008	1,356	687	44
2009	720	373	32
2010	1,021	629	57
2011	712	486	43
2012	5,674	2,873	286
Total	37,088	16,196	1,549

Source: CDC, *Morbid. Mortal. Wkly. Rep.*, 62, 513, 2013e.

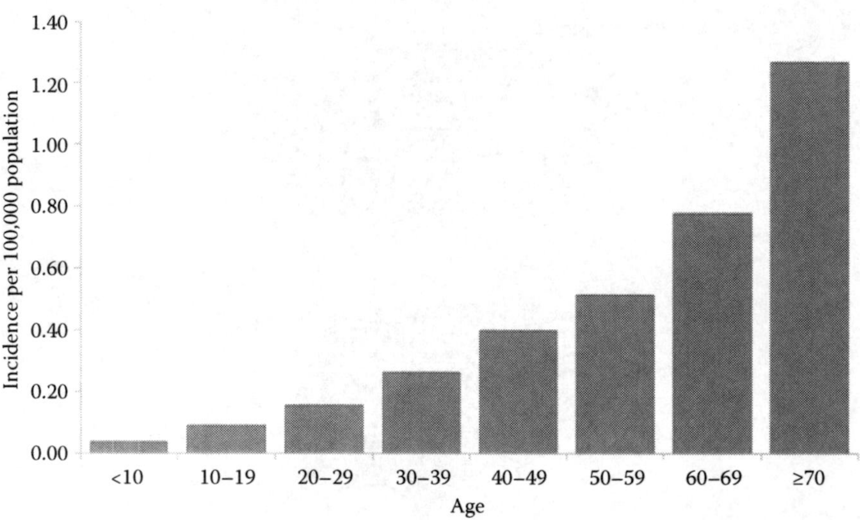

FIGURE 21.4 Incidence of reported cases of WNF that infected the central nervous system among different age groups from 1999 to 2012 in the United States. (From Division of Vector-Borne Diseases, National Center for Emerging and Zoonotic Infectious Diseases (ArboNET Surveillance); CDC, *Morbid. Mortal. Wkly. Rep.*, 59(53), 1, 2012b.)

the arms, legs, or breathing muscles. Many survivors of WNF encephalitis or meningitis suffer from long-term complications, which include fatigue, weakness, and muscle aches; some long-term patients also suffer from depression, tremors, loss of memory, and inability to concentrate (Murray et al. 2010a, 2012b, Rossi et al. 2010, Mayo Clinic 2012).

21.3 WNV INFECTIONS IN ANIMALS

Birds exhibit diverse signs of a WNV infection. The most susceptible birds become lethargic and unaware of their surroundings, have ruffled feathers, cannot hold their heads up, and lack muscle coordination. Most birds that appear ill succumb within a day. Most birds are infectious for only 4–6 days; after which, either the bird's immune system has swept WNV from the blood (although the virus can persist longer in other tissue) or the bird has succumbed (CDC 2003, 2012a, Komar et al. 2003, National Wildlife Health Center 2005).

Some avian species, especially passerines, are the reservoir host for WNV and are infected by mosquito bites. In much of the United States, American robins are the most important amplifying host for WNV (Figure 21.5). Hamer et al. (2009) found that in Chicago, Illinois, 66% of mosquitoes infected with WNV acquired the virus from feeding on just a few avian species, including America robins (responsible for 35% of infected mosquitoes), blue jays (17%), and house finches (15%). In Guatemala, the great-tailed grackle serves the role of a reservoir host. In Russia, the marsh frog can serve as a reservoir host for WNV (Morales-Betoulle et al. 2012, Wheeler et al. 2012).

FIGURE 21.5 Many *Culex* mosquitoes in the United States become infected with WNV after obtaining a blood meal from an infected American robin. (Courtesy of the U.S. Fish and Wildlife Service.)

The vast majority of birds that contract WNV are infected by mosquitoes, but birds can also become infected from a tick bite or through close association with an infected bird (e.g., if they share the same cage or roost). More than 300 bird species, including domestic chickens, turkeys, geese, and ducks, are known to have been infected with WNV (CDC 2013c). Susceptibility to WNV varies among species. For example, mortality rates from experimental infections of WNV transmitted by mosquitoes were high in American crows (100%), blue jays (75%), and fish crows (75%), while there were few if any mortalities among mourning doves, European starlings, and feral pigeons. Reports of dead crows are often the first sign of an outbreak of WNV in an area (Komar et al. 2003, Kilpatrick 2011). For this reason, public health agencies monitor avian die-offs.

Since its arrival in North America, WNV has killed millions of wild birds and caused population declines of more than 50% in American crows, blue jays, chickadees, titmice, wrens, and thrushes in different regions of North America (Sidebar 21.3). The rapid movement of WNV across the United States and Western Hemisphere was facilitated by the dispersal and migration of birds while infected with WNV. In Europe, several outbreaks of WNV have been linked to migratory birds. Humans also may have helped spread WNV by inadvertently transporting infected mosquitoes in ships or planes (Malkinson et al. 2002, Marra et al. 2004, Brown et al. 2012).

SIDEBAR 21.3 IS WNV RESPONSIBLE FOR LARGE-SCALE DECLINES IN SOME BIRD POPULATIONS IN NORTH AMERICA? (LADEAU ET AL. 2007)

Since the 1999 arrival of WNV in New York City, millions of North American birds have become infected with the virus and died. This does not mean that bird populations have declined due to WNV: most avian species have high reproductive rates, and the increase in WNV mortality rates may result in lower mortality rates from other causes. To test if WNV has caused avian populations to decline, LaDeau et al. (2007) predicted that population declines would occur after 1999 in those avian species known to be susceptible to WNV (American crow, fish crow, blue jay, and tufted titmouse) but not in those species known to be unsusceptible (downy woodpecker, mourning dove, gray catbird, northern mockingbird, and wood thrush). The authors examined 26 years of data from the Audubon Christmas bird counts and found that three of the four susceptible species (American crow, blue jay, and tufted titmouse) experienced population declines after 1999 and one susceptible species did not (fish crow). The American crow had the greatest population decline (Figure 21.6). In contrast, none of the five unsusceptible species experienced a population decline after 1999. Additionally, three avian species with moderate vulnerability to WNV declined during this period (American robin, house wren, and chickadee). These data indicate that WNV has caused some bird populations to decline.

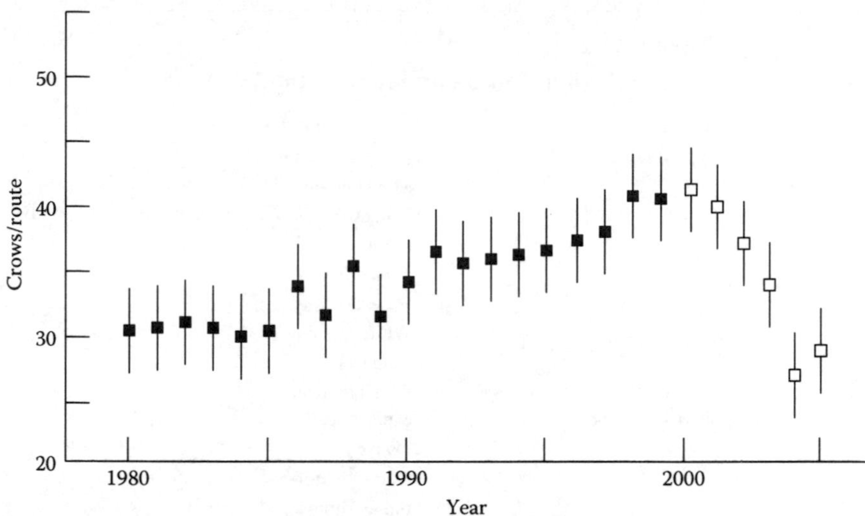

FIGURE 21.6 Changes in the U.S. population of American crows since 1980 based on the mean number of crows seen per route on the U.S. Breeding Bird Survey. Data before the arrival of WNV into the United States in 1999 are shown as open squares, and data since then are shown as closed squares. (From LaDeau, S.L. et al., *Nature*, 447, 710, 2007.)

WNV has been detected (either its RNA or antibodies against it) in a few reptiles and many free-ranging mammals, but most of them are dead-end hosts for the virus (Table 21.2). Horses are particularly susceptible to WNV. For this reason, horses are a good sentinel species for monitoring the appearance of WNV in an area because they often become ill before humans. Clinical signs of WNV neuroinvasive disease in horses include muscle and muzzle twitches or tremors (similar to what a horse does when trying to get flies off its body), aimless wandering, stiff walking, and problems with balance. One or both sides of the face may be paralyzed so that one ear is erect, and the other droops or the tongue may hang out of the mouth. As the illness progresses, horses may experience convulsions or lose the ability to stand. Some horses collapse on their front legs so they look as if they were praying. Most infected horses do not become ill, but most of those that develop clinical signs are killed by the infection (Marra et al. 2004, National Wildlife Health Center 2005, Farfán-Ale et al. 2006).

21.4 HOW HUMANS CONTRACT WNF

Almost everyone with WNF has been infected from the bite of an infectious mosquito that earlier fed on an infected bird. WNV has been recovered from 62 different mosquito species in North America. However, *Culex* mosquitoes are the primary vectors because they are highly susceptible to infection, feed frequently on passerine birds, and are able to survive long enough to transmit the virus. Many *Culex* are opportune feeders and will feed frequently on mammals. In some mosquito species, females can

TABLE 21.2
Mammals that Can Be Infected by the WNV

Order	Species
Artiodactyla	Domestic cattle
	Mountain goat
	Domestic sheep
	Llama
	Alpaca
	Mule deer
	White-tailed deer
	Reindeer
	Buru babirusa
Perissodactyla	Mule
	Donkey
	Domestic horse
	Indian rhinocero
Proboscidea	Asian elephant
Lagomorpha	European rabbit
Rodentia	Black-tailed prairie dog
	Gray squirrel
	Fox squirrel
	American red squirrel
	Eastern chipmunk
	Yellow-necked field mouse
	White-footed mouse
	House mouse
	Bank vole
	Black rat
	Hispid cotton rat
Carnivora	Domestic dog
	Wolf
	Red fox
	Domestic cat
	Snow leopard
	Striped skunk
	Raccoon
	Virginia opossum
	Harbor seal
	Red panda
	Black bear
	Brown bear
Primates	Senegal bushbaby
	Ring-tailed lemur
	Olive baboon
	Rhesus macaque
	Pigtail macaque
	Barbary macaque
	Baboon

(*Continued*)

TABLE 21.2 (*Continued*)
Mammals that Can Be Infected by the WNV

Order	Species
Chiroptera	Northern long-eared myotis
	Little brown myotis
	Big brown bat
	Leschenault's rousette

Sources: Hubálek, Z. and Halouzka, J., *Emerg. Infect. Dis.*, 5, 643, 1999; Ludwig et al. 2002; National Wildlife Health Center, West Nile virus, http://www.usgs.gov/disease_information/west_nile_virus/ 2005; Bentler, K.T. et al., *Am. J. Trop. Med. Hyg.*, 76, 173, 2007; Dietrich, G. et al., *Vector-Borne Zoonotic Dis.*, 5, 288, 2005.

pass WNV vertically to their offspring (Anderson and Main 2006). *Culex restuans* and *C. pipiens* (Figure 21.7) are responsible for up to 80% of human infections of WNV in the northeastern United States. In the western United States, several mosquito species transmit WNV, while *C. quinquefasciatus* and *C. nigripalpus* are the primary vectors in the southeastern United States. WNV has also been isolated from 10 tick species. In the lab, soft-bodied ticks of the genus *Argas* can transmit WNV to chickens, but it is unclear how often these ticks are a vector for WNV (Zeller and Schuffenecker 2004, Kilpatrick et al. 2005, Merck 2010, Reisen 2013).

Human-to-human spread of WNV does not occur through casual contact or kissing. It is possible to contract WNF through a blood transfusion or organ transplant, but this is very rare in the United States and Canada where blood donors are screened for the virus. Women, who are infected when pregnant, can infect their fetus, and nursing mothers can infect their babies through breast-feeding; this, however, is uncommon.

21.5 MEDICAL TREATMENT

There is no specific treatment for WNF, and most people recover on their own. Nonetheless, hospitalization is required for people with WNV encephalitis or meningitis so that they can receive the supportive care they need, including intravenous fluids. The CDC recommends seeking medical attention immediately whenever someone has a severe headache or exhibits confusion, because these symptoms may indicate encephalitis or meningitis. Such cases are the most serious and monitoring these patients is critical. The CDC recommends that pregnant women and nursing mothers talk to their doctors if they suspect they have WNF (CDC 2003, 2012a).

There is no human vaccine against WNV, but several groups are trying to develop one. There is a vaccine for horses that can protect them from WNV, and it is available for use in the United States and Canada. Israel has developed a vaccine for use in domestic geese that is 94% effective when two injections are

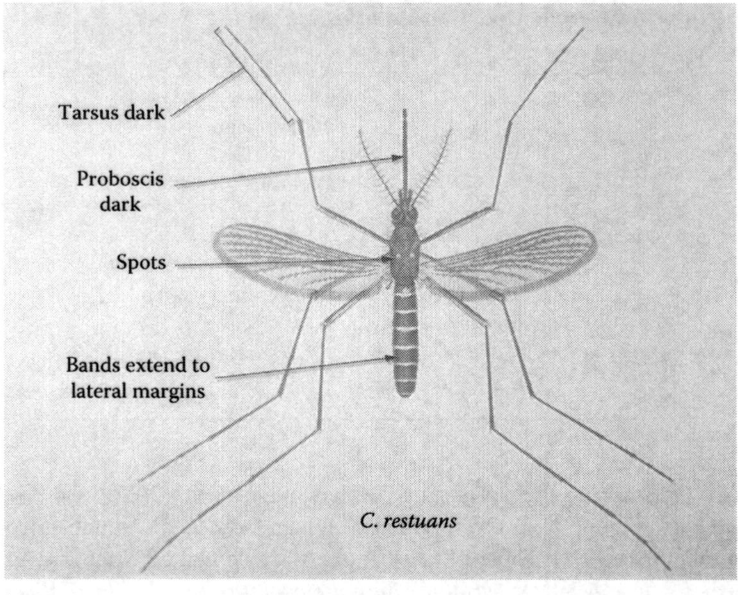

(a)

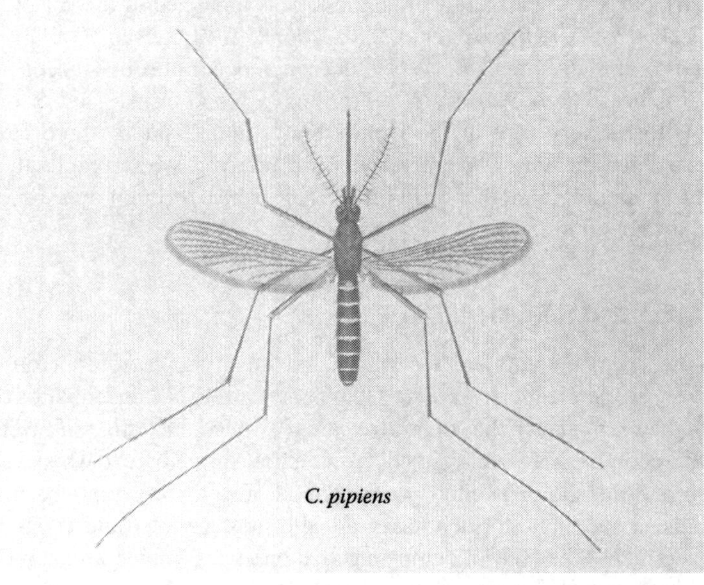

(b)

FIGURE 21.7 (a) *C. restuans* and (b) *C. pipiens* mosquitoes are the main vector for WNV in the eastern United States. (Courtesy of the CDC.)

given spaced 2 weeks apart. Vaccines for geese are not currently available in either Canada or the United States (Merck 2010).

21.6 WHAT PEOPLE CAN DO TO REDUCE THE RISK OF CONTRACTING WNF

In the United States and Canada, WNF is a seasonal disease that occurs mostly during summer when mosquitoes are active and people spend time outdoors. The best way to avoid WNF is to prevent mosquito bites by wearing long-sleeved shirts, long pants, and socks and spraying bare skin with DEET insect repellents when outdoors. Window screens should be checked to make sure that there are no holes and that they fit snugly in the window sills.

The risk of contracting WNF increases with exposure to *Culex* mosquitoes, especially during a WNF outbreak. In North America, the prevalence of WNF is higher in urbanized and agricultural areas where the mosquito species (e.g., *C. pipiens complex* and *C. tarsalis*) that transmit WNV are more abundant. Urban mosquito abundance can be reduced by draining water in buckets, barrels, flowerpots, and children's wading pools. Tire swings should have a hole drilled in the bottom so that water will not accumulate in them. Rain gutters should be checked to ensure that water is draining properly. Water in bird baths should be replaced weekly (Kilpatrick 2011).

Water levels in open marshes can be managed to reduce mosquito densities during outbreaks (Figure 21.8). Mosquito larvae are more likely to hatch in marshes

FIGURE 21.8 Mosquito control districts have the responsibility of managing mosquito populations during outbreaks of WNV. (Courtesy of H.E. Stark and the CDC.)

where there are small depressions of water, but these can be drained with shallow ditches. These ditches also allow fish access to these water pools, and many small fish eat mosquito larva. Mosquito control districts also use insecticides to control either larva or adult mosquitoes.

One of the first signs of a WNV outbreak is finding dead birds, especially American crows or other species known to be susceptible to the disease. In 89% of WNV outbreaks in 2002, the virus was detected in birds prior to its detection in humans (CDC 2003). For this reason, local health authorities often test dead birds for signs of the WNV. Once an outbreak occurs, its intensity is related to the abundance of infected mosquitoes. For this reason, sampling of mosquitoes is the best tool for quantifying the infection risk in an area (CDC 2003).

21.7 OTHER DISEASES CAUSED BY FLAVIVIRUSES

WNV is only one of several viruses in the Flaviviridae family. Other members of this family of single-stranded RNA viruses that produced human disease in the United States or Canada include yellow fever virus, dengue virus, St. Louis encephalitis virus, Powassan encephalitis virus, and deer tick virus. These viruses will be described next. Japanese encephalitis virus will also be covered, although this virus is not endemic in North America (WHO 2007).

21.7.1 Yellow Fever (YF)

For the last 400 years, YF has been the most infamous viral disease in North and South America; its infamy resulted from the loss of human life after the YF virus, and its mosquito vector reached the New World from Africa in the 1600s as a result of the slave trade and the concurrent introduction of the vector mosquito, *Aedes aegypti*. During the next three centuries, deadly epidemics swept through North and South America. These epidemics were common in the tropics but not limited to them; YF epidemics also occurred each summer in the cities of New Orleans, Savannah, Charleston, Philadelphia, New York, and Boston (Sidebar 21.4).

The disease was named YF because the virus often caused jaundice so that the skin and eyes of YF patients appeared yellow. Working in Cuba, Walter Reed and his colleagues discovered that YF was caused by a virus and transmitted among humans by *A. aegypti*, which was subsequently named the YF mosquito. Once it was realized how the pathogen was transmitted, control efforts aimed at *A. aegypti* were implemented in many countries. These efforts, along with vaccination programs, produced a dramatic reduction in the number and severity of YF outbreaks. By 1925, YF was limited in the Americas to small parts of Brazil. However, control efforts were relaxed during the recent decades, and the YF virus made a comeback in both the Americas and Africa (Gubler 2004).

YF is currently endemic in tropical areas of Africa and South America where each year there are an estimated 200,000 YF cases in humans and 30,000 deaths (Figure 21.10). YF virus is not endemic in the United States and Canada and is a threat primarily to people who travel to tropical countries where it is endemic.

SIDEBAR 21.4 YF EPIDEMIC OF 1793 HITS PHILADELPHIA (ARNEBECK 2008 AND *PHILADELPHIA THE GREAT EXPERIMENT: EPISODE TWO: 1793*)

In 1793, the population of Philadelphia, Pennsylvania, was swollen by the arrival of French colonists who had fled a French colony in the Caribbean due to a slave revolt. It is likely that the ships carrying them also carried the YF virus and the YF mosquito. One of the first victims in Philadelphia was one of these French refugees who had resided near the city's wharfs (Figure 21.9). More victims quickly followed, and Dr. Benjamin Rush, who was present during Philadelphia's earlier epidemic of YF, realized that the city was facing a new YF epidemic. No one at the time, however, realized that diseases were caused by bacteria or viruses and that mosquitoes were the vector for YF. Instead, Dr. Rush blamed the epidemic on a shipment of coffee that had putrified on the wharf. The city's medical society recommended that the city stop the tolling of church bells, stop public burials of the dead, clean streets and wharves, and explode gun power to increase the amount of oxygen in the city. Treatment for YF patients included bleeding and purging. Many of the sick were abandoned because people feared that if they cared for patients or got too close to them, they, too, would become infected. With the death toll mounting, panic spread through the city, and 20,000 of the city's 50,000 inhabitants fled. The epidemic continued until cold weather in November killed the mosquitoes. By then, the official death toll in Philadelphia had reached 3,881.

People are infected by the bite of an infected mosquito. Fever, backaches, headaches, shivering, nausea and vomiting are the initial symptoms of YF and begin after an incubation period of three to 6 days. In most patients, the symptoms wane after a few days, but 15% of patients experience a second phase of the infection, which begins within 24 hours of the initial symptoms disappearing. This phase is more serious, and clinical signs include jaundice and abdominal pain as the pathogen infects the liver. Bleeding may occur from the mouth, nose, eyes, or stomach. Blood may occur in vomit and in feces. Half of the patients who enter this second phase die within 10–14 days; the other half recover. YF is difficult to diagnose, especially during the initial phase, because the symptoms are similar to other diseases. Diagnoses are usually made following the isolation of the YF virus from the patient or the detection of antibodies against YF virus in the patient's blood (WHO 2013).

Several other mosquitoes can transmit the YF virus in addition to *A. aegypti. A. africanus*, *A. luteocephalus*, *A. opok*, and *A. simpsoni* are vectors in Africa as are *Haemagogus* mosquitoes in South America. Monkeys and humans are the reservoir hosts for the YF virus. African primates infected with the YF virus develop only a mild infection, but YF viral infections can be fatal in South American primates, such as howler monkeys, squirrel monkeys, owl monkeys, and spider monkeys. Presumably South American primates are more susceptible than African primates

FIGURE 21.9 An engraving of Philadelphia's wharf where the YF epidemic of 1793 began. (Engraving by William Birch, from "The City of Philadelphia, 1800," Independence National Historical Park via WikiMedia Commons.)

because the latter coevolved with the virus and are more immune to it (Gubler 2004, Weissenböck 2012).

Antibiotics are ineffective against the YF virus, and most care is directed at making the patient comfortable and providing supportive care to prevent respiratory failure, dehydration, and high fevers. Such care can reduce the mortality rate among YF patients, but it is unavailable in many poor regions of South America and Africa where YF is prevalent.

The YF virus is maintained through two transmission cycles: one is a jungle cycle, and the second is the urban cycle (Figure 21.11). In tropical jungles, the YF virus is maintained by circulating among monkeys; people who live or work in the forest may become infected with the YF virus as an offshoot of this transmission cycle. The jungle cycle results in a slow but steady stream of people with YF, albeit without large epidemics. The second YF transmission cycle occurs in urban areas where many people are unvaccinated, leading to YF epidemics. The urban cycle begins when travelers from rural areas bring the YF virus into a city. The epidemic often starts at a single site and spreads outward as urban mosquitoes spread the virus from one person to another. In urban areas, the main mosquito vector is the YF mosquito, and humans are the reservoir host.

Mosquito control programs have proven effective in eliminating the YF mosquito from many urban areas, but this has not been accomplished in tropical rainforests

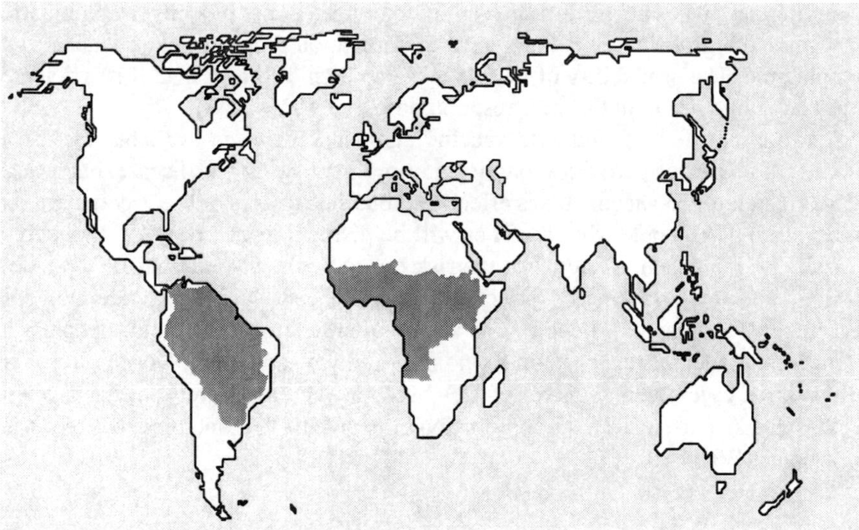

FIGURE 21.10 YF is endemic in parts of tropical South America and Africa. These areas are shown in black on the map; in these endemic areas, the WHO recommends that people become vaccinated against YF. Epidemics of YF occur outside these areas but in a more sporadic manner than in the endemic areas. (Data from WHO, Yellow Fever. Fact Sheet 100, World Health Organization, Geneva, Switzerland, 2013.)

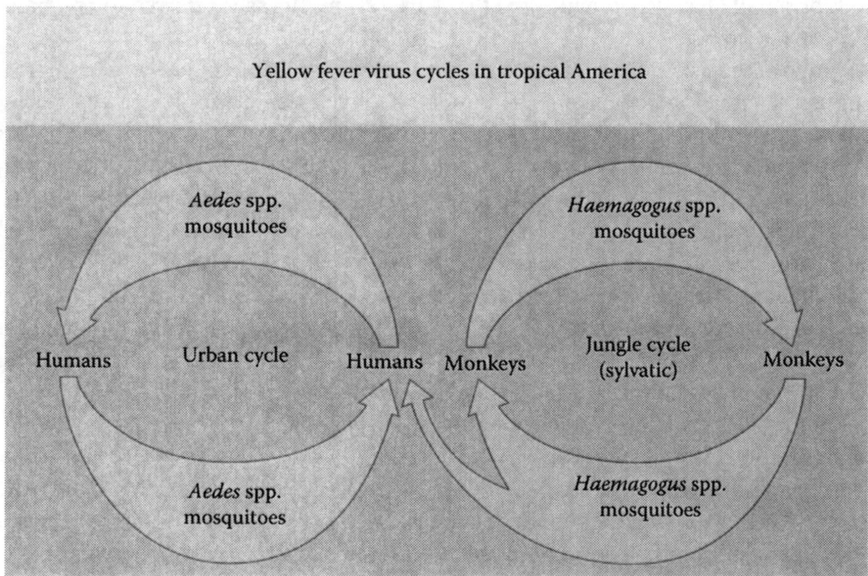

FIGURE 21.11 YF has both an urban disease cycle and a jungle cycle. In urban areas, humans are the amplifying host, and *Aedes* mosquitoes are the primary vector. In tropical jungles, monkeys are the amplifying host, and *Haemagogus* mosquitoes are the main vector. (Courtesy of the CDC.)

where there are wild monkeys. When a YF epidemic occurs in a city, a combination of mosquito control efforts and large-scale vaccination programs can reduce or halt the epidemic. The probability of halting a YF epidemic is increased if it is detected early and public health authorities respond quickly (WHO 2013).

There is an effective and safe vaccine against YF that is available for use in humans. A single dose provides long-term immunity against the disease, but it takes 7–10 days before the vaccine takes effect. Serious side effects of the vaccine are rare but do occur. For people who live in or will be visiting areas where YF is endemic, the risk of YF is much greater than the risk posed by the vaccine. Still, WHO does not recommend the vaccine for pregnant women (except during an epidemic), children under 9 months old, people with severe allergies to egg protein, people with severe immunodeficiency, or people who have a thymus disorder. Despite the availability of the vaccine, the number of YF cases has increased over the last couple of decades due, in part, to a smaller proportion of people in YF endemic areas receiving the vaccine (WHO 2013).

21.7.2 Dengue and Dengue Hemorrhagic Fever

This disease is a leading cause of illness and death in tropic and subtropical countries. Mild forms of the disease are called dengue, dengue fever, dandy fever, or breakbone fever. More severe forms are called hemorrhagic dengue, dengue hemorrhagic fever, dengue shock syndrome, Philippine hemorrhagic fever, Thai hemorrhagic fever, and Singapore hemorrhagic fever (Halstead 1994). In this book, the mild form will be called dengue, and the more severe illness is referred to as dengue hemorrhagic fever.

Dengue is caused by four closely related serotypes of the dengue virus: dengue 1, dengue 2, dengue 3, and dengue 4. An infection by one dengue serotype provides the patient lifelong immunity against that particular serotype but not against the others. The dengue virus probably evolved in Southeast Asia where it was maintained by a transmission cycle involving mosquitoes and monkeys. All four dengue viruses occur there, but dengue 1 and dengue 2 spread to Africa and from there to North and South America, perhaps as early as the 1600s. Dengue epidemics have occurred throughout the tropics and reached as far north as Philadelphia, Pennsylvania, which had a dengue epidemic in 1793. Over a million people in and around Houston and Galveston, Texas, were stricken with dengue in 1922. That same year a dengue epidemic was reported in Florida (Halstead 1994, Luby 1994, Gubler 2004).

Scientists learned during the beginning of the twentieth century that the pathogen responsible for dengue was transmitted by the YF mosquito (*A. aegypti*). Mosquito control programs almost eradicated this mosquito from North and South America in the 1950s. Nonetheless, dengue has reemerged a worldwide problem since the 1960s when support for mosquito control programs waned, and *A. aegypti* returned to most of its former range in Central and South America. Before 1970, only nine countries had experienced epidemics of dengue hemorrhagic fever; the number exceeded 100 countries by 2010 when locally acquired cases of dengue were first documented in Europe. Currently, 100 million people are infected with the dengue virus, and 9 million people develop dengue fever annually around the world. Half of

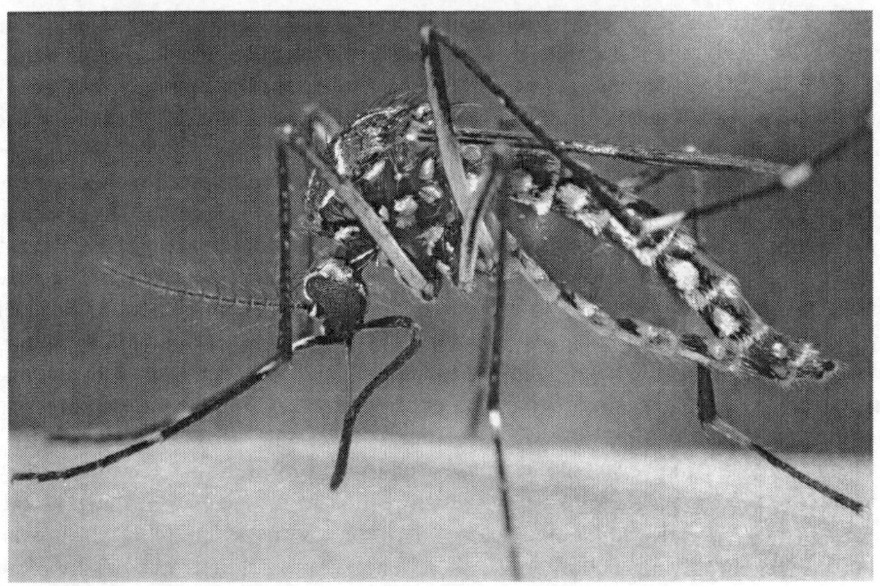

FIGURE 21.12 Photo of the *A. aegypti* mosquito. (Courtesy of James Gathany and the CDC.)

the cases occur in Southeast Asia. In the Western Hemisphere, there are 1.6 million cases of dengue, and 49,000 of these result in dengue hemorrhagic fever (Gubler 2004, WHO 2008, 2012b).

In 2011, there were 243 diagnosed cases of dengue in the United States, almost all were travelers who acquired the pathogen while abroad. Exceptions included a local outbreak in Hawaii, which sickened five people, and one in Florida, which involved seven patients who acquired the disease locally. A year later, two patients in Miami and two near Orlando acquired dengue locally, indicating the dengue virus may be endemic in Florida. One study found that 5% of residents in Key West, Florida, have antibodies against the virus. There is a risk of a large outbreak of dengue in U.S. cities in Florida and along the Gulf of Mexico because the mosquitoes that can transmit dengue virus are common there (Gubler 2004, McKenna 2012, CDC 2013a,b).

In the Western Hemisphere, the most important vector is the mosquito *A. aegypti*. This mosquito is native to Africa but has spread to North and South America and Asia (Figure 21.12). Historically, this mosquito lived in the tree canopy where it obtains blood meals from monkeys and lays its eggs inside walls of tree cavities. Eggs can remain dormant for months until these cavities fill with rainwater and the eggs hatch. *A. aegypti* has expanded its range by adapting to areas in towns and cities where it uses humans as a host and water-holding containers to lay eggs. For this reason, dengue mostly occurs in urban and semiurban areas where the dengue virus is maintained in a human–mosquito–human transmission cycle.

Another mosquito vector for the dengue virus is *A. albopictus*. This species is native to Asia, but it has spread to North and South America in the 1980s, possibly through the importation of used tires. Several other mosquito species transmit the dengue virus in the jungles of Asia and Africa. Dengue virus can replicate in the mosquito's gut, brain, salivary glands, and reproductive organs without harming the mosquito. After the salivary glands are infected, the mosquito can transmit the virus to primates. Once infected, a mosquito can remain infected for life (Halstead 1994, Monath 1994, CDC 2013b).

People develop dengue 3–8 days after being bitten by an infected mosquito. The dengue virus makes blood capillaries very permeable, causing bleeding. The principal symptoms of dengue are a fever of over 100°F (38°C) and at least two of the following: severe headache, severe pain behind the eyes, joint pain, muscle ache, or bone pain. These symptoms usually last less than a week. After this initial period, the fever and symptoms may wane for a few hours or days to be replaced by nonitchy rash of red and purple spots. The rash often begins on the hands and feet, spreading to arms, legs, and trunk. A diagnosis of dengue is confirmed by isolating the virus from blood or body tissue or detecting antibodies specific to the dengue virus (Monath 1994).

Sometimes as the fever and symptoms of the initial period are waning, they are replaced with excessive bleeding, which results in bruises, nosebleeds, bloodshot eyes, bleeding from the gums, blood in vomit, and dark feces that result from intestinal bleeding. This condition is called dengue hemorrhagic fever, and its first 24–48 hours is critical. Excessive bleeding during this period can lead to failure of the circulatory system, shock, and death. Patients with shock may have cold, clammy skin, sweaty hands and feet, restlessness, a rapid pulse, low blood pressure, and rapid or labored breathing. The case fatality rate of dengue hemorrhagic fever is 20% if left untreated but declines to less than 1% with medical care. In Latin America and Asia, dengue hemorrhagic fever has become a leading cause of hospitalization and death of children. Given how rapidly dengue shock can develop, the CDC warns people to go immediately to the emergency room of a hospital if any of the following warning signs appear 3–7 days after the first symptoms appear: severe abdominal pain or persistent vomiting; red spots or patches on the skin; bleeding from the nose or gums; vomiting blood; black tarry feces; drowsiness or irritability; pale, cold, or clammy skin; or difficulty breathing (Halstead 1994, CDC 2012a,b, 2013b).

The risk of dengue developing into dengue shock syndrome is increased when a patient has a second infection of dengue produced by a different serotype of the dengue virus than the first infection. Babies less than 1 year old are also at risk, especially if they passively acquired dengue antibodies at birth (Halstead 1994).

There is no specific treatment for dengue; antibiotics are ineffective against the virus. People who think they may have dengue should avoid aspirin or ibuprofen as these drugs can make the bleeding worse. Patients should drink plenty of fluids and seek medical attention. Dengue hemorrhagic fever requires critical care management; patients may need rapid blood transfusions and careful management of their fluids and electrolytes to replace losses caused by disease (CDC 2012a,b).

There is no vaccine against dengue. Most preventative steps in areas where the dengue virus is endemic involve eliminating water sources that *A. aegypti* mosquitoes use to lay eggs. People can reduce the risk of being bitten by these mosquitoes by using an insect repellent and wearing long-sleeved shirts, long pants, socks, and shoes. Mosquitoes can transmit the dengue virus from one person to another, warranting extra caution when a member of the household has dengue.

21.7.3 St. Louis Encephalitis (SLE)

This disease first attracted attention during the summer of 1933 when a large epidemic that occurred in St. Louis, Missouri, resulted in 1095 reported cases and 201 deaths. Concurrent epidemics occurred in Illinois, Kansas, and Kentucky (Figure 21.13). Since then, there have been over 41 SLE outbreaks in the United States and Canada, each involving more than 20 diagnosed cases. The largest epidemic of SLE in North America occurred in 1975 when there were over 1800 diagnosed cases that ranged from Ontario to Texas. The SLE epidemic began again the next year but was limited to the midwest and south. The same strain of the SLE virus appeared again in 1977 but was isolated to rural parts of Florida (Muckenfuss et al. 1934, Luby 1994).

There is an average 193 confirmed cases annually in the United States, but the number of cases in the United States varies greatly among years (range from three cases annually to 1967), depending upon whether an outbreak occurs that year. Ten patients were diagnosed with SLE in 2010, six in 2011, and three in 2012 (CDC 2012b, 2013e). SLE virus occurs in all states except for Hawaii and Alaska but is most prevalent in the midwestern and southern states. One recent epidemic was in Florida from 1990 to 1991, in which 226 people were diagnosed with SLE, and 11 patients died (Day 2001).

Most people infected with SLE virus are asymptomatic. Those who do become ill typically experience only a mild fever and headache, which begin 4–21 days after the patient was infected. Others develop encephalitis and/or meningitis with a case fatality rate of 5%–15%. Clinical signs that the SLE virus has infected the central nervous system include disorientation, tremors, an unsteady walk, and confusion. The tremors often involve the eyes, eyelids, tongue, and lips. Patients are frequently

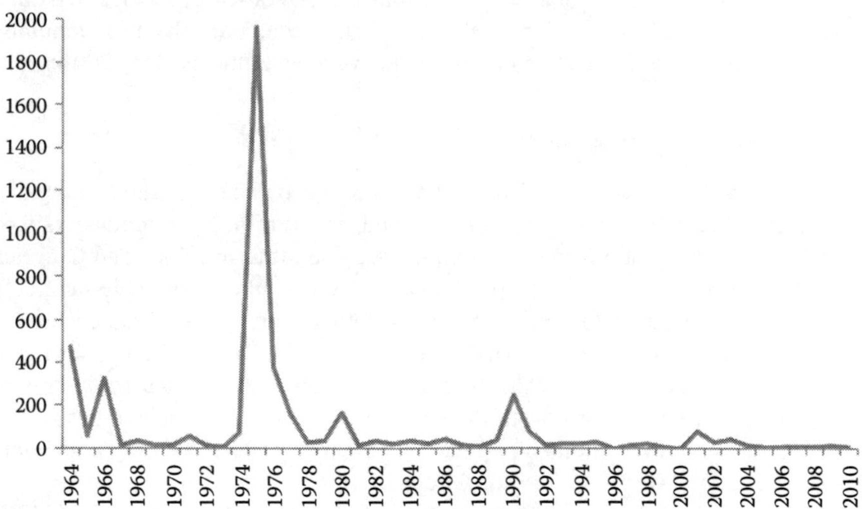

FIGURE 21.13 U.S. distribution of neuroinvasive cases of SLE from 1964 through 2010. (From CDC, Saint Louis Encephalitis epidemiology and geographic distribution. http://www.cdc.gov/sle/technical/epi.html, 2011.)

disoriented as to time and place and have difficulty doing simple mental tasks such as adding and subtracting. The risk of being infected with the SLE virus does not change with a person's age. However, the risk of developing encephalitis or meningitis is highest among people who are over 60 years old. The case fatality also increases with age; 35% of patients over 65 years old succumb to the virus. Recovery is often a lengthy process and can last for months. During recovery, patients complain of chronic fatigue and an inability to concentrate or perform ordinary tasks. Ten percent of SLE patients in the United States suffer long-term neurological problems. An SLE virus infection results in lifelong immunity to the virus (Luby 1994, CDC 2005).

The mosquito species that transmit the pathogen vary across the United States and Canada. *C. nigripalus* is the principal vector in Florida, *C. pipiens* in the midwest and Canada, and *C. quinquefasciatus* and *C. tarsalis* in the west. In some areas, multiple mosquito species may be involved in SLE virus transmission. In urban areas, house sparrows, house finches, and European starlings are reservoir hosts. In rural areas, American robins, common grackles, and mourning doves are reservoir hosts. Birds infected with the SLE virus usually do not become ill (Day 2001, Mahmood et al. 2004, CDC 2005). SLE virus has been detected in black-tailed jackrabbits, snowshoe hares, mountain cottontails, and American black bear. But mammals are considered a dead-end host for the virus (Borne and Fox 2010).

SLE outbreaks in California follow heavy snows and spring rains that promote flooding. In the Midwest, SLE outbreaks occur when rivers flood over their banks and in Texas when rains create playa lakes. In all of these cases, the outbreaks are tied to conditions favorable to the mosquito vectors. In Florida, SLE outbreaks may be linked to the nesting success of amplifying hosts (Luby 1994). SLE outbreaks in Florida sometimes happen within a year or two of a hard winter freeze that kills plants across large tracts of land. This allows for an explosion of annual seed-bearing plants, providing an abundance of food for nesting mourning doves, common grackles, and other ground-feeding birds. The abundance of nestlings and young birds that lack immunity to SLE virus then provides the conditions conducive to an outbreak (Day 2001).

21.7.4 Japanese Encephalitis (JE)

The JE virus was isolated in 1935 from the brain of a patient who died of the disease. During World War II, this virus concerned Allied military commanders who worried that troops set to invade Japan might become infected (Hoke and Gingrich 1994). JE virus occurs across a vast part of eastern and southern Asia (Figure 21.14). JE epidemics first occurred in Japan in the late 1800s; China in the 1930s; Korea in the 1940s; northern Vietnam and Thailand in the 1960s; Burma, Bangladesh, and northern India in the 1970s; and the Torres Strait Islands and the Australian mainland in the 1990s. The reasons for the spread of the JE virus are unclear, but factors include the increase in the human population, acreage irrigated for rice production, and pig production (WHO 2007, Erlanger et al. 2009).

There have been only four confirmed cases of JE in the United States from 1992 to 2008; all occurred among travelers returning from Asia. JE virus is the most widespread of all of the arboviral viruses that produce encephalitis, with approximately 50,000 human cases reported annually and 10,000 deaths. Human infection rates

FIGURE 21.14 Map showing the distribution of JE. (From Hills, S.L. et al., *Japanese Encephalitis. Yellow Book*, CDC, Atlanta, GA, 2012; Provided Courtesy of the CDC.)

are much higher because less than 1% of infected humans develop acute encephalitis. In rural areas of Asia where the JE virus is endemic, JE is a common illness of childhood, and almost everyone has been exposed to the disease by the time they are adults (Hoke and Gingrich 1994, WHO 2007, CDC 2009).

Symptoms of JE in humans are sudden fever, headache, and vomiting; they begin after a 5- to 15-day incubation period. Usually JE is a mild disease in which gastrointestinal symptoms predominate. After a week, the symptoms disappear and the recovery is uneventful. In some patients, the infection spreads to the central nervous system, causing encephalitis, meningitis, or meningoencephalitis. For such patients, the fatality rate is less than 10% but is higher among children. About 30% of JE patients who develop encephalitis suffer long-term neurological problems requiring extensive long-term care. Clinical signs that the JE virus has become neuroinvasive are cogwheel rigidity (i.e., jerky movements), twitching, and involuntary movement of the limbs. Seizures may occur, especially in children. Some JE patients develop acute flaccid paralysis, resulting in weakness or paralysis of one or more limbs despite a normal level of consciousness. The condition occurs more often in the legs than the arms, and paralysis is often asymmetrical involving one arm or leg. Flaccid paralysis of facial muscles creates a masklike appearance (WHO 2007, CDC 2009, Hills et al. 2012).

C. tritaeniorhynchus mosquitoes are the main vector; maintenance hosts are ciconiiform birds, especially black-crowned night herons, little egrets, intermediate egrets, and other wading birds (Buescher et al. 1959). JE virus does not cause illness in these species. Both domestic and feral hogs are the amplifying hosts for the pathogen (WHO 2007, Weissenböck 2012).

JE virus is maintained by a cycle between culicine mosquitoes and waterbirds that serve as the amplifying host, along with feral and domestic hogs. Dogs, sheep, horses, mules, water buffalos, and chickens can become infected with the JE virus, but they and humans are considered dead-end hosts. Other than primates, pigs and horses are the only species known to become ill after being infected with the JE virus. In these animals, the JE virus causes fatal encephalitis in horses and abortion in pigs. During outbreaks, up to 33% of sows can abort their litter. JE virus causes major economic losses for farmers in areas where there are high densities of pigs. Both horses and pigs usually become ill before humans during JE outbreaks, making them good sentinel species (Hoke and Gingrich 1994).

JE is a rural disease in Asia and the western Pacific and is most common in areas where there are irrigated rice fields and adjacent villages with domestic pigs. The JE virus is transmitted primarily by *Culex* mosquitoes, especially by *C. tritaeniorhynchus*, a species that lays its eggs in rice fields. In temperate and subtropical countries, outbreaks occur primarily in late summer and end with the cool weather of fall. In tropical countries, sporadic cases occur throughout the year, but the incidence increases during the monsoon (CDC 2005, 2009).

A human vaccine (Ixiaro) against JE is widely used throughout Asia, including Japan, China, and India (the JE-VAX vaccine is no longer available). It also is available in the United States for people traveling to Asia. The U.S. Advisory Committee on Immunization Practices recommends the JE vaccine for travelers who will spend more than 1 month in an endemic area during the JE virus transmission season. The vaccine is not recommended for short-term travelers whose visit will be restricted to urban areas or outside of the season when JE occurs (Fischer et al. 2010, Hills et al. 2012).

The incidence of JE in Japan, Taiwan, Korea, and China has declined by modifying the cultivation of rice, increasing the use of pesticides, and vaccinating children. Pigs are vaccinated against JE to increase farm productivity as well as part of an integrated approach to curtail JE outbreaks. Physically separating humans from amplifying hosts (e.g., waterbirds and pigs) will lower the risk to humans. Where this is not possible, making houses mosquito proof, using mosquito netting where people are sleeping, wearing long-sleeved shirts and long pants, and applying mosquito repellent are beneficial. Spraying areas with insecticides by ground or aircraft can reduce mosquito populations, but such applications have to be repeated, and many developing countries cannot afford the cost (Hoke and Gingrich 1994, WHO 2007).

21.7.5 Powassan Encephalitis (POW) and Deer Tick Virus (DTV)

The POW virus and DTV are two closely related species of *Flavivirus*. They are unique among North American *Flavivirus* species because they are transmitted by ticks rather than mosquitoes. Other tick-borne flaviviruses occur in Europe and Asia (e.g., European tick-borne encephalitis virus, Siberian tick-borne encephalitis virus, and Far Eastern tick-borne encephalitis virus); these viruses result in several thousand human cases annually. In Eurasia, these tick-borne flaviviruses are transmitted by several species of the *Ixodes* ticks, and small mammals serve as the amplifying hosts. Most tick-borne encephalitis patients in Asia and Europe are infected by a tick bite, but some patients became infected from consuming raw milk from infected cows, goats, and sheep. Women who become infected during pregnancy can infect their fetuses (CDC 2013d).

In Eurasia, most people infected with a tick-borne encephalitis virus have no symptoms or only a mild illness, but the virus can become neuroinvasive. For these latter patients, mortality rates are 1%–2%, and an additional 10%–20% will experience long-term or permanent neurological problems. A vaccine is available in parts of Asia and Europe where the virus is endemic, but some children have an adverse reaction to the vaccine. The vaccine is not available in the United States (CDC 2013d).

There are two lineages of the POW virus in North America: the DTV and the POW virus. The two viruses are serologically indistinguishable but differ genetically. Both lineages probably evolved from a Eurasian virus that may have been introduced to North America by mammals moving across the Bering Land Bridge during the last Ice Age or by migrating birds. Interestingly, the POW virus was reintroduced into Asia during the twentieth century, perhaps when American mink were exported to Russia to support their fur industry (Ebel 2010). The two lineages are broadly distributed in Canada and the United States (Figure 21.15). The POW virus was named after Powassan, Ontario, Canada, where the virus was first discovered in 1958 when it was isolated from the brain of a 5-year-old child who died of encephalitis. The DTV was first isolated in 1997.

The blacklegged tick transmits the DTV; this virus was found in 0.4% of ticks collected from Connecticut and Massachusetts. Female blacklegged tick can transmit the pathogen to their eggs (Telford et al. 1997). The white-footed mice are the amplifying host. There have been only three confirmed human cases of DTV

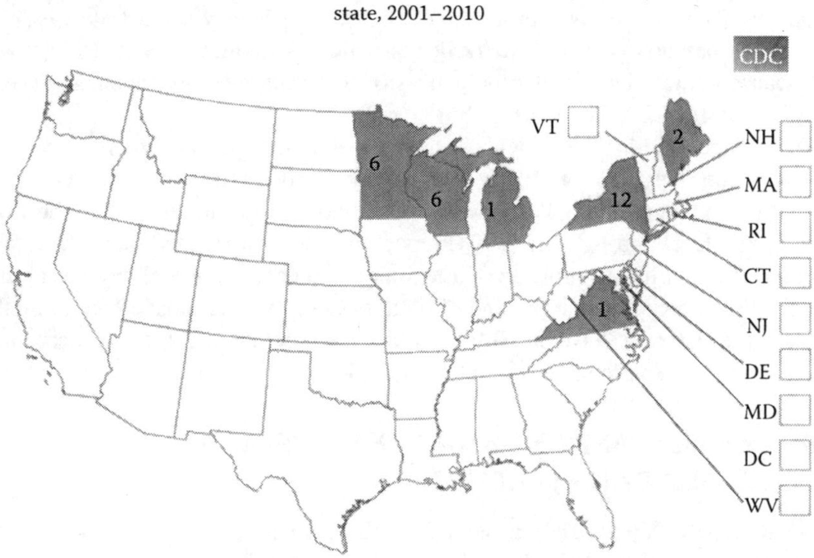

FIGURE 21.15 U.S. distribution of human cases on Powassan virus infections that became neuroinvasive from 2001 through 2010. (From CDC, Powassan virus neuroinvasive disease cases reported by state, 2001–2010. http://www.cdc.gov/powassan/statistics.html, 2012.)

encephalitis in North America. The first case was in Ontario, Canada, in a patient who died of a pulmonary embolism 5 weeks after the onset of symptoms. The second was a 62-year-old man from New York state who developed meningoencephalitis and died 16 days later. The third was a 77-year-old man from New York state who developed encephalitis after being bitten by a tick. He suffered severe neurologic dysfunction and died 8 months after infection (Tavakoli et al. 2009, El Khoury et al. 2013).

The POW lineage has two transmission cycles that allow it to persist in North America. One transmission cycle involves woodchuck ticks as vectors and woodchucks and striped skunks as reservoir hosts; the other transmission cycle centers on the squirrel ticks and squirrels. In some areas, half of all woodchucks have antibodies against the POW virus. The POW virus produces mild or no signs in their hosts. Woodchuck ticks are found mainly in woodchuck burrows, and squirrel ticks are found in squirrel nests. Both rarely bite humans, which is one reason for the paucity of humans stricken with POW. The Rocky Mountain wood tick and *I. spinipalpus* also can transmit the POW virus. The Rocky Mountain wood tick, and probably other tick species, can transmit the POW virus to their eggs (Artsob et al. 1986, Artsob 1989, Ebel 2010, Raval et al. 2012).

Early symptoms of infection by the DTV and POW virus in humans are fever, headaches, stiff neck, and muscle aches and dizziness. These viruses can become neuroinvasive with symptoms of dizziness, loss of coordination, confusion, memory loss, seizures, and altered mental states. When POW viral infections result in meningitis or encephalitis, the case fatality rate is 10%, and about half of the survivors suffer long-term neurological problems. In 2012, there were seven POW cases that developed into a neuroinvasive disease in the United States; these occurred in Minnesota (four cases), Wisconsin (two cases), and Pennsylvania (one case). Six of the seven patients were hospitalized; none died. Although DTV and POW virus rarely cause human disease, the incidences of these diseases are increasing (Hardy 1994, Ebel 2010, CDC 2013d,e).

The average age of POW patients is 13 years old; most patients are male, probably because young men are more likely than women to spend time outdoors where they are exposed to ticks (Hardy 1994). Most POW patients become ill during the spring and summer. Ebel and Kramer (2004) report that the blacklegged tick can transmit the POW virus within 15 minutes of attachment. Hence, a thorough body search for ticks should be completed at the end of outdoor activity. The most effective method to reduce the risk of a POW or DTV infection is to prevent tick bites. Specific techniques to avoid ticks are provided in Chapter 13 on Lyme Disease.

21.8 ERADICATING WNF AND OTHER DISEASES CAUSED BY FLAVIVIRUSES

At one time, the YF virus had been eradicated in much of the world with the exception of countries where the virus was able to maintain itself by infecting wild primates. More recently, the pathogen has made a resurgence due to fewer people being vaccinated and reduced funding for mosquito control. We currently lack the ability to eradicate WNV and most other flaviviruses because they circulate among wild animals.

LITERATURE CITED

Anderson, J. F. and A. J. Main. 2006. Importance of vertical and horizontal transmission of West Nile virus by *Culex pipiens* in the northeastern United States. *Journal of Infectious Diseases* 194:1577–1579.

Arnebeck, B. 2008. A short history of yellow fever in the US. http://bobarnebeck.com/history.html (accessed June 13, 2013).

Artsob, H. 1989. Powassan encephalitis. In: T. P. Monath, editor. *The Arboviruses: Epidemiology and Ecology*. CRC Press, Boca Raton, FL, pp. 29–49.

Artsob, H. L. Spence, C. Th'ng, V. Lampotang et al. 1986. Arbovirus infections in several Ontario mammals, 1975–1980. *Canadian Journal of Veterinary Research* 50:42–46.

Asnis, D. S., R. Conetta, A. A. Teixeira, G. Waldman, and B. A. Sampson. 2000. The West Nile virus outbreak of 1999 in New York: The Flushing Hospital experience. *Clinical Infectious Diseases* 30:413–418.

Bentler, K. T., J. S. Hall, J. J. Root, K. Klenk et al. 2007. Serologic evidence of West Nile virus exposure in North American mesopredators. *American Journal of Tropical Medicine and Hygiene* 76:173–179.

Borne, D. and N. Fox. 2010. St. Louis encephalitis (SLE) in bears and lagomorphs. http://wildpro.twycrosszoo.org/S/00dis/viral/SLE_Bears_lagomorphs.htm (accessed June 10, 2013).

Brown, E. B. E., A. Adkin, A. R Fooks, B. Stephenson et al. 2012. Assessing the risks of West Nile virus-infected mosquitoes from transatlantic aircraft: Implication for disease emergence in the United Kingdom. *Vector-Borne and Zoonotic Diseases* 12:310–320.

Buescher, E. L., W. F. Scherer, H. E. McClure, J. T. Moyer et al. 1959. Ecologic studies of Japanese encephalitis virus in Japan: IV. Avian infection. *American Journal of Tropical Medicine and Hygiene* 8:689–697.

CDC. 2003. *Epidemic/Epizootic West Nile Virus in the United States: Guidelines for Surveillance, Prevention, and Control*. Third revision. Centers of Disease Control and Prevention, Division of Vector-Borne Infection Diseases, Atlanta, GA.

CDC. 2005. Information on arboviral encephalitides. http://www.cdc.gov/ncidod/dvbid/arbor/arbdet.htm (accessed June 2, 2013).

CDC. 2009. Japanese encephalitis among three U.S. travelers returning from Asia, 2003–2008. *Morbidity and Mortality Weekly Report* 58:737–742.

CDC. 2011. Saint Louis Encephalitis epidemiology and geographic distribution. http://www.cdc.gov/sle/technical/epi.html (accessed December 5, 2013).

CDC. 2012a. West Nile virus. http://www.cdc.gov/westnile/index.html (accessed December 21, 2013).

CDC. 2012b. Summary of notifiable diseases—United States, 2010. *Morbidity and Mortality Weekly Report* 59(53):1–111.

CDC. 2012c. Powassan virus neuroinvasive disease cases reported by state, 2001–2010. http://www.cdc.gov/powassan/statistics.html (accessed June 8, 2013).

CDC. 2013a. Summary of notifiable diseases—United States, 2011. *Morbidity and Mortality Weekly Report* 60(53):1–117.

CDC. 2013b. Dengue. http://www.cdc.gov/dengue/ (accessed June 10, 2013).

CDC. 2013c. Species of dead birds in which West Nile virus has been detected, United States, 1999–2012. http://www.cdc.gov/westnile/resources/pdfs/Bird%20Species%201999-2012.pdf (accessed December 21, 2013).

CDC. 2013d. Tick-borne encephalitis, fact sheet. http://www.cdc.gov/ncidod/dvrd/spb/pdf/Tick-borne_Encephalitis_Fact_Sheet.pdf (accessed June 10, 2013).

CDC. 2013e. West Nile virus and other arboviral diseases—United States, 2012. *Morbidity and Mortality Weekly Report* 62:513–517.

CDC. 2013f. West Nile Virus annual maps and data. http://www.cdc.gov/westnile/statsMaps/finalMapsData/index.html (accessed May 13, 2013).

Day, J. F. 2001. Predicting St. Louis encephalitis virus epidemics: Lessons from recent, and not so recent, outbreaks. *Annual Review of Entomology* 46:111–138.

Dietrich, G., J. A. Montenieri, N. A. Panella, S. Langevin et al. 2005. Serologic evidence of West Nile virus infection in free-ranging mammals, Slidell Louisiana, 2002. *Vector-Borne and Zoonotic Diseases* 5:288–292.

Ebel, G. D. 2010. Update on Powassan virus: Emergence of a North American tick-borne flavivirus. *Annual Review of Entomology* 55:95–110.

Ebel, G. D. and L. B. Kramer. 2004. Short report: Duration of tick attachment required for transmission of Powassan virus by deer ticks. *American Journal of Tropical Medicine and Hygiene* 71:268–271.

El Khoury, M. Y., R. C. Hull, P. W. Bryant, K. L. Escuyer et al. 2013. Diagnosis of acute deer tick virus encephalitis. *Clinical Infectious Diseases* 56(4):e40–e47.

Erlanger, T. E., S. Weiss, J. Keiser, J. Utzinger, and K. Wiedenmayer. 2009. Past, present, and future of Japanese encephalitis. *Emerging Infectious Diseases* 15:1–7.

Farfán-Ale, J. A., B. J. Blitvich, N. L. Marlenee, M. A. Loroño-Pino et al. 2006. Antibodies to West Nile virus in asymptomatic mammals, birds, and reptiles in the Yucatan Peninsula of Mexico. *American Journal of Tropical Medicine and Hygiene* 74:908–914.

Fischer, M., N. Lindsey, J. E. Staples, and S. Hills. 2010. Japanese encephalitis vaccines: Recommendations of the Advisory Committee on Immunization Practices (ACIP). *Morbidity and Mortality Weekly Report, Recommendations and Reports* 59(RR01):1–27.

Gomes, B., E. Kioulos, A. Papa, A. P. G. Almeida et al. 2013. Distribution and hybridization of *Culex pipiens* forms in Greece during the West Nile virus outbreak of 2010. *Infection, Genetics and Evolution* 16:218–225.

Gubler, D. J. 2004. The changing epidemiology of yellow fever and dengue, 1900 to 2003: Full circle? *Comparative Immunology, Microbiology and Infectious Diseases* 27:319–330.

Halstead, S. B. 1994. Dengue and dengue hemorrhagic fever. In G. W. Beran, editor. *Handbook of Zoonoses*, 2nd edition. CRC Press, Boca Raton, FL, pp. 89–99.

Hamer, G. L., U. D. Kitron, T. L. Goldberg, J. D. Brawn et al. 2009. Host selection by *Culex pipiens* mosquitoes and West Nile virus amplification. *American Journal of Tropical Medicine and Hygiene* 80:268–278.

Hardy, J. L. 1994. Arboviral zoonoses of North America. In: G. W. Beran, editor. *Handbook of Zoonoses*, 2nd edition. CRC Press, Boca Raton, FL, pp. 185–200.

Hills, S. L., R. J. Nett, and M. Fischer. 2012. *Japanese Encephalitis. Yellow Book*. CDC, Atlanta, GA.

Hoke Jr., C. H. and J. B. Gingrich. 1994. Japanese encephalitis. In: G. W. Beran, editor. *Handbook of Zoonoses*, 2nd edition. CRC Press, Boca Raton, FL, pp. 59–69.

Hubálek, Z. and J. Halouzka. 1999. West Nile fever—A reemerging mosquito-borne viral disease in Europe. *Emerging Infectious Diseases* 5:643–650.

Kilpatrick, A. M. 2011. Globalization, land use, and the invasion of West Nile virus. *Science* 334:323–327.

Kilpatrick, A. M., L. D. Kramer, S. R. Campbell, E. O. Alleyne et al. 2005. West Nile virus risk assessment and the bridge vector paradigm. *Emerging Infectious Diseases* 11:425–429.

Kilpatrick, A. M., M. A. Meola, R. M. Moudy, and L. D. Kramer. 2008. Temperature, viral genetics, and the transmission of West Nile virus by *Celex pipiens mosquitoes*. *PloS Pathogens* 4(6): e1000092.

Komar, N., S. Langevin, S. Hinten, N. Nemeth et al. 2003. Experimental infection of North American birds with the New York 1999 strain of West Nile virus. *Emerging Infectious Diseases* 9:311–322.

LaDeau, S. L., A. M. Kilpatrick, and P. P. Marra. 2007. West Nile virus emergence and large-scale declines of North American bird populations. *Nature* 447:710–713.

Luby, J. P. 1994. St. Louis encephalitis. In: G. W. Beran, editor. *Handbook of Zoonoses*, 2nd edition. CRC Press, Boca Raton, FL, pp. 47–58.

Ludwig, G. V., P. P. Calle, J. A. Mangiafico, B. L. Raphael, et al. 2002. An outbreak of West Nile virus in a New York City captive wildlife population. *American Journal of Tropical Medicine and Hygiene* 67:67–75.

Mahmood, F., R. E. Chiles, Y. Fang, C. M. Barker, and W. Reisen. 2004. Role of nestling mourning doves and house finches as amplifying hosts of St. Louis encephalitis virus. *Journal of Medical Entomology* 41:965–972.

Malkinson, M., C. Banet, Y. Weisman, S. Pokamunski et al. 2002. Introduction of West Nile virus in the Middle East by migrating white storks. *Emerging Infectious Diseases* 8:392–397.

Marra, P. P., S. Griffing, C. Caffrey, A. M. Kilpatrick et al. 2004. West Nile virus and wildlife. *BioScience* 54:393–402.

Mayo Clinic. 2012. West Nile virus. http://www.mayoclinic.org/health/west-nile-virus/DS00438 (accessed May 15, 2013).

McKenna, M. 2012. Dengue, aka "breakbone fever," is back. http://www.slate.com/articles/Health_and_science/pandemics/2012/12/dengue_fever_in_united_states_breakbone_fever_outbreaks_florida_texas_and.html (accessed October 1, 2013).

MedicineNet.com. 2013. http://MedicineNet.com/dengue_fever/article.htm#dengue_fever_facts (accessed May 27, 2014).

Merck. 2010. *Merck Veterinary Manual*, 10th edition. Merck, Whitehouse Station, NJ.

Monath, T. P. 1994. Dengue: The risk to developed and developing countries. *Proceedings of the National Academy of Sciences, USA* 91:2395–2400.

Morales-Betoulle, M. E., N. Komar, N. A. Panella, D. Alvarez et al. 2012. West Nile virus ecology in a tropical ecosystem in Guatemala. *American Journal of Tropical Medicine and Hygiene* 88:116–126.

Muckenfuss, R. S., C. Armstrong, and L. T. Webster. 1934. Etiology of the 1933 epidemic of encephalitis. *Journal of the American Medical Association* 103:731–733.

Murray, K., C. Walker, E. Herrington, J. A. Lewis et al. 2010a. Persistent infection with West Nile virus years after initial infection. *Journal of Infectious Diseases* 201:2–4.

Murray, K. O., E. Mertens, and P. Despres. 2010b. West Nile virus and its emergence in the United States of America. *Veterinary Research* 41:67.

National Wildlife Health Center. 2005. West Nile virus. http://www.usgs.gov/disease_information/west_nile_virus/ (accessed December 23, 2013).

Papa, A., K. Danis, A. Baka, A. Bakas et al. 2010. Outbreak of West Nile virus infections in humans in Greece, July–August 2010. *Euro Surveillance* 15(34):19644.

Raval, M., M. Singhal, D. Guerrero, and A. Alonto. 2012. Powassan virus infection: Case series and literature review from a single institution. *BMC Research Notes* 5(594):1–6.

Reisen, W. K. 2013. Ecology of West Nile virus in North America. *Viruses* 5:2079–2105.

Rossi, S. L., T. M. Ross, and J. D. Evans. 2010. West Nile virus. *Clinical Laboratory Medicine* 30:47–65.

Tavakoli, N. P., H. Wang, M. Dupuis, R. Hull et al. 2009. Fatal case of deer tick virus encephalitis. *New England Journal of Medicine* 360:2099–2107.

Taylor, R. M., T. H. Work, H. S. Hurlbut, and F. Rizk. 1956. A study of the ecology of West Nile virus in Egypt. *American Journal of Tropical Medicine and Hygiene* 5:579–620.

Telford III, S. R., P. M. Armstrong, P. Katavolos, I. Foppa et al. 1997. A new tick-borne encephalitis-like virus infecting New England deer ticks, *Ixodes dammini. Emerging Infectious Diseases* 3:165–170.

Weissenböck, H. 2012. Other flaviviruses. In: D. Gavier-Widén, J. P. Duff, and A. Meredith, editors. *Infectious Diseases of Wild Mammals and Birds in Europe*. Wiley-Blackwell, West Sussex, U.K., pp. 142–145.

Wheeler, S. S., M. P. Vineyard, L. W. Woods, and W. K. Reisen. 2012. Dynamics of West Nile virus persistence in house sparrows (*Passer domesticus*). *PLoS Neglected Tropical Diseases* 6(10):e1860.

WHO. 2007. *Manual for the Laboratory Diagnosis of Japanese Encephalitis Virus Infection*. World Health Organization, Geneva, Switzerland.

WHO. 2008. *The Global Burden of Disease: 2004 Update*. World Health Organization, Geneva, Switzerland.

WHO. 2012a. *Dengue and Severe Dengue*. Fact sheet 117. World Health Organization, Geneva, Switzerland.

WHO. 2012b. *Handbook for Clinical Management of Dengue*. World Health Organization, Geneva, Switzerland.

WHO. 2013. *Yellow Fever. Fact Sheet 100*. World Health Organization, Geneva, Switzerland.

Zeller, H. G. and I. Schuffenecker. 2004. West Nile virus: An overview of its spread in Europe and the Mediterranean Basin in contrast to its spread in the Americas. *European Journal of Clinical Microbiology and Infectious Diseases* 23:147–156.

22 Colorado Tick Fever and Human Diseases Caused by Bunyaviruses

La Crosse Encephalitis, California Encephalitis, Cache Valley Virus, and Jamestown Canyon Virus

Colorado tick fever was initially called mountain fever. The etiology of Colorado tick fever was initially considered to be "rarified mountain air."

Paraphrased from Silver et al. (1961) and Loge (2000)

22.1 INTRODUCTION AND HISTORY

Colorado tick fever (CTF) is a disease caused by a double-stranded RNA virus of the same name, belonging to the genus *Coltivirus*, family Reoviridae (Figure 22.1). La Crosse encephalitis, California encephalitis, Cache Valley virus, and Jamestown Canyon virus are caused by viruses of the family Bunyaviridae. They will also be covered in this chapter.

CTF has afflicted white men since their arrival in the Rocky Mountains and even earlier for Native Americans (Sidebar 22.1). In 1855, a physician named Ewing described a disease with symptoms and a course of development similar to CTF. Initially, CTF was believed to be a mild form of Rocky Mountain spotted fever. Then in the 1930s and 1940s, F. E. Becker reported that the illness was a separate disease and transmitted by the Rocky Mountain wood tick. The CTF virus was first isolated from a human in 1944. Common names for CTF were mountain fever, mountain tick fever, and American tick fever (Silver et al. 1961, Riley and Spruance 2000, Klasco 2002, Attoui et al. 2005). There are between 200 and 400 reported cases of CTF in the United States annually, but the actual incidence of CTF is considerably higher because this disease is easily confused with other viral infections and often is not diagnosed (Eads and Smith 1983, CDC 2013b, Cranshaw and Peairs 2013).

The distribution of CTF is similar to the distribution of its vector: the Rocky Mountain wood tick (Figure 22.4). The CTF virus occurs across the western

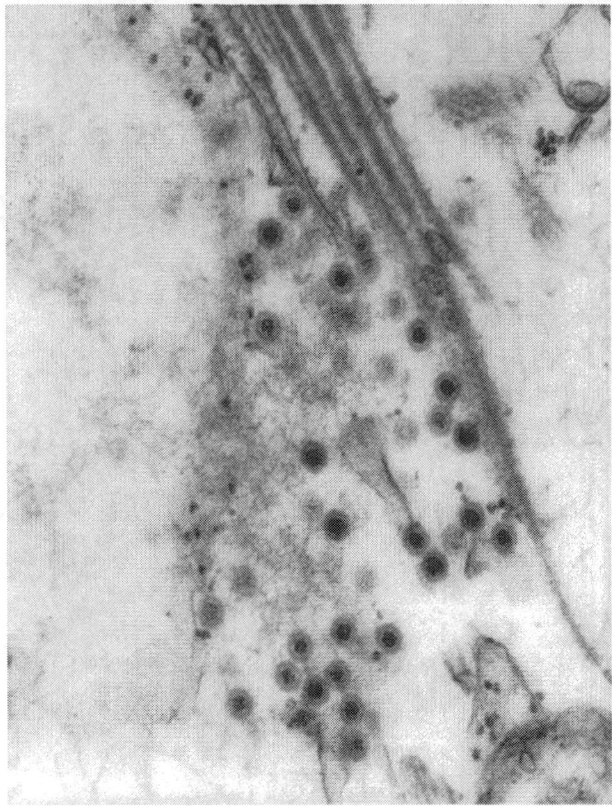

FIGURE 22.1 Transmission electron micrograph of several CTFs. The virus particles are the circles with a dark interior. (Photo courtesy of the CDC.)

United States and Canada (British Columbia and Alberta); yet more than 90% of CTF patients acquired the disease in just three states: Colorado, Montana, and Utah. Typical habitat for the virus and its tick vector occurs at 4,000–10,000 ft (1,219–3,048 m) in elevation and includes sagebrush, juniper, or pine vegetation. CTF is a seasonal disease with 90% of cases occurring between April and July when Rocky Mountain wood ticks are active and seeking hosts (Eklund et al. 1955, Emmons 1988, McLean et al. 1993, Riley and Spruance 2000, Brackney et al. 2010).

The Eyach virus occurs in Europe and is closely related to the CTF virus. The Eyach virus causes fever and encephalitis in humans and is transmitted by the European sheep tick and European rabbit tick. The European rabbit is the main amplifying host for this virus, and it has been isolated from mice, deer, domestic sheep, and domestic goats (Chastel et al. 1984, Attoui et al. 2005).

22.2 SYMPTOMS IN HUMANS

Symptoms of CTF typically begin 4 days after the tick bite, but the incubation period can lengthen to as long as 3 weeks. The initial symptoms include fever,

SIDEBAR 22.1 CTF AND THE LEWIS AND CLARK EXPEDITION (LOGE 2000)

The early American explorers and settlers of the western United States often referred to an illness called mountain fever, now known as CTF. The first recorded case may have been on the Lewis and Clark Expedition (Figures 22.2 and 22.3). On July 25, 1805, the expedition had reached the Three Forks of the Missouri in what is today west central Montana. But before these explorers could continue their journey to the Pacific Ocean, Captain William Clark became so incapacitated from illness that the expedition came to a halt. His symptoms included high fever, chills, fatigue, bone and muscle aches, and a loss of appetite. Following accepted medical practices of their day, Captain Clark was given five laxative or purging pills. The pills worked as planned and kept him awake all night responding to their effects. After 4 days of illness, Captain Clark reported that he felt better but was still weak. Due to a lack of knowledge that bacteria and viruses cause illness, his sickness was attributed to "wading in cold water when overheated." Today, scholars suspect that he was suffering from CTF. They note that Clark's symptoms are similar to those of CTF; he was in an area where the CTF virus is endemic, and Clark became ill during the season when ticks were active.

FIGURE 22.2 Portraits of Meriwether Lewis (left) and William Clark (right) by the artist Charles Willson Peale.

FIGURE 22.3 Lewis and Clark on the Lower Columbia by Charles Marion Russell.

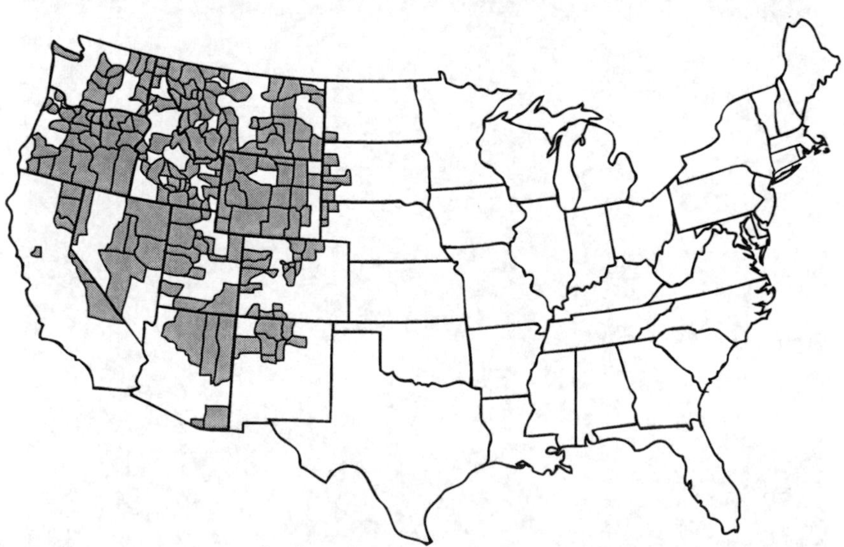

FIGURE 22.4 Map showing the distribution of Rocky Mountain wood tick. (Courtesy of the CDC.)

SIDEBAR 22.2 CTF VEXES MORMON SETTLERS (LOGE 2000)

Mormon pioneers and settlers provided vivid descriptions of a local disease they called mountain fever and known today as CTF (Figure 22.5). The following quote comes from William McBride, during his 1850 trip to Utah.

> At dawn this morning E. W. Summy and Lewis Mitchell applied to me for medications. They were severely and suddenly attacked last night after partaking of a hearty supper. Their disease is what the Mormons have termed mountain fever, being a disease particular to this region. Many patients complain of a most violent pain in the head and eyes with dimness of vision, pain in the back and limbs, great lassitude, alternations of chill and fever, nausea and vomiting, constipation. Such are the most prominent symptoms. The pain from this disease is very great!
>
> I prescribed for them and this evening they are some easier. I saw a company of nine men today which have been compelled to lay at Green River [Wyoming] a week, five of their men having the mountain fever at one time—it is very common at the river.

chills, excessive sweating, muscle aches, joint pains, joint stiffness, and headaches. A spotted rash occurs only in 5%–12% of CTF patients (Sidebar 22.2). Half of CTF patients experience a double-humped (i.e., saddle back) fever that is characterized by an initial fever that lasts 2–3 days, followed by a period of remission of equal length, and then a second period of fever, which lasts 2–3 days and is often worse than the first fever. Clinical signs of CTF involve low white blood cell counts, swollen lymph node, anorexia, chills, nausea, vomiting, and fatigue. Eighteen percent of CTF patients require hospitalization. In 5%–10% of patients, the infection can spread to the central nervous system within a week of the onset of disease; this is more common in children than adults. Severe headache, sensory impairment, neck stiffness, and light sensitivity are indications that the CTF virus has become neuroinvasive. Such infections can develop into meningitis, encephalitis, or meningoencephalitis. In rare cases, patients may go into a coma and die; fortunately, no fatalities from CTF have been reported in the United States during the past few years (Riley and Spruance 2000, Romero and Simonsen 2008). Other rare complications of CTF are pneumonia, myocarditis, and hepatitis. Most patients make a full recovery within 2 weeks, but some patients experience weakness, malaise, and depression that endure for months. A long convalescent period may result from the ability of the CTF virus to persist in human tissue for several months (Silver et al. 1961, Calisher 1994, CDC 2013b).

CTF is difficult to distinguish from other illnesses because its symptoms are nonspecific. The appearance of a saddleback fever and the absence of a hemorrhagic rash common to Rocky Mountain spotted fever may indicate CTF. Confirmation of the presumptive diagnosis is made by isolating the virus from a patient or by detecting antibodies against the virus, but it may take a couple of weeks after the onset of CTF before antibodies can be detected (Calisher et al. 1985, Dana 2009).

FIGURE 22.5 Photo of Mormon settlers traveling across the West around 1879 to reach the Salt Lake Valley. (Photo by C. W. Carter and provided by the U.S. National Archives and Records Administration.)

22.3 CTF VIRUS INFECTIONS IN ANIMALS

Typically, animals are infected with the CTF virus after being bitten by an infected tick. Many small mammal species serve as amplifying hosts, including golden mantled ground squirrels, Columbian ground squirrels, Richardson's ground squirrels, yellow-pine chipmunks, least chipmunks, deer mice, and bushy-tailed woodrats. Further, the virus can infect larger wild mammals, such as porcupines, mountain cottontails, snowshoe hares, black-tailed jackrabbits, yellow-bellied marmots, coyotes, mule deer, and elk. Horses and domestic sheep are subject to infection as well (Spruance and Bailey 1973, Emmons 1988, Cranshaw and Peairs 2013). In Rocky Mountain National Park, McLean et al. (1993) found an average of 26 Rocky Mountain wood ticks on each adult porcupine (Figure 22.6), and 85% of adult porcupines had antibodies against the CTF virus.

The CTF virus has been isolated from at least eight tick species. After a tick feeds upon an infected mammal, the CTF virus replicates in the tick's midgut and is disseminated to other tissues, including the tick's salivary gland where the CTF

FIGURE 22.6 Photo of a North American porcupine. (Courtesy of the U.S. Fish and Wildlife Service.)

virus undergoes further replication. Once the CTF virus becomes abundant in the salivary glands, the tick can transmit the virus to other mammals by feeding upon them. Infected ticks harbor the virus for life and can infect their eggs with the virus (Carey et al. 1980, Emmons 1988, Calisher 1994).

22.4 HOW HUMANS CONTRACT CTF

The Rocky Mountain wood tick (Figure 22.7) transmits the CTF virus to people. This tick generally occurs above 4000 ft (1219 m) elevation. Anyone who lives or travels in areas of the western United States and Canada above this elevation can acquire CTF. Campers, hikers, and people who work outdoors have a higher risk of being infected than the general population in the Rocky Mountain states (Klasco 2002, Günther and Haglund 2005). Most people are infected during late spring and early summer. Among CTF patients, males outnumber females two to one, presumably due to males spending more time outdoors than females. For the same reason, the highest incidence of CTF infections occurs in persons between 20 and 30 years old. While the Rocky Mountain wood tick is the only species known to transmit the virus to humans, some speculation exists that the American dog tick may transmit the virus to humans on occasion. Humans are a dead-end host for the CTF virus, and there is no evidence of natural person-to-person transmission. However, a few people have contracted CTF from a blood transfusion when the blood donor was infected. Also, women who develop CTF while pregnant can transmit the virus to their fetuses, but this is rare (Riley and Spruance 2000, Brackney et al. 2010).

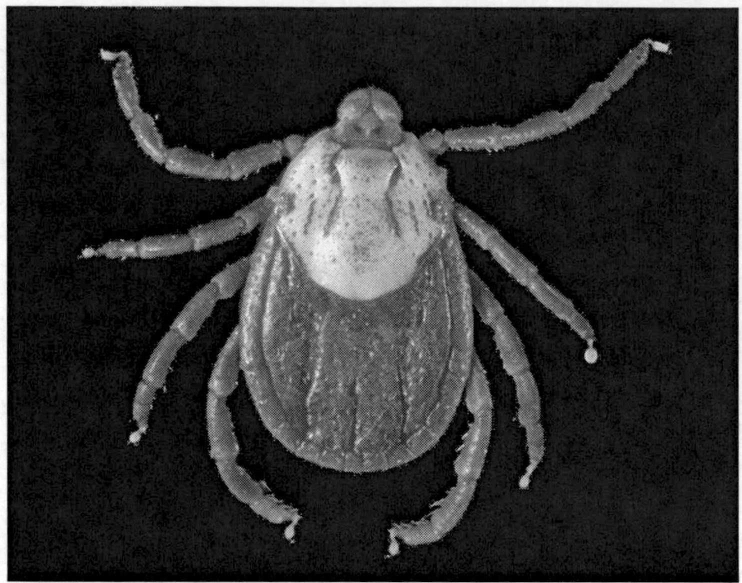

FIGURE 22.7 Photo of a Rocky Mountain wood tick. (Courtesy of the CDC and Christopher Paddock.)

22.5 MEDICAL TREATMENT

No antiviral medication is effective against the CTF virus. Treatment for CTF patients is supportive and includes acetaminophen or nonsteroidal anti-inflammatory medications for pain and fever control; aspirin should be avoided because it might exacerbate any bleeding. Diagnosis of CTF in patients usually occurs when the CTF virus or antibodies against it have been isolated from blood. A spinal tap may be necessary for patients suffering from meningitis or encephalitis (Emmons 1988, Günther and Haglund 2005, Dana 2009).

22.6 WHAT PEOPLE CAN DO TO REDUCE THE RISK OF CONTRACTING CTF

Prevention of CTF requires limiting exposure to ticks. For persons exposed to tick-infested habitats, a careful body inspection and removal of any crawling or attached ticks provide an important method of preventing the disease. Repellents can be used to discourage tick attachment. Permethrin repellents can be sprayed on boots and clothing and will last for several days. Repellents containing DEET can be applied to the skin but will last only a few hours before reapplication is necessary. More information on avoiding ticks and reducing tick populations is available in Chapter 13. There is no vaccine against the CTF virus that can be used either for animals or humans; fortunately, most people who have been infected once by the CTF virus develop lifelong immunity against the virus (Dana 2009, CDC 2013b, Cranshaw and Peairs 2013).

22.7 OTHER BUNYAVIRUSES THAT CAUSE HUMAN ILLNESS IN THE UNITED STATES OR CANADA

There are several viruses in the Bunyaviridae family and *Orthobunyavirus* genus that produce human illness worldwide. In the United States and Canada, these viruses include the LAC virus, California encephalitis virus, Cache Valley virus, and Jamestown Canyon virus (JCV) (Figure 22.8). These viruses are single-stranded RNA and are spherical in shape. Collectively, these viruses are called the California encephalitis serogroup because they have similar antigens. They are also similar in that they all can cause encephalitis, meningitis, or meningoencephalitis in humans. In 2011, 137 people in the United States were diagnosed with an illness caused by a virus in this serogroup, and the virus became neuroinvasive in 120 of the patients (CDC 2013a). Each virus in this serogroup will be described in turn.

22.7.1 La Crosse Encephalitis (LAC)

The LAC virus was first detected in 1963 and named after the site of their discovery: La Crosse, Wisconsin. LAC was initially limited primarily in the upper midwest states of Wisconsin, Minnesota, Iowa, Illinois, Indiana, and Ohio. More recently, the range of the disease has extended into the eastern states of West Virginia, Virginia, and North Carolina and in the southern states of Mississippi

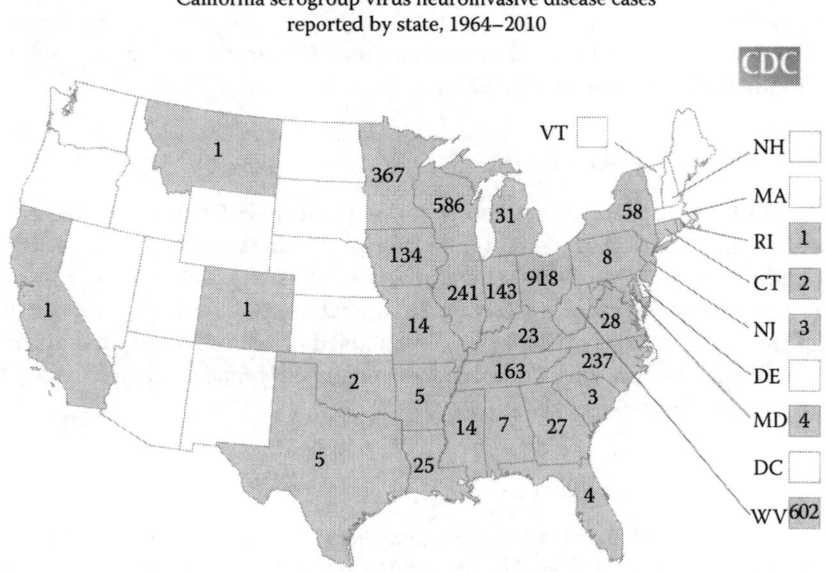

FIGURE 22.8 U.S. distribution of human cases of California serogroup viruses that developed into encephalitis, meningitis, or meningoencephalitis from 1964 through 2010. Most cases are caused by the La Crosse virus, but some cases are caused by the California encephalitis virus, Cache Valley virus, and Jamestown Canyon virus. (From CDC, La Crosse encephalitis epidemiology and distribution. http://www.cdc.gov/lac/tech/epi.html, 2011.)

and Alabama. Most LAC patients are less than 18 years old. LAC patients become infected after being bitten by an infected mosquito, and most cases occur during the summer when mosquitoes are active. About 75 cases of LAC are reported annually to the CDC, but many physicians do not order the tests to confirm an infection of the LAC virus, owing to a lack of a specific treatment for LAC infections. Hence, the cases are often diagnosed as viral encephalitis caused by an unidentified virus rather than LAC encephalitis. In 2011, 130 LAC cases were reported to the CDC: 15% developed meningitis and 72% encephalitis. In 2012, 78 cases were reported: 18% developed meningitis and 72% encephalitis. More than 90% of LAC patients required hospitalization. One LAC patient died in 2011 and another in 2012 (CDC 2005, 2013c).

Most people infected with the LAC virus are asymptomatic or develop influenza-like symptoms of malaise, fever, headache, nausea, and vomiting. In most people, the disease is self-limiting and lasts for only a few days. Sometimes, the virus infects the nervous system and develops into encephalitis. Most cases of LAC encephalitis occur in children less than 16 years old; clinical signs include seizures, coma, and paralysis. Case fatality rate is less than 1% (CDC 2005, 2013c).

The LAC virus maintains itself by infecting the mosquito *Aedes triseriatus*. This mosquito is active during the day and specializes in biting small mammals living in tree holes. Chipmunks and tree squirrels are amplifying hosts. When suckling mice were experimentally infected, the LAC virus first replicated in muscle before invading the central nervous system and killing the mice (Janssen et al. 1984). *A. triseriatus* mosquitoes are able to infect their eggs, and the virus can overwinter inside the eggs. The following year after the eggs hatch, the adult mosquito that develops from infected eggs can spread the virus to chipmunks, squirrels, or humans although humans are dead-end hosts (CDC 2005).

22.7.2 California Encephalitis

The California encephalitis virus was first isolated in 1943 from mosquitoes caught in Kern County, California. A short time later, three human cases of encephalitis in the Central Valley of California were attributed to this virus (Hammon and Reeves 1952). For over two decades, there were no known cases in humans; then in 1966, a man from Marin County, California, complained of blurred vision and dizziness and was diagnosed as having California encephalitis. He made a complete recovery within a month (Eldridge et al. 2001).

22.7.3 Cache Valley Virus

This virus was named after the site of its discovery: Cache Valley, Utah. Its distribution, however, is much broader. The virus has been isolated from mosquitoes across North America: 4 Canadian provinces, 22 U.S. states, Mexico, and Jamaica. The virus has been isolated from many different mosquitoes, including fourteen species of *Aedes*, seven *Anopheles*, four *Psorophora*, three *Culex*, two *Culiseta*, and one *Coquillettidia*. Despite its wide distribution and the numerous mosquito species that can transmit it, human cases of Cache Valley virus are rare. Only three cases are known: all three developed meningitis or encephalitis, and one patient succumbed (Sexton et al. 1997, Campbell et al. 2006, Nguyen et al. 2013).

Horses, cattle, sheep, and dogs have become infected with the Cache Valley virus. Antibodies against the virus were found in a high proportion of these animals in coastal Virginia and Maryland. In Texas, 19% of domestic sheep had antibodies against the virus. If domestic sheep become infected during their first trimester of pregnancy, the virus may cross the placenta resulting in fetal death, stillbirths, or birth defects. The amplifying hosts for this virus are wild ungulates, especially white-tailed deer; among white-tailed deer in Minnesota, North Dakota, Wisconsin, and Texas, 88%–100% had antibodies against the virus. A high proportion of elk, caribou, and raccoons from endemic areas also had antibodies against the virus (Edwards 1994, de la Concha-Bermejillo 2003).

22.7.4 Jamestown Canyon Virus (JCV)

This virus occurs throughout the United States. A diverse group of mosquitoes serve as vectors for this pathogen. White-tailed deer are amplifying hosts. The distribution of people in Michigan with antibodies against the JCV closely matches the distribution of white-tailed deer. In Indiana, 2%–15% of humans have antibodies against JCV, and their locations correlate with the prevalence of JCV antibodies in white-tailed deer (Grimstad et al. 1986, Boromisa and Grimstad 1987, Zamparo et al. 1997, Andreadis et al. 2008). Such is not the case in Connecticut where deer are abundant throughout the state and 21% of the deer are seropositive. White-tailed deer fawns born to females infected with JCV exhibit paralysis and respiratory distress (Grimstad 1994).

Studies in New England have found that up to 12% of the human population has antibodies against JCV. Most people infected with the virus are asymptomatic or have a mild illness; nonetheless, meningoencephalitis is a possibility. Fifteen human cases of JCV have been reported to the CDC between 2004 and 2011, and 11 of the patients had a moderate or severe case of meningoencephalitis. Two JCV cases were reported in 2012. Adults are more susceptible to JCV than children. JCV patients often contract the infection during late spring or early summer when the adult mosquitoes that transmit JCV hatch are active (Grimstad et al. 1982, Lowell et al. 2011, CDC 2013c).

22.8 ERADICATING CTF AND OTHER DISEASES CAUSED BY BUNYAVIRUSES FROM A COUNTRY

Wildlife species serve as reservoir hosts for CTF virus and other bunyaviruses. Hence, they will be difficult to eradicate from large areas.

LITERATURE CITED

Andreadis, T. G., J. F. Anderson, P. M. Armstrong, and A. J. Main. 2008. Isolation of Jamestown Canyon virus (Bunyaviridae: *Orthobunyavirus*) from field-collected mosquitoes (Diptera: Culicidae) in Connecticut, USA: A ten-year analysis, 1997–2006. *Vector-Borne and Zoonotic Diseases* 8:175–188.

Attoui, H., F. M. Jaafar, P. de Micco, and X. de Lamballerie. 2005. Coltiviruses and seadornaviruses in North America, Europe, and Asia. *Emerging Infectious Disease* 11:1673–1679.

Boromisa, R. D. and P. R. Grimstad. 1987. Seroconversion rates to Jamestown Canyon virus among six populations of white-tailed deer (*Odocoileus virginianus*) in Indiana. *Journal of Wildlife Diseases* 23:23–33.

Brackney, M. M., A. A. Marfin, J. E. Staples, L. Stallones et al. 2010. Epidemiology of Colorado tick fever in Montana, Utah, and Wyoming, 1995–2003. *Vector-Borne and Zoonotic Disease* 10:381–385.

Calisher, C. H. 1994. Medically important arboviruses of the United States and Canada. *Clinical Microbiology Reviews* 7:89–116.

Calisher, C. H., J. D. Poland, S. B. Calisher, and L. A. Warmoth. 1985. Diagnosis of Colorado tick fever virus infection by enzyme immunoassays for immunoglobulin M and G antibodies. *Journal of Clinical Microbiology* 22:84–88.

Campbell, G. L., J. D. Mataczynski, E. S. Reisdorf, J. W. Powell et al. 2006. Second human case of Cache Valley virus disease. *Emerging Infectious Diseases* 12:854–856.

Carey, A. B., R. G. McLean, and G. O. Maupin. 1980. The structure of a Colorado tick fever ecosystem. *Ecological Monographs* 50:131–151.

CDC. 2005. Information on arboviral encephalitides. http://www.cdc.gov/ncidod/dvbid/arbor/arbdet.htm (accessed June 2, 2013).

CDC. 2011. La Crosse Encephalitis epidemiology and distribution. http://www.cdc.gov/lac/tech/epi.html (accessed June 30, 2013).

CDC. 2013a. Summary of notifiable diseases—United States, 2011. *Morbidity and Mortality Weekly Report* 60(53):1–117.

CDC. 2013b. *Tickborne Diseases of the United States: A Reference Manual for Health Care Providers*. Centers for Disease Control and Prevention, Atlanta, GA.

CDC. 2013c. West Nile virus and other arboviral diseases—United States, 2012. *Morbidity and Mortality Weekly Report* 62:513–517.

Chastel, C., A. J. Main, A. Couatarmanac'h, G. LeLay et al. 1984. Isolation of *Eyach* virus (Reoviridae, Colorado tick fever group) from *Ixodes ricinus* and I. *ventalloi* ticks in France. *Archives of Virology* 82:161–171.

Cranshaw, W. S. and F. B. Peairs. 2013. *Colorado Ticks and Tick-Borne Diseases*. Colorado State University Extension, Division of the Office of Engagement, Fort Collins, CO.

Dana, A. N. 2009. Diagnosis and treatment of tick infestation and tick-borne diseases with cutaneous manifestation. *Dermatologic Therapy* 22:293–326.

de la Concha-Bermejillo, A. 2003. Cache Valley virus is a cause of fetal malformation and pregnancy loss in sheep. *Small Ruminant Research* 49:1–9.

Eads, R. B. and G. C. Smith. 1983. Seasonal activity and Colorado tick fever virus infection rates in Rocky Mountain wood ticks, *Dermacentor andersoni* (Acari: Ixodidae), in north-central Colorado, USA. *Journal of Medical Entomology* 20:49–55.

Edwards, J. F. 1994. Cache Valley virus. *Veterinary Clinics of North America: Food Animal Practice* 10:515–524.

Eklund, C. M., G. M. Kohls, and J. M. Brennan. 1955. Distribution of Colorado tick fever and virus-carrying ticks. *Journal of American Medical Association* 157:335–337.

Eldridge, B. R., C. Glaser, R. E. Pedrin, and R. E. Chiles. 2001. The first reported case of California encephalitis in more than 50 years. *Emerging Infectious Diseases* 7:451–452.

Emmons, R. W. 1988. Ecology of Colorado tick fever. *Annual Reviews of Microbiology* 42:49–64.

Grimstad, P. R. 1994. California group viral infections. In: G. W. Beran, editor. *Handbook of Zoonoses, Section B: Viral*. CRC Press, Boca Raton, FL, pp. 71–79.

Grimstad, P. R., C. H. Calisher, R. N. Harroff, and B. B. Wentworth. 1986. Jamestown Canyon virus (California serogroup) is the etiologic agent of widespread infection in Michigan humans. *American Journal of Tropical Medicine and Hygiene* 35:376–386.

Grimstad, P. R., C. L. Shabino, C. H. Calisher, and R. J. Waldman. 1982. A case of encephalitis in a human associated with a serologic rise to Jamestown Canyon virus. *American Journal of Tropical Medicine and Hygiene* 31:1238–1244.

Günther, G. and M. Haglund. 2005. Tick-borne encephalopathies: Epidemiology, diagnosis, treatment and prevention. *CNS Drugs* 19:1009–1032.

Hammon, W. McD. and W. C. Reeves. 1952. California encephalitis virus, a newly described agent. *California Medicine* 77:303–309.

Janssen, R., F. Gonzalez-Scarano, and N. Nathanson. 1984. Mechanism of bunyavirus virulence. Comparative pathogenesis of a virulent strain of La Crosse and avirulent strains of Tahyna virus. *Laboratory Investigations: A Journal of Technical Methods and Pathology* 50:447–455.

Klasco, R. 2002. Colorado tick fever. *Medical Clinics of North America* 86:435–440.

Loge, R. V. 2000. Illness at Three Forks: Captain William Clark and the first recorded case of Colorado tick fever. *Montana* 50(2):2–15.

Lowell, J., D. P. Higgins, M. Drebot, K. Makowski, and J. E. Staples. 2011. Human Jamestown Canyon virus infection—Montana, 2009. *Morbidity and Mortality Weekly Report* 60:652–655.

McLean, R. G., A. B. Carey, L. J. Kirk, and D. B. Francy. 1993. Ecology of porcupines (*Erethizon dorsatum*) and Colorado tick fever virus in the Rocky Mountain National Park 1957–1977. *Journal of Medical Entomology* 30:236–238.

Nguyen, N. L., G. Zhao, R. Hull, M. A. Shelly et al. 2013. Cache Valley virus in a patient diagnosed with aseptic meningitis. *Journal of Clinical Microbiology* 51:1966–1969.

Riley, D. K. and S. L. Spruance. 2000. Colorado tick fever. In: B. A. Cunha, editor. *Tickborne Infectious Diseases: Diagnosis and Management*. CRC, Boca Raton, FL.

Romero, J. R. and K. A. Simonsen. 2008. Powassan encephalitis and Colorado tick fever. *Infectious Disease Clinics of North America* 22:545–550.

Sexton, D. J., P. E. Rollin, E. B. Breitschwerdt, G. R. Corey et al. 1997. Life-threatening Cache Valley virus infection. *New England Journal of Medicine* 336:547–549.

Silver, H. K., G. Meiklejohn, and C. H. Kempe. 1961. Colorado tick fever. *American Journal of Diseases of Children* 101:30–36.

Spruance, S. L. and A. Bailey. 1973. Colorado tick fever: A review of 115 laboratory confirmed cases. *Archives of Internal Medicine* 131:288–293.

Zamparo, J. M., T. G. Andreadis, R. E. Shope, and S. J. Tirrell. 1997. Serologic evidence of Jamestown Canyon virus infection in white-tailed deer populations from Connecticut. *Journal of Wildlife Diseases* 33:623–627.

23 Hantaviruses

Hantavirus Pulmonary Syndrome (HPS) has turned out to be a newly identified, but not a 'new' disease.

CDC (2012a)

23.1 INTRODUCTION AND HISTORY

Hantaviruses are a group of viruses in the family Bunyaviridae, which are single-stranded RNA viruses and spherical in shape (Figure 23.1). All hantaviruses belong to the genus *Hantavirus* (one of five genera in the Bunyaviridae family). The *Hantavirus* genus includes more than 40 known species and probably many more that are not yet described. These viruses are grouped into four serotypes. Hantaviruses are the causative agents for two diseases in humans: hantavirus pulmonary syndrome (HPS), which is found in the Americas, and hemorrhagic fever with renal syndrome (HFRS), which occurs primarily in Eurasia (Heyman et al. 2012).

Diseases with symptoms similar to HFRS were described in Chinese texts over a thousand years ago. Based on similarities in symptoms, HFRS has been described in different parts of the world and given various names in the past. During World War I (1914–1918), troops fighting in Europe called it trench nephritis. In the 1930s, it was called Puumala fever in Finland, and Japanese troops stationed in Manchuria referred to it as Songo fever. In the 1950s, more than 3000 United Nations troops fighting in the Korean War became ill with HFRS, known to them as Korean hemorrhagic fever (Heyman et al. 2012).

HFRS occurs mainly in Europe and Asia and is caused by at least five viral species *H. hantaan*, *H. seoul*, *H. dobrava-belgrade*, *H. saaremaa*, and *H. puumala* (hantavirus species are usually named after the place where they are first detected). The Hantaan virus is widely distributed in eastern Asia, particularly in China, Russia, and Korea. The Seoul virus is found worldwide. Dobrava-Belgrade virus is found in the Balkans; Saaremaa is found in central Europe and Scandinavia; and Puumala virus is found in Scandinavia, western Europe, and western Russia (Shakespeare 2009, Zhang et al. 2010, Vaheri et al. 2011, 2013, Heyman et al. 2012).

Annually, 150,000–200,000 people develop HFRS worldwide. Since 1950, there have been over 1.5 million human cases with over 46,000 deaths in China alone. In Europe, over 10,000 cases are diagnosed annually with the highest number of patients occurring in Scandinavia, Belgium, and France (Figure 23.2). However, the actual number of Europeans infected with a hantavirus is five to ten times higher than the number of reported cases. In Europe, 1%–5% of the general population is seropositive for hantavirus; this rate climbs above 20% in Finland, Bosnia, and Herzegovina. In recent years, the annual incidence rate for HFRS has been declining

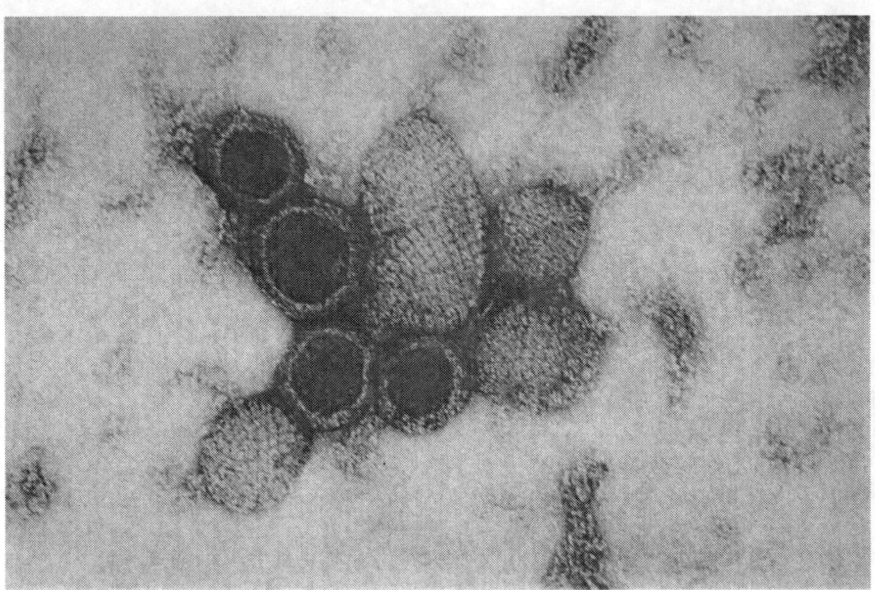

FIGURE 23.1 A photo taken with a transmission electron microscope of the Sin Nombre hantavirus. (Courtesy of M. L. Marintin, E. L. Palmer, and the CDC.)

in China but increasing in some European countries (Shakespeare 2009, Zhang et al. 2010, Vaheri et al. 2011, 2013, Heyman et al. 2012).

Hantaviral disease was initially believed to be limited to the Eastern Hemisphere until an outbreak of a mysterious pulmonary illness occurred in 1993 in the southwestern United States in the Four Corners region where the states of Arizona, Colorado, New Mexico, and Utah all meet. First, a 21-year-old, healthy Navajo man, who was experiencing shortness of breath, was transported to a New Mexican hospital where he suddenly died. Local doctors soon learned that the victim's fiancée had died a few days earlier with similar symptoms. This realization sparked an intensive effort to determine the cause of the deaths. An investigation in the Four Corners region identified five more healthy adults who had all died earlier with similar symptoms. Their tissue samples were sent to the CDC where virologists determined the causative agent to be an unknown species of hantavirus. Because all other hantaviruses were associated with rodents, efforts were made to trap rodents in and around buildings where victims had lived or worked, as well as in the general area. Tissue samples from 1700 rodents were analyzed; 30% of all deer mice were seropositive against hantavirus. Other rodent species were also infected, although less frequently than deer mice. By November 1993, the specific hantavirus that caused the Four Corners outbreak was isolated and named the Muerto Canyon virus. Later, the name was changed to Sin Nombre virus. The new disease was called HPS (CDC 1993, 2012a, Duchin et al. 1994).

While most HPS disease in the United States is caused by Sin Nombre virus, other hantavirus species cause the same disease elsewhere in the United States.

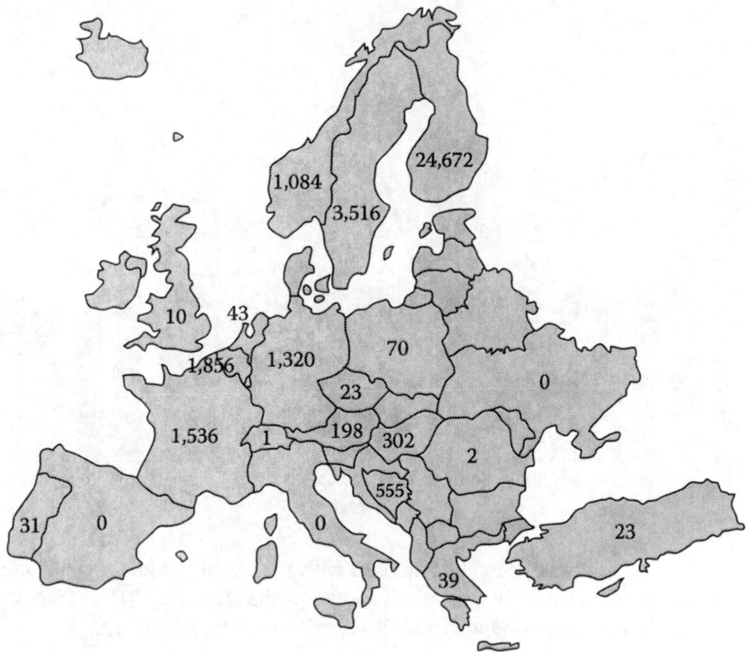

FIGURE 23.2 Number of hantavirus cases in Europe from 1990 to 2006. (Data from Heyman, P. et al., Bunyavirus infections, in: D. Gavier-Widén, J. P. Duff, and A. Meredith, editors, *Infectious Diseases of Wild Mammals and Birds in Europe*, Wiley-Blackwell, West Sussex, U.K., pp. 241–248, 2012.)

In June 1993, a Louisiana bridge inspector developed HPS. The patient's tissues were tested for the presence of antibodies against hantavirus, leading to the discovery of a new hantavirus, named Bayou virus. In late 1993, a 33-year-old Florida man was diagnosed with a nonfatal case of HPS caused by another hantavirus, named the Black Creek Canal virus. Another person in New York developed HPS; this time, the hantavirus was named New York-1 (CDC 2012a, 2013a).

From 1993 to 2011, there were 587 human cases of HPS in the United States (Figure 23.3). Cases were reported in 34 states with 95% of cases occurring west of the Mississippi River. All HPS patients in Canada and most in the United States were infected with the Sin Nombre virus; most patients (70%) lived in rural areas. The mean age of HPS patients in the United States was 37 years (range from 6 to 83 years). Few children developed HPS, for reasons unknown. Most HPS were male (61%), probably because more men were more employed in occupations that exposed them to rodents. Native Americans accounted for about 18% of cases—a rate higher than other ethnic group, probably because many Native Americans lived in the Four Corners region (CDC 2012a, 2013b, Knust et al. 2012).

HPS is more prevalent in South America than in North America with cases occurring in Argentina, Chile, Uruguay, Paraguay, Brazil, and Bolivia. Andes virus causes HPS in Argentina and Chile. Other hantavirus species that cause

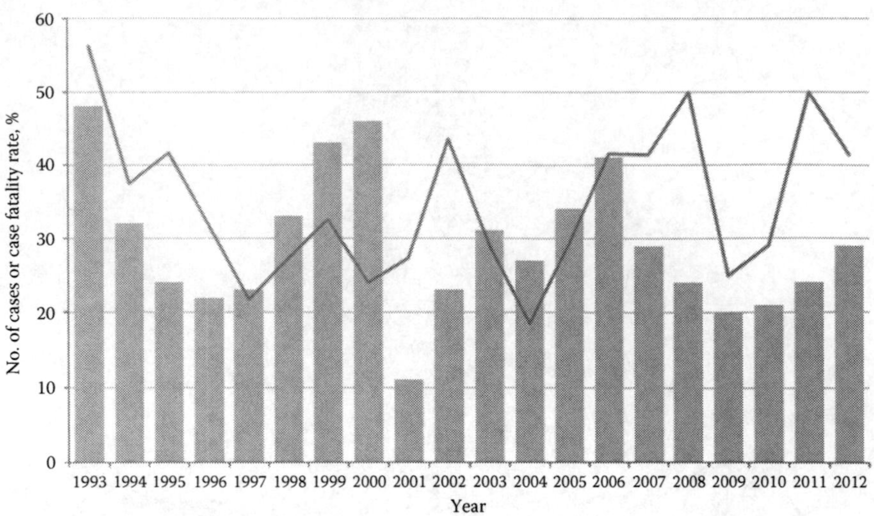

FIGURE 23.3 The number of people diagnosed with HPS in the United States annually are shown by the bars, and the case fatality rate is shown by the black line. (Data from the CDC, Hantavirus, http://www.cdc.gov/hantavirus/, 2012a, accessed July 26, 2012.)

HPS in Argentina include the Bermejo, Hu39694, Lechiguanas, Maciel, Oran, and Pergamino. In Bolivia, HPS is a product of the Bermejo and Laguna Negra hantaviruses. In Paraguay, HPS is caused by Laguna Negra hantavirus. In Brazil, hantaviruses responsible for HPS include Araraquara, Castelo dos Sonhos, and Juquitiba. In 1999, an outbreak in Panama marked the first cases of HPS identified in Central America, and a new hantavirus, Choclo, was identified. It is likely that hantaviruses occur in all countries of the Americas, owing to the broad geographic distribution of sigmodontine rodents (CDC 2012a, 2013a).

23.2 SYMPTOMS IN HUMANS

As the disease's name suggests, the main clinical signs of HFRS involve internal bleeding, high fever, and infection of the renal system. In the most serious cases, the infection of the renal system results in kidney destruction and death. Fatality rates among HFRS patients hover around about 1% (Zhang et al. 2010).

Symptoms of HFRS usually develop between 1 and 3 weeks after infection, but they can be delayed up to 8 weeks. Sudden, intense headaches, back and abdominal pain, fever, chills, nausea, and blurred vision are some of the early symptoms. Individuals may exhibit flushing of the face, inflammation or redness of the eyes, or a rash. Later clinical signs may include low blood pressure, vascular leakage, shock, and kidney failure. Disease severity varies depending upon the virus causing the infection; Hantaan and Dobrava hantaviruses produce the most severe cases of HFRS. In Europe, 5% of hospitalized HFRS patients infected with the Puumala hantavirus required dialysis or prolonged care in the intensive care unit, as did 16%–48% of hospitalized patients infected with the Dobrava hantavirus. Long-term consequences include hormonal, renal, and cardiovascular problems; complete recovery can require months. Case fatality rates range from 5% to 15% for HFRS produced by Hantaan and Dobrava hantaviruses to less than 1% for infections involving the Saaremaa and Puumala hantaviruses (CDC 2011, Vaheri et al. 2011, Heyman et al. 2012).

HPS differs from HFRS because the infection occurs mainly in the pulmonary system, not the renal system. HPS has two different stages, which begin after a 2- or 3-week incubation period. During the first stage, symptoms are similar to influenza and include fatigue, fever, headaches, dizziness, and chills. Muscle aches occur especially in the thighs, hips, and back. About half of all HPS patients experience nausea, vomiting, diarrhea, and abdominal pain. The second stage occurs 4–10 days after the first signs of illness and can begin abruptly. Clinical signs during the second stage include coughing, shortness of breath, and difficulty breathing, owing to the lungs filling with fluid. These clinical signs result because the hantavirus attacks the capillaries in the lung, causing them to leak and fill the lung with fluid. A drop in blood pressure follows, as the heart becomes less efficient. These lung and heart problems can quickly become life threatening. The fatality rate for HPS patients in the United States and Canada is over 30%. Hence, the Mayo Clinic recommends anyone experiencing flu-like symptoms and who has recently been exposed to rodents should immediately see a physician. They should also tell their physician about any potential exposure to rodents or to areas where there were rodent feces or urine (Mertz et al. 2006, Mayo Clinic 2011, CDC 2013a).

23.3 HANTAVIRUS INFECTIONS IN ANIMALS

In Europe, HFRS in humans is caused by five different hantaviruses: Puumala, Dobrava-Belgrade, Saaremaa, Tula, and Seoul hantaviruses. European reservoir hosts for these hantaviruses are bank vole for Puumala; yellow-necked field mouse, striped field mouse, and Black Sea field mouse for Dobrava-Belgrade; striped field mouse for Saaremaa; and common vole for Tula. The Seoul hantavirus was inadvertently introduced into Europe from Asia and now occurs in cities where there are large populations of black rats and brown rats. Thus far, Saaremaa, Tula, and Seoul hantaviruses rarely cause human illness in Europe. Instead, the vast majority of HFRS patients in Europe are infected with the Puumala and Dobrava-Belgrade hantaviruses.

The distribution of these hantaviruses in Europe is similar to the distribution of their reservoir hosts, and incidence of HFRS in humans is highest in those areas where the reservoir hosts are common (Figure 23.4). For example, the bank vole is the reservoir host of the Puumala hantavirus and is a forest-dwelling vole. Thus, the risk of someone being infected with a Puumala hantavirus increases with the proportion of the land covered with forests, especially the old-growth moist forests that bank voles prefer. Furthermore, annual changes in the HFRS incidence in humans are related to annual population cycles of those rodents that serve as reservoir hosts for hantaviruses. In northern parts of Europe, both bank vole populations and the incidence of HFRS are cyclic, peaking every 3–4 years. In warmer parts of Europe, annual incidence of HFRS varies with the yearly variation in the

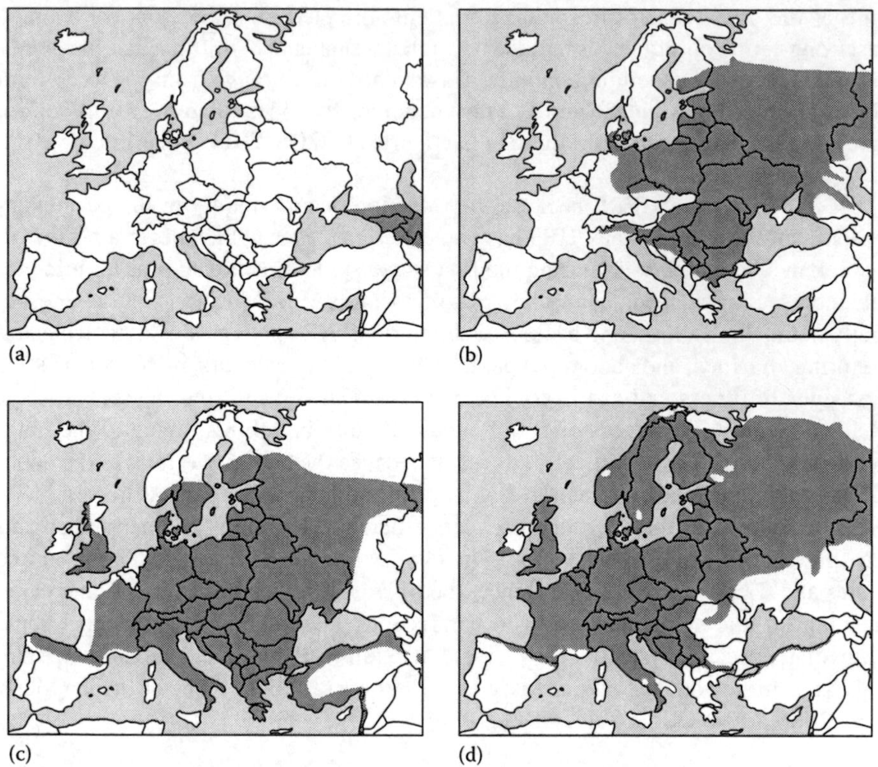

FIGURE 23.4 European distribution of rodents that serve as reservoir hosts for hantaviruses. (a) Black Sea field mouse, (b) striped field mouse, (c) yellow-necked mouse, and (d) bank vole. (Data from Vaheri, A. et al., *Rev. Med. Virol.*, 23, 35, 2013.)

nut production by beech and oak trees; a heavy nut crop increases rodent populations (CDC 2007, 2011, Vaheri et al. 2013).

In Asia, HFRS in humans is caused by Hantaan, Amur, and Seoul hantaviruses. The striped field mouse is the reservoir host for the Hantaan hantavirus, the Korean field mouse for Amur hantavirus, and black rat and brown rat for Seoul hantavirus. Several other hantavirus species have been isolated from other species of voles, mice, and shrews, but these hantaviruses are not known to cause human illness (Zhang et al. 2010, Vaheri et al. 2013).

Rats and mice of the family Muridae, subfamily Sigmodontinae, are the reservoir hosts for all hantaviruses known to cause HPS. There are at least 430 species of them, and they are widespread across North and South America. Generally, a single species serves as the reservoir host for each hantavirus. The North American deer mouse is the reservoir host for Sin Nombre virus, which causes most HPS cases in the United States, and annual changes in incidence of HPS among humans are correlated with deer mouse populations (Sidebar 23.1). The deer mouse is common throughout North America and lives in habitat types ranging from desert to alpine tundra. As a result, the Sin Nombre virus occurs in all of these areas. About 10% of deer mice tested

SIDEBAR 23.1 WHAT HAPPENED IN 1993 THAT CAUSED THE OUTBREAK OF HPS IN THE FOUR CORNERS OF THE UNITED STATES (HJELLE AND GLASS 2000, YATES ET AL. 2002)

Rodents in the Four Corners region of the United States probably were infected with hantavirus for a long time. Yet, HPS was not known to occur in the region until 1993 when suddenly an HPS outbreak struck 54 people living in the area. In the following years, less than five people were infected annually with HPS. This raised the question of why the outbreak occurred in 1993. One hypothesis was that deer mouse densities may have increased that year; in Europe, the incidence of HFRS in Europe was related to rodent numbers. Perhaps the same was true for HPS. Field research in the Four Corners region supported the hypothesis because the region had been in a multiyear drought. Then, in early 1993, heavy snows and rainfall helped drought-stricken plants and animals to revive and proliferate. The area's deer mice, with ample food, reproduced so rapidly that there were 10 times more mice in May 1993 than in May of 1992. With so many mice, it was more likely that mice and humans would come into contact and hantavirus would be transmitted from mice to humans.

In most years, precipitation in the Four Corners region is sparse, but rainfall and snow increase when there is an El Niño southern oscillation, which occurred in 1991–1992. This led to the prediction that outbreaks of HPS should occur shortly after an El Niño. To test this hypothesis, researchers examined what happened during the next El Niño southern oscillation, which occurred in 1997–1998. As predicted, there was an outbreak of HPS in the region in 1998–1999 with the number of HPS patients increasing fivefold. Furthermore, HPS patients were clustered in those areas experiencing the greatest increase in precipitation.

throughout North America show evidence of infection with Sin Nombre virus. The white-footed mouse is the reservoir host for the New York hantavirus; the cotton rat is the reservoir host for the Black Creek Canal hantavirus, and the rice rat is the reservoir host for the Bayou hantavirus (Figure 23.5). Nearly the entire continental United States falls within the range of one or more of these host species (Figure 23.6). Other hantaviruses, isolated from several other sigmodontine rodent species in North America, are not known to cause human disease (CDC 2007, 2012a).

Other species of small mammals can be infected with a hantavirus but are much less likely to transmit the virus to other animals or humans. Domestic cats, domestic dogs, and red foxes have tested positive for hantavirus antibodies, but these are all dead-end hosts. There are no known cases where one of these animals transmitted a hantavirus to a person (CDC 2007, 2012a, Vaheri et al. 2011, Heyman et al. 2012).

Most rodents become infected with hantavirus after weaning, and they shed hantavirus in their saliva, urine, and feces for weeks, months, or life. Transmission from one rodent to another may happen by physical contact; male rodents are more likely to be infected than females, so perhaps transmission results from fights or other

FIGURE 23.5 Photo of a deer mouse (a), white-footed mouse (b), cotton rat (c), and rice rat (d). These species are the reservoir hosts for hantaviruses in North America. (All Courtesy of the CDC; cotton rat taken by James Gathany.)

aggressive behavior. Hantaviruses are stable outside of their hosts and remain infectious for days inside rodent dens and burrows. Hence, direct transmission from one rodent to another is not required; an infected rodent can contaminate a den, burrow, or feeding site and transmit the hantavirus to another rodent that enters a contaminated area at a later date. Hantaviruses do not cause overt illness in their reservoir hosts, but deer mice infected with the Sin Nombre hantavirus have a higher mortality rate than uninfected mice (Kallio et al. 2006, Luis et al. 2012, Vaheri et al. 2013).

23.4 HOW HUMANS CONTRACT HFRS OR HPS

Hantaviruses shed in rodent urine and droppings are stable at room temperature for up to 10 days. Sweeping or a wind gust can cause contaminated dust particles to become airborne where people can inhale them and become infected. People can also become infected if bitten by an infected rodent or when rodent urine or fecal material gains entry into broken skin or the mucous membranes of the mouth, nose, or eyes.

The hantavirus species that cause HPS in the United States have not been transmitted from one person to another. However, the Andes hantavirus that exists in South America can be spread between people. Persons who camp or live in buildings infested with rodents carrying hantavirus are at risk of HPS (Sidebar 23.2). Any activity that brings a person into contact with rodent droppings, urine, saliva, or nesting materials places them at risk. Most HPS patients (80%) in the Four Corners

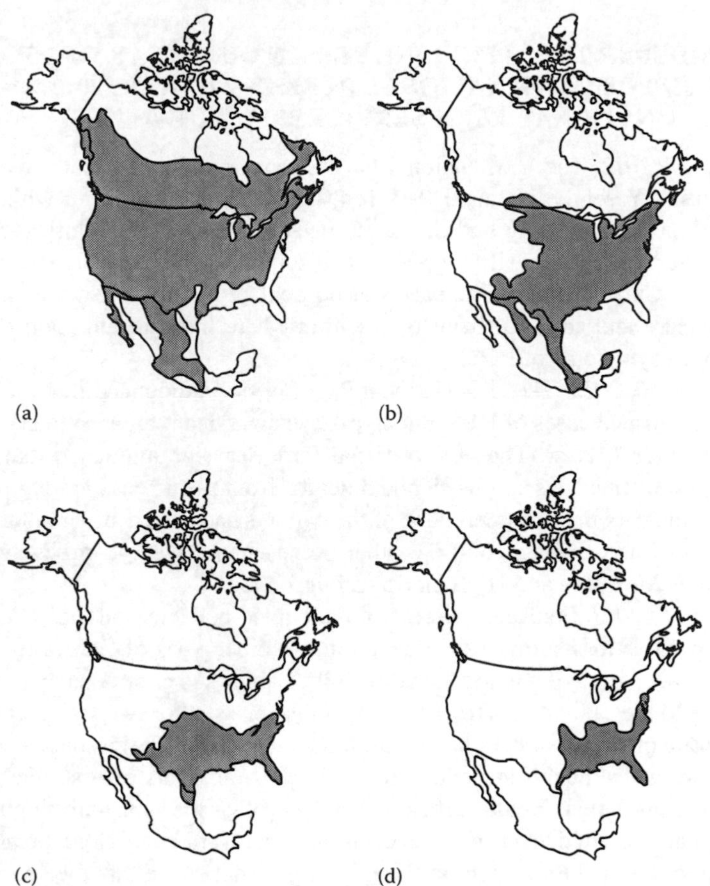

FIGURE 23.6 Distribution of (a) deer mouse, (b) white-footed mouse, (c) cotton rat, and (d) rice rat in North America. (From CDC, Rodents in the United States that carry hantavirus, http://www.cdc.gov/hantavirus/rodents/, 2012c.)

region in 1997–1998 were likely infected while they were inside a building contaminated with mouse feces and urine. This suggests that outdoor activities, such as hiking or gardening, do not place people at risk as much as living in a building infested with deer mice (Hjelle and Glass 2000, Martinez et al. 2005, Mayo Clinic 2011, CDC 2012a).

23.5 MEDICAL TREATMENT

HPS patients should be immediately transferred to an emergency department or intensive care unit for close monitoring because they may experience respiratory distress and need help breathing. Supportive care requires careful management of the patient's fluid and electrolyte levels. Dialysis may be needed to correct severe fluid overload (Mayo Clinic 2011, CDC 2013a).

SIDEBAR 23.2 TIMELINE FOR AN OUTBREAK OF HPS IN YOSEMITE NATIONAL PARK IN 2012 (CDC 2012B, NATIONAL PARK SERVICE 2012, BROWN 2013)

August 27, 2012. The U.S. National Park Service announced that three people visiting Yosemite National Park in 2012 had been diagnosed with HPS; a fourth person probably had the same disease (Figure 23.7). Unfortunately, two of the people died. All four people had stayed at the park's Signature Tent Cabins in Curry Village. The park started contacting other visitors who had stayed there and advised them to seek immediate medical attention if they developed symptoms of HPS.

August 30, 2012. The U.S. National Park Service announced that the number of confirmed cases of HPS among park visitors had increased to six.

November 1, 2012. The U.S. National Park Service announced that there were 10 confirmed cases of HPS and 3 deaths from the disease among people visiting the park that summer. Nine of the patients had stayed at Signature Tent Cabins in Curry Village, and the other person had stayed 15 miles away at Tuolumne Meadows and High Sierra campgrounds.

May 25, 2013. Fourteen percent of deer mice collected at Curry Village tested positive for hantavirus, a rate similar to other parts of California. Why so many people staying in the Signature Tent Cabins had contracted hantavirus was due to the cabins' construction. The tents consisted of two walls: an outer wall made of canvas and an inner wall made of drywall. In the space between these two walls, health inspectors found copious amounts of mice feces and shredded insulation. Health officials concluded that the unusually high morbidity rates resulted from mice and humans being in such close proximity. Yosemite National Park spent millions trying to make sure that their facilities were rodent proof.

Ribavirin is an antiviral drug and has been used to treat HPS; unfortunately, it is ineffective against the Sin Nombre hantavirus that causes HFRS in the United States and Canada. It is not available in the United States or Canada for the treatment of hantavirus. Yet, it has some effectiveness against the Andes hantavirus that occurs in South America (Mertz et al. 2004, Safronetz et al. 2011).

23.6 WHAT PEOPLE CAN DO TO REDUCE THEIR RISK OF CONTRACTING HFRS OR HPS

In the 1990s, the Chinese government developed an effective vaccine against the Hantaan and Seoul hantaviruses and has distributed about two million vaccine doses annually in areas where HFRS is endemic. Hantavirus vaccines are also available in Korea. The Chinese government used poisons and traps in residential areas to reduce rat populations and an educational program to inform the public about the disease and the importance of keeping rat populations under control.

FIGURE 23.7 Photo of Half Dome, a famous mountain in Yosemite National Park. Yosemite Valley is the dark, forested area to the left of it. (Courtesy of Carroll Ann Hodges and U.S. Geological Survey.)

These programs have been effective in reducing the incidence of HFRS in China. In Europe, there are no large-scale programs to control mouse populations as a means of reducing the incidence of HFRS. There are no vaccines available in the United States or Europe to protect people from hantaviruses, but research is being conducted to develop one (Custer et al. 2003, Zhang et al. 2010, Heyman et al. 2012, Vaheri et al. 2013).

The incidence of HPS varies considerably across years and is positively correlated with annual changes in the density of deer mice. These observations suggest that reducing rodent densities in human communities can reduce the incidence of HPS in humans and studies have demonstrated that having rodents in a home greatly increases the risk of infection (Heyman et al. 2012).

Several methods are effective in reducing rodent numbers inside homes. Rodents can be excluded by sealing their access holes in homes and removing brush, lumber, or other items from the yard that rodents can use for shelter. Because mice eat grains and other seeds, items made with grains, such as crackers, cookies, bread, noodles, breakfast cereal, and pet food, should be kept in rodent-proof containers. Mice traps can be baited with peanut butter, nuts, or grain (do not use cheese because mice rarely eat it). However, trapping mice will be ineffective as long as they have access to food because any trapped mice will quickly be replaced by others (Conover 2002).

People should avoid contact with rodent urine, droppings, saliva, and nesting materials, and they should thoroughly wash their hands after touching a contaminated surface. Entering or cleaning buildings that have been closed for an extended period increases the risk of being infected with hantavirus. Therefore, closed rooms and buildings should be thoroughly ventilated before entry. Care should be taken when cleaning rooms or buildings where rodent feces and urine have accumulated. Contaminated surfaces should be wet down with alcohol, disinfectants (e.g., 3% Lysol), or 10% bleach for at least 5 minutes. However, disinfectants may be ineffective in areas where there are large amounts of accumulated organic matter, such as animal bedding or feces. Surfaces should then be wiped with a wet towel rather than sweeping or vacuuming so that less dust will become airborne. After all dust and organic material have been removed, clean surfaces should be disinfected a second time to kill any remaining virus (Mayo Clinic 2011).

Rodent populations are highest where grain is abundant, and people who work in such areas have a higher risk of being infected. Primary among these occupations are grain and dairy farmers, feedlot, and grain millworkers. Field biologists and people who trap rodents also have a high risk as well as people who clean rooms or building that have remained closed for a period of time (CDC 2002, 2012a).

23.7 ERADICATING HFRS OR HPS FROM A COUNTRY

The rodent species that serve as reservoir hosts for hantaviruses are very abundant and widely distributed. As long as these hantaviruses circulate in these rodents, it will be impossible to eradicate these diseases from a country. Instead, efforts to reduce the incidence of HFRS and HPS among humans are centered on preventing humans from becoming infected by reducing rodent numbers in and around houses and educating people of steps they can take to reduce their risk of infection.

LITERATURE CITED

Brown, E. 2013. Yosemite makeover seeks to keep hantavirus at bay. *Los Angeles Times* (May 25, 2013).

CDC. 1993. Hantavirus infection—Southwestern United States: Interim recommendations for risk reduction. *Morbidity and Mortality Weekly Report* 42(RR-11):1–13.

CDC. 2002. Hantavirus pulmonary syndrome—United States: Updated recommendations for risk reduction. *Morbidity and Mortality Weekly Report* 51(RR09):1–12.

CDC. 2007. Hantavirus pulmonary syndrome (hantavirus) and animals. http://www.cdc.gov/healthypets/diseases/hantavirus.htm (accessed July 26, 2013).

CDC. 2011. Hemorrhagic fever with renal syndrome (HFRS). http://www.cdc.gov/hantavirus/hfrs/index.html (accessed July 26, 2012).

CDC. 2012a. Hantavirus. http://www.cdc.gov/hantavirus/ (accessed July 26, 2012).

CDC. 2012b. Outbreak of hantavirus infection in Yosemite National Park. http://www.cdc.gov/hantavirus/outbreaks/yosemite–national-park-2012.html (accessed December 22, 2013).

CDC. 2012c. Rodents in the United States that carry hantavirus. http://www.cdc.gov/hantavirus/rodents/ (accessed November 3, 2013).

CDC. 2013a. Hantavirus pulmonary syndrome (HPS). http://www.cdc.gov/hantavirus/hps/ (accessed December 13, 2013).

CDC. 2013b. U.S. HPS cases, by reporting states. http://www.cdc.gov/hantavirus/surveillance/reporting-state.html (accessed July 26, 2013).

Conover, M. R. 2002. *Resolving Human–Wildlife Conflicts: The Science of Wildlife Damage Management*. CRC Press, Boca Raton, FL.

Custer, D. M., E. Thompson, C. S. Schmaljohn, T. G. Ksiazek, and J. W. Hooper. 2003. Active and passive vaccination against hantavirus pulmonary syndrome with Andes virus M genome segment-based DNA vaccine. *Journal of Virology* 77:9894–9905.

Duchin, J. S., F. T. Koster, C. J. Peters, G. L. Simpson et al. 1994. Hantavirus pulmonary syndrome: A clinical description of 17 patients with a newly recognized disease. *New England Journal of Medicine* 330:949–955.

Heyman, P., E. P. J. Gibbs, A. Meredith. 2012. Bunyavirus infections. In: D. Gavier-Widén, J. P. Duff, and A. Meredith, editors. *Infectious Diseases of Wild Mammals and Birds in Europe*. Wiley-Blackwell, West Sussex, U.K.

Hjelle, B. and G. E. Glass. 2000. Outbreak of hantavirus infection in the Four Corners region of the United States in the wake of the 1997–1998 El Niño southern oscillation. *Journal of Infectious Diseases* 181:1569–1573.

Kallio, E. R., J. Klingström, E. Gustafsson, T. Manni et al. 2006. Prolonged survival of Puumala hantavirus outside the host: Evidence for indirect transmission via the environment. *Journal of General Virology* 87:2127–2134.

Knust, B., A. MacNeil, and P. E. Rollin. 2012. Hantavirus pulmonary syndrome clinical findings: Evaluating a surveillance case definition. *Vector-Borne and Zoonotic Diseases* 12:393–399.

Luis, A. D., R. J. Douglass, P. J. Hudson, J. N. Mills, and O. N. Bjørnstad. 2012. Sin Nombre hantavirus decreases survival of male deer mice. *Oecologia* 169:431–439.

Martinez, V. P., C. Bellomo, J. San Juan, D. Pinna et al. 2005. Person-to-person transmission of Andes virus. *Emerging Infectious Diseases* 11:1848–1853.

Mayo Clinic. 2011. Hantavirus pulmonary syndrome. http://www.mayoclinic.org/health/hantavirus-pulmonary-syndrome/DS00900 (accessed September 1, 2013).

Mertz, G. J., B. Hjelle, M. Crowley, G. Iwamoto et al. 2006. Diagnosis and treatment of new world hantavirus infections. *Current Opinion in Infectious Diseases* 19:437–442.

Mertz, G. J., L. Miedzinski, D. Goade, A. T. Pavia et al. 2004. Placebo-controlled, double-blind trial of intravenous ribavirin for the treatment of hantavirus cardiopulmonary syndrome in North America. *Clinical Infectious Diseases* 39:1307–1313.

National Park Service. 2012. Hantavirus in Yosemite. http:// www.nps.gov/yose/planyourvisit/hantafaq.htm (accessed July 26, 2013).

Safronetz, D., E. Haddock, F. Feldmann, H. Ebihara, and H. Feldmann. 2011. In vitro and in vivo activity of ribavirin against Andes virus infections. *PLoS One* 6(8):e23560.

Shakespeare, M. 2009. *Zoonoses*, 2nd edition. Pharmaceutical Press, London, U.K.

Vaheri, A., H. Henttonen, L. Voutilainen, J. Mustonen et al. 2013. Hantavirus infections in Europe and their impact on public health. *Reviews in Medical Virology* 23:35–49.

Vaheri, A., J. N. Mills, C. F. Spiropoulou, and B. Hjelle. 2011. Hantaviruses. In: S. R. Palmer, L. Soulsby, P. R. Torgerson, and D. W. G. Brown, editors. *Oxford Textbook of Zoonoses*, 2nd edition. Oxford University Press, New York, pp. 307–322.

Yates, T. L., J. N. Mills, C. A. Parmenter, T. G. Ksiazek et al. 2002. The ecology and evolutionary history of an emergent disease: Hantavirus pulmonary syndrome. *BioScience* 52:989–998.

Zhang, Y.-Z., Y. Zou, Z. F. Fu, and A. Plyusnin. 2010. Hantavirus infections in humans and animals, China. *Emerging Infectious Diseases* 16:1195–1203.

24 Influenza

Waterbirds, to an influenza researcher, are more than majestic swans and charming mallards. They are instead stealthy vectors of novel influenza viruses, some of nature's bioterrorist agents, chauffeuring dangerous microbes from place to place....

Smith (2013a)

With a highly pathogenic virus like H5N1, identifying which birds have the virus is very easy; if your chickens lie upside down in the pen then they probably have H5N1.

Ron Fouchier (Eramas Medical Center, Netherlands)

24.1 INTRODUCTION AND HISTORY

Influenza or flu is caused by three species of single-stranded RNA viruses: influenza virus C, influenza virus B, and influenza virus A, which will be referred to in this book as types C, B, and A viruses (Figures 24.1 and 24.2). Humans are reservoir hosts for type C and type B viruses, and these are referred to as human influenza viruses. Types C and B viruses rarely infect other species except that type B viruses are also found in marine mammals. Type C viruses are common in humans and usually cause no symptoms or only a mild illness. Type B viruses cause sporadic outbreaks of respiratory disease in humans, especially in children. Type A viruses infect birds, especially waterbirds, but some strains have developed the ability to infect humans and other mammals including horses, hogs, and marine mammals. Type A viruses cause flu epidemics every winter that sicken and kill people. In some years, type A viruses mutate into a more lethal virus that can easily spread from one person to another. When this happens, a worldwide pandemic (i.e., a worldwide epidemic) may occur, resulting in great loss of life.

Influenza pandemics have happened for centuries. The first pandemic that clearly fits the description of influenza occurred in 1580; the name "influenza" predates that pandemic. The disease was named by the Italians in the 1400s, and influenza is translated as "influenced by the stars"; the disease was so named because of the belief that influenza epidemics were caused by an ominous star pattern. When a pandemic will occur, it cannot be predicted, but an average of three occur every century (WHO 2005a, Atkinson et al. 2012, CDC 2012).

Type A viruses have two main proteins (antigens) on their outer surfaces: hemagglutinin (H) antigen and neuraminidase (N) antigen; each influenza virus is further divided into subtypes based on which copy of the two antigens it possesses. For instance, H5N1 and H1N1 are two different subtypes that differ because they possess different H antigens. In total, there are 16 known H antigens and 9 N antigens and, therefore, 144 potential subtypes of the type A virus. Human immune systems recognize influenza viruses by their antigens so that when a person is vaccinated or

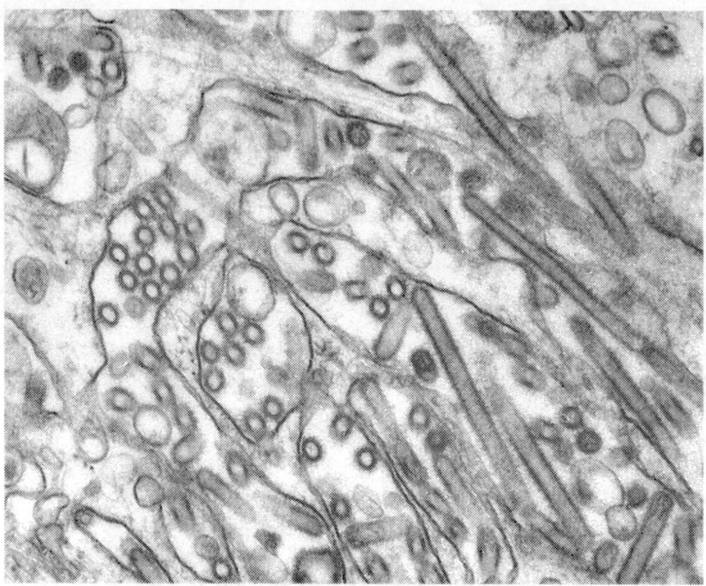

FIGURE 24.1 A highly magnified transmission electron micrograph of an influenza virus. (Courtesy of Cynthia Goldsmith, Jacqueline Katz, Sherif Zaki, and the CDC.)

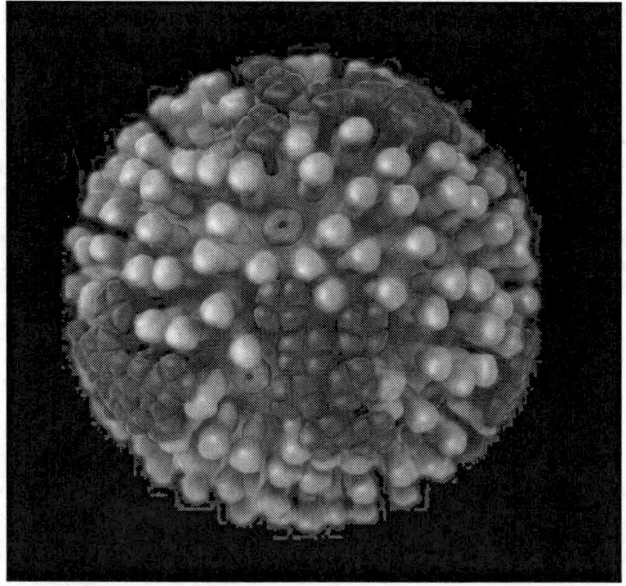

FIGURE 24.2 A 3D portrayal of an influenza virus. (Courtesy of Dan Higgins and the CDC.)

had a prior infection, the immunity does not extend to subtypes with different antigens (WHO 2005a, OSHA 2006, CDC 2011a, Reperant et al. 2012).

Type A viruses are notorious for their ability to mutate and change the proteins on their outer surfaces. They can do so using two different methods. The first occurs when two influenza viruses exchange genetic material, a process called antigenic shift, which can create new combinations of the H and N proteins. The second method is called antigenic drift; it results from point mutations to the virus's RNA and results in minor changes to its proteins, including the H and A antigens. Antigenic shift and drift continually create new viral strains. In some cases, these changes allow the virus to become more virulent to its host species or be able to infect a new avian or mammalian species. In either case, an influenza outbreak may result. Further, antigen changes create problems for medical authorities because doctors cannot predict accurately which influenza viruses will pose the greatest threats to people during an upcoming flu season.

Additionally, type A viruses are grouped into low and highly pathogenic strains. Poultry, domestic ducks, and domestic geese infected with a low pathogenic strain typically experience mild symptoms, such as ruffled feathers or a decrease in egg production. Highly pathogenic strains cause serious illnesses and death in birds and are so lethal that they have no reservoir host. Instead, they result when a low pathogenic strain mutates into a highly pathogenic strain (Newman et al. 2010, CDC 2011b, Reperant et al. 2012). From 1994 to 2005, there were several outbreaks of highly pathogenic strains, including H5N1 in various parts of the Old World from 1996 onward; H5N2 in Mexico in 1994, Italy in 1997, and Texas in 2004; H7N1 in Italy in 1999; H7N3 in Pakistan and Australia in 1994, Chile in 2002, and Canada in 2003; H7N4 in Australia in 1997; and H7N7 in the Netherlands in 2003 (Olsen et al. 2006).

There is great concern that someday an influenza type A virus will cause a pandemic, resulting in hundreds of millions of people becoming sick and millions dying. To start a pandemic, an influenza virus needs to have three characteristics. First, the virus needs to be novel so that most humans will not have any immunity against it. Second, the virus needs to be able to infect humans and cause illness. Third, the virus needs the ability to be transmitted easily from one person to another. Three pandemics have occurred during the last century (1918–1919, 1957–1958, and 1968–1969).

Influenza pandemics are the disease equivalent of a tsunami or tidal wave. They occur without warning, sweep through an area with devastating speed, and disappear just as fast (WHO 2005a). While an influenza pandemic is unlikely to happen during any given year, they are recurring events and will occur in the future with potentially devastating health consequences. It is instructive to examine pandemics of the last century as well as recent influenza outbreaks that did not develop into a pandemic.

24.1.1 Spanish Influenza Pandemic of 1918 and 1919

This pandemic began close to the end of World War I (Figure 24.3), but it is unclear where it began. The first cases were recorded in March 1918 in both the United States and Europe, and cases were soon being reported across the world. The pandemic is known as Spanish influenza, but this was a misnomer because the pandemic did not begin there, nor was it worse there than elsewhere. Rather, the name arose

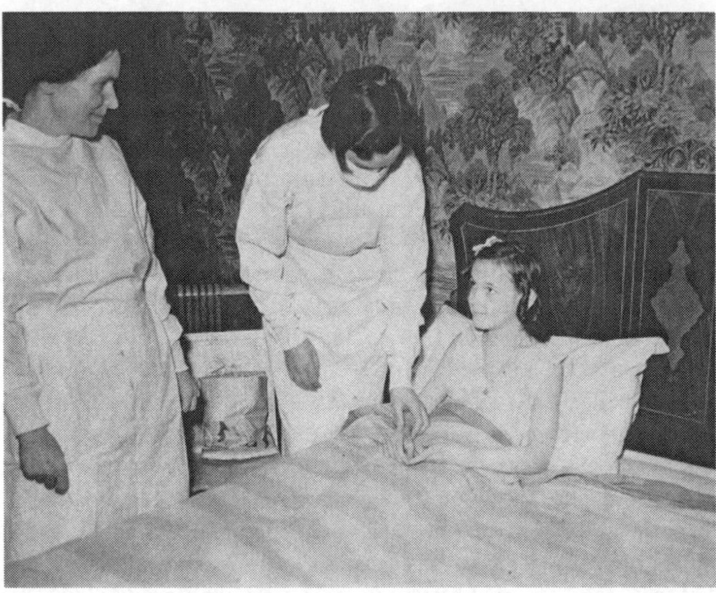

FIGURE 24.3 The Spanish influenza pandemic began during World War I. (Courtesy the U.S. National Library of Medicine.)

because press censorship by all of the warring countries downplayed the pandemic and Spain's free press was the main source of information about the severity of the pandemic. Spain's reward for being an accurate source of information was to have the pandemic named after it, a dubious honor at best (Johnson and Mueller 2002, Taubenberger and Morens 2006).

The first wave of this pandemic occurred during the spring and summer of 1918, with a second wave occurring that fall and a third a year later. By the time the pandemic ended in 1920, 25%–30% of the world population or 500 million people had become ill, and 50,000,000 people (some estimates are 100,000,000) had perished. By way of comparison, all of the battles of World War I claimed 8,000,000 lives. Death usually occurred quickly—within a couple of days after the onset of illness—the cause of death was suffocation resulting from pneumonia or the lungs filling with fluid. Physicians could do little for their patients, and health authorities had few, if any, viable options to stem the epidemic. Quarantines were imposed, and many people began wearing masks in public. In some countries, laws were passed making it a crime to cough or sneeze in public without covering the mouth. This pandemic killed more people than any other single event in human history with the possible exception of World War II when there were approximately 60 million fatalities. It is this pandemic that haunts medical authorities today because they fear that it will be repeated (Johnson and Mueller 2002, Taubenberger and Morens 2006).

Recently, the virus responsible for Spanish influenza pandemic of 1918–1919 was isolated from frozen tissue obtained from three victims who died in 1918 (two U.S. soldiers whose lung tissue was archived and an Inuit woman who was buried in permafrost). When the virus was genetically sequenced, it was identified as the type A virus: H1N1 (Taubenberger et al. 2000).

24.1.2 1957–1958 Influenza Pandemic

This pandemic was caused by H2N2 and began in February 1957 in a single province of China but spread with alarming speed. Within 2 months, it was found throughout China. Six months later, human cases were being diagnosed around the world. In Europe and Asia, the disease hit in September 1957 with the opening of schools. Illness was concentrated in children of school age and people living or working in crowded situations. Some countries restricted international travel and trade, banned public gatherings, and closed schools, but these did little more than delay the onset of influenza by a couple of weeks or months. Fortunately, mortality rates were much lower than the Spanish influenza pandemic of 1918–1919; nonetheless, more than two million people died in the pandemic (WHO 2005a).

24.1.3 1968–1969 Influenza Pandemic

This pandemic was caused by H3N2 and began during the summer of 1968 in southeastern China. It reached Hong Kong by July and had sickened 500,000 people within 2 weeks; but as the virus spread around the world, it moved slower, caused milder symptoms, and killed fewer people than the two earlier pandemics. This probably resulted because the virus contained the N2 antigen as did the 1957–1958 pandemic. Hence, people who were alive in 1968 would have some immunity to the new influenza virus. The epidemic reached the United States in September 1968, carried by U.S. troops returning home from the Vietnam War. An estimated 34,000 people in the United States died from the pathogen. Globally, about 1,000,000 people died during this pandemic (WHO 2005a).

24.1.4 2004 Outbreak of H5N1

The outbreak began in 1996 when a low pathogenic strain of the H5N1 virus mutated into a highly pathogenic strain in Guangdong Province, China. A year later, an outbreak of the same highly pathogenic H5N1 virus started killing poultry in Hong Kong (Figure 24.4). At the same time, 18 people in Hong Kong were infected with the same strain of H5N1, and six of them died. In response, Hong Kong authorities killed 1.3 million chickens to stop the outbreak (Smith 2012). The effort seemed to be successful because there were no more reports of people or poultry being infected with the H5N1 virus for the next couple of years. Then in December 2003, large numbers of chickens started dying in Korea; genetic tests revealed the same highly pathogenic H5N1 virus had reappeared. This time, the virus spread rapidly; by 2004, it was infecting poultry throughout Southeast Asia (Korea, Thailand, Vietnam, Japan, Hong Kong, Cambodia, Laos, Indonesia, China, and Malaysia), and human cases were documented in China, Vietnam, and Thailand. In Asia, more than 120 million poultry were killed by the pathogen or destroyed by public health authorities within 3 months. People throughout Asia started to become ill by the same virus. Most of

FIGURE 24.4 In Asia, small backyard flocks of chickens and ducks can exchange type A viruses with wild birds and humans. (Courtesy of the CDC.)

the influenza patients acquired the virus directly from poultry, but some cases of human-to-human transmission were reported in Thailand (Chan 2002, WHO 2005b, 2013b, CDC 2008, Newman et al. 2010).

Ominously, thousands of free-ranging birds, including ducks, geese gulls, and cormorants, that started dying in China are killed by the virus. Many of the sickened birds were members of migratory species, creating the threat that they would distribute the pathogen beyond Asia. The threat was confirmed in 2006 when Mongolia, Russia, Kazakhstan, Croatia, Poland, Denmark, Turkey, Kuwait, and Egypt all found the pathogen in wild birds. Soon, the viral strain had been detected in birds located in 63 countries; human cases were reported in 15 countries. Most human patients were infected from birds rather than from another person. Fortunately, H5N1 has not created a massive pandemic at the time this book was published because the virus had not developed the ability to spread from one person to another. H5N1 has been called a pandemic in waiting because it may yet acquire that ability. Still, there were 633 confirmed cases of H5N1 in humans and 337 deaths. Another cost of pandemic is the loss of over 250 million poultry, which reduced the amount of protein in the diet of millions of people (WHO 2005b, 2012b, CDC 2008, Newman et al. 2010, Smith 2013a).

24.1.5 2009 Swine Flu Pandemic

In March and April 2009, a new strain of H1N1 emerged in Mexico and the United States where it developed the ability to infect humans and to spread from

FIGURE 24.5 Influenza viruses can infect pigs. (Courtesy of the USDA Agricultural Research Service.)

human to human. It was dubbed "swine flu" because the virus originated from pigs (Figure 24.5). The swine flu virus spread quickly around the world and became the first flu pandemic of the twenty-first century. By June 2009, 74 countries were reporting human cases of H1N1 influenza. Before it ended in August 2010, the virus had spread to virtually all countries. Most of the deaths resulted from viral pneumonia; mortality rates were highest among pregnant women, children under the age of two, and people with chronic lung diseases or weakened immune systems (WHO 2012a, 2013c). The CDC (2010) estimated that approximately 61 million people in the United States were infected by the swine flu virus, 274,000 patients were hospitalized, and 12,470 died.

24.1.6 2013 Epidemic of H7N9

Another strain of avian flu virus (H7N9) was isolated from both pigeons and humans in eastern China. It was first detected in February 2013 among pneumonia patients located in eastern China around the Yangtze River and Shanghai. Similar cases were soon being reported in other Chinese locations hundreds of miles apart. Again, no human-to-human transmission was detected despite intensive surveillance. Chinese authorities moved quickly, killing over 20,000 chickens, ducks, geese, and pigeons and closing down the poultry market in Shanghai. Within a few months, the number of human cases of H7N9 in China exceeded 130 and resulted in over 40 deaths (Holmes 2013, Smith 2013b, WHO 2013b,c,d).

24.2 SYMPTOMS IN HUMANS

During an influenza epidemic or pandemic, most people become infected from another person. After being inhaled, the virus infects epithelial cells in the trachea and bronchi where they replicate. This results in the destruction of the infected cell, and the pathogens are freed to infect other cells. Only about 50% of people infected with the influenza virus will become ill. Symptoms usually begin 2 days after infection but can occur as soon as 1 day or as late as 4 days of postinfection. Symptoms include fever of 101°F–102°F (38°C–39°C) that begins suddenly, a sore throat, nonproductive cough, nasal discharge, headache, lower backaches, and weakness. Influenza symptoms differ from a common cold in that influenza has a more abrupt beginning, fever is common, and influenza patients feel much worse. Influenza symptoms rarely persist for more than 5 days, but patients can feel drained of energy for a couple of weeks. The most common complication of influenza is pneumonia, usually resulting from a secondary infection caused by bacteria (bacterial pneumonia). Viral pneumonia occurs only rarely but has a high mortality rate. Another rare complication is Reye's syndrome, which can be fatal; its clinical signs involve severe vomiting and confusion. Reye's syndrome occurs almost exclusively in children who have influenza and take aspirin. For this reason, aspirin should not be given to children or teenagers (Atkinson et al. 2012).

24.3 INFLUENZA VIRUS INFECTIONS IN ANIMALS

Type A viruses have been detected in over 105 avian species from 26 avian families. Wild Anseriformes (e.g., ducks and geese) and Charadriiformes (e.g., gulls, terns, and wading birds) are reservoir hosts of all subtypes of the influenza A virus. That is, influenza viruses containing each of the 16 H antigens and 9 N antigens have been detected in wild birds (Table 24.1). Of 28,955 dabbling ducks that were tested around the world prior to 2006, 10.1% tested had antibodies to type A viruses as did 1.6% of 1,011 diving ducks tested, 1.9% of 5,009 swans, 1.4% of 14,505 gulls, 1.4% of 1,962 rails, 1.0% of 4,806 geese, 0.9% of 2,521 terns, 0.8% of 2,637 wading birds, 0.4% of 4,500 cormorants, and 0.3% of 1,416 petrels (Olsen et al. 2006).

Most type A viruses, such as H5N1, replicate in the gastrointestinal tracts of ducks, geese, swans, and gulls; birds shed the virus in feces for long periods of time. Low pathogenic influenza viruses can persist from 1 year to the next by infecting birds or by surviving in the environment. The virus can survive for months in feces, on a frozen lake, or in surface water. Waterbirds are then infected when they drink contaminated water or forage on aquatic food. Dabbling ducks, such as mallards, are more likely to be infected than diving ducks, perhaps because dabbling ducks feed on the water surface where the influenza viruses are concentrated, while diving ducks feed at greater depths (Olsen et al. 2006, Reperant et al. 2012).

Domestic ducks, geese, and poultry are also reservoir hosts for type A viruses, and there is an exchange of these viruses between domestic and wild birds. It is unclear exactly how this happens, but one hypothesis is that it occurs when domestic ducks and geese are allowed to feed in flooded rice fields where wild ducks and geese

TABLE 24.1
Outbreaks of Highly Pathogenic Avian Influenza in Birds from 1959 to 2005

Year	Strain	Country	Great Economic Losses?
1959	H5N1	Scotland	No
1963	H7N3	England	No
1966	H5N9	Canada	No
1976	H7N7	Australia	No
1979	H7N7	England	No
1979	H7N7	Germany	No
1983	H5N8	Ireland	No
1983–1985	H5N2	United States	Yes
1985	H7N7	Australia	No
1991	H5N1	England	No
1992	H7N3	Australia	No
1994	H7N3	Australia	No
1994	H7N3	Pakistan	Yes
1994–1995	H5N2	Mexico	Yes
1997	H5N1	Hong Kong	Yes
1997	H5N2	Italy	No
1997	H7N4	Australia	No
1999–2000	H7N1	Italy	Yes
2002	H5N1	Hong Kong	No
2002	H7N3	Chile	No
2003	H7N7	Netherlands	Yes
2004	H5N2	South Africa	No
2004	H5N2	United States	No
2004	H7N3	Canada	Yes
2004	H7N3	Pakistan	No

Sources: WHO, Avian influenza: Assessing the pandemic threat, WHO/CDS/2005.29, World Health Organization, Geneva, Switzerland, 2005a; WHO, *N. Engl. J. Med.*, 353, 1374, 2005b.

Note: Outbreaks that spread to numerous poultry farms and causing great economic losses are noted.

are also foraging. Domestic geese and ducks then transmit the pathogens to chickens and turkeys (Gilbert et al. 2006, Merck 2010, Reperant et al. 2012).

Typically, wild and domestic birds that are infected with influenza viruses do not experience illness or only have a mild respiratory illness, but some strains can be fatal in both wild and captive birds. Sometimes, the mortality rate is significant: the 2005 outbreak of H5N1 decreased the global population of bar-headed geese by 10% (Figure 24.6) and also decreased European populations of mute swan and whooper

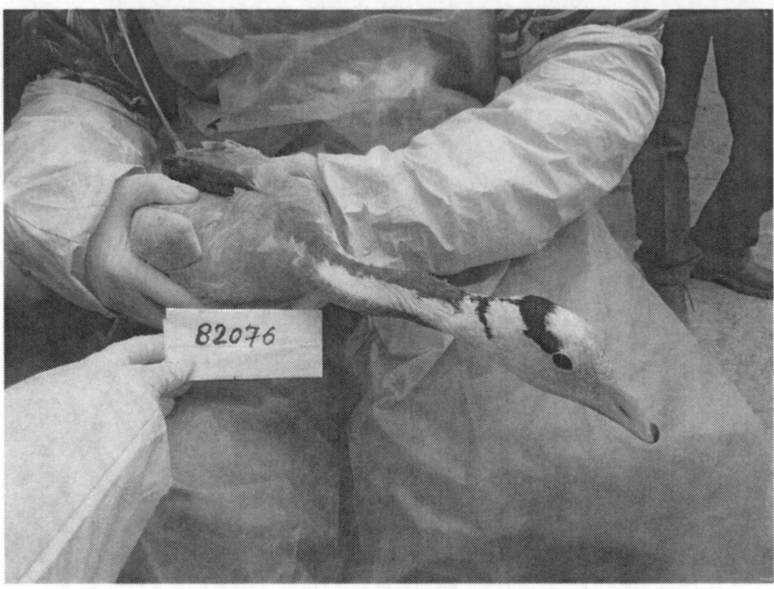

FIGURE 24.6 Outbreaks of H1N1 influenza virus reduced populations of bar-headed geese by 10%. (Photo by Diann Prosser and provided courtesy of the U.S. Geological Survey.)

swan (Olsen et al. 2006). Birds infected with H5N1 virus had breathing problems, swollen heads, and greenish diarrhea. Highly pathogenic viruses often cause disseminated infections involving multiple organs and producing massive internal hemorrhaging. If the infection spreads to the central nervous system, birds can exhibit an unusual posture, drooping wings, loss of muscle coordination, and paralysis. Death often occurs within 48 hours of the onset of clinical signs, and mortality rates of infected birds can approach 100% (Brown et al. 2008, Merck 2010).

Interspecific transmission of the influenza virus occurs between birds and marine animals (Figure 24.7). Both types A and B viruses have been found in several species of marine mammals around the world (Table 24.2). Hundreds of harbor seal died along the New England coast in 1979–1980, 1982–1983, and 2011 from infections of H7N7, H4N4, and H5N1, respectively (Figure 24.8). Other subtypes of type A virus that have been recovered from dead marine mammals from around the world include H1N1, H1N3, H3N3, H3N8, H4N5, H4N6, H13N2, and H13N9. Respiratory problems and a bloody discharge from the nose and mouth are clinical signs of an infected seal. On rare occasions, seals have transmitted the influenza virus to people. In one case, a captive seal that had influenza sneezed into the face of an investigator, who developed conjunctivitis within 2 days of the incident; tests revealed that the seal and trainer were infected by the same viral strain. Other people became ill after conducting autopsies of seals that died of influenza. Although it is very rare for anyone to be infected with the influenza virus from a marine mammal, the ability of type A viruses to jump from marine mammals to humans is troubling. Marine mammals are the only

FIGURE 24.7 Birds often feed near whales; this close association allows for the transmission of type A viruses from infected birds to whales. (Courtesy of the U.S. National Oceanic and Atmospheric Administration.)

species other than humans known to be infected with both types A and B viruses. If these influenza viruses exchange genes, a new virus might evolve that could cause a pandemic (Van Campen and Early 2001, Reperant et al. 2009, Anthony et al. 2012).

Some type A viruses are able to maintain themselves by infecting pigs and have done so both in the United States and around the world. The main viruses that have been circulating in U.S. domestic pigs in recent years are H1N1, H1N2, and H3N2. The H1N1 virus has circulated among U.S. pigs since at least 1930, H1N2 since 1999, and H3N2 since 1998. Influenza viruses can be very contagious in pigs, especially during the fall and winter, and can spread quickly through a pig farm. However, these viruses are not usually fatal in pigs. These same influenza viruses also occur in feral hogs, complicating the task of eradicating influenza viruses at a pig farm because local feral hogs can reinfect them (Figure 24.9). Influenza signs in pigs are similar to human clinical signs and include fever, coughing, difficulty breathing, and nasal discharge. Normally, swine viruses do not infect humans, but sometimes they evolve the ability to do so. For example, the 2009 pandemic was caused by a H1N1 virus that was circulating in pigs (Reperant et al. 2009, 2012, CDC 2012, 2013).

One strain of H7N7 and a strain of H3N8 can maintain themselves by infecting horses. Both were probably transmitted to horses from birds. Clinical signs of influenza in horses include fever, cough, and nasal discharge. Some horses develop pneumonia and myocarditis. Influenza is usually not fatal in horses unless a secondary infection develops (Reperant et al. 2009, 2012, CDC 2012).

TABLE 24.2
Marine Mammals that Have Been Infected with Influenza Type A Virus Based on Serological Evidence or Virus Isolation

Species	Location
Porpoise, Dall's	Northern Pacific and Antarctic oceans
Sea lion	Bering Sea
Sea lion, California	California
Seal, Baikal	Lake Baikal
Seal, Caspian	Caspian Sea
Seal, Kuril harbor	Japan
Seal, gray	North Sea
Seal, Harbor	California, New England, North Sea
Seal, harp	Barents Sea
Seal, hooded	Barents Sea
Seal, northern elephant	California
Seal, ringed	Alaska and Arctic Canada
Seal, South American fur	Uruguay
Whale, belugas	Arctic Canada
Whale, minke	South Pacific
Whale, pilot	New England

Sources: Nielsen, O. et al., *J. Wildl. Dis.*, 37, 820, 2001; Blanc, A. et al., *J. Wildl. Dis.*, 45, 519, 2009; Reperant, L.A. et al., Influenza virus infections, in: D. Gavier-Widen, J.P. Duff, and A. Meredith, editors, *Infectious Diseases of Wild Mammals and Birds in Europe*, Wiley-Blackwell, West Sussex, U.K., 2012, pp. 37–58; Boyce, W.M. et al., *Emerg. Microbes Infect.*, 2, e40, 2013.

24.4 HOW HUMANS CONTRACT INFLUENZA

Most people are infected with the influenza virus from another person. When an influenza patient coughs or sneezes, tiny droplets containing the virus are blown into the air. These can be inhaled by anyone who is near. The droplets also contaminate surfaces. A person who later touches a contaminated surface can become infected when they touch their eyes, nose, or mouth. Outbreaks of influenza in the Northern Hemisphere peak during the late fall and winter months.

Humans can contract influenza from birds, pigs, and seals. Birds and swine infected with the influenza virus produce mucus in their lungs and nasal passages that contain high densities of the virus. These viruses are dispersed in the immediate area via coughing and sneezing; people working with these animals are exposed to the virus. In most cases, people are not infected because the viral strains of animals do not have the ability to infect humans. Sometimes, however, the virus mutates so that it can infect people. The human immune system may have no defenses against

FIGURE 24.8 Infections of type A viruses often sicken and kill harbor seals. (Courtesy of the Woods Hole Science Aquarium and the U.S. National Oceanic and Atmospheric Administration.)

FIGURE 24.9 The presence of type A viruses in feral hogs greatly complicates the effort to keep domestic hog operations free of the influenza virus. (Courtesy of Nova Silvy.)

viruses that previously only infected animals, resulting in severe disease in people. If such a virus develops the ability to spread from human to human, an epidemic may occur (Van Campen and Early 2001, CDC 2011b, 2012).

24.5 MEDICAL TREATMENT

Antiviral drugs can be used to treat influenza or to prevent it. Two are approved by the U.S. Food and Drug Administration and are used in the United States. Both are effective against types A and B viruses and require a prescription. Tamiflu® contains the active ingredient oseltamivir and is approved for adults, children, and infants over 2 weeks old. It is formulated as a dry powder and administered by inhalation. Relenza® contains zanamivir and can be used in adults and children over 7 years old. It should not be used by people with respiratory diseases. It comes in the form of a capsule and is swallowed. Both medicines can shorten the duration of illness and reduce the risk of complications. To be effective, the drugs must be started within 2 days of the onset of symptoms. CDC recommends that the drugs be used to treat patients with confirmed or suspected influenza who are hospitalized or have a severe or progressive illness (CDC 2012).

Influenza patients should rest, drink plenty of liquids, and use a pain reliever, such as acetaminophen or ibuprofen, for aches and pains. Aspirin should not be given to children. The WHO recommends that people seek medical attention if they have difficulty breathing, experience shortness of breath, or have a fever for more than 3 days. Children need medical care if they have labored or fast breathing, convulsions, or a continuing fever. One serious complication of influenza is bacterial pneumonia, which can be treated with antibiotics.

Experts recommend that people should contact a doctor immediately if they think they have influenza within 10 days of handling flu-infected birds or visited an area where there is an influenza outbreak. This is especially true for people who visited a farm or open-air market where there were poultry (CDC 2012, Mayo Clinic 2013).

24.6 WHAT PEOPLE CAN DO TO REDUCE THEIR RISK OF CONTRACTING INFLUENZA

24.6.1 Preventing the Risk That a Person Will Be Infected with the Influenza Virus from Another Person

The best method for people to protect themselves against influenza is to obtain an influenza (flu) vaccination annually. Vaccines against seasonal influenza are designed to protect people from three different influenza viruses: two type A viruses and one type B virus. The actual viruses targeted are changed annually and are based on which type A and B viruses are circulating at that time. One vaccine is the trivalent inactive influenza vaccine, which contains split virus and subunit inactivated virus. The second vaccine uses live attenuated influenza viruses. It is administered in a single dose that is sprayed into the nose (Figure 24.10). Both vaccines target the same type A and B viruses (Atkinson et al. 2012).

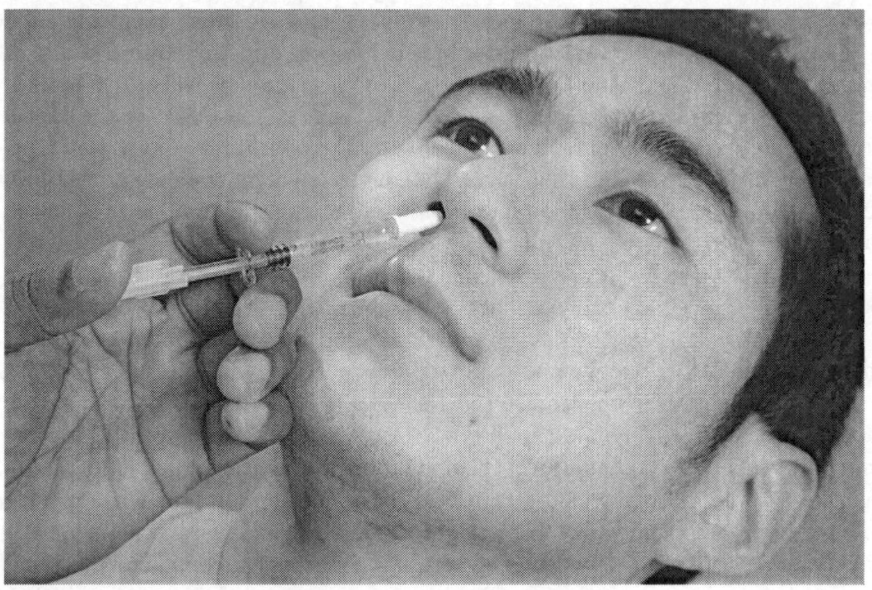

FIGURE 24.10 One type of influenza vaccine is administered through inhalation, making it a favorite of people who are afraid of hypodermic needles. (Courtesy of the CDC.)

People should also cover their nose and mouth with a tissue when coughing or sneezing and dispose the used tissue. People should wash their hands thoroughly and often with soap and water and avoid touching their eyes, nose, and throat. Wearing a surgical mask and using alcohol-based hand sanitizers can help protect people from becoming infected during an influenza epidemic. To avoid infecting others, influenza patients should stay home and avoid contact with others. The Mayo Clinic reports that most healthy people who have influenza and have only mild symptoms do not need to seek medical attention. Nonetheless, people who have a higher risk of developing influenza-related complications should see a doctor. These people include children under the age of five or people over 65, pregnant women, people who have a weakened immune system, and those who suffer from a chronic health problem (e.g., heart or liver disease, diabetes, asthma, or emphysema) or are obese (CDC 2012, Mayo Clinic 2013).

24.6.2 Preventing the Evolution of New Strains of Type A Viruses That Can Infect Humans or Are Highly Pathogenic

Avian flu outbreaks often start in Asia because of the area's high densities of humans and poultry. For example, China has more than one billion people and 16 billion chickens. Many rural households throughout Asia maintain a flock of chickens and ducks. This close association between humans and domestic birds increases the risk that an influenza virus in poultry develops the ability to infect human. The public health agencies across the world are constantly on the alert for outbreaks of avian influenza A viruses in wild birds, poultry, or humans. This monitoring allows local governments to quickly eliminate infected flocks, stop the movement of poultry that may have been exposed to the virus, and close markets where live birds are sold. WHO has created the Global Outbreak Alert and Response Network, which is a collaboration of various governmental health agencies, institutions (e.g., Red Cross, Red Crescent, and medical colleges), and humanitarian organizations. Its mission is to pool global resources for the rapid identification, confirmation, and response to outbreaks of international importance (WHO 2005a).

When the influenza virus is detected in domestic birds, the infected flock and any other flocks that had contact with it are culled and quarantine practices are imposed to contain the virus. For instance, countries in the European Union prohibit the movement of all live poultry and poultry products from within 2 miles (3 km) of where an infected bird was located; bird hunting is banned within this same area (Reperant et al. 2012). While quarantines or travel bans may only slow the arrival of influenza virus into a country rather than prevent it, the additional time allows health agencies to prepare (e.g., creating more vaccines). Furthermore, influenza pandemics are normally so contagious that health facilities are quickly overwhelmed by the sick. Any steps that can slow the pathogen's spread through a population can reduce the number of patients requiring medical attention at any one point in time (OSHA 2006, CDC 2011a,b, 2012).

24.6.3 Reducing the Risk That a Person Will Acquire Influenza from Birds or Pigs

According to the CDC, the best way for a person to avoid becoming infected during an outbreak in poultry is to avoid exposure to sick or dead birds. Poultry farmers and workers should use personal protective equipment. They can take prophylactic antiviral medication when an outbreak occurs in chickens. It is possible, but rare, for a person to become infected directly from wild birds (CDC 2011c, 2012, WHO 2013a,b).

People cannot become infected with the type A virus from eating properly cooked eggs, chicken, or pork although people who handle live or dead birds or pigs or who eat raw or uncooked poultry or pork can be infected. Cooks should use a food thermometer to cook meat to at least 165°F (74°C). Cooks and food handlers should also wash hands with warm water and soap before and after handling raw poultry and eggs. Cooking and cutting utensils should be cleaned regularly (OSHA 2006, CDC 2013).

Antiviral drugs (Tamiflu and Relenza) can be used to prevent influenza but are not as effective as vaccinations. Still, these drugs are 70%–90% effective in preventing influenza. The CDC does not recommend the widespread or routine use of these antiviral drugs to prevent influenza because of the danger that influenza viruses will develop immunity to them. Their use, however, may be warranted during an influenza outbreak in a nursing home, long-term care facility, and hospital. Treatment should continue until 1 week after the last known influenza case has occurred (CDC 2012).

24.7 ERADICATING INFLUENZA FROM A COUNTRY

Influenza viruses pose a constant threat to cause a new pandemic. To prevent this, public health agencies across the world are constantly on the alert for outbreaks of avian influenza viruses in wild birds, poultry, or humans. When one occurs, governments need to be able to react quickly and decisively to prevent the virus from spreading.

LITERATURE CITED

Anthony, S. J., J. A. St. Leger, K. Pugliares, H. S. Ip et al. 2012. Emergence of fatal avian influenza in New England harbor seals. *MBio* 3(4):e00166-12.

Atkinson, W., J. Hamborsky, and S. Wolfe. 2012. *Epidemiology and Prevention of Vaccine-Preventable Diseases*, 12th edition. CDC, Washington, DC.

Blanc, A., D. Ruchansky, M. Clara, F. Achaval et al. 2009. Serologic evidence of influenza A and B viruses in South American fur seals (*Arctocephalus australis*). *Journal of Wildlife Diseases* 45:519–521.

Boyce, W. M., I. Mena, P. K. Yochem, F. M. D. Gulland et al. 2013. Influenza A (H1N1)pdm09 virus infection in marine mammals in California. *Emerging Microbes and Infections* 2:e40.

Brown, J. D., D. E. Stallknecht, and D. E. Swayne. 2008. Experimental infection of swans and geese with highly pathogenic avian influenza virus (H5N1) of Asian lineage. *Emerging Infectious Diseases* 14:136–142.

CDC. 2008. Avian influenza: Current H5N1 situation. http://www.cdc.gov./flu/avian/outbreaks/current.htm (accessed July 10, 2013).

CDC. 2010. Updated CDC estimates of 2009 H1N1 influenza cases, hospitalization and deaths in the United States, April 2009–April 20, 2010. CDC, Atlanta, GA.

CDC. 2011a. Influenza type A viruses and subtypes. http://www.cdc.gov/flu/avianflu/infuenza-a-virus-subtypes.htm (accessed July 10, 2013).

CDC. 2011b. Key facts about avian influenza (bird flu) and highly pathogenic avian influenza A (H5N1) virus. http://www.cdc.gov/flu/avian/gen-info/facts.htm (accessed July 10, 2013).

CDC. 2011c. Transmission of avian influenza A viruses between animals and people. http://www.cdc.gov/flu/avianflu/virus-transmission.htm (accessed July 10, 2013).

CDC. 2012. Seasonal flu (influenza). http://www.cdc.gov/flu/avianflu/ (accessed December 23, 2013).

CDC. 2013. Avian influenza A (H7N9) virus. http://www.cdc.gov/flu/avianflu/h7n9-virus.htm (accessed July 10, 2013).

Chan, P. K. S. 2002. Outbreak of avian influenza A (H5N1) virus infection in Hong Kong in 1997. *Clinical Infectious Diseases* 34(Supplement 2):S58–S64.

Conover, M. R. 2002. *Resolving Human-Wildlife Conflicts: The Science of Wildlife Damage Management*. CRC Press, Boca Raton, FL.

Gilbert, M., P. Chaitaweesub, T. Parakamawongsa, S. Premashthira et al. 2006. Free-grazing ducks and highly pathogenic avian influenza, Thailand. *Emerging Infectious Diseases* 12:227–234.

Holmes, D. 2013. The world waits for H7N9 to yield its secrets. *Lancet Infectious Disease* 13:477–478. http://www.cnn.com/2013/04/05/world/asia/china-bird-flu (accessed July 10, 2013).

Johnson, N. P. A. S. and J. Mueller. 2002. Updating the accounts: global mortality of the 1918–1920 "Spanish" influenza pandemic. *Bulletin of the History of Medicine* 76:105–115.

Mayo Clinic. 2013. Influenza (flu): Risk factors. http://www.mayoclinic.org/health/influenze/DS00081/DSECTION=risk-factors (accessed September 1, 2013).

Merck. 2010. *Merck Veterinary Manual*, 10th edition. Merck, Whitehouse Station, NJ.

Newman, S. H., J. Siembieda, R. Koch, T. McCracken et al. 2010. FAO EMPRES Wildlife unit fact sheet: Wildlife and H5N1 HPAI virus—Current knowledge. Animal Production and Health Division, Food and Agriculture Organization of the United Nations, New York.

Nielsen, O., A. Clavijo, and J. A. Boughen. 2001. Serologic evidence of influenza A infection in marine mammals of Arctic Canada. *Journal of Wildlife Diseases* 37:820–825.

Olsen, B., V. J. Munster, A. Wallensten, J. Waldenström et al. 2006. Global patterns of influenza A virus in wild birds. *Science* 312:384–388.

OSHA. 2006. OSHA guidance update on protecting employees from avian flu (avian influenza) viruses. http://www.osha.gov (accessed July 10, 2013).

Reperant, L. A., A. D. M. E. Osterhaus, and T. Kuiken. 2012. Influenza virus infections. In: D. Gavier-Widen, J. P. Duff, and A. Meredith, editors. *Infectious Diseases of Wild Mammals and Birds in Europe*. Wiley-Blackwell, West Sussex, U.K., pp. 37–58.

Reperant, L. A., G. F. Rimmelzwaan, and T. Kuiken. 2009. Avian influenza viruses in mammals. *Revue Scientifique et Technique* 28:137–159.

Smith, T. C. 2012. Holy influenza, Batman! http://www.scienceblogs.com/aetiology/2012/02/28/holy-influenza-batman/ (accessed July 10, 2013).

Smith, T. C. 2013a. Nature's bioterrorist agents: Just how bad is the new bird flu. http://www.slate.com/articles/health_and_science/medical_examiner/2013/04/the_h7n9_bird_flu_in_china_how_dangerous_is_it.html (accessed December 23, 2013).

Smith, T. C. 2013b. What's up with H7N9, the new avian influenza? http://www.scienceblogs.com/aetiology/2013/04/10/whats-up-with-h7n9-the-new-avian-influenza/ (accessed July 10, 2013).

Taubenberger, J. K. and D. M. Morens. 2006. 1918 influenza: The mother of all pandemics. *Emerging Infectious Diseases* 12:15–22.

Taubenberger, J. K., A. H. Reid, and T. G. Fanning. 2000. The 1918 influenza virus: a killer comes into view. *Virology* 274:241–245.

Van Campen, H. and G. Early. 2001. Orthomyxovirus and paramyxovirus infections. In: E. S. Williams and I. K. Barker, editors. *Infectious Diseases of Wild Mammals*, 3rd edition. Iowa State University Press, Ames, IA, pp. 271–279.

WHO. 2005a. Avian influenza: Assessing the pandemic threat, WHO/CDS/2005.29. World Health Organization, Geneva, Switzerland.

WHO. 2005b. Avian influenza A (H5N1) infection in humans. *New England Journal of Medicine* 353:1374–1385.

WHO. 2012a. Global alert and response (GAR): What is the pandemic (H1N1) 2009 virus? http://www.who.intt/csr/disease/swineflu/frequently_asked_questions/about_disease/en/index.html (accessed September 1, 2013).

WHO. 2012b. H5N1 avian influenza: timeline of major events. www/who.int/influenza/H5N1_avian_influenza_update_20121217b.pdf timeline. (accessed July 10, 2013).

WHO. 2013a. Avian influenza—Fact sheet. http://www.who.int/mediacentre/factsheets/avian_influenza/en/ (accessed July 10, 2013).

WHO. 2013b. China H7N9 joint mission report. Executive summary. http://www.who.int/influenza/human_animal_interface/en/ (accessed July 10, 2013).

WHO. 2013c. Cumulative number of confirmed human cases of avian influenza A (H5N1) reported to WHO. http://www.who.int/influenza/human_animal_interface/H5N1_cumulative_table_archives/en/index.html (accessed July 10, 2013).

WHO. 2013d. Human infection with avian influenza A (H7N9) virus—Update. http://www.who.int/csr/don/2013_07_04/en/index.html (accessed July 10, 2013).

Section V

Fungal Diseases

25 Cryptococcosis

Because *Cryptococcus* is common in the environment, most people probably breathe in small amounts of microscopic, airborne spores every day. Sometimes these spores cause symptoms of a respiratory infection.

CDC (2010)

25.1 INTRODUCTION AND HISTORY

Cryptococcosis is a human disease caused by the fungal species in the genus *Cryptococcus*. In Latin, the name means hidden spheres: a name selected because the fungi are both small and spherical (Figure 25.1). The species was isolated from a German patient during the 1890s. *Cryptococcus* are yeastlike, monomorphic fungi, which reproduce by budding. The spherical yeasts are enclosed by a thick capsule. There are over 30 species in the genus, but nearly all human cases of cryptococcosis result from just two species: *C. neoformans* and *C. gattii*. Both species are found in soils across the world. *C. neoformans* is associated with soil that has been enriched with bird feces, especially pigeon feces. *C. neoformans* cryptococcosis is rare among healthy people; the annual incidence rate is between 0.4 and 1.3 cases per 100,000, but the rate is much higher among people with HIV/AIDS. Among them, the annual incidence is 200–700 cases per 100,000 among people in the United States (CDC 2012a, 2012b).

C. gattii is found in soil where eucalyptus trees grow (Mitchell and Perfect 1995). The annual incidence rate for *C. gattii* cryptococcosis is 1 per 100,000 people in Australia where there is a plethora of eucalyptus trees (Ellis and Pfeifer 1990, Chen et al. 2012). It has also been detected in other tropical and subtropical countries where eucalyptus trees are grown such as Spain, Italy, South America, and parts of Africa and Asia. It also occurs in southern California (Cafarchia et al. 2006).

Most people across the world probably inhale *Cryptococcus* on a daily basis with no adverse effect, but some people become ill after inhaling the pathogen. In sub-Saharan Africa, cryptococcal meningitis is the leading cause of death among HIV/AIDS patients but is uncommon among healthy people. Worldwide, there are a million new cases of cryptococcal meningitis annually. While most cases occur in Africa (700,000 annually), there are 70,000 new cases annually in the Western Hemisphere, 28,000 in Europe and Central Asia, and 134,000 in the rest of Asia. There are 625,000 fatalities annually from cryptococcal meningitis across the world. Patients who survive this disease can have long-lasting or permanent neurological problems (Mitchell and Perfect 1995, CDC 2010, 2012b).

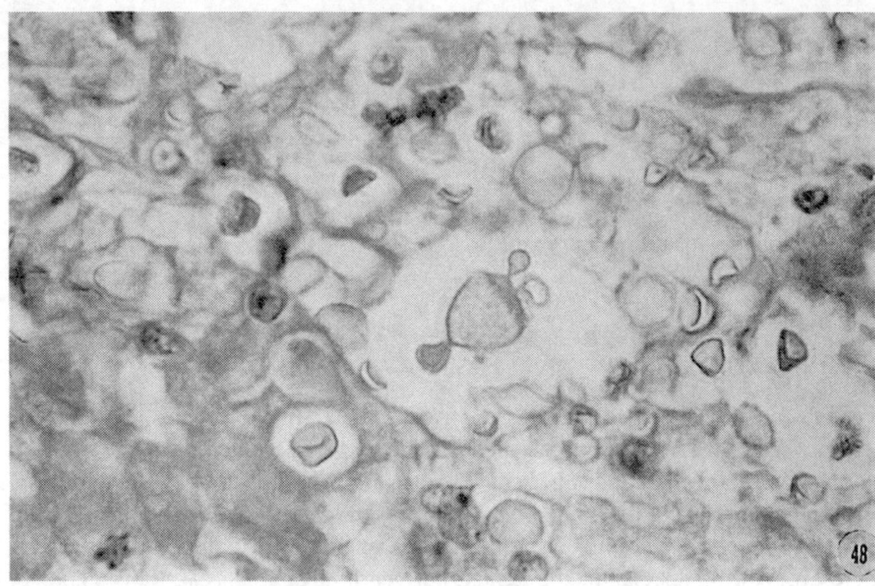

FIGURE 25.1 A micrograph showing *C. neoformans* that appear in the photo as grains of sand. (Courtesy of Lucille Georg and the CDC.)

25.2 SYMPTOMS IN HUMANS

In some people, *Cryptococcus* are able to infect their lungs when inhaled and produce symptoms similar to flu. Common symptoms are fever, cough, chest pains, and shortness of breath. In some patients, especially those with a weakened immune system, the fungus spreads from the lungs to other organs, such as the bones, joints, skin, digestive tract, and, most serious of all, meninges of the central nervous system resulting in cryptococcal meningitis, which produces symptoms of headache, neck ache, sensitivity to light, nausea, lethargy, confusion, and unconsciousness. *Cryptococcus* can also cause a latent infection during which the pathogen remains dormant in the body and does not cause disease until later (Mitchell and Perfect 1995, CDC 2010).

A *Cryptococcus* infection can be diagnosed by culturing the fungus from a blood or tissue sample. A faster diagnosis can be made using a microscopic examination or testing for antibodies against *Cryptococcus*. These methods, however, cannot distinguish between *C. neoformans* and *C. gattii* infections. Doing that requires specialized laboratory tests that are available through state or federal health agencies (Mitchell and Perfect 1995, CDC 2010).

There are a few differences in cryptococcosis caused by *C. gattii* and *C. neoformans*. *C. gattii* cryptococcosis is more likely to strike healthy people, while most patients with *C. neoformans* cryptococcosis have a weakened immune system resulting from AIDS or cancer. Pneumonia is more likely to result in patients infected with *C. gattii* than *C. neoformans* (Harris et al. 2011, Chen et al. 2012).

During 1999, an unusual genotype of *C. gattii* was found in Vancouver Island, Canada, where there were no eucalyptus trees (Sidebar 25.1). Later, it was detected

SIDEBAR 25.1 AN OUTBREAK OF *C. GATTII* CRYPTOCOCCOSIS IN VANCOUVER ISLAND, CANADA (KIDD ET AL. 2004, LESTER ET AL. 2004, MACDOUGALL ET AL. 2007)

C. gattii cryptococcosis has been regarded as a disease of Australia and other tropical and subtropical countries where eucalyptus trees occur. This view has changed since an outbreak of *C. gattii* began in Vancouver Island during 1999 and spread to Washington and Oregon, areas where eucalyptus trees are uncommon. From 2004 until 2011, approximately 100 people in the United States have been documented with *C. gattii* infections; 83 of them were located in Washington and Idaho, and the Vancouver strain was isolated from 78. The annual incidence of *C. gattii* among people living in Vancouver Island from 1999 to 2003 was between 0.9 and 3.7 cases per 100,000 residents, a rate much higher than the 0.1 incidence rate in Australia where *C. gattii* is endemic. Most cases have occurred along the east coast of Vancouver Island within the coastal Douglas-fir biogeoclimatic zone. This zone is limited to a narrow range from sea level to 500 ft (152 m) elevation (Figure 25.2), and *C. gattii* has been detected in the area's air, soil, freshwater, and seawater. *C. gattii* has been isolated from dogs, cats, and an unusual group of British Columbia wildlife including eastern gray squirrels and Dall's porpoises (Figures 25.2 and 25.3).

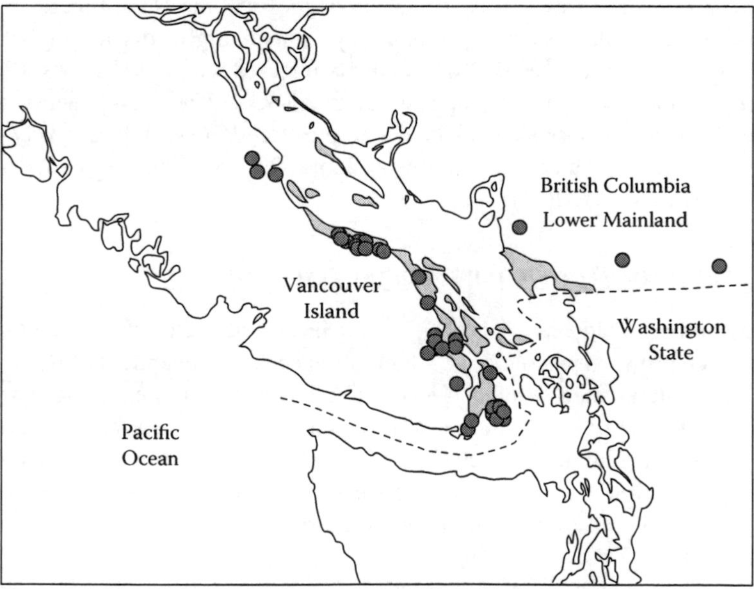

FIGURE 25.2 Map of the southwest corner of British Columbia and Vancouver Island. The coastal Douglas-fir biogeoclimatic zone is shown in gray and the black dots represent where human cases of cryptococcosis were detected prior to 2002. (Data from MacDougall, L. et al., *Emerg. Infect. Dis.*, 13, 42, 2007; Kidd, S.E. et al., *Proc. Natl. Acad. Sci.*, 101, 17258, 2004.)

FIGURE 25.3 Photo of a Dall's porpoise. (Courtesy of the NOAA National Marine Fisheries Service, Southwest Fisheries Science Center, Protected Resources Division.)

elsewhere in British Columbia, Canada, and in the United States by 2004. There have only been about 100 cases of *C. gattii* cryptococcosis in the United States between 2004 and 2011, but there is concern that it may be an emerging disease in the United States and Canada (Harris et al. 2011). Patients infected with the Vancouver stain of *C. gattii* were more likely to develop respiratory disease (75%) than patients infected with the *C. gattii* strain associated with eucalyptus trees (36%) and less likely to have symptoms involving the central nervous system (37% and 90%, respectively; Kidd et al. 2004, Harris et al. 2011).

25.3 *CRYPTOCOCCUS* INFECTIONS IN ANIMALS

Feral and domestic pigeons are both reservoir hosts and vectors for *C. neoformans*. The pathogen grows in pigeon feces, which often accumulate under nests and roosts, and in soil contaminated by pigeon feces. The link between the pathogen and bird feces results because birds excrete it in feces and because avian feces contain high concentrations of creatinine, which *C. neoformans* can digest but other yeast species cannot (Burek 2001). *C. neoformans* has also been isolated from the intestines of common buzzards and common kestrels, suggesting that birds of prey can also serve as reservoir hosts for the pathogen.

The encapsulated cells of *Cryptococcus* become airborne easily when dry, especially when the soil or fecal material is disturbed. Animals normally become infected with *Cryptococcus* by inhaling it but the pathogen can also infect skin wounds. *Cryptococcus* cells are small enough to penetrate the alveolus of the lungs. In healthy birds, inhalation of *Cryptococcus* does not cause illness, or the infection is limited to the upper respiratory tract, including the throat and nasal cavity, causing only minor signs, such as nasal discharge. In birds with a weakened immune system, however, an infection can develop in the lungs and disseminate to other body organs with lethal consequences (Rosario et al. 2005, Burek 2001, Cafarchia et al. 2006).

A wide variety of wild mammals, such as primates, bats, dolphins, and shrews, can be infected with *C. neoformans* or *C. gattii*. *Cryptococcus* infections in mammals normally produce a generalized form of disease, but they can develop into a progressive respiratory disease with fever, nasal discharge, chronic cough, and weight loss. *Cryptococcus* can further cause meningoencephalitis in mammals, which can be fatal (Spencer et al. 1993, Ajello and Padhey 1994, Burek 2001, Merck 2010, Cafarchia 2012). *C. gattii* causes disease in koalas; this phenomenon is interesting because both the pathogen and koalas are associated with eucalyptus trees and are endemic to Australia. Most infections occur in the respiratory tract of koalas, but in 37% of koalas, the infection spreads to the central nervous system (Krockenberger et al. 2003).

Among livestock and companion animals, *Cryptococcus* infections are most common in domestic cats but also occur in dogs, horses, sheep, goats, and cattle. Clinical signs in cats include skin lesions, sneezing, nasal discharge, and a swelling over the bridge of the nose. The pathogen can spread to the central nervous system of cats, causing changes in mood, circling behavior, or seizures. Retinal detachment and blindness can result if the eyes become infected. In cows, infections of *Cryptococcus* occur in the mammary tissue and the lymph nodes that drain them; the infections can cause a loss of appetite and a reduction of milk production but are not fatal in cattle. In horses, *Cryptococcus* infects the respiratory tract and can result in growths within the nasal cavities; horses can die if the infection spreads to their central nervous system. *Cryptococcus* infections in companion animals and livestock can be treated with antibiotics. Fluconazole or itraconazole are usually the antibiotics of choice, but amphotericin B can also be administered (Merck 2010).

25.4 HOW HUMANS CONTRACT CRYPTOCOCCOSIS

People become infected by inhaling the fungus. *Cryptococcus* is so common in soils across the world that most people inhale it on a daily or weekly basis. Obviously, most people do not become ill from doing so; sometimes, however, the pathogen infects the respiratory tract and results in a flu-like illness. The risk of developing cryptococcosis varies based on the virulence of the pathogen, the amount inhaled, and the person's ability to fight off the infection. Cryptococcosis is not a contagious disease; a person cannot become infected through contact with an infected person.

25.5 MEDICAL TREATMENT

The CDC (2012a, 2012b) recommends that people see their doctor if they think they may have cryptococcosis. The disease can be treated with antifungal medicines, such as fluconazole or itraconazole, but treatment takes at least 6 months and sometimes much longer. Cryptococcal meningitis can be treated with amphotericin B usually in combination with flucytosine. Treating HIV/AIDS patients with highly active antiretroviral treatment (HAART) has reduced their vulnerability to cryptococcal meningitis by strengthening their immune systems. HAART treatments are less available in developing countries. A more cost-effective approach being used in Africa is to take a small blood sample from HIV-infected people and test it for antibodies against *Cryptococcus*. The test's results are available in only 10 minutes, and the test detects more than 95% of all infections, even in people that have not yet developed cryptococcosis. When the tests are positive, patients are given fluconazole orally to eliminate the infection before it can develop into cryptococcal meningitis. Among HIV/AIDS patients, the mortality rate from cryptococcal meningitis is about 12% in the United States and Canada and above 50% in sub-Saharan Africa (Perfect et al. 2010, CDC 2010, 2012b).

25.6 WHAT PEOPLE CAN DO TO REDUCE THE RISK OF CONTRACTING CRYPTOCOCCOSIS

People infected with HIV or with blood ailments, such as diabetes, leukemia, or Hodgkin's disease, have an elevated risk of contracting cryptococcosis and for the disease developing into cryptococcal meningitis. Consequently, these people should avoid contact with birds and areas where bird feces might accumulate. In developing countries where cryptococcal meningitis is prevalent, access to health care needs to be improved so that HIV patients can be routinely screened for *Cryptococcus*, and those who are infected can obtain antifungal drugs (Ajello and Padhye 1994).

People who raise pigeons obviously have a high risk of developing cryptococcosis. Occupational workers who have an enhanced risk of contracting cryptococcosis are building inspectors, demolition workers, bridge inspectors, street sweepers, poultry farmers, and pest control workers.

Pigeon feces should not be allowed to accumulate. Areas where they have accumulated should be cleaned by professionals trained in the removal of hazardous waste. Use of a respirator and mask can reduce one's risk of inhaling pathogens when working at hazardous sites. People should consult the U.S. National Institute for Occupational Safety and Health for information about which masks are effective and how to use them. Before starting work at a site that may contain *Cryptococcus*, the soil at roost sites should be sprayed thoroughly with water and kept wet. Roost sites that have accumulations of pigeon feces can be decontaminated with a 5% solution of sodium hypochlorite (Ajello and Padhye 1994).

25.7 ERADICATING CRYPTOCOCCOSIS FROM A COUNTRY

Cryptococcus is so common in soils that most people inhale it on a daily or weekly basis. The ubiquity of this pathogen makes it possible to eradicate. Across the world, most victims of this pathogen have weakened immune systems, often due to HIV infections. Efforts to protect people from cryptococcosis are directed at preventing HIV/AIDS.

LITERATURE CITED

Ajello, L. and A. A. Padhye. 1994. Systemic mycoses. In: G. W. Beran, editor-in-chief. *Handbook of Zoonoses*, 2nd edition. CRC Press, Boca Raton, FL, pp. 483–504.

Burek, K. 2001. Mycotic diseases. In: E. S. Williams and I. K. Barker, editors. *Infectious Diseases of Wild Mammals*. Iowa State University, Ames, IA, pp. 514–531.

Cafarchia, C. 2012. Yeast infections. In: D. Gavier-Widén, J. P. Duff, and A. Meredith, editors. *Infectious Diseases of Wild Mammals and Birds in Europe*. Wiley-Blackwell, West Sussex, U.K., pp. 462–465.

Cafarchia, C., D. Romito, R. Iatta, A. Camarda et al. 2006. Role of birds of prey as carriers and spreaders of *Cryptococcus neoformans* and other zoonotic yeasts. *Medical Mycology* 44:485–492.

CDC. 2010. *Cryptococcal Meningitis: A Deadly Fungal Disease among People Living with HIV/AIDS*. CDC, National Center for Emerging and Zoonotic Infections Diseases, Atlanta, GA.

CDC. 2012a. *C. gattii* cryptococcosis. http://www.cdc.gov/fungal/cryptococcosis-gattii/information.html (accessed July 7, 2013).

CDC. 2012b. *C. neoformans* cryptococcosis. http://www.cdc.gov/fungal/cryptococcosis-neoformans/information.html (accessed July 7, 2013).

Chen, S. C.-A., M. A. Slavin, C. H. Heath, E. G. Playford et al. 2012. Clinical manifestations of *Cryptococcus gattii* infection: Determinants of neurological sequelae and death. *Clinical Infectious Diseases* 55:789–798.

Ellis, D. H. and T. J. Pfeiffer. 1990. Natural habitat of *Cryptococcus neoformans* var. *gattii*. *Journal of Clinical Microbiology* 28:1642–1644.

Harris, J. R., S. R. Lockhart, E. Debess, N. Marsden-Haug et al. 2011. *Cryptococcus gattii* in the United States: Clinical aspects of infection with an emerging pathogen. *Clinical Infectious Diseases* 53:1188–1195.

Kidd, S. E., F. Hagen, R. L. Tscharke, M. Huynh et al. 2004. A rare genotype of *Cryptococcus gattii* caused the cryptococcosis outbreak on Vancouver Island (British Columbia, Canada). *Proceedings of the National Academy of Science* 101:17258–17263.

Krockenberger, M. B., P. J. Canfield, and R. Malik. 2003. *Cryptococcus neoformans* var. *gattii* in koala (*Phascolarctos cinereus*): A review of 43 cases of cryptococcosis. *Medical Mycology* 41:225–234.

Lester, S. J., N. J. Kowalewich, K. H. Bartlett, M. B. Krockenberger et al. 2004. Clinicopathologic features of an unusual outbreak of cryptococcosis in dogs, cats, ferrets and a bird: 38 cases (January 2003 to July 2003). *Journal of the American Veterinary Medical Association* 225:1716–1722.

MacDougall, L., S. E. Kidd, E. Galanis, S. Mak et al. 2007. Spread of *Cryptococcus gattii* in British Columbia, Canada, and detection in the Pacific Northwest, USA. *Emerging Infectious Diseases* 13:42–50.

Merck. 2010. *Merck Veterinary Manual*, 10th edition. Merck, Whitehouse Station, NJ.

Mitchell, T. G. and J. R. Perfect. 1995. Cryptococcosis in the era of AIDS—100 years after the discovery of *Cryptococcus neoformans*. *Clinical Microbiology Reviews* 8:515–548.

Perfect, J. R., W. E. Dismukes, F. Dromer, D. L. Goldman et al. 2010. Clinical practice guidelines for the management of cryptococcal disease: 2010 update by the Infectious Diseases Society of America. *Clinical Infectious Diseases* 50:291–322.

Rosario, I., M. Hermoso de Mendoza, S. Déniz, G. Soro et al. 2005. Isolation of *Cryptococcus* species including *C. neoformans* from cloaca of pigeons. *Mycoses* 48:421–424.

26 Histoplasmosis

> [Histoplasmosis] is highly endemic in the Ohio and Mississippi river valleys of the United States. An estimated 40 million people in the United States have been infected with *H. capsulatum*, with 500,000 new cases occurring each year.
>
> **Kurowski and Ostapchuk (2002)**

26.1 INTRODUCTION AND HISTORY

Histoplasmosis is an infectious disease caused by the fungus *Histoplasma capsulatum*. The fungus lives in soil but also survives by infecting animals and people. Other names for the disease include cave sickness and Ohio Valley disease. There are three varieties of *H. capsulatum*. One of them (*H. capsulatum* variety *farciminosum*) only infects horses, donkeys, and mules; this variety occurs in Africa, the Middle East, and Asia. *H. capsulatum* variety *duboisii* only occurs in Africa where it infects humans and baboons. The third variety (*H. capsulatum* variety *capsulatum*) has a worldwide distribution and infects humans and numerous animal species (Ajello and Padhye 1994). In this chapter, only the third variety will be discussed; whenever we mention *H. capsulatum*, we are referring only to *H. capsulatum* variety *capsulatum*.

H. capsulatum is a dimorphic fungus that has two life forms. The first form is a mycelia (mold) stage that occurs when the fungus is growing in soil at ambient temperatures (Figure 26.1). The mycelia produce two types of spores: microconidia, which are small spores (2–5 μm in diameter), and macroconidia, which are large spores (8–15 μm) and have thick walls and outgrowths. When growing in soil, many of the mycelia protrude from the ground and produce spores. Because of this, many spores are released into the air, especially when the soil surface is disturbed. If these spores are inhaled by a susceptible host, the warmer temperatures inside the lungs cause *H. capsulatum* to develop into the second form, which has the appearance of oval or round yeast cells with diameters of 1–5 μm, and regenerate by budding (Figure 26.2). The host's immune cells respond to the yeast cells in the lungs by engulfing them (phagocytosis) and carrying them to the lymph nodes. Once there, the yeasts can travel in the blood to many other organs and infect them (Burek 2001, Kauffman 2007, CDC 2013).

H. capsulatum was first identified in the Panama Canal Zone by an American physician Samuel T. Darling, during an autopsy of a patient. Darling believed that the patient had died from a disease caused by a protozoan infection. In 1912, a physician in Germany recognized that the microorganism described by Darling was not a protozoa, but a yeast. Hence, this disease was first called "Darling's disease" despite Darling's misdiagnosis. Then, in the 1930s, two physicians, Dodd and Tompkins, diagnosed histoplasmosis in a sick child. The pathogen was isolated, cultured, and identified as a fungus by De Monbreun at Vanderbilt University. The same pathogen

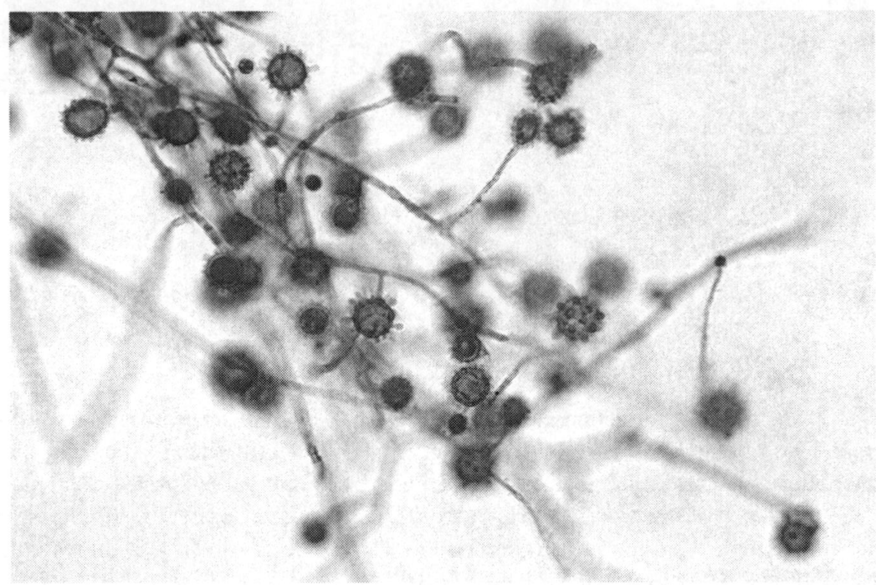

FIGURE 26.1 A micrograph showing the mycelium life form of *H. capsulatum*. The round objects with projections sticking out from them are macrospores. (Courtesy of Libero Ajello and the CDC.)

was first isolated from wild mammals during 1947 and from soil during 1948 (Kauffman 2007, Fischer et al. 2009, Harnalikar et al. 2012).

H. capsulatum has a worldwide distribution but some regions are more heavily infested than others. In the United States, histoplasmosis is prevalent in the Ohio–Mississippi–Missouri river valleys and in the American Midwest (Figure 26.3). In Canada, the St. Lawrence River Valley is heavily infested. In Latin America, histoplasmosis is common in Mexico, Guatemala, Venezuela, and Peru (Wilson 1991, Ajello and Padhye 1994).

The incidence of histoplasmosis in humans is unknown because most people do not seek medical attention or the disease is misdiagnosed as flu. Between 50% and 80% of people who live in areas where *H. capsulatum* is common in the environment will become infected at some point in their lives, but few will realize it (CDC 1995). Histoplasmosis is not a notifiable disease in the United States, so data on its incidence are scarce. According to Kurowski and Ostapchuk (2002) and the University of Maryland Medical Center (2013), approximately 500,000 people are infected with *H. capsulatum* each year, and an estimated 40 million people in the United States have been infected with H. *capsulatum* at some time in their lives.

26.2 SYMPTOMS IN HUMANS

More than 90% of individuals infected with the *H. capsulatum* do not exhibit symptoms. For the few who do, symptoms begin 3–17 days after exposure to the fungus. People become infected with *H. capsulatum* after inhaling its spores, so

(a)

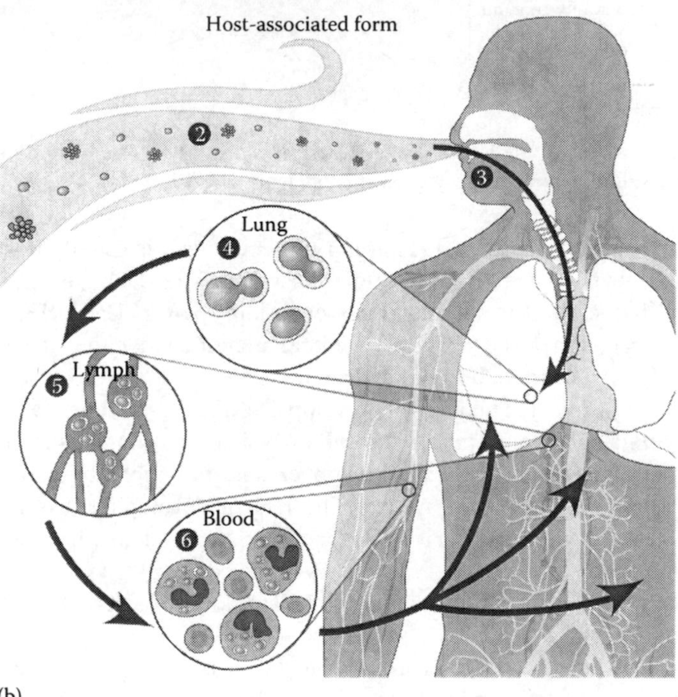

(b)

FIGURE 26.2 In the environment, *H. capsulatum* exists as a mold with filaments (a). The filaments or mycelia produce both large and small spores that become airborne. When inhaled by a susceptible host, the spores develop into a yeastlike form and reproduce by budding (b). From the lungs, the pathogens are transported to the lymph nodes by immune cells. Once there, they are carried by the blood to other body organs. (From CDC, Histoplasmosis, http://www.cdc.gov/fungal/histoplasmosis/, 2013.)

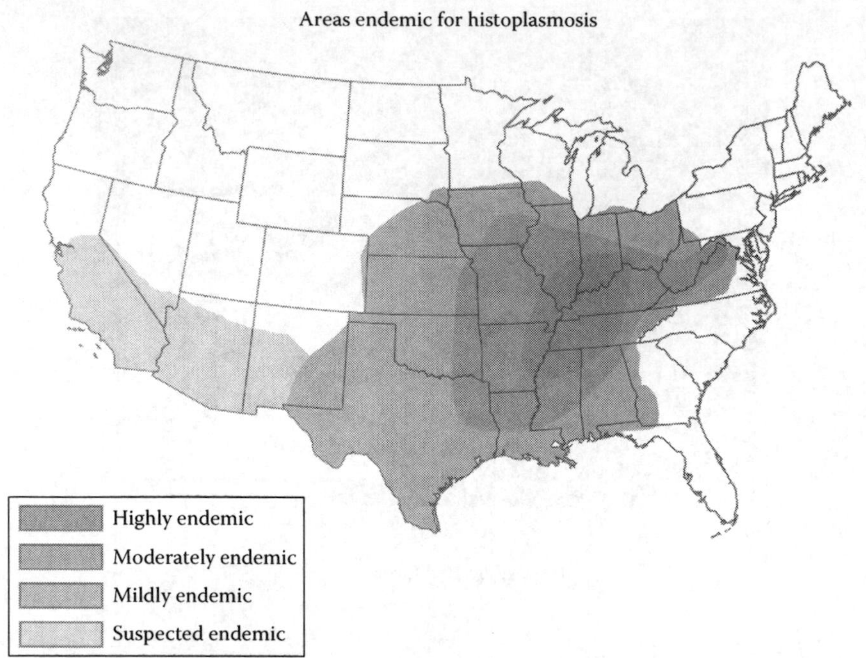

FIGURE 26.3 Areas in the United States where *H. capsulatum* is endemic. (From CDC, Histoplasmosis, http://www.cdc.gov/fungal/histoplasmosis/, 2013.)

histoplasmosis usually starts as a respiratory infection and is usually mistaken for a common cold or flu. Symptoms are often nonspecific (headache, fever, dry cough, and nausea) but some patients may experience joint pain (CDC 2013). The Mayo Clinic (2012) recommends people seek medical attention when they develop flu-like symptoms after exposure to bird or bat droppings.

One percent of infected people will develop a serious, sometimes fatal, condition called disseminated histoplasmosis; it results when *H. capsulatum* spreads from the lungs to other organs in the body including liver, spleen, lymph nodes, bone marrow, eyes, skin, and central nervous system. The risk of developing disseminated disease is greatest for people who have weakened immune systems. In areas where the *H. capsulatum* is common, 10%–25% of HIV-infected people have experienced disseminated histoplasmosis (Kauffman 2007, Dylewski 2011, Harnalikar et al. 2012, Mayo Clinic 2012).

Diagnosis of histoplasmosis is usually made by isolating *H. capsulatum* from blood or tissue and culturing the fungus. Histoplasmosis also can be diagnosed by testing urine and skin or examining a small sample of infected tissue under a microscope (Cafarchia et al. 2012).

26.3 *HISTOPLASMA CAPSULATUM* INFECTIONS IN ANIMALS

The primary reservoir of *H. capsulatum* is soil contaminated with bird and bat droppings. The fungus can survive in the soil for years and thrives in caves where the soil has been enriched by bat guano (Lenhart et al. 2004). *H. capsulatum* infects dogs, cats, bats, carnivores, marsupials, rodents, and ungulates. Clinical signs of disease caused by *H. capsulatum* are rare in all species except cats and dogs. Animals are infected by inhalation, creating an infection in the respiratory tract and thoracic lymph nodes. The pathogen can spread throughout the body including the liver, spleen, bone marrow, and skin. If the skin is involved, nodular lesions can result. Clinical signs of infection in animals vary depending upon which organs have become infected. The digestive system often is the primary infection site in dogs. Clinical signs in dogs include fever, weight loss, persistent diarrhea, increased respiratory rate, and chronic cough. The disease can be fatal in dogs. In cats, clinical signs include fever, anorexia, weight loss, and difficulty breathing. In both cats and dogs, conjunctivitis and retinal detachment can occur if the disease spreads to the eyes (Menges et al. 1967, Brömel and Sykes 2005, Merck 2010).

H. capsulatum has been isolated from over 25 bat species. The course of disease in bats is similar to that in humans. Infection usually results from inhalation of the pathogen and begins in the lungs. From there, the fungus can spread to other organs, including the liver, spleen, and gastrointestinal tract. Infected bats shed *H. capsulatum* in the feces and carry it on their wings, feet, and body surface. Bats are known to infect the guano-enriched soil under their roost and colonies (Tesh and Schneidau 1966).

Blackbirds (e.g., starlings, grackles, red-winged blackbirds, and cowbirds) form large roosts in the fall and winter containing tens of thousands of birds. They also forage and loaf during winter in flocks of hundreds or thousands at dairy farms and livestock feedlots where the birds consume grain intended for the livestock. Roosts and feeding sites are visited daily during the fall and winter, and the soil soon contains enough bird feces that *H. capsulatum* can thrive. Likewise, starlings and pigeons are a vector for *H. capsulatum* in urban and suburban areas (Lenhart et al. 2004).

26.4 HOW HUMANS CONTRACT HISTOPLASMOSIS

Vegetative forms of *H. capsulatum* grow in soil enriched with the feces of birds or bats. Such soil exists at sites where there are or have been large concentrations of birds such as roost sites, bird breeding colonies, or poultry farms. *H. capsulatum* can remain in the soil for years so that abandoned roosts and colonies can be a source of infection although no evidence of the roost or colony remains. In areas where *H. capsulatum* is endemic, the spores are released constantly into the air from contaminated soil, and human cases of histoplasmosis occur regularly. The incidence of histoplasmosis varies little from 1 year to the next, but local outbreaks of histoplasmosis can also occur when activities, such as plowing, soil excavation, building construction, or demolition, disturb the ground surface, releasing large quantities of *H. capsulatum* into the air. Wind can move the spores for kilometers, meaning that people distant from a bird roost can develop histoplasmosis. Farmers, bulldozer operators, bridge workers, poultry-farm workers, pigeon fanciers, demolition crews, construction crews, and landfill employees all have an increased risk of being exposed to *H. capsulatum* spores and developing histoplasmosis (Huhn et al. 2005, Wheat et al. 2007). *H. capsulatum* can also thrive in soil enriched with

SIDEBAR 26.1 EXPLORATION OF A CAVE RESULTS IN AN OUTBREAK OF HISTOPLASMOSIS (LOTTENBERG ET AL. 1979)

During February 1973, 29 people on a church outing explored a limestone cave located in central Florida. Two weeks later, one of the participants, an 18-year-old woman, sought medical attention for a respiratory illness that resulted in shortness of breath and cyanosis (i.e., a bluish skin color resulting from a lack of oxygen). The following day, another participant, an 18-year-old male, visited the local hospital with similar symptoms. Both were diagnosed with acute pulmonary histoplasmosis. Further investigation revealed that 23 of the 29 people on the outing had symptoms of cough, afternoon fever and sweats, chest pains, and shortness of breath on exertion. Histoplasmin skin tests were conducted on 24 individuals who were part of the group, and 18 people tested positive for past exposure to the fungus. Pulmonary histoplasmosis is often found in people who spend time in caves where bats roost (Figure 26.4). The experiences of this youth group were unique only in the number of people infected simultaneously.

bat feces. For this reason, people who visit bat caves have an elevated risk of developing histoplasmosis (Sidebar 26.1).

The incidence of histoplasmosis is highest in humans aged 15 and 34 years old because these people are more likely to be engaged in occupations or activities that increase their exposure to the pathogen. Histoplasmosis is more prevalent in males than females, but there are no sexual differences in predisposition to developing histoplasmosis. Rather, the differences between males and females in the incidence to the disease are due to sexual differences in exposure. There is no evidence of direct transmission of histoplasmosis from one animal to another or

FIGURE 26.4 People who explore bat caves have an increased risk of contracting histoplasmosis. (Courtesy of the U.S. Department of Energy, Idaho National Laboratory.)

from one animal to a human. Instead, concurrent outbreaks of histoplasmosis in both humans and animals occur when both have been exposed to the same source of the pathogen (Wittler 2007).

The risk of any one person developing histoplasmosis increases with the infection dose because inhaling many spores at any one time can overwhelm the person's natural resistance. Infants, young children, the elderly, and people who have chronic obstructive pulmonary disease are at increased risk of histoplasmosis developing into a severe disease. Finally, the risk of histoplasmosis developing into disseminated histoplasmosis increases in patients who have a weakened immune system.

26.5 MEDICAL TREATMENT

Mild lung infections of histoplasmosis will generally resolve themselves within a month, and treatment is usually not necessary. When symptoms continue beyond a month, itraconazole is often prescribed for patients. For patients with severe acute pulmonary histoplasmosis, amphotericin B may be given intravenously each day for 1 or 2 weeks followed by oral doses of itraconazole for 12 weeks. Complications caused by histoplasmosis are treated as needed. People can develop histoplasmosis more than once in their lives, but prior infections provide some immunity; hence, reinfections are less likely to be serious than the first infection (Wheat 2006, Wheat et al. 2007).

26.6 WHAT PEOPLE CAN DO TO REDUCE THE RISK OF CONTRACTING HISTOPLASMOSIS

There is no vaccine to prevent histoplasmosis; instead, people with weakened immune systems should avoid areas with accumulations of bird or bat droppings. Areas with accumulations of bird or bat droppings should be cleaned by professional companies that specialize in the removal of hazardous waste. Use of a respirator and mask can reduce one's risk of inhaling *H. capsulatum* spores when working at hazardous sites. People should consult the U.S. National Institute for Occupational Safety and Health for information about which masks are effective and how to use them. Before starting work at a site that may contain *H. capsulatum* spores, the soil at roost sites should be sprayed thoroughly with water and kept wet. This will reduce the number of spores that are released into the air. Ajello and Padhye (1994) reported that small areas containing *H. capsulatum* can be decontaminated with a 5% solution of sodium hypochlorite.

The best method to prevent histoplasmosis is to prevent the accumulation of bird or bat feces. Thus, when birds or bats are found roosting in a building, immediate action should be taken to exclude them from the building. Care needs to be taken not to trap birds or bats inside buildings by making sure that all of the bats or birds have exited the building before the last exit is sealed up. A one-way door can also be installed that allows birds and bats to exit a building but prevents them from returning (Conover 2002). Birds and bats should not be excluded when baby chicks or bats are in the building. Birds and bats are protected by state and federal law. Therefore, it is wise to check with both federal and state wildlife agencies before disturbing the animals. More information about how to exclude bats from buildings is available elsewhere (Conover 2002).

26.7 ERADICATING HISTOPLASMOSIS FROM A COUNTRY

In some parts of the United States and the world, *H. capsulatum* is common in soil enriched by avian feces and will be difficult to eradicate from countries where it is endemic. The fungus lives in soil but also survives by infecting animals and people. One of the best approaches to reducing histoplasmosis is to not allow birds to roost in areas where there is an elevated risk of people being exposed to the pathogen. As one example, an elementary school playground was littered with feces from thousands of starlings roosting in the school's trees. The problem was solved by cutting down some of the trees.

LITERATURE CITED

Ajello, L. and A. A. Padhye. 1994. Systemic mycoses. In: G. W. Beran, editor-in-chief. *Handbook of Zoonoses*, 2nd edition. CRC Press, Boca Raton, FL, pp. 483–504.

Brömel, C. and J. E. Sykes. 2005. Histoplasmosis in dogs and cats. *Clinical Techniques in Small Animal Practice* 20:227–232.

Burek, K. 2001. Mycotic diseases. In: E. S. Williams and I. K. Barker, editors. *Infectious Diseases of Wild Mammals*. Iowa State University Press, Ames, IA, pp. 514–531.

Cafarchia, C., K. Eatwell, D. S. Jansson, C. U. Meteyer, and G. Wibbelt. 2012. Other fungal infections. In: D. Gavier-Widén, J. P. Duff, and A. Meredith, editors. *Infectious Diseases of Wild Mammals and Birds in Europe*. Wiley-Blackwell, West Sussex, U.K., pp. 466–475.

CDC. 1995. Histoplasmosis—Kentucky, 1995. *Morbidity and Mortality Weekly Report* 44:701–703.

CDC. 2013. Histoplasmosis. http://www.cdc.gov/fungal/histoplasmosis/ (accessed July 6, 2013).

Conover, M. R. 2002. *Resolving Human-Wildlife Conflicts: The Science of Wildlife Damage Management*. CRC Press, Boca Raton, FL.

Dylewski, J. 2011. Acute pulmonary histoplasmosis. *Canadian Medical Association Journal* 183:e1090.

Fischer, G. B., H. Mocelin, C. B. Severo, F. de Mattos Oliveira et al. 2009. Histoplasmosis in children. *Pediatric Respiratory Reviews* 10:172–177.

Harnalikar, M., V. Kharkar, and U. Khopkar. 2012. Disseminated cutaneous histoplasmosis in an immunocompetent adult. *Indian Journal of Dermatology* 57:206–209.

Huhn, G. D., C. Austin, M. Carr, D. Heyer et al. 2005. Two outbreaks of occupationally acquired histoplasmosis: More than workers at risk. *Environmental Health Perspectives* 113:585–589.

Kauffman, C. A. 2007. Histoplasmosis: A clinical and laboratory update. *Clinical Microbiology Reviews* 20:115–132.

Kurowski, R. and M. Ostapchuk. 2002. Overview of histoplasmosis. *American Family Physician* 66:2247–2253.

Lenhart, S. W., M. P. Schafer, M. Singal, and R. A. Hajjeh. 2004. *Histoplasmosis: Protecting Workers at Risk*. National Institute for Occupational Safety and Health, Washington, DC.

Lottenberg, R., R. H. Waldman, L. Ajello, G. L. Hoff et al. 1979. Pulmonary histoplasmosis associated with exploration of a bat cave. *American Journal of Epidemiology* 110:156–161.

Mayo Clinic. 2012. Histoplasmosis. http://www.mayoclinic.org/health/histoplasmosis/DS00517/Dsection=symptoms (accessed July 10, 2013).

Menges, R. W., M. L. Furcolow, R. T. Habermann, and R. J. Weeks. 1967. Epidemiologic studies on histoplasmosis in wildlife. *Environmental Research* 1:129–144.

Merck. 2010. *Merck Veterinary Manual*, 10th edition. Merck, Whitehouse Station, NJ.

Tesh, R. B. and J. D. Schneidau, Jr. 1966. Experimental infection of North American insectivorous bats (*Tadarida brasiliensis*) with *Histoplasma capsulatum. American Journal of Tropical Medicine and Hygiene* 15:544–540.

University of Maryland Medical Center. 2011. Histoplasmosis. http//:www.umm.edu/health/medical/altmed/condition/histoplasmosis.html (accessed December 23, 2013).

Wheat, L. J. 2006. Histoplasmosis: A review for clinicians from non-endemic areas. *Mycoses* 49:274–282.

Wheat, L. J., A. G. Freifeld, M. B. Kleiman, J. W. Baddley et al. 2007. Clinical practice guidelines for the management of patients with histoplasmosis: 2007 update by the Infectious Diseases Society of America. *Clinical Infectious Diseases* 45:807–825.

Wilson, M. E. 1991. *A World Guide to Infections: Diseases, Distribution, Diagnosis.* Oxford University Press, Oxford, U.K.

Wittler, R. R. 2007. Histoplasmosis. In: L. C. Garfunkel, J. Kaczorowski, and C. Christy, editors. *Pediatric Clinical Advisor.* Mosby Elsevier, Philadelphia, PA, pp. 277–278.

Section VI

Prions

27 Creutzfeldt–Jakob Disease in Humans, Chronic Wasting Disease in Cervids, Mad Cow Disease in Cattle, and Scrapie in Sheep and Goats

Prion diseases, also known as transmissible spongiform encephalopathies (TSEs), are a group of animal and human brain diseases that are uniformly fatal.... Prion diseases attracted much attention and public concern after an outbreak of bovine spongiform encephalopathy (BSE) occurred among cattle in many European countries and scientific evidence indicated the foodborne transmission of BSE to humans.

Belay and Schonberger (2005)

27.1 INTRODUCTION AND HISTORY

Prion diseases are a group of diseases that are caused by prions; the term prion is an abbreviation of "proteinaceous infectious particles." Prions are proteins that occur in cells but can fold abnormally and become misshapen. These misshapen prions have the ability to cause similar proteins to fold abnormally and turn into misshapened prions. Hence, a few misshapened prions in a cell can develop into many, and their accumulation in brain tissue causes cell death. Prions are absorbed by adjacent nerve cells, and the process is repeated. When enough nerve cells die in the brain, tiny holes develop, which give brain tissue the appearance of a sponge when viewed through a microscope (Figure 27.1). Thus, the name for this group of diseases is transmissible spongiform encephalopathy (TSE), which means a disease of the brain that causes it to appear in the form of a sponge. The process of converting normal proteins into prions is very slow, and it can take years between when an animal or human is infected and when it first exhibits symptoms; ultimately, however, enough nerve cells are destroyed that normal brain function is impaired. Prion diseases are always fatal providing

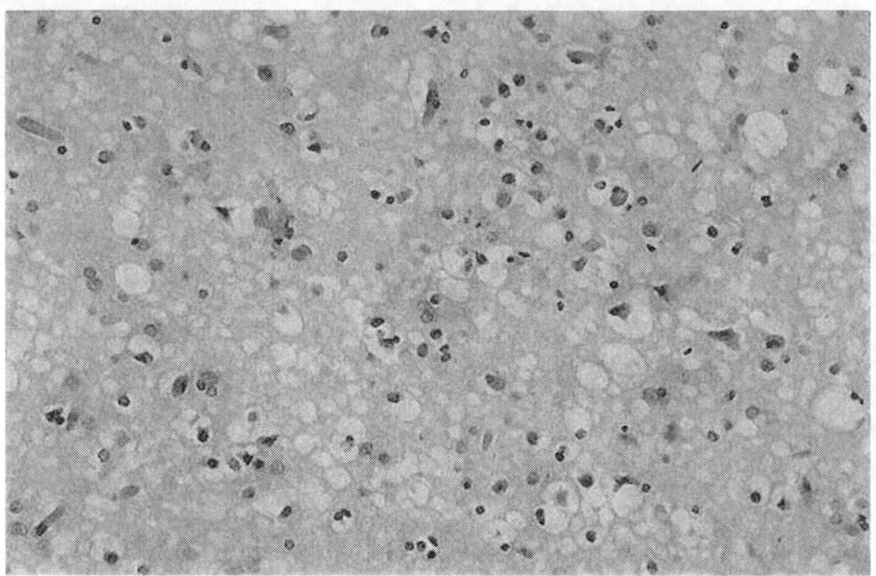

FIGURE 27.1 Brain tissue from an animal with a TSE has tiny holes that give the tissue the appearance of a sponge when viewed under a microscope. (Courtesy of Ermlas Belay and the CDC.)

that the infected animal or person lives long enough. Yet, prions do not produce a response by the immune system because they are not recognized as a foreign agent (Londhe et al. 2012).

Prion diseases occur in a variety of animals as well as humans. Prion diseases include scrapie, which occurs in sheep; bovine spongiform encephalopathy, which occurs in cattle; transmissible mink encephalopathy in captive mink; feline spongiform encephalopathy in cats; chronic wasting disease in deer elk and moose; and Creutzfeldt–Jakob disease and variant Creutzfeldt–Jakob disease, both of which occur in humans (Prusiner 1982, Krauss et al. 2003).

Scrapie was first reported in England during 1732 and has since spread throughout most of the world with the exception of Australia, New Zealand, and a few other countries. The first U.S. record of scrapie occurred during 1947 in Michigan among sheep that had been imported from England via Canada (Travis and Miller 2003). Although related, bovine spongiform encephalopathy took longer to be identified. It was first detected among cattle in the United Kingdom during 1986 and was clinically described 2 years later. The disease was given the common name of mad cow disease by the British Press, and we will use this term because it is easy to remember (Wells et al. 1987, Nathanson et al. 1999, Travis and Miller 2003).

Classical Creutzfeldt–Jakob disease was described during the 1920s by two German neurologists, Creutzfeldt and Jakob, and is named after them. It occurs worldwide but is not an infectious disease. Rather, the disease results from a spontaneous transformation of normal proteins into abnormal prions (85% of cases) or from an inherited mutation of the gene (15% of cases). These inherited forms include fatal familial insomnia, familial Creutzfeldt–Jakob disease, and Gerstmann–Straussler–Scheinker syndrome (Belay 1999, CDC 2010).

Variant Creutzfeldt–Jakob disease was first detected during 1996 when 10 patients with the disease were diagnosed in the United Kingdom. These patients were much younger than most patients with classical Creutzfeldt–Jakob disease, and their symptoms were different. Epidemiologists realized that most patients became infected from consuming infected beef and that a causal relationship existed between mad cow disease and variant Creutzfeldt–Jakob (CDC 2010, Holman et al. 2010).

Chronic wasting disease in deer was first described by researchers at the University of Wyoming in 1967; since then, chronic wasting disease has been diagnosed in wild deer, elk, and moose across the United States and in Saskatchewan and Alberta, Canada. Chronic wasting disease is the only TSE that has been detected in wild animals (Williams and Young 1980, Travis and Miller 2003).

27.2 SYMPTOMS IN HUMANS

Classical Creutzfeldt–Jakob disease is the most common TSE in humans, but it is still a rare disease. The annual incidence rate in both the United States and the world is about one case of Creutzfeldt–Jakob disease per million people each year. In the United States, between 150 and 350 cases occur annually (Figure 27.2). Early clinical signs include abnormal reflexes, tremors, uncoordinated muscles, difficulty walking, and speech abnormalities. Nearly all (90%) patients exhibit muscle twitching. Patients may be confused, depressed, or agitated. As the disease progresses, clinical signs include memory loss and dementia. During the final stages of the disease, patients become immobile and mute. Once clinical signs develop, the disease

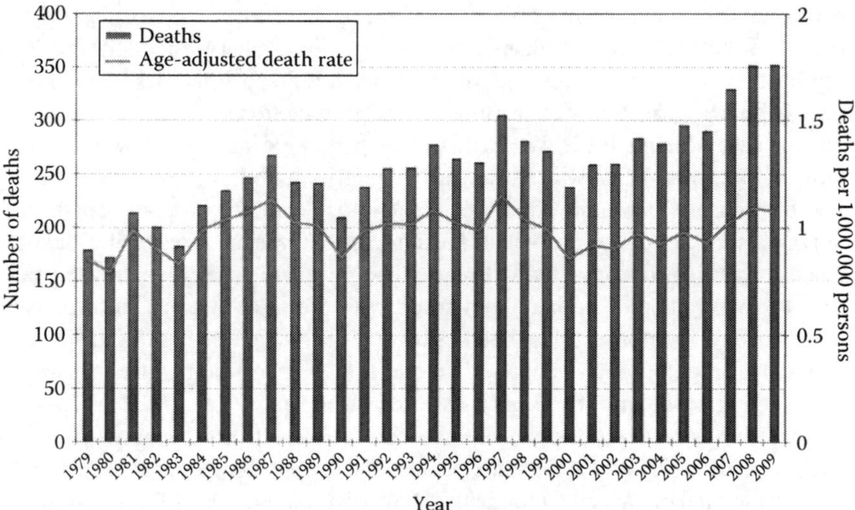

FIGURE 27.2 Number of U.S. deaths annually from Creutzfeldt–Jakob disease and the annual fatality rate per one million persons. (From CDC, vCJD (variant Creutzfeldt-Jakob disease), http://www.cdc.gov/ncidod/dvrd/cjd/, 2010.)

rapidly progresses to dementia; 85% of patients die within a year of developing clinical signs (Belay 1999, WHO 2006, CDC 2012).

Variant Creutzfeldt–Jakob disease differs from the classical form of the disease in that most patients with the variant form were infected from consuming meat from an animal with mad cow disease (Sidebar 27.1). Most cases occurred in the United

SIDEBAR 27.1 HISTORY OF MAD COW DISEASE OUTBREAK IN THE UNITED KINGDOM (WILL ET AL. 1996, BROWN ET AL. 2001, CUMMINGS 2011, GAVIER-WIDÉN 2012)

The outbreak of bovine spongiform encephalopathy or mad cow disease began in the United Kingdom during the 1980s when an increase in the price of protein led cattle producers in the United Kingdom to replace soybean and fish meal as a protein supplement with a meat and bone meal products from livestock carcasses. Unfortunately, some of these carcasses are now thought to have been cattle that were infected with mad cow disease; thus, the disease spread. Mad cow disease was first detected in UK cattle during 1986. A year later, John Wilesmith was asked to investigate the cause of the disease, which had been named mad cow disease by the press. He determined that it resulted from the use of contaminated meat and bone meal in cattle feed. His findings resulted in a ban on the use of bovine offal in feed during 1988. At the time, the infectious agent responsible for mad cow disease was not considered to result in human illness because scrapie only caused disease in sheep and goats. However, on May 6, 1990, the U.K. Department of Health learned that a new TSE had been diagnosed in a domestic cat. This was a bombshell because it indicated that the assumption that TSEs could not occur in species other than livestock was false. Under increasing public concern about the health implications of mad cow disease, the U.K. Health Department issued a press release on May 16, 1990, reaffirming that beef was safe to eat. Sadly, this was not the case; during 1995, two British teenagers were diagnosed with a previously unknown TSE, which was subsequently named variant Creutzfeldt–Jakob disease. On March 20, 1996, the UK government officially reported during a speech before the British Parliament that variant Creutzfeldt–Jakob disease was linked to mad cow disease. The government also acknowledged that 10 young people had been diagnosed with the fatal disease. In response to the mad cow outbreak, the United Kingdom killed nearly 185,000 diseased cattle and preemptively slaughtered an additional 4.5 million cattle that were asymptomatic but may have been exposed to the pathogen. These steps to remove the prions responsible for mad cow disease from the human foods were, nonetheless, too late for some people. The United Kingdom reported that 163 people died of variant Creutzfeldt–Jakob disease between 1999 and March 2008. The final death toll is still not known; the incubation period for variant Creutzfeldt–Jakob disease is so long that some people who consumed UK beef before 1995 may still develop the variant Creutzfeldt–Jakob disease.

Kingdom after 1996, but there were three cases of variant Creutzfeldt–Jakob disease in the United States from 1996 to 2011 (Prusiner et al. 1999, Krauss et al. 2003, CDC 2010, 2012).

Symptoms of the variant Creutzfeldt–Jakob disease usually do not appear until after an incubation period of about 16 years. Depression, fits of rage, difficulty walking, and involuntary movements are some of the early clinical signs. Patients with the variant Creutzfeldt–Jakob disease are more likely than patients with classical forms of the disease to exhibit pronounced behavioral changes during the onset of the disease and a rapid loss of muscle coordination. The total time from the onset of symptoms to death is much longer with the variant form (13–14 months) than with the classical form (4–5 months). Both the variant and classical forms of the disease are always fatal, but the median age of death for people with variant Creutzfeldt–Jakob disease is 28 years old, versus 68 for classical Creutzfeldt–Jakob disease. Diagnosis of the disease can only be confirmed through a brain biopsy or autopsy. However, a presumptive diagnosis of variant Creutzfeldt–Jakob can be made when a patient exhibits the symptoms of the disease, and all other neurologic diseases have been eliminated (Prusiner et al. 1999, Krauss et al. 2003, CDC 2010, 2012).

27.3 PRION DISEASES IN ANIMALS

Scrapie affects domesticated sheep and goats and captive herds of moufflon. In the United States, scrapie is most prevalent in black-faced breeds of sheep (e.g., Hampshire, Suffolk, and their crosses); but in other countries, scrapie is equally common in other breeds. There is no evidence that scrapie prions cause disease in humans nor are there any reports of scrapie in free-ranging populations of wild sheep or goats. Infected ewes can transmit the disease to their lambs and to other adult sheep that consume infected placentas or amniotic fluids (Pattison et al. 1972, Prusiner et al. 1999, Gatti et al. 2002, Ecroyd et al. 2004, Gavier-Widén 2012).

Clinical signs of scrapie typically begin after an incubation period of 2–5 years. Early signs of scrapie include impaired social behavior, ear and head tremors, a vacant stare, and a strong propensity to scratch in the absence of a skin infection. Later clinical signs include loss of wool, skin damage, emaciation, and loss of motor skills especially in the hind legs. Sheep with scrapie can appear uncoordinated when running and may adopt a bunny-hop type of gait or one resembling a high-stepping horse. Ultimately, infected sheep become prostrate, and death results usually 1–3 months after the illness begins. At present, live sheep from the United States are banned from export to Europe and China because they may have scrapie; it is estimated that scrapie costs U.S. sheep ranchers \$20–\$25 million annually in lost sales (Laplanche et al. 1999, Merck 2010).

Mad cow disease was first detected in cattle during 1986, and during the next decade, an outbreak of the disease occurred among UK cattle (Sidebar 27.1). The list of countries that have detected mad cow disease in their cattle is large and includes most European countries, Japan, Canada, the United States, and Israel. By 2012, 18 cattle in Canada had been diagnosed with mad cow. Only four cases of mad cow disease have been detected among U.S. cattle: one in Washington State during 2003, one in Texas during 2005, one in Alabama during 2006, and one in California during

2012. The first infected cow was an animal imported from Canada, while the others were born in the United States. In each case, the infected animal was destroyed and its carcass disposed of in a manner to denature the prions or keep them out of the food chain (Travis and Miller 2003, CDC 2013).

Two forms of mad cow disease are now recognized. Classical mad cow disease is the strain responsible for the UK outbreak, and cows with this strain become infected from eating contaminated feed. Atypical mad cow disease occurs in older cattle and is not associated with contaminated feed. It may result from a spontaneous mutation in the infected animal's DNA or protein. The 2003 infected cow in Washington State had classical mad cow disease, while the last three infected cows in the United States had atypical mad cow disease (CDC 2013).

Although few cattle are infected with mad cow disease, the economic impact of the disease is enormous. After the infected cow was detected in the United States during 2003, U.S. exports of beef plunged as numerous countries enacted bans, and U.S. consumers became wary of eating beef. This reduction in the demand for beef resulted in U.S. ranchers, feedlots, and meatpacking companies losing billions of dollars. A decade later, some of these bans on U.S. cattle still remain. For example, Japan still does not allow the import of beef from U.S. cattle that are over 30 months old because older cattle are more likely to have mad cow disease (Strom 2012, Strom and Tabuchi 2013).

Most cattle with mad cow disease were infected when offal or other products from an infected animal were used in their feed—a practice that was banned in most countries during the 1990s. Since then, the number of infected cattle has declined significantly, but the disease still occurs in cattle, and it is unclear how cattle are currently being infected (Cohen et al. 2004). Besides humans, the mad cow prion has caused disease in several zoo animals including eight species from the family Bovidae (zebu, bison, common eland, kudu, nyala, scimitar-horned oryx, Arabian oryx, and gemsbok), three species of the order of primates (brown lemur, mongoose lemur, and rhesus macaque), and several species of felids (Kirkwood and Cunningham 1994, 1999, Bons et al. 1999, Nathanson et al. 1999, Seuberlich et al. 2006, Gavier-Widén 2012).

Transmissible mink encephalopathy was first described on a mink ranch in Wisconsin during 1947. Outbreaks of transmissible mink encephalopathy are rare but have occurred during 1961, 1963, and during 1985 on mink farms in the United States. Since the mid-1980s, outbreaks have been reported at mink farms located in Canada, Finland, Germany, and former republics of the Soviet Union. All cases of transmissible mink encephalopathy have occurred among farm-raised mink that had been fed meat from infected animals. The disease has not been observed among wild mink. Feeding mink commercial food rather than meat from dead livestock has greatly decreased the rate of disease in mink farms. The incubation period for transmissible mink encephalopathy is 8–12 months; clinical signs include sleepiness, lack of muscle coordination, and a propensity to bite. Many infected mink exhibit the behavior of flipping their tails over their backs in a manner similar to the tail posture of tree squirrels (Laplanche et al. 1999, Brown et al. 2001, Merck 2010).

Feline spongiform encephalopathy is a fatal TSE that occurs in domestic cats and was first detected during the 1990s in the United Kingdom. All of the sick cats had been fed pet food made from cattle with mad cow disease. By 2010, the disease was diagnosed in at least 89 domestic cats in the United Kingdom and a few more cats located throughout Europe. Infected cats are hypersensitive to sound or tactile stimuli and lose muscle coordination. Few domestic cats have been diagnosed with feline spongiform encephalopathy following a 1990 ban on the use of the spleen, brain, and spinal cord tissue in pet foods. Feline spongiform encephalopathy was also detected in captive cheetahs, tigers, lions, cougars, ocelots, Asian gold cat, and leopard cat that had been fed meat from a mad cow. The total number of captive felids that developed the disease numbered less than 100 animals. After the disease was detected, its prevalence diminished by changing the diet of zoo animals (Sigurdson and Miller 2003, Eiden et al. 2010).

Chronic wasting disease is the only TSE that has been detected in free-ranging wildlife. The disease was first described among captive mule deer in Colorado during the 1960s, but the disease was not diagnosed as a TSE until 1978. Soon thereafter, it was also detected in captive elk located in Wyoming (Williams and Young 1980). By 1981, it was diagnosed among free-ranging elk, mule deer, and white-tailed deer living in a localized area within northeastern Colorado and southeastern Wyoming (Figure 27.3), but once surveillance was expanded, it was detected in other states. By 2005, the area where wild deer or elk are infected with chronic wasting disease had expanded to Utah, New Mexico, Nebraska, South Dakota, Wisconsin, Illinois, New York, and Saskatchewan, Canada (Figure 27.4). During 2007, chronic wasting disease was detected in a moose. By 2013, the prion has been found in free-ranging deer and elk in 18 U.S. states and in Alberta and Saskatchewan, Canada (Figure 27.5). Prevalence increases slowly, but in some areas, more than 30% of deer have been infected. To date, chronic wasting disease has not been detected among free-ranging scavengers (e.g., raccoons, coyotes, striped skunks, and Virginia opossum) within the United States or Canada. It also has not been detected among wild or farm-raised deer in Europe (Sigurdson and Miller 2003, U.S. Geological Survey 2007, Jennelle et al. 2009).

Deer and elk with chronic wasting disease shed prions through their saliva and feces. Prions also exist in the carcasses of infected animals. Other deer and elk become infected through direct contact with an infected animal or indirectly through the inadvertent consumption of prions in food or water from the environment. Chronic wasting disease prions can persist for years in soil and bedding material; their durability in the environment increases the risk that a healthy deer or elk will consume them and become infected (Sigurdson and Miller 2003, Travis and Miller 2003, Holland 2004).

Chronic wasting disease is uncommon in young deer and elk because the incubation period for the disease is usually 18–24 months. The disease also is more prevalent among males than females. Behavioral signs of chronic wasting disease can include an increase in aggression and solitary behavior and a loss of fear of humans. Infected deer and elk may carry their head and ears lower and stand with their legs further apart than normal. They may engage in repetitive behaviors, such as walking in circles. Neurological problems include head tremors, hind limb paralysis, and an inability to rise to their feet. As the disease progresses, infected animals waste away (hence the name for the disease) as they reduce their food intake and become emaciated (Figure 27.6). During late stages, prions occur throughout the body including

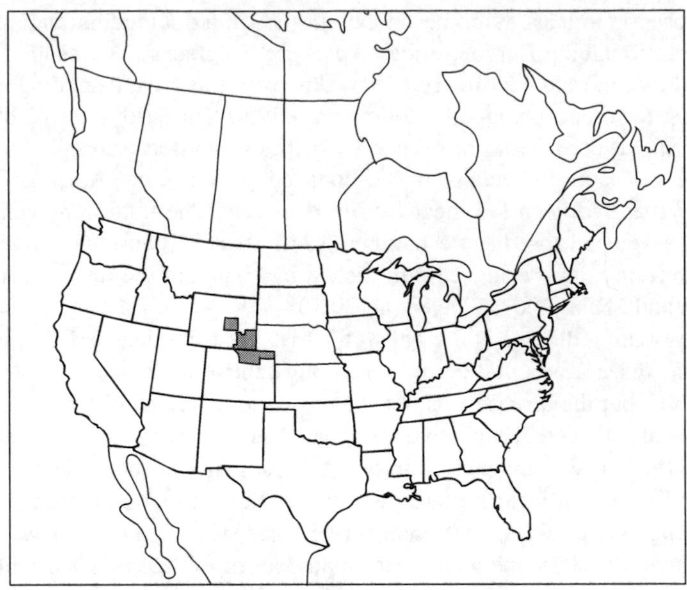

FIGURE 27.3 Distribution of chronic wasting disease in North America in 2000. (Data from U.S. Geological Survey, Chronic wasting disease, U.S. Geological Survey, Fact sheet 2007–3070, U.S. Department of the Interior, Madison, WI, 2007.)

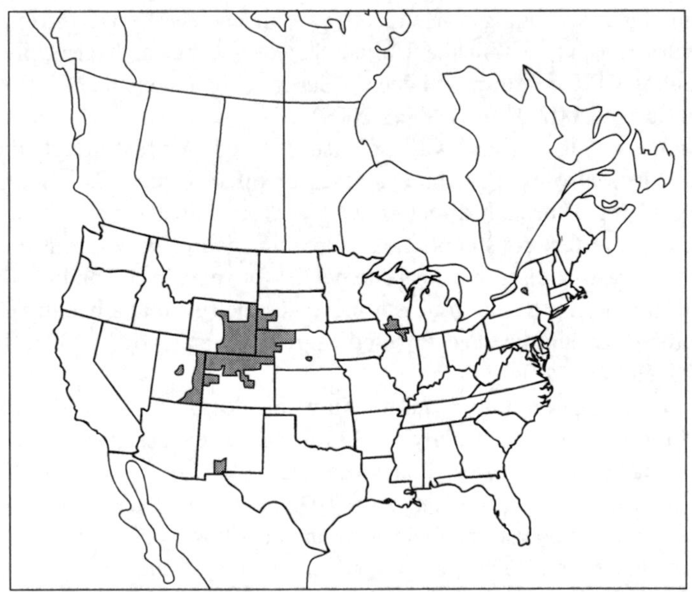

FIGURE 27.4 Distribution of chronic wasting disease in North America in 2005. (Data from U.S. Geological Survey, Wildlife Health Bulletin #05-01, U.S. Department of the Interior, Madison, WI, 2005.)

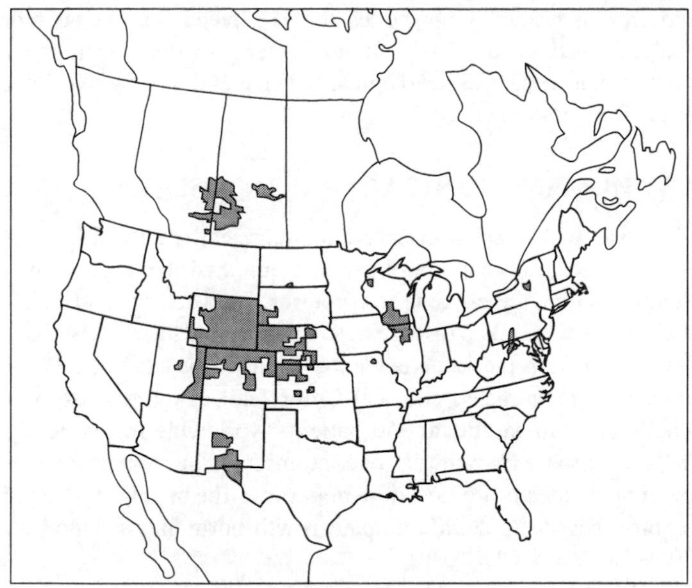

FIGURE 27.5 Distribution of chronic wasting disease in North America in 2013. (Data from U.S. Geological Survey, Chronic wasting disease (CWD), http://www.nwhc.usgs.gov/disease_information/chronic_wasting_disease/, 2013.)

FIGURE 27.6 This white-tailed deer in Wisconsin has chronic wasting disease and has become emaciated. (Courtesy of Donald Savoy and the U.S. Geological Service.)

the brain and nervous system, lymph nodes, tonsils, spleen, muscle, fat, blood, saliva, and antler velvet. Death occurs 20–25 months after infection (Williams and Young 1980, Laplanche et al. 1999, U.S. Geological Service 2007, Gavier-Widén 2012, U.S. National Wildlife Heath Center 2013).

27.4 HOW HUMANS CONTRACT PRION DISEASES

Most people (85%) with a prion disease have classical Creutzfeldt–Jakob disease, which is not contagious. Instead, these patients obtained the disease due to a spontaneous mutation of one of their genes or proteins. Between 5% and 15% of patients with classical Creutzfeldt–Jakob disease have an inherited form of the disease, which often strikes people from the same extended family. For example, 17 members of one Brazilian family contracted classical Creutzfeldt–Jakob disease over a period of three generations. An additional 250 patients worldwide inadvertently acquired Creutzfeldt–Jakob disease from the use of contaminated instruments in neurosurgery or from receiving an infected cornea, dura mater (i.e., the membrane surrounding the brain and central nervous system), human growth hormone, or blood transfusions (Belay 1999, WHO 2006, CDC 2012).

Kuru is a TSE disease of humans that occurs only in one part of New Guinea and was transmitted through ritual cannibalistic practices that involved consuming the brain and other human tissue of dead relatives as a way of honoring their dead (Sidebar 27.2). The prevalence of kuru declined markedly after people realized how people developed kuru, and the government outlawed the practice (Ricketts 1997, Prusiner et al. 1999).

Variant Creutzfeldt–Jakob disease is another human TSE that resulted when people consumed beef from an animal that had mad cow disease. This disease was detected in the United Kingdom during 1996 and was much worse in the United Kingdom than in the United States; by 2009, there were 168 patients in the United Kingdom versus four in the United States. This difference resulted because mad cow disease was more prevalent among UK cattle than U.S. cattle. The number of animals with mad cow disease is miniscule today as is the risk of developing this disease from eating beef today. During 2010, for instance, there was only one cow that was detected with mad cow disease in Canada and none in the United States. Yet the incubation period for variant Creutzfeldt–Jakob disease is so long that some people who consumed UK beef products prior to 1996 may still develop the disease. There are no reported cases of humans being infected with a TSE from consuming meat from sheep with scrapie or from deer and elk with chronic wasting disease (Detwiler and Pritchett 2004, CDC 2013).

27.5 MEDICAL TREATMENT

There are no vaccines or cures for either the classical or variant form of Creutzfeldt–Jakob disease, and no specific therapy is known that can slow or stop the progression of a prion disease or to prevent fatalities. Medical treatment for patients with Creutzfeldt–Jakob disease involves treating disease's adverse symptoms and making the patient comfortable.

SIDEBAR 27.2 ERADICATING KURU DISEASE BY STOPPING CANNIBALISM (PRUSINER ET AL. 1999, COLLINGE ET AL. 2006)

Kuru disease is a TSE that was first documented during the early 1900s among cannibalistic tribes in New Guinea, although it was not identified as a TSE until much later. The Fore people of Papua New Guinea had a ritualistic funeral practice of preparing and consuming the brain and other tissue from deceased relatives (Figures 27.7 and 27.8). This practice resulted in the spread of kuru disease when the brains of people that died of kuru where consumed by relatives. During the 1950s and 1960s, kuru had become so common that it was the leading cause of death among Fore women. Between 1957 and 2004, over 2,700 people died from kuru (13% of the total Fore population of 20,000). According to oral history of the Fore people, the disease did not occur prior to the 1920s. Kuru in the local language meant "shake with fear," so named because tremors were a common trait of the disease. Other early symptoms included headaches, joint pains, involuntary movements, unsteady gait, and slurred speech. The symptoms are distinct enough that the disease is easily recognizable by the local community. Later signs included dementia and an inability to eat. The disease was invariably fatal with most patients dying within a year when symptoms developed.

The transmission of kuru was stopped when the practice of mortuary feasting ceased during the 1950s. Thereafter, the incidence of kuru began to decline, but some people still develop kuru due to its exceptionally long incubation period. Eleven people were diagnosed with kuru during 1996 to 2004; all of them were born before 1950 and participated in mortuary feasting as children. For these 11 patients, the minimum incubation period ranged from 34 to 41 years. Kuru has not been detected in anyone born after 1960 suggesting that kuru may be eradicated in the upcoming decades.

27.6 WHAT PEOPLE CAN DO TO REDUCE THEIR RISK OF CONTRACTING A PRION DISEASE

The risk of any one person contracting a prion disease is extremely rare. The risk of someone developing Creutzfeldt–Jakob disease from a medical procedure or tissue transplant has been reduced considerably through the use of stringent donor selection procedures that exclude donors who may have a TSE. Current blood donor laws make it illegal for people with a TSE to donate blood. Persons who have lived in the United Kingdom for more than 3 months from 1980 to 1996 are not allowed to donate in the United States (WHO 2004).

To date, only variant Creutzfeldt–Jakob disease has been conclusively shown to be transmitted from animals to humans, primarily through the consumption of meat or other products from infected cattle. All developed countries have extensive surveillance programs to monitor the prevalence of mad cow disease in their cattle populations. These programs are administered in Canada by the Canadian Food Inspection

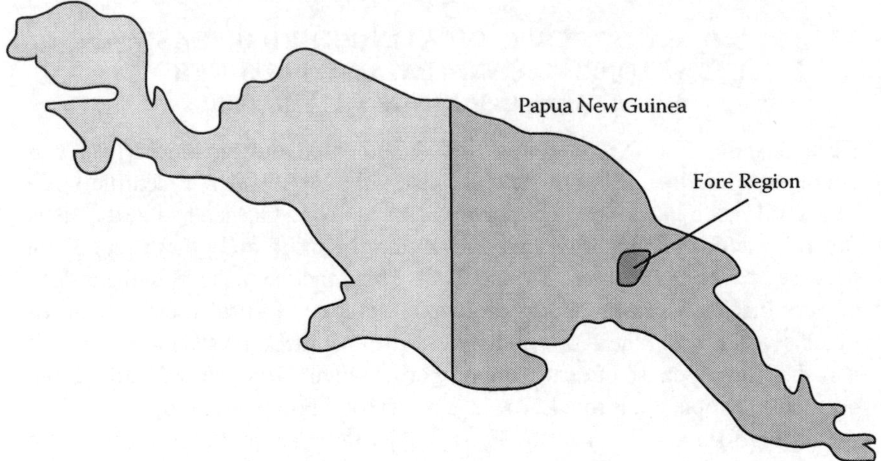

FIGURE 27.7 Location of the 20,000 members of the Fore people in Papua New Guinea.

FIGURE 27.8 Photo of the Fore people of Papua New Guinea. The photo shows a man holding the body of a young man that has died from kuru. (Courtesy of the U.S. National Library of Medicine and WHO.)

and in the United States by the USDA's Animal and Plant Health Inspection Service (APHIS). Most countries, including the United States and Canada, have restricted the addition of certain animal proteins from all animal feeds, pet foods, and fertilizers. These restrictions have greatly reduced the risk of TSEs in cattle, domestic cats, captive mink, and zoo animals (CDC 2013).

Lambs can contract scrapie from their mothers, and other sheep can contract it if their premises are contaminated or if they consume infected placentas. Some sheep are genetically more susceptible to scrapie, and many sheep producers in the United States and Canada are culling these sheep from their herds. The USDA's APHIS has instituted a scrapie eradication program with the goal of substantially eliminating the disease in the United States by 2017. The program includes (1) prohibition against using protein from ruminants in sheep rations; (2) official identification of all breeding sheep, all sheep over 18 months of age, and certain classes of goats when they change ownership; and (3) scrapie testing of targeted mature sheep and goats sold for slaughter. Whenever an infected animal is detected, it is killed and all animals in the herd from which it originated are quarantined and tested. All exposed animals that are genetically susceptible to scrapie are depopulated or permanently restricted. By 2008, this program had reduced prevalence of the disease to 0.03% of U.S. sheep (Merck 2010).

Prions responsible for chronic wasting disease in deer and elk differ from prions that cause human disease. Still, scientist cannot rule out the possibility that the prion responsible for chronic wasting disease could spread to humans. The U.S. Geological Service (2007) recommends that deer and elk hunters take simple precautionary measure to reduce the risk by (1) not shooting animals that appear sick, lethargic, or emaciated; (2) not eating the spinal cord, brain, spleen, lymph nodes, or other internal organs; and (3) wearing disposable gloves while field dressing animals and thoroughly cleaning knives or saws used in the process. Most wildlife agencies in states where chronic wasting disease is endemic will test deer and elk carcasses for hunters without charge and inform them if the animal they shot is infected (Williams et al. 2002, Travis and Miller 2003, U.S. Geological Survey 2007).

The primary goal for managing chronic wasting disease is to stop its spread to new areas. In states where chronic wasting disease is absent, wildlife managers often try to prevent the introduction of the disease by not allowing the importation of live deer and elk and not allowing hunters to bring whole deer carcasses into their home states. Several states have taken steps to protect wild cervids from contracting the disease from captive cervids. These steps can include routine testing for chronic wasting disease of all carcasses of captive cervids that died when the animals were over a year old and complete depopulation of any captive herds where any infected animal is detected. Many states require that captive facilities be enclosed by two fences to both prevent the nose-to-nose contact between wild animals and to stop a captive animal from escaping (VerCauteren et al. 2007).

Supplemental feeding or baiting of deer and elk may increase the spread of chronic wasting disease by concentrating sick and healthy animals together. Hence, feeding deer and elk is discouraged or outlawed in areas where chronic wasting disease is found (Thompson et al. 2008). Some states have culled their deer herds after chronic wasting disease was detected in localized areas with mixed results. In Wisconsin, chronic wasting disease was first detected in a wild deer during 2002. The state immediately began an aggressive program to test deer and to reduce deer numbers where the disease was endemic. The goal was to eliminate quickly the disease from the state, but the program was unsuccessful. Ten years later, people in Wisconsin are polarized over the issue of what should be done in response to the disease.

Some want the deer herds culled and others want them to be as large as possible (Grear et al. 2006, Weiss 2011).

Prions are very durable and can survive common methods of disinfection and incineration, unless very high temperatures are used to incinerate carcasses (Brown et al. 2004, WHO 2004). They can survive in soil for years. Their durability enhances their ability to persist long enough in the environment to infect new animals (Brown and Gajdusek 1991, Georgsson et al. 2006, Johnson et al. 2006, Seidel et al. 2007).

27.7 ERADICATING PRION DISEASES FROM A COUNTRY

Government actions have almost eliminated the threat of kuru and variant Creutzfeldt–Jakob disease as human diseases. Eradicating scrapie and chronic wasting disease in animal populations will be a much greater challenge. Fortunately, the prions responsible for these diseases have not been detected in humans.

LITERATURE CITED

Belay, E. D. 1999. Transmissible spongiform encephalopathies in humans. *Annual Review of Microbiology* 53:283–314.

Belay, E. D. and L. B. Schonberger. 2005. The public health impact of prion diseases. *Annual Review of Public Health* 26:191–212.

Bons, N., N. Mestre-Frances, P. Belli, F. Cathala, D. C. Gajdusek, and P. Brown. 1999. Natural and experimental oral infection of nonhuman primates by bovine spongiform encephalopathy agents. *Proceedings of the National Academy of Sciences United States of America* 96:4046–4051.

Brown, P. and D. C. Gajdusek. 1991. Survival of scrapie virus after three years' interment. *Lancet* 337:269–270.

Brown, P., E. H. Rau, P. Lemieux, B. K. Johnson et al. 2004. Infectivity studies of both ash and air emissions from simulated incineration of scrapie-contaminated tissues. *Environmental Science and Technology* 38:6155–6160.

Brown, P., R. G. Will, R. Bradley, D. M. Asher, and L. Detwiler. 2001. Bovine spongiform encephalopathy and variant Creutzfeldt-Jakob disease: Background, evolution, and current concerns. *Emerging Infectious Diseases* 7:6–16.

CDC. 2010. vCJD (variant Creutzfeldt-Jakob disease). http://www.cdc.gov/ncidod/dvrd/cjd/ (accessed December 23, 2013).

CDC. 2012. CJD (Creutzfeldt-Jakob disease, classic). http://www.cdc.gov/ncidod/dvrd/cjd/index.htm (accessed August 1, 2013).

CDC. 2013. CJD (Bovine spongiform encephalopathy, or mad cow disease). http://www.cdc.gov/ncidod/dvrd/bse/ (accessed December 23, 2013).

Cohen, J. T., K. Dugger, G. M. Gray, S. Kreindel et al. 2004. A simulation model for evaluating the potential for spread of bovine spongiform encephalopathy in animals or to people. In: B. K. Nunnally and I. S. Krull, editors. *Prions and Mad Cow Disease.* Marcel Dekker, New York, pp. 61–123.

Collinge, J., J. Whitfield, E. McKintosh, J. Beck et al. 2006. Kuru in the 21st century—An acquired human prion disease with very long incubation periods. *Lancet* 367:2068–2074.

Cummings, L. 2011. Considering risk assessment up close: The case of bovine spongiform encephalopathy. *Health, Risk and Society* 13:255–275.

Detwiler, L. A. and B. Pritchett. 2004. Actions to prevent bovine spongiform encephalopathy from entering the United States. In: B. K. Nunnally and I. S. Krull, editors. *Prions and Mad Cow Disease.* Marcel Dekker, New York, pp. 125–135.

Ecroyd, H., P. Sarradin, J. L. Dacheux, and J. L. Gatti. 2004. Compartmentalization of prion isoforms within the reproductive tract of the ram. *Biology of Reproduction* 71:993–1001.

Eiden, M., C. Hoffmann, A. Balkema-Buschmann, M. Müller et al. 2010. Biochemical and immunohistochemical characterization of feline spongiform encephalopathy in a German captive cheetah. *Journal of General Virology* 91:2874–2883.

Gatti, J. L., S. Métayer, M. Moudjou, O. Andréoletti et al. 2002. Prion protein is secreted in soluble forms in the epididymal fluid and proteolytically processed and transported in seminal plasma. *Biology of Reproduction* 67:393–400.

Gavier-Widén, D. 2012. Transmissible spongiform encephalopathies. In: D. Gavier-Widén, J. P. Duff, and A. Meredith, editors. *Infectious Diseases of Wild Mammals and Birds in Europe*. Wiley-Blackwell, West Sussex, U.K., pp. 489–496.

Georgsson, G., S. Sigurdarson, and P. Brown. 2006. Infectious agent of sheep scrapie may persist in the environment for at least 16 years. *Journal of General Virology* 87:3737–3740.

Grear, D. A., M. D. Samuel, J. A. Langenberg, and D. Keane. 2006. Demographic patterns and harvest vulnerability of chronic wasting disease infected white-tailed deer in Wisconsin. *Journal of Wildlife Management* 70:546–553.

Holland, S. D. 2004. Overview of transmissible spongiform encephalopathy in cervids in the United States. In: B. K. Nunnally and I. S. Krull, editors. *Prions and Mad Cow Disease*. Marcel Dekker, New York, pp. 137–150.

Holman, R. C., E. D. Belay, K. Y. Christensen, R. A. Maddox et al. 2010. Human prion diseases in the United States. *PLoS One* 5(1):e8521.

Jennelle, C. S., M. D. Samuel, C. A. Nolden, D. P. Keane et al. 2009. Surveillance for transmissible spongiform encephalopathy in scavengers of white-tailed deer carcasses in the chronic wasting disease area of Wisconsin. *Journal of Toxicology and Environmental Health, Part A* 72:1018–1024.

Johnson, C. J., K. E. Phillips, P. T. Schramm, D. McKenzie et al. 2006. Prions adhere to soil minerals and remain infectious. *PLoS Pathogens* 2:e32.

Kirkwood, J. K. and A. A. Cunningham 1994. Epidemiological observations on spongiform encephalopathies in captive wild animals in the British Isles. *Veterinary Record* 135:296–303.

Kirkwood, J. K. and A. A. Cunningham. 1999. Scrapie-like spongiform encephalopathies (prion diseases) in nondomesticated species. In: M. E. Fowler and R. E. Miller, editors. *Zoo and Wild Animal Medicine*. W.B. Saunders, Philadelphia, PA, pp. 662–669.

Krauss, H., A. Weber, M. Appel, B. Enders et al. 2003. *Zoonoses: Infectious Diseases Transmissible from Animals to Humans*, 3rd edition. ASM Press, Washington, DC.

Laplanche, J. L., N. Hunter, M. Shinagawa, and E. Williams. 1999. Scrapie, chronic wasting disease, and transmissible mink encephalopathy. In: S. B. Prusiner, editor. *Prion Biology and Diseases*. Cold Spring Harbor Laboratory Press, Cold Spring Harbor, NY, pp. 393–429.

Londhe, M. S., N. K. Mahajan, R. P. Gupta, and R. M. Londhe. 2012. Review of prion diseases in animals with emphasis to bovine spongiform encephalopathy. *Veterinary World* 5:443–448.

Merck. 2010. *Merck Veterinary Manual*, 10th edition. Merck, Whitehouse Station, NJ.

Nathanson, N., J. Wilesmith, G. A. Wells, and C. Griot. 1999. Bovine spongiform encephalopathy and related diseases. In: S. B. Prusiner, editor. *Prion Biology and Diseases*. Cold Spring Harbor Laboratory Press, Cold Spring Harbor, NY, pp. 431–463.

Pattison, I. H., M. N. Hoare, J. N. Jebbett, and W. A. Watson. 1972. Spread of scrapie to sheep and goats by oral dosing with fetal membranes from scrapie-infected sheep. *Veterinary Record* 90:465–468.

Prusiner S. B. 1982. Novel proteinaceous infectious particles cause scrapie. *Science* 216:136–144.

Prusiner, S. B. 1996. Molecular biology and pathogenesis of prion diseases. *Trends in Biochemical Sciences* 21:482–487.

Prusiner, S. B., M. R. Scott, S. J. DeArmond, and G. Carlson. 1999. Transmission and replication of prions. In: S. B. Prusiner, editor. *Prion Biology and Diseases*. Cold Spring Harbor Laboratory Press, Cold Spring Harbor, NY.

Ricketts, M. N. 1997. Is Creutzfeldt–Jakob disease transmitted in blood? Is the absence of evidence of risk evidence of the absence of risk? *Canadian Medical Association Journal* 157:1367–1370.

Seidel, B., A. Thomzig, A. Buschmann, M. H. Groschup et al. 2007. Scrapie agent (strain 263K) can transmit disease via the oral route after persistence in soil over years. *PLoS One* 2:e435.

Seuberlich, T., C. Botteron, C. Wenker, V. Café-Marçal et al. 2006. Spongiform encephalopathy in a miniature zebu. *Emerging Infectious Diseases* 12:1950–1953.

Sigurdson, C. J. and M. W. Miller. 2003. Other animal prion diseases. *British Medical Bulletin* 66:199–212.

Strom, S. 2012. Case of mad cow disease is found in U.S. *New York Times* (April 25, 2012). http://www.nytimes.com/2012/04/25/health/case-of-mad-cow-disease-is-found-in-us.html?_r=0 (accessed May 20, 2014).

Strom, S. and H. Tabuchi. 2013. A break for embattled ranchers. *New York Times* (January 28, 2013). http://www.nytimes.com/2013/01/29/business/global/japan-to-ease-restrictions-on-us-beef.html (accessed May 20, 2014).

Thompson, A. K., M. D. Samuel, and T. R. Van Deelen. 2008. Alternate feeding strategies and potential disease transmission in Wisconsin white-tailed deer. *Journal of Wildlife Management* 72:416–421.

Travis, D. and M. Miller. 2003. A short review of transmissible spongiform encephalopathies, and guidelines for managing risks associated with chronic wasting disease in captive cervids in zoos. *Journal of Zoo and Wildlife Medicine* 34:125–133.

U.S. Geological Survey. 2007. Chronic wasting disease. U.S. Geological Survey, Fact sheet 2007-3070. U.S. Department of the Interior, Madison, WI.

U.S. Geological Survey. 2013. Chronic Wasting Disease (CWD). http://www.nwhc.usgs.gov/disease_information/chronic_wasting_disease/ (accessed July 27, 2013).

U.S. National Wildlife Health Center. 2013. Frequently asked questions concerning chronic wasting disease. USGS National Wildlife Health Center, Madison, WI.

VerCauteren, K. C., M. J. Lavelle, N. W. Seward, J. W. Fischer, and G. E. Phillips. 2007. Fence-line contact between wild and farmed cervids in Colorado: Potential for disease transmission. *Journal of Wildlife Management* 71:1594–1602.

Weiss, J. 2011. Wisconsin's CWD road isn't one to follow. *The Post-Bulletin* (March 12, 2011). Rochester, MN. http://www.postbulletin.com/news/special_report/wisconsin-s-cwd-road-isn-t-one-to-follow/article_cfe413e6-33b2-5fe3-ba3d-6c4278279a26.html?mode=jqm (accessed May 20, 2014).

Wells, G. A. H., A. C. Scott, C. T. Johnson, R. F. Gunning et al. 1987. A novel progressive spongiform encephalopathy in cattle. *Veterinary Record* 121:419–420.

WHO. 2004. *Laboratory Biosafety Manual*, 3rd edition. World Health Organization, Geneva, Switzerland.

WHO. 2006. WHO guidelines on tissue infectivity distribution in transmissible spongiform encephalopathies. World Health Organization, Geneva, Switzerland.

Will, R. G., J. W. Ironside, M. Zeidler, S. N. Cousens et al. 1996. A new variant of Creutzfeldt-Jakob disease in the UK. *Lancet* 347:921–925.

Williams, E. S., M. W. Miller, T. J. Kreeger, R. H. Kahn, and E. T. Thorne. 2002. Chronic wasting disease of deer and elk. *Journal of Wildlife Management* 66:551–563.

William, E. S. and S. Young. 1980. Chronic wasting disease of captive mule deer: A spongiform encephalopathy. *Journal of Wildlife Diseases* 16:89–98.

Section VII

Parasites

28 Baylisascariasis and Raccoon Roundworms

A single adult female worm may produce an estimated 115,000 to 877,000 eggs per day and an infected raccoon can shed as many as 45,000,000 eggs daily. In light of the relatively low infectious dose of *B. procyonis* (estimated to be ≤5000 eggs) and the viability of the eggs in the environment for months to years, the infection potential is not insubstantial.

Sorvillo et al. (2002)

28.1 INTRODUCTION AND HISTORY

Baylisascaris is a genus of roundworms, which occur in the intestines of a number of mammals. There are nine species, each of which has a small number of mammal species that serve as its reservoir hosts. In North America, raccoons are the definitive host (i.e., the host in which the parasite reaches sexual maturity and reproduces) for *Baylisascaris procyonis*; skunks for *Baylisascaris columnaris*; black, grizzly, and polar bears for *Baylisascaris transfuga*; mustelids for *Baylisascaris devosi*; and rodents for *Baylisascaris laevis*. In definitive hosts, female roundworms reach sexual maturity and produce eggs that are shed in the animal's feces. All of these roundworm species can potentially infect humans, but *B. procyonis* is the only one known to cause serious diseases in humans; hence, this chapter will focus on raccoon roundworms (Figure 28.1).

Raccoons often use latrines when defecating. These latrines are often located at the base of trees or on top of raised horizontal structures, such as fallen logs, barn lofts, tree stumps, boulders, or woodpiles. Raccoon feces and the eggs of raccoon roundworms accumulate in these latrines. Humans infected with these parasites can develop a disease referred to as baylisascariasis. To avoid confusion, the term baylisascariasis will be used when referring to the human illness, while the term raccoon roundworms will be utilized when referring to the parasite itself (Kazacos 2001, CDC 2012, Bauer 2013).

28.2 SYMPTOMS IN HUMANS

Baylisascariasis is rare in humans, but it is a serious disease when it occurs. Humans are stricken with baylisascariasis after consuming the eggs of raccoon roundworms. Ingested eggs hatch in the digestive system; the larvae then burrow through the lining of the intestines and are carried by the bloodstream to different organs. The growing larvae start to migrate through organs, muscles, and other tissue. In doing so, the larvae cause damage and inflammation. Initial symptoms

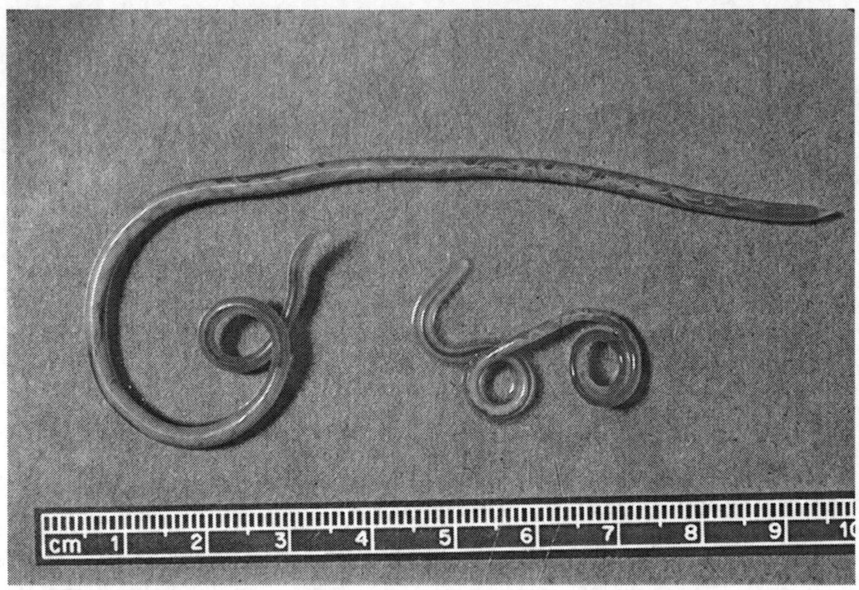

FIGURE 28.1 Photo of adult raccoon roundworms recovered from the intestines of a raccoon. (Courtesy of Kevin Kazacos, Purdue University, West Lafayette, IN.)

of baylisascariasis begin 2–4 weeks after the eggs are ingested and include nausea, lethargy, and loss of muscle coordination. Symptoms vary based on the number of migrating larvae inside a person and which organs the larvae invade. Neurological problems can result from an invasion of the brain or central nervous system, a condition called neural larva migrans. Symptoms of this include lethargy, irritability, slurred speech, minor changes in vision, and a loss of coordination or muscle control. Symptoms can quickly worsen and progress to seizures, paralysis, tremors, coma, and death. Neural larva migrans has a case fatality rate of 36% in the United States, and many of the survivors are left in a vegetative state or have long-term neurological problems. At least 23 people in North America have been diagnosed with neural larva migrans as of 2013; most were children under the age of two (CDC 2012, Center for Food Security and Public Health 2012, Bauer 2013).

Another disease called ocular larva migrans results when roundworm larvae invade the eye; dozens of U.S. patients have been diagnosed with ocular larva migrans. Clinical signs include photophobia and inflammation of one eye rather than both; blindness can result. Visceral larva migrans develops when the larvae invade tissue and internal organs, including the lung, heart, or liver, producing abdominal pain. Invasion of the lung can cause chest pains, cough, and pulmonary damage. Cutaneous larva migrans occurs when the larvae invade the skin causing a macular skin rash, often on the trunk or face. The prevalence of visceral or cutaneous larva migrans in the United States is unknown. Most people infected with raccoon roundworms do not have symptoms or have only a mild illness. Such people may not seek medical attention, and those who do may not be diagnosed as having baylisascariasis (Sorvillo et al. 2002, CDC 2012, Center for Food Security and Public Health 2012).

28.3 RACCOON ROUNDWORMS IN ANIMALS

Raccoons are a definitive host for raccoon roundworms (Figure 28.2). Raccoons are native to North and Central America and are widely distributed from Mexico and Canada. Habitat changes by humans have allowed raccoons to expand their range northward and to occupy parts of North America that previously were inhospitable for them (Figure 28.3). As raccoons invaded new areas, they carried raccoon roundworms with them. Human cases have been documented in Massachusetts, New York, Pennsylvania, Michigan, Minnesota, Illinois, Missouri, Louisiana, Oregon, and California (CDC 2012). Raccoons were deliberately introduced in both Europe and Japan for pets or for the fur industry and have become established in both areas. Some of these raccoons were infected with raccoon roundworm, and the parasite is now endemic in Europe. In Germany, up to 71% of wild raccoons are infected with raccoon roundworms. In Japan, raccoon roundworms have been found among captive raccoons, but not among Japan's wild raccoons (Kazacos 2001, Bauer 2013, Page 2013).

Juvenile raccoons are more likely to be infected than adults and to have higher levels of infestation (juveniles have an average of 40–50 roundworms, while adults have 12–22). The number of roundworms per raccoon often declines during the winter in the northern United States and Canada but then increases again

FIGURE 28.2 Raccoons are the reservoir host for raccoon roundworms. (Courtesy of the U.S. Fish and Wildlife Service, Willapa, WA.)

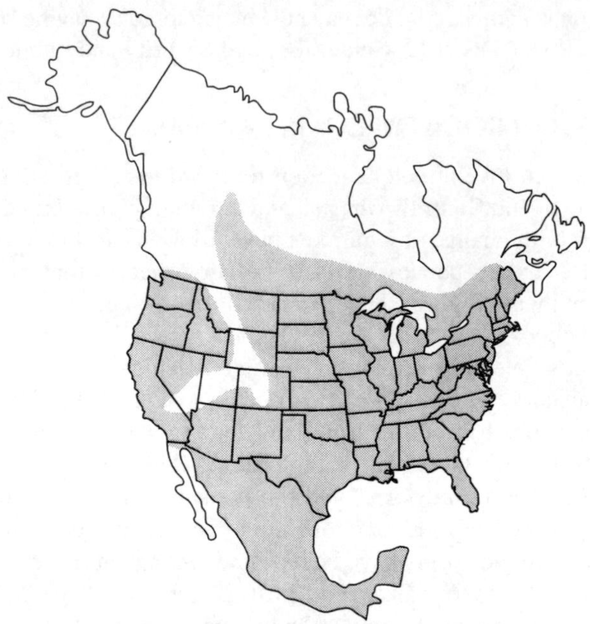

FIGURE 28.3 The range of raccoons in North America.

during the spring and summer. Many raccoons in the northeastern and midwestern sections of the United States are infected with them, but this is not true in southern states (Table 28.1).

Juvenile raccoons usually become infected with raccoon roundworm eggs when foraging, visiting a raccoon latrine, or grooming (roundworm eggs readily adhere to fur). After being ingested, the eggs hatch in the raccoon's digestive system. In raccoons, the roundworms stay there and do not invade other tissue. The roundworms become an adult while in the intestines and start to produce eggs that are excreted in the raccoon's feces. Raccoon roundworms do not cause illness in raccoons except for the rare occasion when the adult roundworms become so numerous that they block the intestine (Kazacos 2001).

A single adult roundworm can produce 100,000–900,000 eggs daily, and a raccoon may be infected with several female roundworms giving it the potential to shed up to 45 million eggs every day. The eggs are not initially infectious, but after the embryos inside the eggs have developed for a couple weeks, the eggs are ready to hatch as soon as they are consumed (Figure 28.4). Raccoon roundworm eggs can survive for years in soil or other moist environments—long after all of the fecal material has decomposed (Kazacos 2001, Sorvillo et al. 2002).

Many species of mammals and birds become infected when they consume roundworm eggs; rodents and rabbits are especially susceptible. These animals serve as intermediate hosts for the roundworm (Figure 28.5). Several species of rodents, other small mammals, and ground-foraging birds visit raccoon latrines searching for food, such as insects drawn to the feces or undigested seeds

TABLE 28.1
Percent of Raccoons or Raccoon Feces Found to Be Infected with Raccoon Roundworms in Different Parts of Canada and the United States

Location	Percentage Infected	Number Examined
Nova Scotia	8	219
Nova Scotia	7	236
Nova Scotia	7	491
New York	68	429
New York	20	277
New Jersey	34	137
Maryland	30	304
Georgia	1	110
Georgia	0	100
North Carolina	0	148
South Carolina	0	128
Alabama	0	371
Tennessee	8	253
Indiana	72	1425
Indiana	74	391
Indiana	29	218
Indiana	15	219
Illinois	82	310
Illinois	86	100
Michigan	0	256
Wisconsin	51	213
Minnesota	61	163
Minnesota	66	109
South Dakota	12	250
Kansas	44	128

Source: Kazacos, K.R., *Baylisascaris procyonis* and related species, in: W.M. Samuel, M.J. Pybus, and A.A. Kocan, editors, *Parasitic Diseases of Wild Animals*, 2nd edition, Iowa State University Press, Ames, IA, pp. 301–341, 2001.
Note: Each study is listed on a separate line.

contained in the raccoon feces. These animals inadvertently consume the eggs of raccoon roundworms along with any food they find. In intermediate hosts, the roundworm larvae do not stay within the intestine but pass through it to invade other tissue (e.g., liver, lungs, heart, pancreas, kidneys, and muscles) where they grow and then encapsulate themselves. Some of these encapsulated roundworms are large enough to be seen with the naked eye. Clinical signs of visceral larva migrans among intermediate hosts include lethargy and ruffled feathers in birds and a rough hair coat in mammals. Some of the larvae invade the brain of intermediate hosts

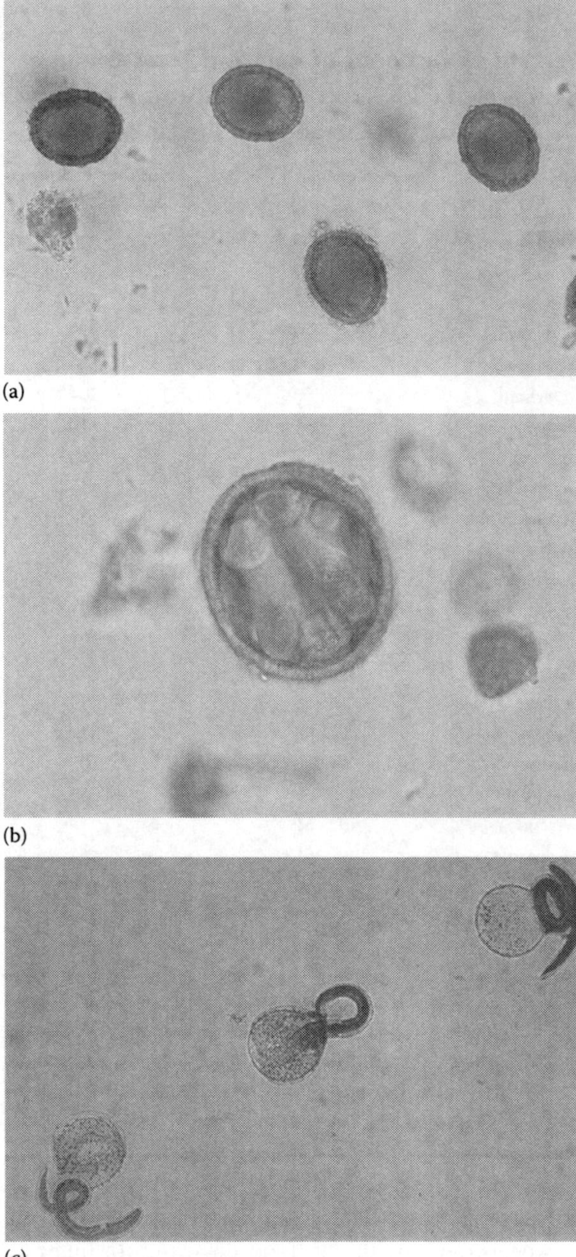

FIGURE 28.4 Raccoon roundworm eggs from raccoon feces (the larva inside the eggs has not yet developed to the point where the eggs are infective) (a). An older egg that contains an infective larva (the coiled worm inside the egg is shown in cross section) (b). Larvae as they are emerging from their eggs (c). (First two photos: courtesy of Kevin Kazacos, Purdue University, West Lafayette, IN; last photo: courtesy of the CDC, Atlanta, GA.)

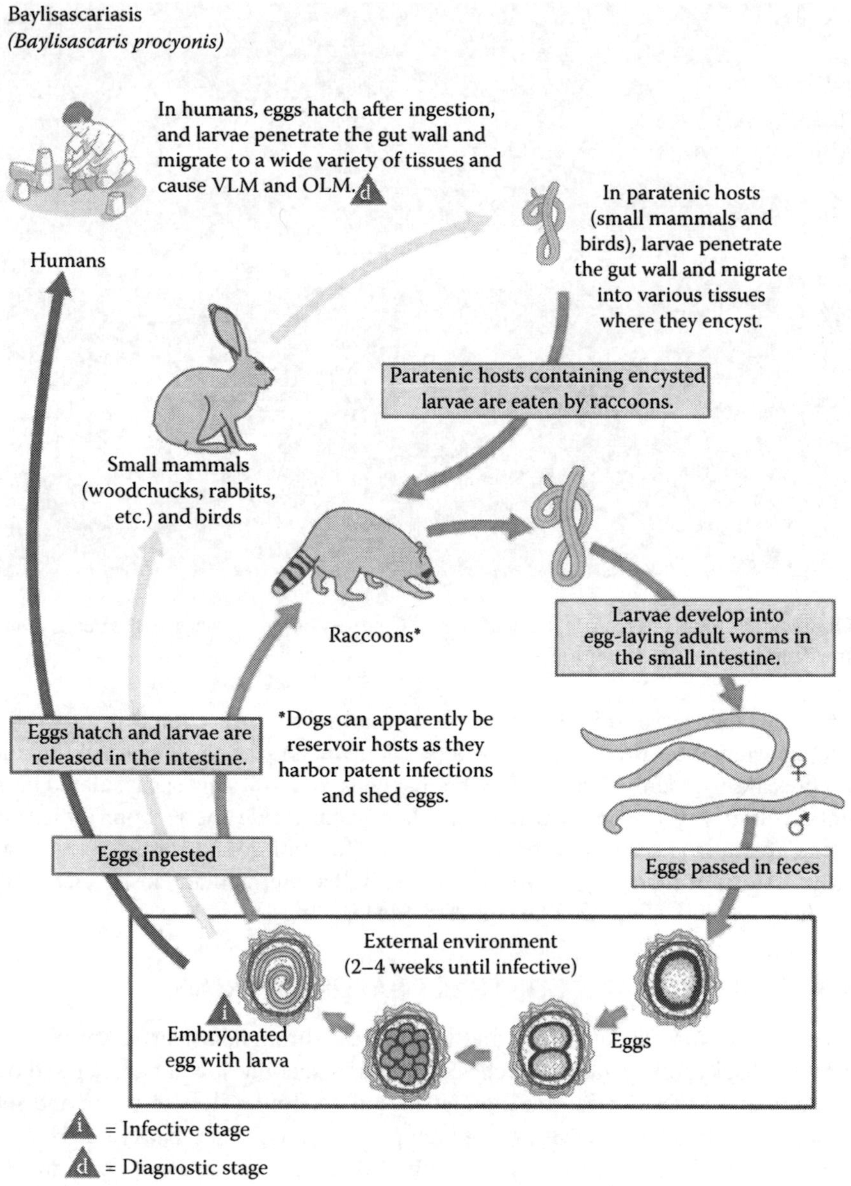

Baylisascariasis
(Baylisascaris procyonis)

In humans, eggs hatch after ingestion, and larvae penetrate the gut wall and migrate to a wide variety of tissues and cause VLM and OLM. d

In paratenic hosts (small mammals and birds), larvae penetrate the gut wall and migrate into various tissues where they encyst.

Humans

Paratenic hosts containing encysted larvae are eaten by raccoons.

Small mammals (woodchucks, rabbits, etc.) and birds

Raccoons*

Larvae develop into egg-laying adult worms in the small intestine.

Eggs hatch and larvae are released in the intestine.

*Dogs can apparently be reservoir hosts as they harbor patent infections and shed eggs.

♀

♂

Eggs ingested

Eggs passed in feces

External environment (2–4 weeks until infective)

i

Embryonated egg with larva

Eggs

i = Infective stage

d = Diagnostic stage

FIGURE 28.5 Life cycle of raccoon roundworms. (Courtesy of Alexander J. da Silva, Melanie Moser, and the CDC, Atlanta, GA.)

and cause neurological problems, such as lethargy, circling behavior, tremors, prostestation, and coma. The severity of neural larva migrans in intermediate hosts will depend on the number of roundworms in the brain, their location, and travel paths through the brain. In a small bird or mammal, a single larva in the brain is often fatal (Kazacos 2001, Page et al. 2001, Evans 2002, Sorvillo et al. 2002).

FIGURE 28.6 Photo of an Allegheny woodrat. (Courtesy of Joe Kosack and the Pennsylvania Game Commission, Harrisburg, PA.)

Animals that serve as intermediate hosts transmit raccoon roundworms to adult raccoons when they are consumed by a raccoon when raccoons either kill them for food or scavenge their carcasses. If either happens, roundworms encapsulated in the animal's flesh will also be ingested and then hatch inside the raccoon's digestive system, completing the roundworm's life cycle. Raccoon roundworms can be a significant source of mortality for species that serve as intermediate hosts, such as the Allegheny woodrat (Figures 28.6 and 28.7; Sidebar 28.1).

28.4 HOW HUMANS CONTRACT BAYLISASCARIASIS

Raccoons live in rural, suburban, and urban areas throughout North America and venture in backyards, garages, porches, and homes seeking food or shelter and defecating in all of these areas. Raccoon latrines are common in both urban and suburban areas with latrine densities ranging up to six latrines per backyard. Humans become infected after ingesting soil or other material that has been contaminated by raccoon feces. This often happens when people put their fingers in their mouth. Children are more likely to do this than adults and therefore are more at risk of becoming ill. Over half of the reported human cases were children under 3 years of age. Baylisascariasis is not contagious: the parasite cannot be transmitted from one person to another (Sorvillo et al. 2002, Roussere et al. 2003, Page 2013).

Domestic dogs typically become infected with raccoon roundworms without showing any signs of illness. Some raccoon roundworms stay inside a dog's intestine and can shed eggs in the dog's feces although this happens infrequently. Hence, people should avoid contact with dog feces, and dogs should be dewormed by a

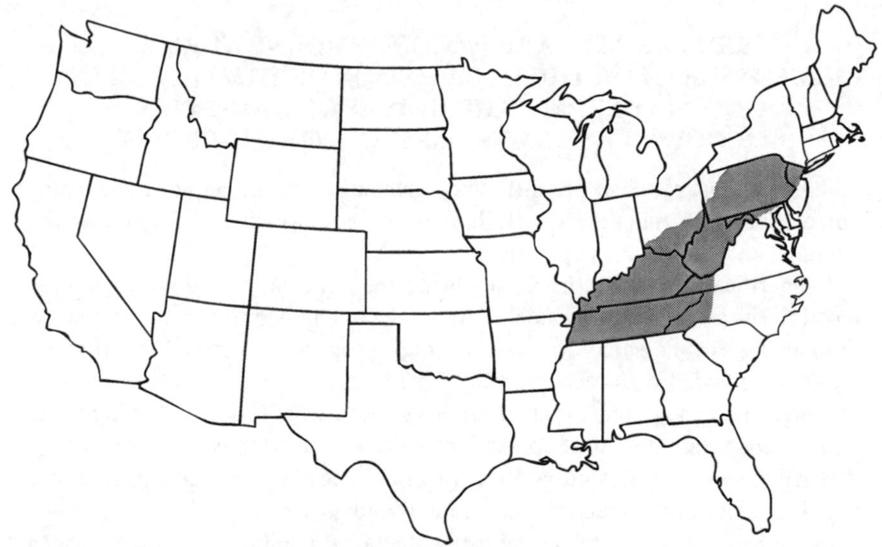

FIGURE 28.7 Range map of the Allegheny woodrat.

veterinarian at regular intervals to kill raccoon roundworms and other intestinal parasites (Center for Food Security and Public Health 2012).

Along with raccoons, kinkajous are members of the family Procyonidae; kinkajous live in the rainforests of Central and South America, and they can serve as definitive hosts for the raccoon roundworm (Figure 28.8). It is less clear if other members of the Procyonidae, such as coatis and ringtails, can serve as definitive hosts for this parasite. Some people keep kinkajous as pets, and these captive animals can be infected with raccoon roundworms. If infected, pet kinkajous can shed roundworm eggs in their feces, posing a risk of spreading the parasite to people. Hence, people should avoid contact with kinkajou feces, and pet kinkajous should be dewormed at regular intervals by a veterinarian (Overstreet 1970, Kazacos et al. 2011).

Veterinarians, fur trappers, taxidermists, wildlife biologists, animal control operators, and wildlife rehabilitators have an increased risk of developing baylisascariasis. Raccoons and raccoon roundworms both have a broad range in North America (CDC 2012).

28.5 MEDICAL TREATMENT

No drugs are totally effective against raccoon roundworms, but the drug of choice is albendazole, which is usually given orally for 10–20 days to stop the migration of larva through the body. If albendazole is not available, mebendazole or ivermectin can be substituted. Inflammation can be reduced by using corticosteroids. Antibiotics should be administered as soon as possible as their effectiveness increases when initiated within 3 days of exposure. Unfortunately, there is a time lag of 2–3 weeks between the beginning of neural larva migrans and the onset of clinical signs; and by

SIDEBAR 28.1 ARE RACCOON ROUNDWORMS RESPONSIBLE FOR THE EXTIRPATION OF THE ALLEGHENY WOODRAT FROM SOME NEW ENGLAND STATES? (MCGOWAN 1993, SMYSER ET AL. 2013A, PAGE 2013)

Allegheny woodrats live on cliffs and talus slopes found along the Appalachian Mountains and Interior Highlands of the northeastern United States (Figures 28.6 and 28.7). Populations of the Allegheny woodrat have declined in the northeastern part of its range during the last century and become extirpated from several states. Concomitantly, raccoon populations have increased dramatically in the same areas, and a high proportion of raccoons in these areas are infected with raccoon roundworms. Scientists hypothesize that the increase in raccoon populations is a detriment to woodrats by infecting them with raccoon roundworms. Support for the hypothesis comes from the finding that Allegheny woodrats often cache raccoon feces in their middens and that woodrats often are infected with raccoon roundworms.

An effort was made to reestablish Allegheny woodrats into their former range in New York by releasing disease-free Allegheny woodrats along talus slopes that historically were occupied by woodrats. Unfortunately, all of the released animals and other offspring died. Before they died, 11 woodrats were captured, and all had raccoon roundworms in their central nervous system. Four of the woodrats exhibited abnormal behavior when live trapped, and such behavior would increase their risk of being killed by predators even if neural larva migrans did not kill them outright. This unsuccessful attempt to reintroduce the Allegheny woodrats provided further support for the hypothesis that woodrats suffer when raccoons and raccoon roundworms move into their habitat. Efforts to reintroduce woodrats in Indiana were more successful, in part, because anthelmintic baits (i.e., baits containing a chemical that kills parasitic worms) were dispersed to reduce the prevalence of raccoon roundworms.

then, damage is irreparable. For patients with meningoencephalitis of an unknown cause, baylisascariasis may be the cause, and treatment with albendazole and corticosteroids should begin promptly. This is especially true for patients who are young children. Laser photocoagulation has been used successfully to kill ocular larva and to prevent further damage (Perlman et al. 2010, CDC 2013). The Mayo Clinic recommends that anyone who suspects that they have baylisascariasis should seek immediate medical attention and inform the attending physician about any recent exposure to raccoons or their feces.

28.6 WHAT PEOPLE CAN DO TO REDUCE THEIR RISK OF CONTRACTING BAYLISASCARIASIS

No vaccines have been developed that can prevent infections of raccoon roundworm. Therefore, the best way to prevent baylisascariasis is to avoid contact with raccoons

FIGURE 28.8 Kinkajous are in the same family as raccoons, and they also can shed the eggs of raccoon roundworms in their feces. (This drawing of a kinkajou is from the *Dictionnaire Universel d'Histoire Naturelle* by Alcide d'Orbigny, which was published in 1849.)

or their feces. Children should learn to wash their hands thoroughly with soap and water after playing outside and to keep their fingers out of their mouths. They should be taught to avoid raccoons and not play in areas where raccoon feces or contaminated soil may exist. Outdoor play areas should be inspected daily for raccoon fecal material. Raccoons and kinkajous should not be kept as pets (Figure 28.9), or their owners should have them dewormed at regular intervals by a veterinarian (Overstreet 1970, Kazacos et al. 2011).

Raccoons are drawn to homes in search of food or shelter. Raccoons are wild animals and should not be fed. Dog or cat food should not be left in areas where raccoons can obtain it; garbage should be kept in raccoon-proof containers. Homes and other buildings are often used by raccoon for shelter. Chimneys should be capped so raccoons cannot use them for shelter. Holes that raccoon can use to gain entry into a garage or building should be covered. Crawl spaces beneath buildings should be secured. Woodpiles, thick brush, and other dense cover that raccoons can use for den sites should be removed (Conover 2002). Sandboxes should be kept covered to prevent raccoons from using them for a latrine (Figure 28.10).

Raccoon feces should be removed and latrines sterilized (Sidebar 28.2). Removal of the top 2–4 in. (5–10 cm) of soil at latrines help decrease egg numbers, but some viable eggs will remain despite rigorous cleanup measures. The CDC recommends that feces and soil contaminated with their fecal material be removed with a shovel or inverted plastic bag and then disposed by burning, burying, or putting it in the garbage for disposal at a landfill. Most chemical

FIGURE 28.9 Keeping raccoons as a pet is illegal in most states and increases the risk of being infected with raccoon roundworms. (Courtesy of Tony DeNicola.)

disinfectants do not kill roundworm eggs, but high temperatures will. Thus, sites contaminated with raccoon roundworm eggs should be sterilized with boiling water or a propane torch; the local fire department, however, should be contacted before using a propane torch to learn how to use it safely and any restrictions on its use. Disposable gloves and rubber boots should be worn. A N95-rated respiratory should be used if cleaning in a confined space. After finishing, boots, shovels, and other tools used in the cleaning operation should be disinfected, clothes cleaned in hot water, and hands washed with soap and warm water (Center for Food Security and Public Health 2012, CDC 2013).

Raccoons like to defecate in water and sometimes use swimming pools for this purpose. While chlorine will kill most bacteria in pools, it does not kill roundworm eggs. People should not use a contaminated pool until it is thoroughly cleaned. To clean a pool, it should be filtered for at least a day, and the pool filter should then be

FIGURE 28.10 Sandboxes should be kept covered when not in use so raccoons do not defecate in them. (Courtesy of Pieter Bos.)

SIDEBAR 28.2 CAN THE PREVALENCE OF RACCOON ROUNDWORMS BE REDUCED IN RACCOONS AND THEIR FECES BY A NONLETHAL APPROACH? (PAGE ET AL. 2011, SMYSER ET AL. 2013B)

Many people are fond of raccoons and are opposed to killing them to control raccoon roundworms. Kristen Page and others tested the effectiveness of a nonlethal approach. Their study involved comparing eight untreated sites to eight treated sites where all raccoon latrines were removed and sterilized with a propane torch. Anthelmintic baits were then distributed monthly in the treated sites with the number of distributed baits being dependent on the local density of raccoons.

Before treatments began, raccoon roundworm eggs were found in 33% of all latrines. After the anthelmintic baits were distributed, the prevalence of eggs declined more than threefold in treatment plots, but prevalence did not change in untreated plots. One year after the onset of treatments, the proportion of white-footed mice infected with raccoon roundworm larva differed significantly between the treated sites (27% of mice infected) and untreated sites (38%). The results of this study indicate that the prevalence of raccoon roundworms can be reduced by using this nonlethal approach.

backwashed. People cleaning the filter should wear disposable gloves. The discarded filter, gloves, and other contaminated material should be double bagged in plastic garbage bags and thrown away. Alternatively, the pool can be drained and cleaned before being refilled (CDC 2013).

28.7 ERADICATING BAYLISASCARIASIS FROM A COUNTRY

Raccoons have successfully adapted to human-dominated environments and are common in rural, suburban, and urban areas. It will be impossible to eradicate *Baylisascaris* until raccoons cease to be a reservoir host. Unfortunately, we do not know how to eliminate raccoon roundworms from raccoons.

LITERATURE CITED

Bauer, C. 2013. Baylisascariosis—Infections of animals and humans with 'unusual' roundworms. *Veterinary Parasitology* 193:404–412.

CDC. 2012. *Baylisascaris*. http://www.cdc.gov/parasites/ (accessed December 23, 2013).

CDC. 2013. *Baylisascaris* infections. Factsheets. CS233728 and CS222686. http://www.cdc.gov/parasites/baylisascaris/ (accessed September 15, 2013).

Center for Food Security and Public Health. 2012. Baylisascariasis. Iowa State University, Ames, IA.

Conover, M. R. 2002. *Resolving Human-Wildlife Conflicts: The Science of Wildlife Damage Management*. CRC Press, Boca Raton, FL.

Evans, R. H. 2002. *Baylisascaris procyonis* (Nematode: Ascarididae) larva migrans in free-ranging wildlife in Orange County, California. *Journal of Parasitology* 88:299–311.

Kazacos, K. R. 2001. *Baylisascaris procyonis* and related species. In: W. M. Samuel, M. J. Pybus, and A. A. Kocan, editors. *Parasitic Diseases of Wild Animals*, 2nd edition. Iowa State University Press, Ames, IA, pp. 301–341.

Kazacos, K. R., T. P. Kilbane, K. D. Zimmerman, T. Chavez-Lindell et al. 2011. Raccoon roundworms in pet kinkajous—Three states, 1999 and 2010. *Morbidity and Mortality Weekly Report* 60:302–305.

McGowan, E. M. 1993. Experimental release and fate study of the Allegheny woodrat (*Neotoma magister*). New York Federal Aid Project W-166-E. Albany, NY.

Overstreet, R. M. 1970. *Baylisascaris procyonis* (Sefanski and Zarnowski, 1951) from kinkajou, *Potos flavus*, in Colombia. *Proceedings of the Helminthological Society of Washington* 37:192–193.

Page, K., J. C. Beasley, Z. H. Olson, T. J. Smyser et al. 2011. Reducing *Baylisascaris procyonis* roundworm larvae in raccoon latrines. *Emerging Infectious Diseases* 17:90–93.

Page, L. K. 2013. Parasites and the conservation of small populations: The case of *Baylisascaris procyonis*. *International Journal for Parasitology: Parasites and Wildlife* 2:203–210.

Page, L. K., R. K. Swihart, and K. R. Kazacos. 2001. Seed preferences and foraging by granivores at raccoon latrines in the transmission dynamics of the raccoon roundworm (*Baylisascaris procyonis*). *Canadian Journal of Zoology* 79:616–622.

Perlman, J. E., K. R. Kazacos, G. H. Imperato, R. U. Desai et al. 2010. *Baylisascaris procyonis* neural larva migrans in an infant in New York City. *Journal of Neuroparasitology* 1(N100502):1–5.

Roussere, G. P., W. J. Murray, C. B. Raudenbush, M. J. Kutilek et al. 2003. Raccoon roundworm eggs near homes and risk for larva migrans disease, California communities. *Emerging Infectious Diseases* 9:1516–1522.

Smyser, T. J., S. A. Johnson, L. K. Page, C. M. Hudson, and O. E. Rhodes Jr. 2013a. Use of experimental translocations of Allegheny woodrat to decipher causal agents of decline. *Conservation Biology* 27:752–762.

Smyser, T. J., L. K. Page, S. A. Johnson, C. M. Hudson et al. 2013b. Management of raccoon roundworm in free-ranging raccoon populations via anthelmintic baiting. *Journal of Wildlife Management* 77:1372–1379.

Sorvillo, F., L. R. Ash, O. G. W. Berlin, J. Yatabe et al. 2002. *Baylisascaris procyonis*: An emerging helminthic zoonosis. *Emerging Infectious Diseases* 8:355–359.

29 Trichinellosis

And the swine, though he divide the hoof, and be clovenfooted, yet he cheweth not the cud; he is unclean to you. Of their flesh shall ye not eat, and their carcase shall ye not touch; they are unclean to you.

Leviticus 11:7–8 (King James Version of The Bible)

Forbidden to you are: dead meat, blood, the flesh of swine....

The Qur'an (5:3)

The presence of *Trichinella spiralis* among sylvatic carnivores represents the Sword of Damocles for domestic pigs living in the same area.

Murrell and Pozio (2000)

29.1 INTRODUCTION AND HISTORY

Trichinellosis is a human disease caused by the larvae of several roundworm species within the genus *Trichinella*. There are at least eight species of *Trichinella*, but most human disease is the result of *T. spiralis*: a pathogen that now occurs around the world due to humans inadvertently moving infected pigs around the world. Animals and humans are infected by consuming meat containing the encysted larvae of *Trichinella* (Figure 29.1). Once ingested, the digestive enzymes in the stomach dissolve the capsule surrounding the larvae, freeing them to burrow into the cells lining the intestines. Once there, the larvae grow quickly and become adults of about 0.1 in. (3 mm) within a week (Figure 29.2). Females start producing live larvae as soon as they become adults. Most adult roundworms do not survive for more than 4 weeks before they are killed by the host's immune system; however by then, each adult female can produce between 500 and 1500 live larvae (Figures 29.3). These second-generation larvae are not shed in feces but instead burrow through the cells of the intestine and are swept into lymphatic systems and bloodstream, which carries them throughout the body. They then burrow through the capillaries and into cells, seeking a striated muscle cell, which they need to survive. Larvae may select cells to invade at random, resulting in cell destruction. Once invaded by a larva, a muscle cell does not die but is altered by the larva into a nurse cell where the larva grows and encapsulates 18–20 days later. Once encapsulated, the larva can survive for months to years. If the infected animal or its carcass is consumed by a predator, omnivore, or scavenger, the encysted larva will complete its life cycle by infecting the digestive system of the new animal and maturing into an adult (Despommier 1990, Capó and Despommier 1996, Despommier et al. 2005).

Trichinellosis is an ancient disease of humans; *Trichinella* larvae have been found in a 3200-year-old Egyptian mummy. Both the Bible and the Qur'an prohibit

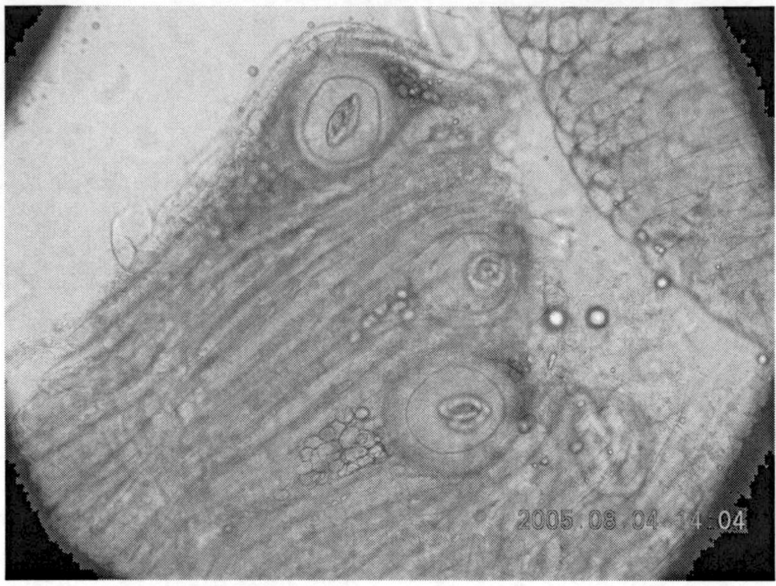

FIGURE 29.1 Photomicrograph showing three *Trichinella* larvae encapsulated in muscle tissue. (Courtesy of Lorraine McIntyre and the British Columbia Centre for Disease Control.)

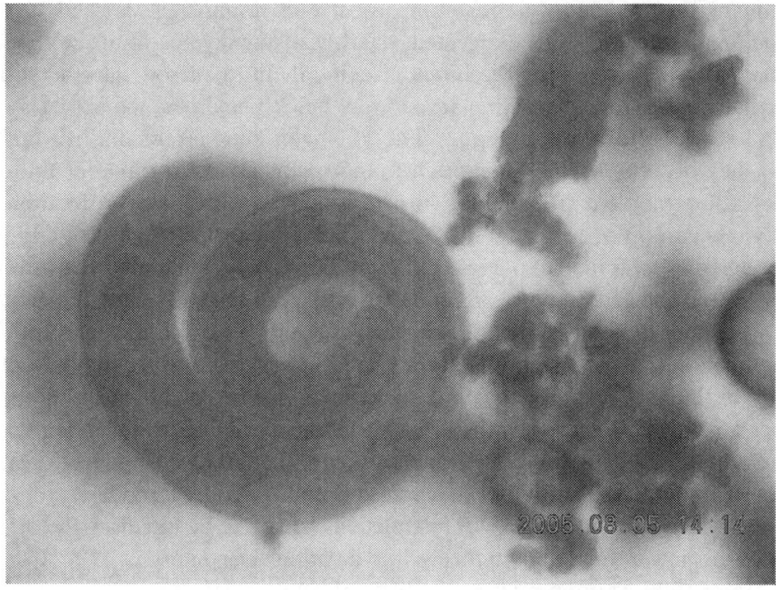

FIGURE 29.2 Photomicrograph of a *Trichinella* larva that has emerged from being encapsulated in muscle tissue. (Courtesy of Lorraine McIntyre and the British Columbia Centre for Disease Control.)

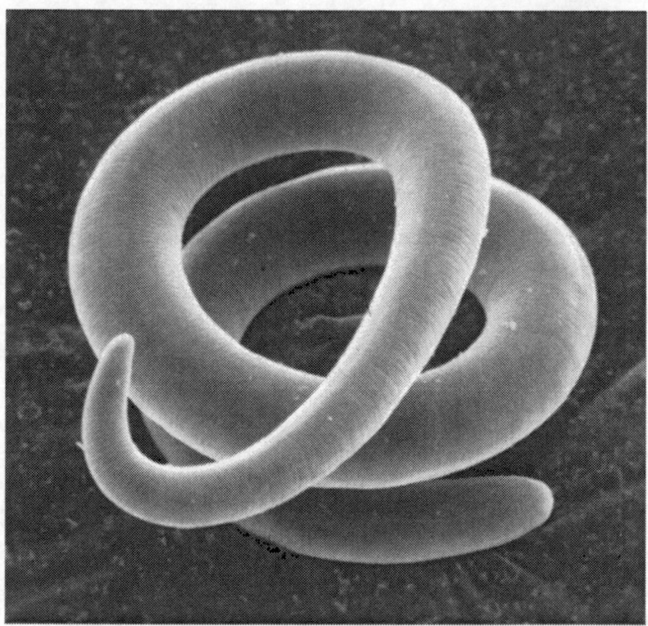

FIGURE 29.3 Photo of an adult *Trichinella* taken through a scanning electron microscope. (Courtesy of Dickson Despommier and trichinella.org.)

the consumption of pork, and these prohibitions protect humans from the risk of trichinellosis. During the Middle Ages, the disease was known as the English sweat or sweating plague; during the 1800s, it was called dandy fever in the West Indies. It was not until the 1820s that the encapsulated larvae of *Trichinella* were described and the cause of the disease was not determined until the 1850s. But even after these discoveries, the disease continued to plague humans. From 1860 to 1880, thousands of people in Germany were stricken with trichinellosis, and more than 500 people died from the disease (Blancou 2001, Neghina et al. 2012).

Trichinellosis, which is also called trichinosis, used to be common in the United States. Before World War II, prevalence of *Trichinella* in humans in the United States reached 36%, but the disease became less common after that. From 1947 to 1951, there was an average of 393 human cases and 57 deaths reported annually in the United States. More recently, the annual incidence declined to 10–30 cases in the United States with a similar number in Canada (Figure 29.4). Worldwide, an estimated 10,000 people develop trichinellosis annually; the case fatality rate is 0.2% (Zimmermann 1970, Roy et al. 2003, Pozio 2007, Gottstein et al. 2009, Hall et al. 2012, CDC 2013).

29.2 SYMPTOMS IN HUMANS

In humans, an infection of *Trichinella* can be separated into two phases: an intestinal phase and a muscular phase. The intestinal phase occurs within 2 days of ingesting infected meat. Symptoms are nonspecific and include nausea, vomiting, diarrhea,

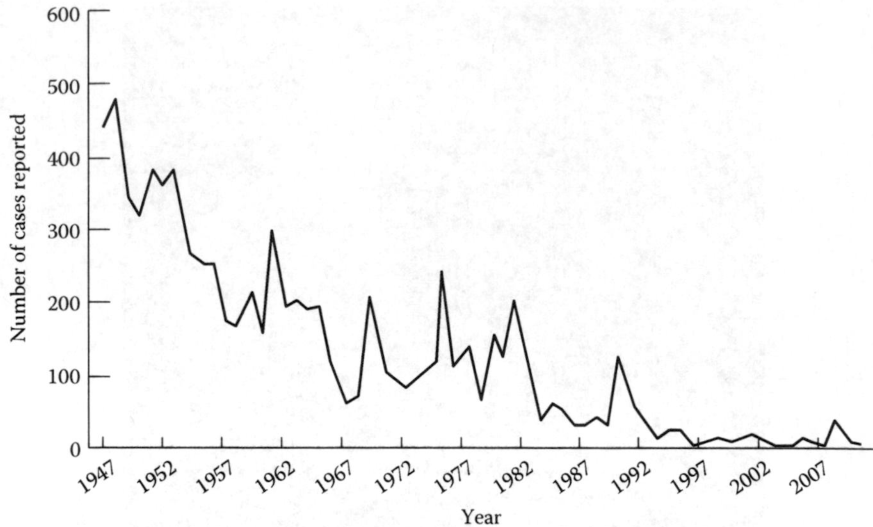

FIGURE 29.4 Annual number of trichinellosis cases in the United States that were reported to the CDC. (From Kennedy, E.D. et al., Trichinellosis surveillance—United States, 2002–2007. *Morbid. Mortal. Wkly. Rep.* 58(SS09), 1, 2009.)

and abdominal pain and result from ingested larva burrowing into the mucosa of the small intestine (Figure 29.5). The second phase begins in 1–2 weeks as the ingested roundworms start to produce larvae of their own. It is the migration of this second-generation of larva through body tissue that produces the symptoms usually associated with trichinellosis. These symptoms include headaches, fever, swelling of the face and eyes, coughing, and an itchy skin rash. Muscles are often painful when pressed. Some patients experience pain when moving the eye muscles. Serious clinical signs can develop, such as myocarditis, encephalitis, breathing problems, and the formation of blood clots. Fatalities from trichinellosis occur during this period (i.e., 3–5 weeks after the initial infection) and often result from heart failure or encephalitis. The illness wanes 6–8 weeks after infection when the production of migrating larvae ceases and the larvae already in the muscle tissue finish encapsulating. Fatigue and muscle pain can continue for a longer period (Murrell and Bruschi 1994, Dupouy-Camet et al. 2002, Despommier et al. 2005).

Severity of illness varies, based on the species of *Trichinella* causing the infection, the number of larvae, and the host's susceptibility to infestation. The minimum number of larvae necessary for causing illness in humans is believed to be between 70 and 100. Infections by *T. spiralis* are often more serious than infections caused by other species because each adult female *T. spiralis* produces more larvae. The encysted larvae in human muscle tissue usually die within 2–5 years and then may become calcified. Prior infections of *Trichinella* impart some immunity to the parasite so that patients do not become as ill on subsequent infections (Murrell and Bruschi 1994, Dupouy-Camet et al. 2002, CDC 2012).

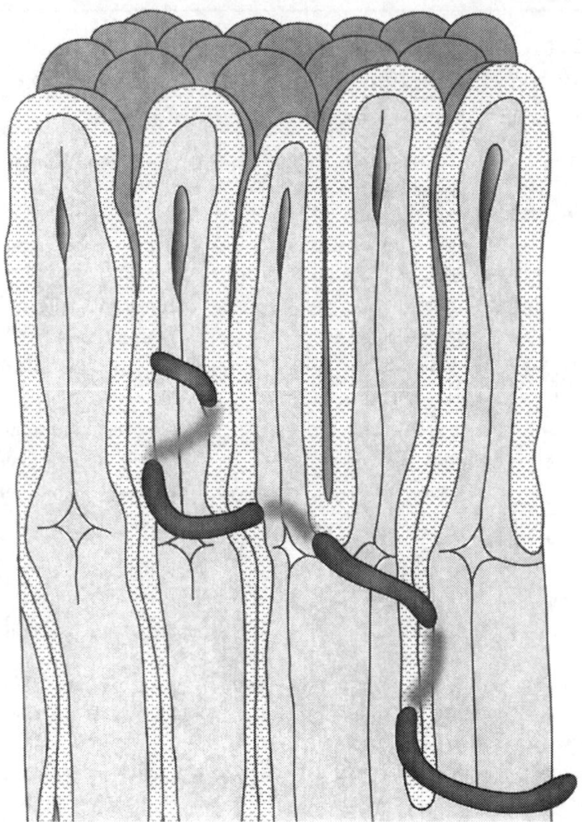

FIGURE 29.5 A drawing of *Trichinella* living within the cells that make up the mucosa of the small intestine. (Adapted from a drawing by J. Karapelou in Despommier, D.D. et al., *Parasitic Diseases*, 5th edn., Apple Trees Productions, New York, 2005.)

29.3 *TRICHINELLA* IN ANIMALS

There are at least eight species and four additional genotypes of *Trichinella* (Table 29.1). The encapsulating group only parasitizes mammals, while the second group parasitizes mammals, birds, and reptiles. This chapter will focus primarily on those species that infect humans in North America.

T. spiralis is responsible for most human cases of trichinellosis across the world. Humans are dead-end hosts for all species of *Trichinella*, but *T. spiralis* is able to maintain itself in two different transmission cycles: a domestic one and a wildlife one. Domestic pigs are the reservoir hosts for the domestic cycle, especially pigs that fed meat or have access to carcasses. Cats, dogs, rats, and wildlife living on farms containing infected pigs can become infected themselves, but *T. spiralis* cannot maintain itself by just infecting these species. Infected rats can spread the parasite to other farms, and the parasite is also transmitted to rats, cats, and dogs when they ingest scraps of infected pork. Horses can become infected if they are fed a

TABLE 29.1
Taxonomy and Distribution of *Trichinella* Species and the Animals that Serve as the Main Source of Human Infections

Trichinella spp.	Distribution	Main Source of Human Infection
Encapsulating Species		
T. britovi	Asia, Europe, Africa	Pigs, horses, wild carnivores
T. nativa	Arctic and subarctic	Wild carnivores
T. nelsoni	Africa	Warthogs, bushpigs
T. murrelli	North America	Horses and bears
T. spiralis	Worldwide	Pigs, horses, and wild mammals
T. genotype T6	United States and Canada	Wild carnivores
T. genotype T8	South Africa	Not known to infect humans
T. genotype T9	Japan	Not known to infect humans
T. genotype T12	Argentina	Not known to infect humans
Nonencapsulating Species		
T. papuae	Southeast Asia	Wild pigs
T. pseudospiralis	Worldwide	Pigs
T. zimbabwensis	Africa	Not known to infect humans

Sources: Based on Pozio, E. and Murrell, K.D., *Adv. Parasitol.*, 63, 367, 2006; Pozio, E., in S.R. Palmer et al., *Oxford Textbook of Zoonoses*, Oxford University Press, New York, pp. 755–766, 2011.

Note: One group of species has larvae that encapsulate themselves in muscle cells, and the other has larvae that lack this ability.

protein supplement that includes infected meat. Other wildlife living near farms with infected pigs also can become infected. This domestic transmission cycle has almost disappeared from commercial livestock farms located in Canada, the United States, and western Europe but still occurs in Mexico, Argentina, Chile, and China. *T. spiralis* can maintain itself through a wildlife transmission cycle where these parasites infect wild carnivores, omnivores, and scavengers and some of their prey species (Pozio and Zarlenga 2005, Gottstein et al. 2009).

T. murrelli infects carnivores across the United States and southern Canada and maintains itself in a wildlife transmission cycle. Humans become infected primarily from eating the raw or undercooked meat from black bears (Sidebar 29.1). An outbreak caused by *T. murrelli* occurred in France when people consumed horse meat that had been imported from the United States (Gottstein et al. 2009).

T. nativa and T. genotype T6 are genetically similar, and both occur in Canada and the northern United States and maintain themselves in a wildlife transmission cycle where the reservoir hosts include grizzly bear, black bear, wolverine, Canada lynx, raccoon, wolf, and fox. A marine transmission cycle also exists that involves polar bears, walruses, and seals in the Arctic and subarctic regions of Canada and Greenland. An interesting feature of these *Trichinella* species is that they can

SIDEBAR 29.1 OUTBREAK OF TRICHINELLOSIS AMONG PEOPLE ATTENDING A COMMUNITY GATHERING IN CALIFORNIA (HALL ET AL. 2012)

On October 27, 2008, there was a community gathering in Humboldt County, California, for a meal that featured fresh bear meat served in several dishes, including one dish with raw chopped meat. The bear had been shot during the hunting season in the mountains east of Humboldt County. In November, a local physician reported to the county health department that he had three trichinellosis patients, all of whom had attended the gathering and consumed bear meat. The state began an investigation and determined that 29 of the 38 people attending the event had a probable or confirmed case of trichinellosis. The only people who consumed bear meat and did not become ill were young children. Everyone else became ill although each person consumed an average of only 1.5 ounces (43 g) of bear meat. The most common symptoms included muscle aches (100% of patients), fatigue (88%), fever (76%), facial edema (67%), and an itchy rash (66%). Symptoms began after an incubation period of 1–38 days (Figure 29.6). Patients were treated with anthelmintic drugs, but most (73%) still had muscle aches in mid-December, 6 weeks after consuming bear meat. All patients made a full recovery. Most patients had antibodies that were specific against *Trichinella*. A paw from the bear contained encapsulated *T. murrelli*, providing further evidence that the patients had trichinellosis. This was the first known outbreak of human trichinellosis in the United States that was caused by this species of *T. murrelli*.

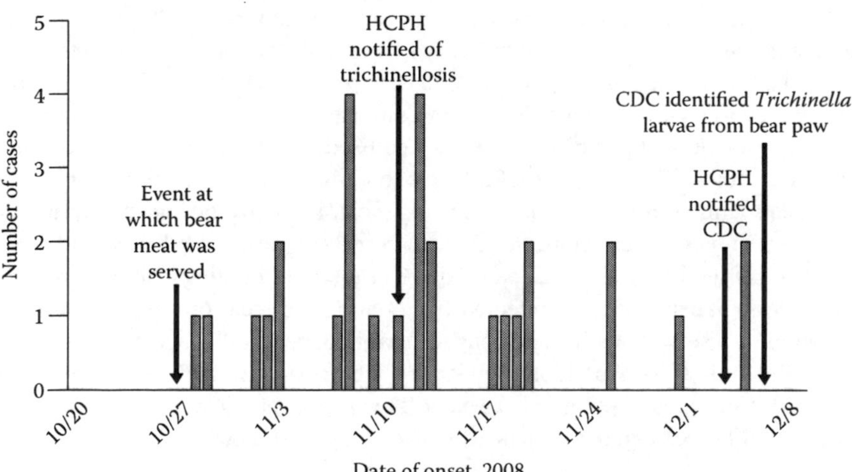

FIGURE 29.6 Onset of trichinellosis symptoms among 29 patients who became infected after consuming bear meat at a community function in California. (Data from Hall, R.L. et al., *Am. J. Trop. Med. Hyg.*, 87, 297, 2012.)

survive freezing temperatures, which explains their presence in northern latitudes. Polar bears are a key part of the marine cycle of the parasite; in some areas, up to 60% of all bears are infected. Arctic foxes, wolves, and fin whales (but not other whale species) also become infected as an offshoot of this marine cycle (Appleyard and Gajadhar 2000, McIntyre et al. 2007).

Ireland is one country where *Trichinella* has been able to maintain itself by just infecting wildlife. The pathogen has not been detected in either pigs or humans for decades; yet 3%–4% of the red foxes in the country have been infected through a fox-to-fox transmission cycle where foxes scavenge infected fox carcasses. In Europe, *Trichinella* can sustain itself by infecting red foxes and raccoon dogs. Many other European carnivores including European badger, beech marten, Eurasian lynx, and Asian black bears can become infected with the parasite, but these predators do not play a major role in sustaining *Trichinella* (Murrell and Pozio 2000, Rafter et al. 2005).

Virtually all wild carnivores and omnivores are susceptible to *Trichinella* infections; infection rates can be high in these animals. As some examples, 55% of cougars collected in Montana, Idaho, and Wyoming were infected, as were 18% in British Columbia (Dworkin et al. 1996). In Pennsylvania, 15% of red foxes, 7% of gray foxes, 6% of mink, 4% of skunks, 3% of opossum, and 3% of raccoons were infected (Schad et al. 1984). Based on samples from across Canada, Smith and Snowdon (1988) reported infection rates of 8% in wolves, 3% in both Arctic fox and red fox, 3% in Canada lynx, 1% in bobcat, 1% in raccoons, and 0.5% in coyotes. In Ontario, 5% of fisher and 3% of American marten were infected (Dick et al. 1986).

29.4 HOW HUMANS CONTRACT TRICHINELLOSIS

Human-to-human transmission of the parasite does not occur. Instead, most trichinellosis patients became infected from raw or undercooked meat. Worldwide, most people develop trichinellosis after consuming raw or undercooked pork. This is especially true in areas where pigs are raised in backyards or under free-ranging conditions.

Local customs and cuisine play a role in how people acquire trichinellosis. Horse meat is a common source of infection in countries where people consume it. For example, over 3000 French and Italians acquired trichinellosis from consuming horse meat from 1975 through 2000. People in China and the Slovak Republic have developed trichinellosis from consuming dog meat. In the Arctic and subarctic, people become infected from consuming meat from walruses, polar bears, and Arctic fox. In the United States and Canada, there is a strong hunting culture, and most trichinellosis patients become infected from eating the meat of wild carnivores or omnivores. These animals include cougar, feral hog, bear, fox, coyote, and raccoon. From 1997 until 2007, 9 of 15 outbreaks of trichinellosis in the United States were attributed to the consumption of bear meat (Dworkin et al. 1996, Proulx et al. 2002, Roy et al. 2003, Pozio and Zarlenga 2005, Kennedy et al. 2009).

29.5 MEDICAL TREATMENT

Treatment for trichinellosis involves administration of anthelmintic drugs, such as mebendazole or albendazole, that kill the adult worms in the intestines. These drugs stop the production of larva and should be used within 3 days of infection to prevent subsequent muscular invasion. However, most patients do not seek medical attention until the second stage of the disease. By then, the larva will have already encapsulated in the patients' muscle cells, where the larva can survive for years. Hence, long periods of anthelmintic treatment may be required. Corticosteroids can be used to reduce the inflammation that accompanies the infection; without them, inflammation and symptoms may worsen when anthelmintic drugs are first used, owing to increased larval deaths. The case fatality rate is up to 5% for patients who do not receive timely treatment and have many larvae invading the muscles. For milder cases, prognosis is good, and symptoms disappear within 2–6 months (Dupouy-Camet et al. 2002, Gottstein et al. 2009). The CDC (2012) recommends seeking medical attention for anyone who may have trichinellosis; patients should notify their physicians if they have eaten raw or undercooked meat.

29.6 WHAT PEOPLE CAN DO TO REDUCE THEIR RISK OF CONTRACTING TRICHINELLOSIS

There is no vaccine against trichinellosis. Instead, preventing trichinellosis in humans is based on a combination of three approaches: (1) educating the public about how to prepare food to avoid the risk of trichinellosis, (2) controlling of *T. spiralis* in domestic pigs, and (3) reducing *Trichinella* infections in wildlife. The best way to avoid trichinellosis is not to consume meat unless it is thoroughly cooked. A food thermometer placed in the middle of the thickest part of meat should be used to determine when meat is adequately cooked rather than relying on the meat's appearance. The CDC (2012) recommends that whole cuts of meat (other than poultry and wild game) should be cooked to at least 145°F (63°C), ground meat (other than poultry and game meat) be cooked to at least 160°F (71°C), and poultry and game meat be cooked to at least 165°F (74°C). Curing, salting, smoking, freezing, or microwaving meat does not consistently kill *Trichinella* (Figure 29.7). Many people in the United States have become ill from consuming homemade jerky or sausage. Meat grinders, knives, or other implements used to prepare meat need to be cleaned thoroughly every time they are used. In the United States and Canada, most trichinellosis patients become infected from eating meat from wild predators or omnivores, especially bears, wild hogs, and walruses (Figure 29.8). Hunters should thoroughly cook meat from these wild animals and caution people who are given some of the game meat to do the same (Dworkin et al. 1996, CDC 2012).

29.6.1 Control of *Trichinella* in Domestic Pigs

The risk of pigs being infected with *Trichinella* is small when pigs are raised in indoor facilities, which are common on large commercial hog facilities in the United States, Canada, and Europe. But this is not true for backyard pigs that are often fed meat scraps or have access to infected rodents and wildlife. The risk of infection for a backyard pig depends, in part, on the prevalence of the parasite among local rodents or wildlife. At one U.S. farm where the pigs were infected, 42% of rats had *Trichinella* infections with larval densities averaging 8219 larvae per ounce of

FIGURE 29.7 Steaks made from a bear may look healthy, but like those shown in this photo harbor numerous *Trichinella* larvae. (Courtesy of Lorraine McIntyre and the British Columbia Centre for Disease Control.)

FIGURE 29.8 Inuits and other people living in the Arctic may develop trichinellosis by consuming undercook meat from walruses. (Photo courtesy of Andrew Trites and the NOAA.)

muscle (293 larvae/g). Rodent control is an important measure: an intensive rat control program involving the use of traps and poisons can reduce the proportion of rats infected with *Trichinella* (Murrell et al. 1987, Leiby et al. 1990). There is no evidence that the parasite can maintain itself for more than a couple of years by infecting just rats; rather, infected rats result from having infected pigs. Still, the presence of infected rats greatly complicates the task of trying to eliminate *Trichinella* from an infected pig farm. At one such farm, all pigs were killed to stop the outbreak and replaced with *Trichinella*-free pigs. Those pigs exposed to rats became infected rapidly, while pigs isolated from rats were still free of the parasites 12 months later (Schad et al. 1987, Pozio and Zarlenga 2005, Papatsiros et al. 2012).

The International Commission on Trichinellosis recommends six steps to reduce the risk of animals becoming infected with *Trichinella* (Murrell and Pozio 2000):

1. Stop all feeding of garbage and meat to pigs and other animals that are susceptible to *Trichinella.*
2. Prevent pigs from having access to animal carcasses of any kind.
3. Dispose of pig, rodent, and carnivore carcasses by incineration, rendering, or burial in such a way that they cannot be dug up by animals.
4. Construct effective barriers that prevent contact among pigs, wild animals, and domestic pets.
5. Control rodents where pigs are raised.
6. Dispose of offal and carcasses of game animal so that they are not accessible to animals. Shallow burial is not sufficient because many wild animals (e.g., bears, wild hogs, raccoons, wolves) will unearth them.

In Europe, meat inspections for *Trichinella* are required for all animals slaughtered for food from pigs, horses, and wild game species that are likely to harbor the parasite. The United States and Canada also employ meat inspectors that examine carcasses for *Trichinella.* These inspections are not designed to prevent human infection but to prevent human illness. In the United States, a food safety issue occurs if more than one to three larvae are found in each gram of meat obtained from a muscle mass where the parasite is most commonly found (e.g., diaphragm or tongue of domestic pigs). Some European countries require the mandatory testing of wild hogs for *Trichinella*, but testing is not required in Canada or the United States. In some areas of the United States and Canada, government agencies offer free testing (Murrell and Pozio 2000, Proulx et al. 2002).

29.6.2 Control of *Trichinella* in Wildlife

Many wild animals become infected as part of the domestic transmission cycle involving domestic pigs, as the reservoir hosts for *Trichinella.* In this situation, infected wildlife are clustered around farms containing infected pigs, and the easiest way to eliminate the parasite from wildlife is to eradicate it in pigs.

Eliminating *Trichinella* in an area where the parasite can maintain itself in a wildlife or marine transmission cycle is difficult to achieve. However, hunters contribute to the problem when they leave offal or carcasses in the field or dispose of them in such a manner that allows wild animals' access. Prevalence of *Trichinella* infection rates was reduced substantially among wild hogs located inside a private hunting preserve after offal from hunter-killed carcasses was incinerated (Worley et al. 1994). In Greenland, hunting practices of the Inuit culture contribute to *Trichinella* infection rates among Arctic foxes and polar bears. Sled dogs are fed the remains of wild animals, including polar bears; because of this, up to 90% of dogs are infected. When the dogs die, their carcasses are often left on the sea ice where their carcasses are consumed by polar bears and other mammals (Murrell and Pozio 2000). By incinerating carcasses or offal of animals that may be infected (e.g., seals, polar bears, walruses, red foxes), the number of infected wild animals can be reduced.

29.7 ERADICATING TRICHINELLOSIS FROM A COUNTRY

Great progress has been made reducing the incidence of trichinellosis in humans by reducing the ability of *Trichinella* to sustain itself through the domestic transmission cycle involving domestic pigs. Eliminating *Trichinella* in countries where the parasite can maintain itself in a wildlife or marine transmission cycle will be difficult to achieve, but progress can be made if people dispose of infected meat so that it is unavailable to dogs, pigs, or wild animals.

LITERATURE CITED

Appleyard, G. D. and A. A. Gajadhar. 2000. A review of trichinellosis in people and wildlife in Canada. *Canadian Journal of Public Health* 91:293–297.

Blancou, J. 2001. History of trichinellosis surveillance. *Parasite (Paris, France)* 8(Supplement 2):16–19.

Capó, V. and D. D. Despommier. 1996. Clinical aspect of infection with *Trichinella* spp. *Clinical Microbiology Reviews* 9:47–54.

CDC. 2012. Parasites—Trichinellosis (also known as trichinosis). Resources for health professionals. http://www.cdc.gov/parasites/trichinellosis/health_professional/ (accessed December 24, 2013).

CDC. 2013. Summary of notifiable diseases—United States, 2011. *Morbidity and Mortality Weekly Report* 60(53):1–117.

Despommier, D. D. 1990. *Trichinella spiralis*: The worm that would be virus. *Parasitology Today* 6:193–196.

Despommier, D. D., R. W. Gwadz, P. J. Hotez, and C. A. Knirsch. 2005. *Parasitic Diseases*, 5th edition. Apple Trees Productions, New York.

Dick, T. A., B. Kingscote, M. A. Strickland, and C. W. Douglas. 1986. Sylvatic trichinosis in Ontario, Canada. *Journal of Wildlife Diseases* 22:42–47.

Dupouy-Camet, J., W. Koeciecka, F. Bruschi, F. Bolar-Fernandez, and E. Pozio. 2002. Opinion on the diagnosis and treatment of human trichinellosis. *Expert Opinion on Pharmacotherapy* 3:1117–1130.

Dworkin, M. S., H. R. Gample, D. S. Zarlenga, and P. O. Tennican. 1996. Outbreak of trichinellosis associated with eating cougar jerky. *Journal of Infectious Diseases* 174:663–666.

Gottstein, B., E. Pozio, and K. Nöckler. 2009. Epidemiology, diagnosis, treatment, and control of trichinellosis. *Clinical Microbiology Reviews* 22:127–145.

Hall, R. L., A. Lindsay, C. Hammond, S. P. Montgomery et al. 2012. Outbreak of human trichinellosis in northern California caused by *Trichinella murrelli. American Journal of Tropical Medicine and Hygiene* 87:297–302.

Kennedy, E. D., R. L. Hall, S. P. Montgomery, D. G. Pyburn, and J. L. Jones. 2009. Trichinellosis surveillance—United States, 2002–2007. *Morbidity and Mortality Weekly Report* 58(SS09):1–7.

Leiby, D. A., C. H. Duffy, K. D. Murrell, and G. A. Schad. 1990. *Trichinella spiralis* in an agricultural ecosystem: Transmission in the rat population. *Journal of Parasitology* 76:360–364.

McIntyre, L., S. L. Pollock, M. Fyfe, A. Gajadhar et al. 2007. Trichinellosis from consumption of wild game. *Canadian Medical Association Journal* 176:449–451.

Murrell, K. D. and F. Bruschi. 1994. Clinical trichinellosis. In: T. Sun, editor. *Progress in Clinical Parasitology*. CRC Press, Boca Raton, FL, pp. 117–150.

Murrell, K. D. and E. Pozio. 2000. Trichinellosis: The zoonosis that won't go quietly. *International Journal for Parasitology* 30:1339–1349.

Murrell, K. D., F. Stringfellow, J. B. Dame, D. A. Leiby et al. 1987. *Trichinella spiralis* in an agricultural ecosystem. II. Evidence for natural transmission of *Trichinella spiralis spiralis* from domestic swine to wildlife. *Journal of Parasitology* 73:103–109.

Neghina, R., R. Moldovan, I. Marincu, C. L. Calma, and A. M. Neghina. 2012. The roots of evil: The amazing history of trichinellosis and *Trichinella* parasites. *Parasitology Research* 110:503–508.

Papatsiros, V. G., S. Boutsini, D. Ntousi, D. Stougiou et al. 2012. Detection and zoonotic potential of *Trichinella* spp. from free-range pig farming in Greece. *Foodborne Pathogens and Disease* 9:536–540.

Pozio, E. 2007. World distribution of *Trichinella* spp. infections in animals and humans. *Veterinary Parasitology* 149:3–21.

Pozio, E. 2011. Trichinellosis. In S. R. Palmer, L. Soulsby, P. R. Torgerson, and D. W. G. Brown, editors. *Oxford Textbook of Zoonoses*. Oxford University Press, New York, pp. 755–766.

Pozio, E. and K. D. Murrell. 2006. Systematics and epidemiology of *Trichinella. Advances in Parasitology* 63:367–439.

Pozio, E. and D. S. Zarlenga. 2005. Recent advances on the taxonomy, systematic and epidemiology of *Trichinella. International Journal for Parasitology* 35:1191–1204.

Proulx, J.-F., J. D. MacLean, T. W. Gyorkos, D. Leclair et al. 2002. Novel prevention program for trichinellosis in Inuit communities. *Clinical Infectious Diseases* 34: 1508–1514.

Rafter, P., G. Marucci, P. Brangan, and E. Pozio. 2005. Rediscovery of *Trichinella spiralis* in red foxes (*Vulpes vulpes*) in Ireland after thirty years of oblivion. *Journal of Infection* 50:61–65.

Roy, S. L., A. S. Lopez, and P. M. Schantz. 2003. Trichinellosis surveillance—United States, 1997–2001. *Morbidity and Mortality Weekly Report* 52(RR06):1–8.

Schad, G. A., C. H. Duffy, D. A. Leiby, K. D. Murrell, and E. W. Zirkle. 1987. *Trichinella spiralis* in an agricultural ecosystem: Transmission under natural and experimentally modified on-farm conditions. *Journal of Parasitology* 73:95–102.

Schad, G. A., D. A. Leiby, and K. D. Murrell. 1984. Distribution, prevalence and intensity of *Trichinella spiralis* infection in furbearing mammals of Pennsylvania. *Journal of Parasitology* 70:372–377.

Smith, H. J. and K. E. Snowdon. 1988. Sylvatic trichinosis in Canada. *Canadian Journal of Veterinary Research* 52:488–489.

Worley, D. E., F. M. Seesee, D. S. Zarlenga, and K. D. Murrell. 1994. Attempts to eradicate trichinellosis from a wild boar population in a private game park (U.S.A.). In: C. W. Campbell, E. Pozio, and F. Frushchi, editors. *Trichinellosis*. ISS Press, Rome, Italy, pp. 611–616.

Zimmermann, W. J. 1970. Trichinosis in the United States. In: S. E. Gould, editor. *Trichinosis in Man and Animals*. Charles C. Thomas, Springfield, IL, pp. 378–400.

30 Swimmer's Itch and Giardiasis

Other common names for swimmer's itch include duck flea, duck lice, duck rash, and pelican itch. These names resulted from the realization that people developed the rash after swimming, bathing or wading in water bodies where these birds are common.

Chamot et al. (1998), CDC (2012)

30.1 INTRODUCTION AND HISTORY

Swimmer's itch is an allergic reaction to microscopic parasites invading the skin. These parasites live in both freshwater and saltwater, and people become infected by swimming or wading in water containing the parasites. The medical term for swimmer's itch is cercarial dermatitis. Swimmer's itch was first described by Cort (1928) in Michigan, and it has been reported on all continents, except Antarctica. Swimmer's itch is caused by several species of parasitic flatworms (family Schistosomatidae), including species of the genera *Schistosoma*, *Austrobilharzia*, *Gigantobilbarzia*, and *Trichobilharzia*, all of which use aquatic snails as intermediate hosts and waterbirds as definitive hosts (Merck 2010). Hence, swimmer's itch occurs in water bodies that have snails and are used by waterbirds (Figure 30.1).

Swimmer's itch should not be confused with schistosomiasis, which is a serious human disease caused by different species of Schistosomatidae. The parasites causing schistosomiasis can survive and reproduce in humans, whereas those that cause swimmer's itch cannot. Schistosomiasis is contagious, and most of the 200 million schistosomiasis patients worldwide became infected by consuming water or food contaminated with minute quantities of fecal material from an infected person. Unlike swimmer's itch, schistosomiasis does not occur in North America, and it will not be covered in this book. Because most species of Schistosomatidae that cause swimmer's itch use birds for hosts, we will refer to these species as avian schistosomes to distinguish them from the species that cause schistosomiasis in humans.

30.2 SYMPTOMS IN HUMANS

Humans are a dead-end host for avian schistosomes; when these parasites burrow into a person's skin, they die there. The burrowing activity can produce an immediate tingling or prickling sensation that lasts about an hour. The burrowing of a single parasite produces a reddish skin bump (i.e., papule) at the burrowing site, which appears within minutes or hours of skin penetration. Usually a multitude of parasites burrow into a single person, causing a rash of red bumps (Figure 30.2), which is accompanied by an intense itching sensation. The severity of the rash and the itching

FIGURE 30.1 Ducks and geese are definitive hosts for the parasites that cause swimmer's itch. (Courtesy of the U.S. Fish and Wildlife Service.)

sensation increases with the number of parasites that burrow into the skin. Because the rash is caused by an allergic reaction, the severity of the rash and the itching sensation can increase after repeated exposure to avian schistosomes (Chamot et al. 1998, Horák and Kolářová 2011). The rash and itching sensation usually end within a few days without medical attention.

Swimmer's itch should not be confused with other rashes. Along ocean beaches, rashes can also be caused by jellyfish. Swimmer's itch is more likely to occur on exposed skin while jellyfish rashes predominate on skin covered by tight-fitting clothing. The rash and symptoms of swimmer's itch are also easily mistaken for those caused by exposure to poison ivy. Most cases of swimmer's itch occur during summer because people are more likely to go swimming when it is hot and the number of avian schistosomes in water peaks during the summer (Wong et al. 1994, Leighton et al. 2000, Coady et al. 2006, Valdovinos and Balboa 2008).

30.3 AVIAN SCHISTOSOME INFECTIONS IN ANIMALS

Avian schistosomes have a complex life cycle involving both sexual and asexual reproduction (Figure 30.3). The adults live in the blood of definitive hosts (waterbirds or mammals) where they engage in sexual reproduction and produce eggs that are excreted in the feces of infected animals. Each egg contains a fully developed larva called a miracidium that will hatch when exposed to water. A miracidium can only survive for a few days and spends this time seeking an aquatic snail to infect, usually a pulmonate snail from the families Lymnaeidae, Physidae, and Planorbidae. Once inside a snail, the miracidium develops into a mother sporocyst,

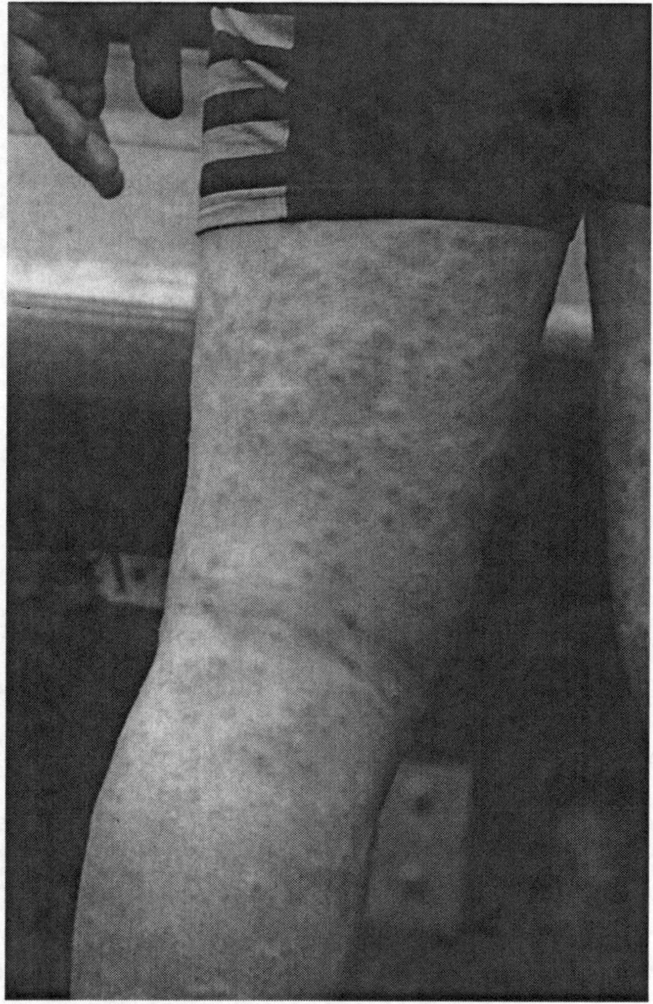

FIGURE 30.2 The rash of swimmer's itch consists of red bumps or papules. (Courtesy of Bernd Sures and the Elsevier Press.)

which produces many daughter sporocysts through asexual reproduction. Each of these daughter sporocysts then produces through asexual reproduction thousands of cercariae, which are the next life stage of the parasite (Chamot et al. 1998, Kolářová et al. 2013).

Cercariae are released into the water from the snail. They have forked tails that are used to move through the water as they seek a vertebrate host to infect (Figure 30.4). It is the cercariae that burrow into human skin causing swimmer's itch. Ducks, mergansers, geese, swans, gulls, and other waterbirds are definitive hosts for most species of avian schistosomes. Cercariae normally penetrate the skin on the feet of waterbirds, burrowing rapidly through their skin and migrating

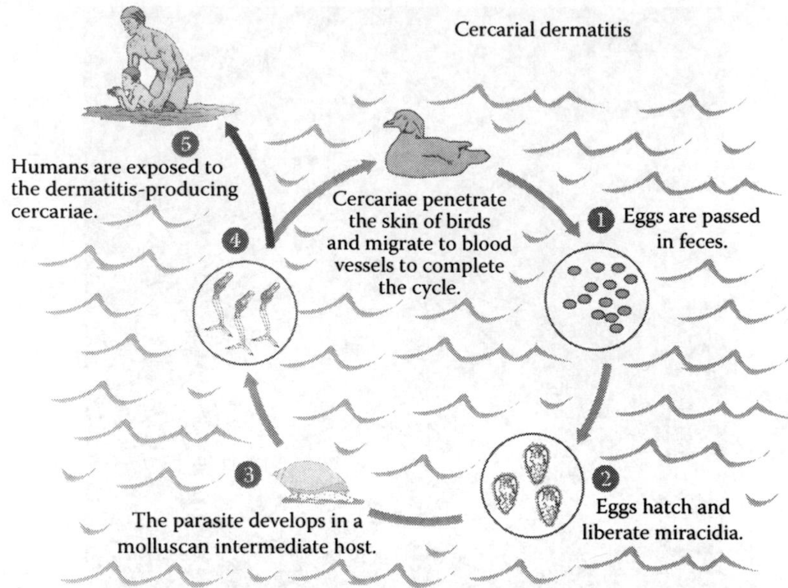

FIGURE 30.3 Life cycle of the parasites that cause swimmer's itch involves aquatic snails that are an intermediate host for the larvae and waterbirds or aquatic mammals that serve as the definitive hosts for the adults. Humans are dead-end hosts for the parasite. (Courtesy of A. J. de Silva, M. Moser, and the CDC.)

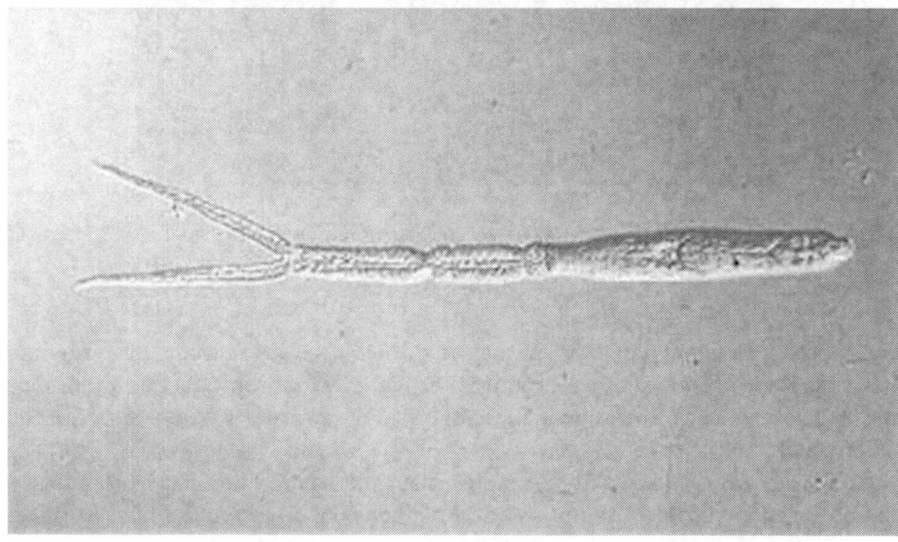

FIGURE 30.4 A micrograph of a cercaria showing the forked tail that the parasitic larva uses to swim through the water. (Courtesy of the DPDx Parasite Image Library.)

through the bloodstream to the viscera—a trip that often takes less than 24 hours. After reaching the viscera, the parasites reach maturity, mate, and start producing eggs. Egg production begins on day 9 after infection and peaks on days 14–16. The eggs that are shed with the host's feces complete the parasite's life cycle. For definitive hosts, infections are usually not fatal or debilitating. Some species of avian schistosomes infect the nasal cavities of birds rather than their viscera, but these parasitic species do not occur in North America (Horák and Kolářová 2011, Kolářová et al. 2013).

A few studies have examined the prevalence of avian schistosomes among birds from North American lakes where swimmer's itch has been a problem. At Flathead Lake, Montana, avian schistosome eggs were found in 87% of fecal samples from common mergansers, 5% from gulls, and 3% from Canada geese (Loken et al. 1995). At Cultus Lake, British Columbia, almost all common mergansers shed avian schistosome eggs in their feces, but this was not true for other avian species such as Canada geese, glaucous-winged gulls, and American coots (Leighton et al. 2000). The prevalence rate was 92% among tundra swans collected in the southwestern United States (Brant 2007). Along the coast of Connecticut, the prevalence rates were 93% in herring gulls, 93% in ring-billed gulls, 82% in great black-backed gulls, 67% in Canada geese, and 28% in double-crested cormorants (Barber and Caira 1995).

Other studies from across the globe have determined the prevalence of avian schistosomes among birds. In Iceland and France, 35% and 60% of aquatic birds were infected, respectively (Kolářová et al. 2013). In a wildlife refuge located in Iran, 79% of northern shovelers and 19% of mallards were infected; other duck species were uninfected (Gohardehi et al. 2013).

Some schistosome species use muskrats, beaver, and raccoons for definitive hosts (Figure 30.5). All these mammals have a propensity to defecate in water; this behavior allows the parasite to complete its life cycle (Bourns et al. 1973, Conover 2007, CDC 2012).

30.4 HOW HUMANS CONTRACT SWIMMER'S ITCH

Swimmer's itch occurs in people who swim or wade in a pond or lake that has both snails that can serve as intermediate hosts and birds or mammals that can serve as definitive hosts. Swimmer's itch is more likely to result when someone swims in a eutrophic lake or pond because these water bodies often harbor large numbers of snails. Streams and oceans can also be infected, but in these water bodies, the parasites are usually not as numerous as they are in ponds and shallow lakes. Cercariae are often concentrated in shallow water where aquatic snails occur and in areas frequented by ducks, geese, or other waterbirds. In Chile, 3% of swimmers at Laguna Chica de San Pedro contracted swimmer's itch, as did 7% of swimmers at Douglas Lake, Michigan (Verbrugge et al. 2004, Valdovinos and Balboa 2008) and 28% of bathers at Lake Leman, Switzerland (Sidebar 30.1). Among rice farmers and people bathing in man-made ponds near a wildlife refuge in Iran, 53% had swimmer's itch (Gohardehi et al. 2013).

FIGURE 30.5 Muskrats are a definitive host for the parasites that cause swimmer's itch. For this reason, people may be at risk for developing swimmer's itch when wading or swimming in swamps and ponds inhabited by muskrats. (Courtesy of the U.S. Fish and Wildlife Service.)

SIDEBAR 30.1 OUTBREAK OF SWIMMER'S ITCH IN GENEVA, SWITZERLAND (CHAMOT ET AL. 1998)

During 1996, Swiss health officials examined the prevalence of swimmer's itch among bathers at four beaches on Lake Leman in Geneva, Switzerland, after the problem received considerable attention in the local press. In Switzerland, swimmer's itch is caused by the parasite *Trichobilharzia ocellata*.

Swiss health officials interviewed 555 bathers 2–7 days after the bathers used Lake Leman and found that 28% had a swimmer's itch skin rash. Of these, 84% reported itching with 46% describing it as uncomfortable, 23% as distressing, and 7% as unbearable. Twenty-six percent complained that the itching had impaired their ability to sleep. Half of the patients reported using drugs for relief; 4% visited a medical doctor. The probability of developing swimmer's itch increased with time spent in the water, barometric pressure, and ambient temperatures during the patients' visit to Lake Leman. People who swam in the morning were more likely to develop swimmer's itch than afternoon or evening swimmers.

Lake-related activities are of great importance to the economy of the communities around Lake Leman. Fifteen percent of people who had experienced swimmer's itch reported that they would not return to Lake Leman or would visit it less often. Swiss health officials concluded that the prevalence of swimmer's itch has had a negative impact on the local economy.

Swimmer's itch is not an infectious disease and cannot spread from one person to another. Avian schistosomes do not occur in swimming pools that have chlorinated water (CDC 2012).

30.5 MEDICAL TREATMENT

Humans are a dead-end host for the parasites that cause swimmer's itch: the cercariae burrow into human skin where they die. Hence, there is little need for drugs to kill them. Instead, medical attention is designed to relieve the symptoms of the allergic reaction. Scratching the rash can result in a secondary infection if bacteria invade through the abraded skin; thus, patients should try to resist the urge to scratch. Most cases of swimmer's itch do not require medical attention, but the itching sensation can become overwhelming and lead to insomnia (Chamot et al. 1998). In such cases, a medical doctor can prescribe corticosteroids or antihistamines to lessen the itching. The CDC (2012) recommends that patients try some of the following actions to relieve discomfort:

1. Use a corticosteroid cream on the rash.
2. Apply an anti-itch lotion.
3. Soak in a colloidal oatmeal bath.
4. Wash in a solution of Epsom salts or baking soda.
5. Place a cool compress on the affected area.
6. Apply a baking soda paste to the rash; the paste is made by adding a small amount of water to baking soda.

30.6 WHAT PEOPLE CAN DO TO REDUCE THEIR RISK OF CONTRACTING SWIMMER'S ITCH

People can prevent swimmer's itch by not swimming in water bodies that contain avian schistosomes. Fishermen can avoid the disease by wearing rubber boots or waders. However, many people are willing to risk infection for the pleasure of swimming on a hot day. For these people, the risk of infection can be reduced, but not eliminated completely, by showering and drying off with a towel as soon as they finish swimming to remove any residual water before the parasites have an opportunity to burrow into the skin. The risk of infection at the same beach can vary from one day to the next. Cercariae can only survive in water for 1–2 days; thus, daily changes in infection rates probably relate to the number of cercariae released into the water that day or the prior day. More cercariae are released on warm days and when it is sunny. Cercariae have a propensity to swim upward and toward light, causing them to accumulate at the surface of the water. Because waterbirds are also at the water surface, these behaviors increase the probability that cercariae will find definite hosts. These behaviors also concentrate parasites where people swim, especially when there is an on-shore wind that blows cercariae toward shore. For the same reason, children are more likely to be infected than adults because children spend more time in shallow water where cercariae are concentrated (Leighton et al. 2000, Verbrugge et al. 2004, Soldánová et al. 2012, Kolářová et al. 2013, Mayo Clinic 2014).

Common insect repellents that contain DEET may help reduce the risk of infection if one applies the repellent prior to swimming (Ramaswamy et al. 2003). Sea Safe®, a cream designed to protect people from jellyfish, and sun-protection lotions containing niclosamide can be effective (Wulff et al. 2007). Public health agencies should erect signs at beaches where avian schistosomes are common to warn swimmers of the risk of swimmer's itch. Public education is needed to inform swimmers of the methods they can employ to reduce their risk of exposure.

One way to mitigate the risk of swimmer's itch is to reduce the populations of aquatic snails that serve as intermediate hosts for the parasites. At popular swimming ponds and lakes, local governments have killed aquatic snails by applying copper sulfate or other chemicals to the water or by using mechanical plows or rototillers that can bury or crush the snails (Leighton et al. 2000, Blankespoor et al. 2001, Soldánová et al. 2012).

Waterbirds are the definitive host for the parasites responsible for swimmer's itch, and efforts have been made to discourage them from loafing or roosting on water bodies where people swim by harassing the birds until the birds leave. State or federal biologists can catch and remove Canada geese where they are a threat to human health. Feeding ducks, geese, and swans should be prohibited at sites where swimmer's itch has been a problem in the past. In Michigan, mergansers and mallards were successfully trapped and treated with praziquantel to kill the parasites that infected them. In areas where mammals serve as the definitive host, trapping has been used to reduce populations of muskrat, beaver, and raccoons. While efforts may reduce populations of aquatic snails, waterbirds, or aquatic mammals, such actions will not eradicate them. Hence, removal efforts must be applied continually (Reimink et al. 1995, Blankespoor et al. 2001).

30.7 ERADICATING AVIAN SCHISTOSOMES FROM A COUNTRY

Eliminating avian schistosomes from a single lake is difficult; eradicating them from an entire country is nearly impossible except in the case of countries consisting of small, oceanic islands. Many infected waterbirds migrate long distances and can reintroduce avian schistosomes back into a water body where the parasites have been eliminated. Thus, eradicating avian schistosomes from a country and keeping them from returning remains a formidable task.

30.8 GIARDIASIS

Giardiasis is a common intestinal illness caused by the parasite *Giardia intestinalis*. This same parasite is also referred to as *G. lamblia* and *G. duodenalis*. Most researchers currently accept that there are six species of *Giardia*, but *G. intestinalis* is the only species that infects humans; it is a zoonotic parasite and can infect several mammalian species. *G. intestinalis* are microscopic parasites with two major life stages. One stage is a cyst (i.e., an egg surrounded by a hard wall), which is able to survive for months in the environment (Figures 30.6 through 30.8). When a cyst is ingested, the acidic environment of the stomach stimulates hatching. The parasite then attaches itself to the epithelial cells lining the intestines

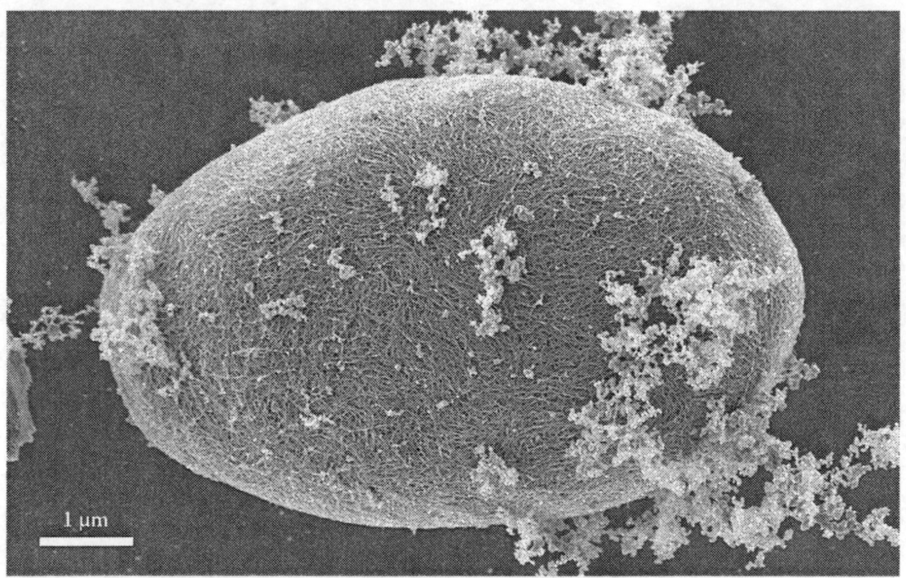

FIGURE 30.6 Scanning electron micrograph showing a *Giardia* cyst. (Courtesy of Stan Erlandsen and the CDC.)

where it grows and starts producing cysts, which are shed with the host's feces (Ryan and Cacciò 2013).

G. intestinalis is divided into eight distinct genetic groups called assemblages. *G. intestinalis* parasites within a group are genetically similar to each other but differ considerably from the parasites in the other assemblages. Further, the assemblages differ as to which mammals they parasitize (Table 30.1). Humans are parasitized by assemblages A and B. These assemblages also parasitize several wild and captive mammals, making them zoonotic parasites. Assemblage A or B have been detected in several species of North American mammals, including beaver, collared peccary, white-tailed deer, elk, muskox, moose, reindeer, muskrat, coyote, red fox, and wolf (Ryan and Cacciò 2013).

Symptoms of giardiasis begin 6–15 days after infection but can be delayed as long as 75 days. The illness is characterized by diarrhea, abdominal pain, gas, bloating, nausea, dehydration, and weight loss. Some patients belch gas that has a foul taste. Usually, symptoms last 2–4 weeks and disappear without medical treatment. Some giardiasis patients (30%–40%) develop a chronic infection that can last over a year during which the patients experience chronic fatigue, weight loss, and intermittent diarrhea. Chronic infections may impair the ability of young children to absorb nutrients, resulting in malnourishment (WHO 2002, 2011, Mayo Clinic 2012).

Giardiasis is contagious, and most people become parasitized by consuming *Giardia* cysts that were shed by another person. This is not surprising given that a person can become infected after consuming as few as 10 cysts and an infected person can shed over a billion cysts daily. The most common route of infection is through person-to-person contact; this is especially true for young children; daycare centers

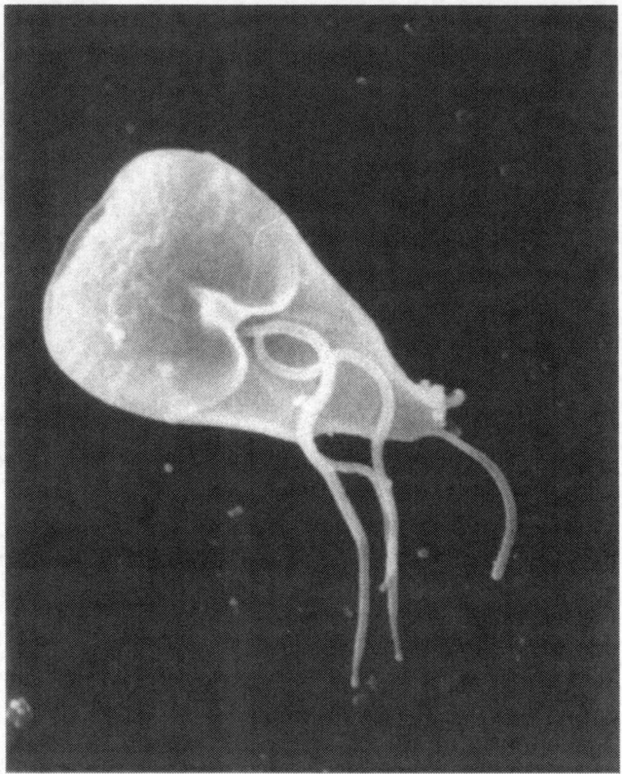

FIGURE 30.7 Scanning electron micrograph showing a free-swimming *Giardia* protozoan. (Courtesy of Janice Haney Carr and the CDC.)

are often the site of transmission because many children at a daycare center may be infected. Water-borne outbreaks occur from drinking water containing *Giardia*; these often result from swimming in contaminated swimming pools, lakes, or streams. Food-borne outbreaks occur as a result of food contamination by food handlers or household members (Ortega and Adam 1997, WHO 2002, 2011, Yoder et al. 2012).

Worldwide, millions of people become infected with *G. intestinalis* each year. The prevalence rate of giardiasis is estimated to be 2% of adults and 8% of children in developed countries and 33% of people in developing countries. In the United States, 19,927 cases of giardiasis were reported to the CDC during 2010, but the actual number of cases is probably much higher. Scallan et al. (2011) estimated that there are more than 75,000 giardiasis patients and two deaths annually in the United States. Giardiasis is more common in the northern half of the United States (Figure 30.9). Children are more likely to become infected than adults. The incidence of giardiasis increases during the summer months, perhaps because this is when most people engage in outdoor recreation (Yoder et al. 2012).

Thompson (2013) considered *Giardia* to be a classic example of a zoonotic parasite that spreads from humans to wildlife, even to wild mammals living in pristine areas. For example, muskox in the Canadian Arctic are infected with *G. intestinalis*

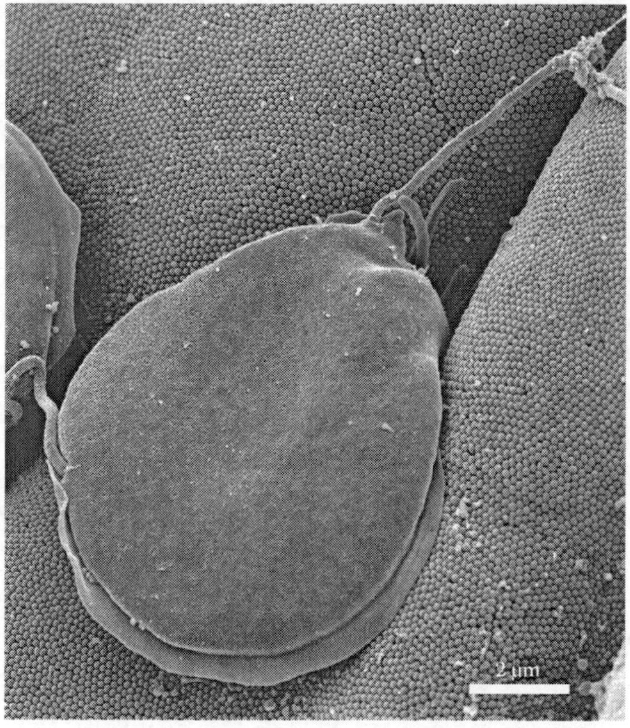

FIGURE 30.8 Scanning electron micrograph showing a *Giardia* protozoan attached to an epithelial cell in the intestine. The small circles under the protozoan are the microvilli on the surface of the epithelial cell. (Courtesy of Stan Erlandawn and the CDC.)

TABLE 30.1
Wild and Captive Mammals Infected by the Eight Assemblages of *Giardia intestinalis*

Assemblage	Hosts
A	Humans, primates, livestock, dogs, cats, wild mammals
B	Humans, primates, dogs, cats, wild mammals
C	Dogs, wild canids
D	Dogs, wild canids
E	Hoofed livestock
F	Cats
G	Rats
H	Marine mammals

Source: Ryan, U. and Cacciò, S.M., *Int. J. Parasitol.*, 43, 943, 2013.

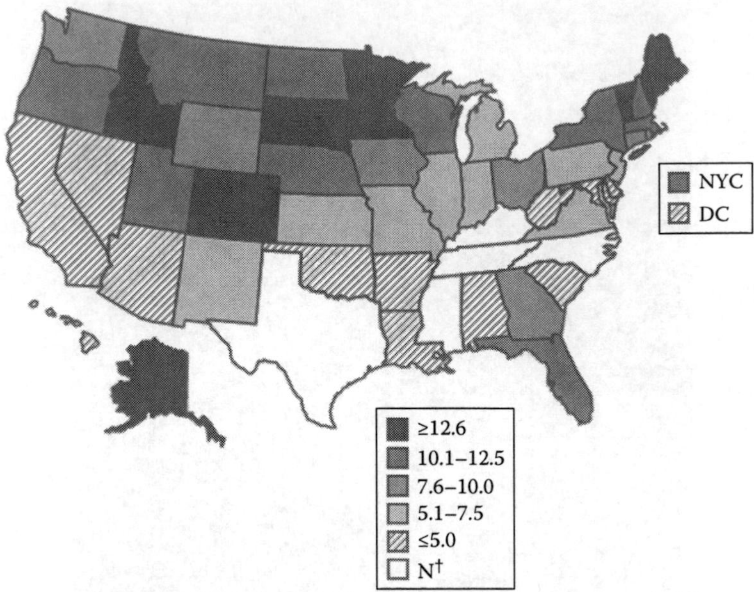

FIGURE 30.9 Incidence of giardiasis per 100,000 population in the United States. (From Yoder, J. S. et al., *Morb. Mortal. Wkly. Rep.*, 61(5), 13, 2012; courtesy of the CDC.)

that are genetically similar to those prevalent in the closest human settlements. Regardless of the parasite's original source, *G. intestinalis* can maintain itself in an area by infecting wild animals. The *Giardia* cysts that are shed by wild mammals can infect humans. For this reason, people drinking from a crystal-clear stream—even in a remote part of the world—can become infected with *Giardia.* Some marine mammals such as seals may have become infected with *Giardia* from a human source, possibly from sewage effluent or surface run-off (Thompson 2013).

There are several drugs that can be used to treat giardiasis such as metronidazole (Flagyl®), nitazoxanide (Alinia®), and tinidazole (Tindamax®). There are no drugs or vaccines that can prevent an infection, but steps can be taken to reduce the risk. These include washing hands regularly, not swallowing water when swimming, and drinking bottled water when in areas where the water supply might be unsafe. *Giardia* parasites are moderately resistant to disinfectants, including chlorine, and can survive freezing temperatures. People should filter surface water or boil it for at least 10 minutes prior to drinking (Ortega and Adam 1997, Mayo Clinic 2012).

LITERATURE CITED

Barber, K. E. and J. N. Caira. 1995. Investigation of the life cycle and adult morphology of the avian blood fluke *Austrobilharzia variglandis* (Trematoda: Schistosomatidae) from Connecticut. *Journal of Parasitology* 81:584–592.

Blankespoor, C. L., R. L. Reimink, and H. D. Blankespoor. 2001. Efficacy of praziquantel in treating natural schistosome infections in common mergansers. *Journal of Parasitology* 87:424–426.

Bourns, T. K., J. C. Ellis, and M. E. Rau. 1973. Migration and development of *Trichobilharzia ocellata* (Trematoda: Schistosomidae) in its duck hosts. *Canadian Journal of Zoology* 51:1021–1030.

Brant, S. V. 2007. The occurrence of the avian schistosome *Allobilharzia visceralis* Kolářová, Rudolfová, Hampl et Skíirnisson, 2006 (Schistosomatidae) in the tundra swan, *Cygnus columbianus* (Anatidae), from North America. *Folia Parasitologica* 54:99–104.

CDC. 2012. Swimmer's itch FAQ. http://www.cdc.gov/parasites/swimmersitch/faqs.html (accessed March 30, 2014).

Chamot, E., L. Toscani, and A. Rougemont. 1998. Public health importance and risk factors for cercarial dermatitis associated with swimming in Lake Leman at Geneva, Switzerland. *Epidemiology and Infection* 120:305–314.

Coady, N. R., P. M. Muzzall, T. M. Burton, R. J. Snider et al. 2006. Ubiquitous variability in the prevalence of *Trichobilharzia stagnicolae* (Schistosomatidae) infecting *Stagnicola emarginata* in three northern Michigan Lakes. *Journal of Parasitology* 92:10–15.

Conover, M. R. 2007. *Predator-Prey Dynamics: The Use of Olfaction*. CRC Press, Boca Raton, FL.

Cort, W. W. 1928. Schistosome dermatitis in the United States (Michigan). *Journal of the American Medical Association* 90:1027–1029.

Gohardehi, S., M. Fakhar, and M. Madjidaei. 2013. Avian schistosomes and human cercarial dermatitis in a wildlife refuge in Mazandaran Province, northern Iran. *Zoonoses and Public Health* 60:442–447.

Horák, P. and L. Kolářová. 2011. Snails, waterfowl and cercarial dermatitis. *Freshwater Biology* 56:779–790.

Kolářová, L., P. Horák, K. Skírnisson, H. Marečková, and M. Doenhoff. 2013. Cercarial dermatitis, a neglected allergic disease. *Clinical Reviews in Allergy and Immunology* 45:63–74.

Leighton, B. J., S. Zervos, and J. M. Webster. 2000. Ecological factors in schistosome transmission, and an environmentally benign method for controlling snails in a recreational lake with a record of schistosome dermatitis. *Parasitology International* 49:9–17.

Loken, B. R., C. N. Spencer, and W. O. Granath, Jr. 1995. Prevalence and transmission of cercariae causing schistosome dermatitis in Flathead Lake, Montana. *Journal of Parasitology* 81:646–649.

Mayo Clinic. 2012. Giardia infections (giardiasis). http://www.mayoclinic.org/diseases-conditions/giardia-infection/basics/definition/con-20024686 (accessed April 2, 2014).

Mayo Clinic. 2014. Swimmer's itch. http://www.mayoclinic.org/diseases-conditions/swimmers-itch/basics/definition/con-20030150 (accessed March 30, 2014).

Merck. 2010. *Merck Veterinary Manual*, 10th edition. Merck, Whitehouse Station, NJ.

Ortega, Y. R. and R. D. Adam. 1997. Giardia: Overview and update. *Clinical Infectious Diseases* 25:545–549.

Ramaswamy, K., Y.-X. He, B. Salafsky, and T. Shibuya. 2003. Topical application of DEET for schistosomiasis. *Trends in Parasitology* 19:551–555.

Reimink, R. L., J. A. DeGoede, and H. D. Blankespoor. 1995. Efficacy of praziquantel in natural populations of mallards infected with avian schistosomes. *Journal of Parasitology* 81:1027–1029.

Ryan, U. and S. M. Cacciò. 2013. Zoonotic potential of *Giardia*. *International Journal for Parasitology* 43:943–956.

Scallan, E., R. M. Hoekstra, F. J. Angulo, R. V. Tauxe et al. 2011. Foodborne illness acquired in the United States—Major pathogens. *Emerging Infectious Diseases* 17:7–15.

Soldánová, M., C. Selbach, M. Kalbe, A. Kostadinova, and B. Sures. 2012. Swimmer's itch: Etiology, impact, and risk factors in Europe. *Trends in Parasitology* 29:65–74.

Thompson, R. C. A. 2013. Parasite zoonoses and wildlife: One health, spillover and human activity. *International Journal for Parasitology* 43:1079–1088.

Valdovinos, C. and C. Balboa. 2008. Cercarial dermatitis and lake eutrophication in south-central Chile. *Epidemiology and Infection* 136:391–394.

Verbrugge, L. M., J. J. Rainey, R. L. Reimink, and H. D. Blankespoor. 2004. Prospective study of swimmer's itch incidence and severity. *Journal of Parasitology* 90:697–704.

WHO. 2002. *Guideline for Drinking-Water Quality*, 2nd edition. World Health Organization, Geneva, Switzerland.

WHO. 2011. *Guideline for Drinking-Water Quality*, 4th edition. World Health Organization, Geneva, Switzerland.

Wong, D. E., T. L. Meinking, L. B. Rosen, D. Taplin et al. 1994. Seabather's eruption: Clinical, histologic and immunologic features. *Journal of the American Academy of Dermatology* 30:399–406.

Wulff, C., S. Haeberlein, and W. Hass. 2007. Cream formulations protecting against cercarial dermatitis by *Trichobilharzia*. *Parasitology Research* 101:91–97.

Yoder, J. S., J. W. Gargano, R. M. Wallace, and M. J. Beach. 2012. Giardiasis surveillance—United States, 2009–2010. *Morbidity and Mortality Weekly Report* 61(5):13–23.

Appendix A: Definition of Medical Terms Used in This Book

Acaricide: A chemical that kills ticks or mites.

Accidental host: A species that can be infected by some bacteria, virus, or other organism but the infecting organism cannot maintain itself by infecting that species. An accidental host is also referred to as a dead-end host (see *Reservoir host* for comparison).

Active infection: An infection in which the disease organism is causing disease symptoms in the patient or host animal.

Acute infection: An infection caused by a pathogen that is of short duration but can be intensive (see *Chronic infection* for comparison).

Aerobic: Requiring oxygen to survive.

Amplifying hosts: A species in which a pathogen can greatly increase its numbers by infecting. However, the pathogen cannot maintain itself by infecting an amplifying host. With viruses, this term is used more commonly than reservoir hosts.

Anaerobic: Not requiring oxygen to survive. An anaerobic organism (such as yeast) can survive without oxygen.

Anorexia: An unwillingness to eat.

Anthrax carcass: A dead animal that was killed by anthrax.

Antigen: A chemical on a bacteria, virus, or other foreign object that provoke the immune to develop specific antibodies against it.

Asymptomatic: An adjective meaning without symptoms. In this book, it means without any sign of infection or disease.

Attenuated strain: A strain of bacteria or virus that is less infectious or virulent. Some vaccines use a live, but attenuated strain to protect animals or people from being infected by a more virulent strain of the same species.

Bell's palsy: A paralysis of facial muscles often on one side of the face due to a disorder of the seventh cranial nerve that controls the facial muscles.

Cardiac: Pertaining to the heart.

Case: As used in this book, a case is a person who has been diagnosed with a particular disease. It only refers to a person; that is, we never refer to a sick animal as a case.

Case fatality rate: The proportion of cases that died from their illness. This is higher than the true fatality rate because people with a minor infection are less likely to seek medical attention and their disease is less likely to be diagnosed.

Causative agent: The agent or thing that is causing the disease or symptoms. With diseases, the causative agent is the bacteria, virus, or parasite that causes the disease or symptoms.

Chronic infection: A disease that persists for long periods of time (see *Acute infection* for comparison).

Conjunctivitis: Inflammation of the outer surface of the eye. This often occurs when capillaries in the eye swell in response to an eye infection or an allergy. This is also called "pinkeye."

Cutaneous: Referring to the skin.

Dead-end host: A species which can be infected by a pathogen but the pathogen cannot maintain itself by infecting the species. Another name for a dead-end host is accidental host.

Definitive host: The animal that a parasite uses to reach sexual maturity and reproduce.

Diagnosis: A confirmation of a patient's disease. A diagnosis often requires that the pathogen be isolated from the patient, the pathogen's DNA being detected in a patient, or the patient has antibodies to the specific pathogen.

Disease organism: The bacteria, virus, or parasite that causes a disease. In this book, disease organism is synonymous with causative agent or pathogen.

Dysentery: Inflammation of the large intestine, especially the colon, that results in bloody diarrhea, fever, and abdominal cramps.

Ectoparasite: A parasite that lives outside of its host and feeds by biting through the skin. In contrast, internal parasites may live inside their hosts' digestive system, blood, muscle, or other body tissue. Ectoparasites include fleas, ticks, and mites.

Edema: Swelling caused by the retention of fluid in body tissue or beneath the skin. Edema often occurs in legs, feet, and ankles.

Encephalitis: An irritation, inflammation, or swelling of the brain, often due to an acute infection of the brain. This is a serious condition and can be fatal.

Encephalopathy: Any degenerative brain disease.

Endemic or endemic area: The area where a pathogen is able to maintain itself (i.e., the area where a disease occurred at the time this book was written in 2014).

Endocarditis: A disease that infects the inner lining of the heart, including the heart valves.

Enteritis: Inflammation of the small intestine.

Epidemic: A disease outbreak among humans.

Epidemiology: The study of how diseases or pathogens are maintained and spread.

Epithelial cells: Cells that form the lining of hollow organs (e.g., intestines) or the skin.

Epizootic hosts: Animal species that may become infected during a disease outbreak. Epizootic hosts have little resistance to the disease organism and die quickly when infected so the disease organism cannot maintain itself in them (sometimes because of a population crash of epizootic host species after an outbreak of a fatal disease). Epizootic hosts species are important during disease outbreaks because the population of the disease organism in the environment is greatly increased through their infection of epizootic hosts.

Eschar: A scab or piece of dead tissue covering a wound that is ultimately sloughed off. It occurs at sites of infection for diseases such as anthrax and spotted fever.

False positive: A test result indicating that an animal has a disease or is infected with bacteria, virus, or other causative agent when in fact the animal neither has the disease nor is infected.

Flaccid paralysis: Loss of muscle control or paralysis. There is a loss of muscle tone; muscles appear weak and flabby. Flaccid paralysis can result from encephalitis or Guillain–Barré syndrome.

Gastroenteritis: Inflammation of the stomach and small intestine which results in vomiting, diarrhea, abdominal pains, and cramps.

Gram-negative bacteria: These are bacteria that appear pink or red when stained with a staining procedure developed by Hans Gram because their outer membrane cannot bind with the violet chemical in the Gram stain. In contrast, Gram-positive bacteria appear purple or violet.

Gram-positive bacteria: These are bacteria that appear dark blue or purple when stained with a staining procedure developed by Hans Gram. In contrast, Gram-negative bacteria appear red or pink. The difference results because Gram-positive bacteria have a chemical on their outer membrane that binds with the violet chemical in the stain, while Gram-negative bacteria lack this chemical. Many species of Gram-negative bacteria cause diseases.

Guillain–Barré syndrome: This is caused when a person's immune system attacks the person's nerves. The first symptoms are weakness and a tingling sensation in the hands and feet. The sensations can quickly spread and result in paralysis.

Incidence rate: The proportion of a human population that contracts a particular disease within a period of one year. Incidence rate is usually expressed as the number of new cases of a disease during a year/100,000 people. We do not use the term when referring to disease in animals (see *Prevalence* for comparison).

Incubation period: The length of time between when a patient became infected with a pathogen and the onset of the first symptoms of illness.

Intravenous: Through the veins.

Jaundice: A yellowing of the skin and eyes. It is often caused when the liver is not functioning correctly.

Latent infection: A condition in which the disease organism is present in a patient or animal host but is not currently producing any symptoms of disease. The danger is that at some time in the future, the latent infection may become an active infection.

Macular rash: A skin rash consisting of red spots that are flat with the surrounding skin (e.g., not raised).

Malaise: A general sense of discomfort or not being well.

Meningitis: An irritation or inflammation of the protective membrane lining the brain and spinal cord often due to an infection of the membrane.

Meningoencephalitis: A medical condition in which the lining of the brain and the brain itself have become inflamed, usually as a result of an infection.

Morbidity: The prevalence or incidence of disease in a human population. It is the probability that a randomly selected person would become ill with a specific disease within a specified period of time.

Myocarditis: Inflammation of the heart muscle, usually due to an infection.

Necrotic enteritis: Enteritis that includes tissue death.

Neuroinvasive: A pathogen that has infected the nervous system. Such infections may results in encephalitis, meningitis, or meningoencephalitis.

Notifiable disease: A disease that medical doctors must report to governmental health agencies whenever they see it in a patient. Notifiable diseases are ones that health agencies want to track for some reason.

Outbreak: An increase of disease in either animals, humans, or both.

Papule: A round area of raised skin that may be pink, red, or brown. Papules are solid (not filled with a liquid) and usually less than 1 cm in width. They often occur at the site of a bite or infection). After a few days, the papule may develop into an eschar.

Patient: As used in this book, a patient is a person who has a particular disease or is infected with a particular pathogen. A patient has to be a person; that is, a sick animal is not referred to as a patient in this book.

Photophobia: An eye pain caused by exposure to bright lights.

Pneumonia: An inflammation of the lungs caused by an infection.

Presumptive diagnosis: Medical doctors make a presumptive diagnosis when a patient's symptoms and history indicate which disease the patient probably has. Antibiotic treatments often begin after a presumptive diagnosis rather than waiting for a diagnosis which may take several days to complete.

Prevalence: The proportion of a population that is infected with a particular disease. For chronic diseases, such as tuberculosis, the prevalence rate can be much higher than the incidence rate because the incidence rate only counts new cases of the disease.

Pulmonary: Involving the lungs.

Reactive arthritis: Swelling and pain in the joints resulting from an infection elsewhere in the body.

Renal: This adjective refers to the kidneys and urinary systems.

Reservoir host: A species that can be infected by some bacteria, virus, or other organism and the infecting organism to able to maintain itself in the environment by infecting that species. In reservoir hosts, the infecting organism often causes only a mild illness or none at all (see *Accidental host* or *Dead-end host* for comparison).

Serological: Pertaining to antibodies. Serological studies are used to determine if an animal or human has antibodies against a particular pathogen. If so, the animal has been exposed to the pathogen in the past but may or may not be currently infected.

Seropositive: Possessing antibodies for antigens from a particular pathogen. A person who is seropositive for a particular pathogen has already been exposed to it and developed antibodies against the pathogen. Being seropositive means that the person either has an active infection by a particular pathogen or had an infection at some time in the past.

Seroprevalence: Proportion of a population that are seropositive.

Serotype: A synonym of serovar.

Serovar: Bacteria or viruses are considered to be of the same serovar if they have the same antigenic makeup. Serotype is a synonym of serovar.

Sputum: Mucous produced by the airways or lungs and often mixed with saliva. It is usually swallowed or spit out.

Trachea: The thin-walled tube made of cartilage extending from the nose and mouth to the lungs through which air reaches the lungs. Another name for the trachea is the windpipe.

Virulent: Bacteria, virus, or parasite that infects another animal and causes illness.

Appendix B: Scientific Names for Species Mentioned in This Book

Alligator, American (*Alligator mississippiensis*)
Alpaca (*Lama pacos*)
Antelope, Lechwe (*Kobus leche*)
Antelope, Pronghorn (*Antilocapra americana*)
Antelope, Roan (*Hippotragus equinus*)
Antelope, Topi (*Damaliscus lunatus*)
Armadillo, Nine-banded (*Dasypus novemcinctus*)
Babirusa, Buru (*Babyrousa babyrussa*)
Baboon, Olive (*Papio anubis*)
Badger, American (*Taxidea taxus*)
Badger, European (*Meles meles*)
Bandicoot, Western Barred (*Perameles bougainville*)
Bat, Big Brown (*Eptesicus fuscus*)
Bat, California Myotis (*Myotis californicus*)
Bat, Eastern Pipistrelle (*Pipistrellus subflavus*)
Bat, Eastern Red (*Lasiurus borealis*)
Bat, Hoary (*Lasiurus cinereus*)
Bat, Keen's Myotis (*Myotis keenii*)
Bat, Little Brown Myotis (*Myotis lucifugus*)
Bat, Long-eared Myotis (*Myotis evotis*)
Bat, Mexican Free-tailed (*Tadarida brasiliensis mexicana*)
Bat, Northern Myotis (*Myotis septentrionalis*)
Bat, Pallid (*Antrozous pallidus*)
Bat, Silver-haired (*Lasionycteris noctivagans*)
Bat, Vampire (*Desmodus rotundus*)
Bat, Western Pipistrelle (*Pipistrellus hesperus*)
Bat, Yuma Myotis (*Myotis yumanensis*)
Bear, American Black (*Ursus americanus*)
Bear, Asian Black (*Ursus thibetanus*)
Bear, Brown (*Ursus arctos*)
Bear, Grizzly (*Ursus arctos horribilis*)
Bear, Polar (*Ursus maritimus*)
Beaver, American (*Castor canadensis*)
Bighorn Sheep (*Ovis canadensis*)
Bison, American (*Bison bison*)
Bison, European (*Bison bonasus*)

Blackbird, Eurasian (*Turdus merula*)
Blackbird, Red-winged (*Agelaius phoeniceus*)
Blackcap, Eurasian (*Sylvia atricapilla*)
Blowfly (Calliphoridae)
Blue Jay (*Cyanocitta cristata*)
Bluebonnet Fly (Calliphoridae)
Boar, Wild (*Sus scrofa*)
Bobcat (*Lynx rufus*)
Buffalo, African (*Syncerus caffer*)
Buffalo, American (*Bison bison*)
Buffalo, European (*Bison bonasus*)
Buffalo, Water (*Bubalus bubalis*)
Bullfinch, Eurasian (*Pyrrhula pyrrhula*)
Bushbaby, Senegal (*Galago senegalensis*)
Bushbuck (*Tragelaphus scriptus*)
Bushpig (*Potamochoerus larvatus*)
Buzzard, Common (*Buteo buteo*)
Camel, One-humped (*Camelus dromedarius*)
Canvasback (*Aythya valisineria*)
Capercaillie, Western (*Tetrao urogallus*)
Cardinal, Northern (*Cardinalis cardinalis*)
Caribou (*Rangifer tarandus*)
Cat, Asian Golden (*Catopuma temminckii*)
Cat, Domestic (*Felis catus*)
Cat, Feral (*Felis catus*)
Cat, Leopard (*Prionailurus bengalensis*)
Catbird, Gray (*Dumetella carolinensis*)
Cattle, Domestic (*Bos taurus*)
Cattle, Feral (*Bos taurus*)
Chaffinch, Common (*Fringilla coelebs*)
Chamois, Alpine (*Rupicapra rupicapra*)
Cheetah (*Acinonyx jubatus*)
Chickadee (*Poecile* spp.)
Chimpanzee, Common (*Pan troglodytes*)
Chipmunk, Cliff (*Tamias dorsalis*)
Chipmunk, Eastern (*Tamias striatus*)
Chipmunk, Least (*Tamias minimus*)
Chipmunk, Uinta (*Tamias umbrinus*)
Chipmunk, Western (*Eutamias* spp.)
Chipmunk, Yellow-pine (*Tamias amoenus*)
Chukar (*Alectoris chukar*)
Cormorant, Double-crested (*Phalacrocorax auritus*)
Cormorant, Great (*Phalacrocorax carbo*)
Cottontail, Desert (*Sylvilagus audubonii*)
Cottontail, Eastern (*Sylvilagus floridanus*)
Cottontail, Mountain (*Sylvilagus nuttallii*)

Cougar (*Puma concolor*)
Cowbird, Brown-headed (*Molothrus ater*)
Coyote (*Canis latrans*)
Coypu (*Myocastor coypus*)
Crane, Sandhill (*Grus canadensis*)
Crane, Whooping (*Grus americana*)
Crocodile, Morelet's (*Crocodylus moreletii*)
Crow, American (*Corvus brachyrhynchos*)
Crow, Carrion (*Corvus corone*)
Crow, Fish (*Corvus ossifragus*)
Deer, European Roe (*Capreolus capreolus*)
Deer, Fallow (*Dama dama*)
Deer, Key (*Odocoileus virginianus clavium*)
Deer, Marsh (*Blastocerus dichotomus*)
Deer, Mule (*Odocoileus hemionus*)
Deer, Red (*Cervus elaphus elaphus*)
Deer, Sika (*Cervus nippon*)
Deer, White-tailed (*Odocoileus virginianus*)
Dog, African Wild (*Lycaon pictus*)
Dog, Domestic (*Canis lupus familiaris*)
Dog, Feral (*Canis lupus familiaris*)
Donkey (*Equus asinus*) = Ass (*Equus asinus*)
Dove, Eurasian Collared (*Streptopelia decaocto*)
Dove, Mourning (*Zenaida macroura*)
Dove, Stock (*Columba oenas*)
Duck, Mallard (*Anas platyrhynchos*)
Duck, Wood (*Aix sponsa*)
Dunlin (*Calidris alpina*)
Eagle, Spanish Imperial (*Aquila adalberti*)
Egret, Intermediate (*Egretta intermedia*)
Egret, Little (*Egretta garzetta*)
Egret, Snowy (*Egretta thula*)
Eland, Common (*Taurotragus oryx*)
Elephant, African (*Loxodonta* spp.)
Elephant, Asian (*Elephas maximus*)
Elk (*Cervus elaphus*)
Ferret, Black-footed (*Mustela nigripes*)
Fieldfare (*Turdus pilaris*)
Finch, House (*Carpodacus mexicanus*)
Flying Squirrel (*Glaucomys* spp.)
Fox, Arctic (*Vulpes lagopus*)
Fox, Bat-eared (*Otocyon megalotis*)
Fox, Gray (*Urocyon cinereoargenteus*)
Fox, Red (*Vulpes vulpes*)
Frog, Blue Mountains Tree (*Litoria citropa*)
Frog, Great Barred (*Mixophyes fasciolatus*)

Frog, Marsh (*Pelophylax ridibundus*)
Fulmar, Northern (*Fulmarus glacialis*)
Gemsbok (*Oryx gazella*)
Gerbil (*Gerbilliscus* spp.)
Giraffe (*Giraffa camelopardalis*)
Goat, Mountain (*Oreamnos americanus*)
Goldfinch, American (*Spinus tristis*)
Goldfinch, European (*Carduelis carduelis*)
Goose, Bar-headed (*Anser indicus*)
Goose, Barnacle (*Branta leucopsis*)
Goose, Cackling (*Branta hutchinsii*)
Goose, Canada (*Branta canadensis*)
Goose, Snow (*Chen caerulescens*)
Goshawk, Northern (*Accipiter gentilis*)
Grackle, Boat-tailed (*Quiscalus major*)
Grackle, Common (*Quiscalus quiscula*)
Grackle, Great-tailed (*Quiscalus mexicanus*)
Greenfinch (*Carduelis chloris*)
Grosbeak, Evening (*Coccothraustes vespertinus*)
Ground Squirrel, Columbian (*Spermophilus columbianus*)
Ground Squirrel, Golden-mantled (*Spermophilus lateralis*)
Ground Squirrel, Richardson's (*Spermophilus richardsonii*)
Groundhog (*Marmota monax*)
Grouse, Ruffed (*Bonasa umbellus*)
Guinea Pig, Domesticated (*Cavia porcellus*)
Gull, Herring (*Larus argentatus*)
Gull, Lesser Black-backed (*Larus fuscus*)
Gull, Ring-billed (*Larus delawarensis*)
Hare (*Lepus* spp.)
Hare, European (*Lepus europaeus*)
Hare, Japanese (*Lepus brachyurus*)
Hare, Mountain (*Lepus timidus*)
Hare, Snowshoe (*Lepus americanus*)
Hartebeest (*Alcelaphus buselaphus*)
Hawk, Red-tailed (*Buteo jamaicensis*)
Hawk, Rough-legged (*Buteo lagopus*)
Hedgehog, African Pygmy (hybrid of *Atelerix albiventris* and *A. algirus*)
Hedgehog, European (*Erinaceus europaeus*)
Heron, Black-crowned Night (*Nycticorax nycticorax*)
Hippopotamus (*Hippopotamus amphibius*)
Hog, Feral (*Sus scrofa*)
Horse (*Equus caballus*)
Horse, Przewalski's (*Equus caballus przewalskii*)
Horsefly (Tabanidae)
Hyrax, Rock (*Procavia capensis*)
Ibis, Glossy (*Plegadis falcinellus*)

Iguana, Green (*Iguana iguana*)
Impala (*Aepyceros melampus*)
Jackal, Black-backed (*Canis mesomelas*)
Jackal, Golden (*Canis aureus*)
Jackal, Side-striped (*Canis adustus*)
Jackdaw, Eurasian (*Corvus monedula*)
Jackrabbit, Black-tailed (*Lepus californicus*)
Jackrabbit, White-tailed (*Lepus townsendii*)
Jay, Blue (*Cyanocitta cristata*)
Kangaroo Rat, California (*Dipodomys californicus*)
Kestrel, American (*Falco sparverius*)
Kite, Black (*Milvus migrans*)
Koala (*Phascolarctos cinereus*)
Kudu (*Tragelaphus* spp.)
Lark, Horned (*Eremophila alpestris*)
Lemur, Brown (*Eulemur fulvus*)
Lemur, Mongoose (*Eulemur mongoz*)
Lemur, Ring-tailed (*Lemur catta*)
Leopard (*Panthera pardus*)
Leopard, Snow (*Uncia uncia*)
Lion (*Panthera leo*)
Lizard, Western Fence (*Sceloporus occidentalis*)
Llama (*Lama glama*)
Loon, Common (*Gavia immer*)
Louse, Human Body (*Pediculus humanus humanus*)
Louse, Human Head (*Pediculus humanus capitis*)
Louse, Human Pubic (*Pthirus pubis*)
Lynx, Canada (*Lynx canadensis*)
Lynx, Eurasian (*Lynx lynx*)
Macaque, Barbary (*Macaca sylvanus*)
Macaque, Rhesus (*Macaca mulatta*)
Mallard (*Anas platyrhynchos*)
Marmot (*Marmota spp.*)
Marmot, Yellow-bellied (*Marmota flaviventris*)
Marten (*Martes spp.*)
Marten, American (*Martes americana*)
Marten, Beech (*Martes foina*)
Martin, House (*Delichon urbica*)
Meadowlark, Eastern (*Sturnella magna*)
Mice (see mouse)
Mink, American (*Mustela vison*)
Mockingbird, Northern (*Mimus polyglottos*)
Mongoose, Indian Gray (*Herpestes edwardsii*)
Mongoose, Slender (*Galerella sanguinea*)
Mongoose, Small Asian (*Herpestes javanicus*)
Mongoose, Yellow (*Cynictis penicillata*)

Monitor, Crocodile (*Varanus salvadorii*)
Monkey, Howler (*Alouatta* spp.)
Monkey, Owl (*Aotus* spp.)
Monkey, Red-bellied Titi (*Callicebus moloch*)
Monkey, Red-tailed (*Cercopithecus ascanius*)
Monkey, Rhesus (*Macaca mulatta*)
Monkey, Spider (*Ateles* spp.)
Monkey, Squirrel (*Saimiri sciureus*)
Moose (*Alces* spp.)
Mosquito, Yellow Fever (*Aedes aegypti*)
Mouflon (*Ovis aries musimon*)
Mouse, Black Sea Field (*Apodemus ponticus*)
Mouse, Cotton (*Peromyscus gossypinus*)
Mouse, Deer (*Peromyscus maniculatus*)
Mouse, House (*Mus musculus*)
Mouse, Korean Field (*Apodemus peninsulae*)
Mouse, Little Pocket (*Perognathus longimembris*)
Mouse, Spinifex Hopping (*Notomys alexis*)
Mouse, Striped Field (*Apodemus agrarius*)
Mouse, Western Harvest (*Reithrodontomys megalotis*)
Mouse, White-footed (*Peromyscus leucopus*)
Mouse, Wood (*Apodemus sylvaticus*)
Mouse, Yellow-necked Field (*Apodemus flavicollis*)
Muskox (*Ovibos moschatus*)
Muskrat (*Ondatra zibethicus*)
Nutria (*Myocastor coypus*)
Nyala (*Tragelaphus angasii*)
Ocelot (*Leopardus pardalis*)
Opossum, Common (*Didelphis marsupialis*)
Opossum, Derby's Woolly (*Caluromys derbianus*)
Opossum, Virginia (*Didelphis virginiana*)
Oryx, Arabian (*Oryx leucoryx*)
Oryx, Scimitar-horned (*Oryx dammah*)
Ostrich (*Struthio camelus*)
Otter, European (*Lutra lutra*)
Owl, Burrowing (*Athene cunicularia*)
Owl, Great Horned (*Bubo virginianus*)
Owl, Little (*Athene noctua*)
Owl, Long-eared (*Asio otus*)
Oystercatcher, Eurasian (*Haematopus ostralegus*)
Panda, Red (*Ailurus fulgens*)
Partridge, Gray (*Perdix perdix*)
Pewee, Western Wood (*Contopus sordidulus*)
Pelican, American White (*Pelecanus erythrorhynchos*)
Pheasant, Ring-necked (*Phasianus colchicus*)
Pig, Domestic (*Sus scrofa*)

Pig, Feral (*Sus scrofa*)
Pigeon, Feral (*Columba livia*)
Pigeon, Wood (*Columba palumbus*)
Polecat, European (*Mustela, putorius*)
Porcupine, North American (*Erethizon dorsata*)
Porpoise, Dall's (*Phocoenoides dalli*)
Porpoise, Harbor (*Phocoena phocoena*)
Rabbit, Brush (*Sylvilagus bachmani*)
Rabbit, Eastern Cottontail (*Sylvilagus floridanus*)
Rabbit, European (*Oryctolagus cuniculus*)
Rabbit, Mountain Cottontail (*Sylvilagus nuttallii*)
Raccoon (*Procyon lotor*)
Raccoon Dog (*Nyctereutes procyonoides*)
Rat, African Grass (*Arvicanthis niloticus*)
Rat, Black (*Rattus rattus*)
Rat, Brown (*Rattus norvegicus*)
Rat, Hispid Cotton (*Sigmodon hispidus*)
Rat, Pacific (*Rattus exulans*)
Rat, Rice (*Oryzomys palustris*)
Redpoll, Common (*Acanthis flammea*)
Redshank, Common (*Tringa totanus*)
Redstart, American (*Setophaga ruticilla*)
Redstart, Common (*Phoenicurus phoenicurus*)
Redwing (*Turdus iliacus*)
Reindeer (*Rangifer tarandus*)
Rhinoceros, Black (*Diceros bicornis*)
Rhinoceros, Indian (*Rhinoceros unicornis*)
Robin, American (*Turdus migratorius*)
Rook (*Corvus frugilegus*)
Rousette, Leschenault's (*Rousettus leschenaultii*)
Sage-grouse, Greater (*Centrocercus urophasianus*)
Sea Lion, California (*Zalophus californianus*)
Sea Lion, Steller (*Eumetopias jubatus*)
Seal, Baikal (*Pusa sibirica*)
Seal, Caspian (*Pusa caspica*)
Seal, Galapagos Fur (*Arctocephalus galapagoensis*)
Seal, Gray (*Halichoerus grypus*)
Seal, Harbor (*Phoca vitulina*)
Seal, Harp (*Pagophilus groenlandicus*)
Seal, Hooded (*Cystophora cristata*)
Seal, Northern Elephant (*Mirounga angustirostris*)
Seal, Northern Fur (*Callorhinus ursinus*)
Seal, Ringed (*Pusa hispida*)
Seal, South American Fur (*Arctocephalus australis*)
Sheep, Bighorn (*Ovis canadensis*)
Sheep, Domestic (*Ovis aries*)

Shrew, Common (*Sorex araneus*)
Siskin, Eurasian (*Spinus spinus*)
Siskin, Pine (*Spinus pinus*)
Skunk, Eastern Spotted (*Spilogale putorius*)
Skunk, Striped (*Mephitis mephitis*)
Sparrow, Eurasian Tree (*Passer montanus*)
Sparrow, Golden-crowned (*Zonotrichia atricapilla*)
Sparrow, House (*Passer domesticus*)
Sparrow, White-throated (*Zonotrichia atricapilla*)
Sparrowhawk, Eurasian (*Accipiter nisus*)
Squirrel, American Red (*Tamiasciurus hudsonicus*)
Squirrel, Belding's Ground (*Spermophilus beldingi*)
Squirrel, California Ground (*Spermophilus beecheyi*)
Squirrel, Eastern Gray (*Sciurus carolinensis*)
Squirrel, Eurasian Red (*Sciurus vulgaris*)
Squirrel, Flying (*Glaucomys* spp.)
Squirrel, Fox (*Sciurus niger*)
Squirrel, Ground (*Spermophilus* spp.)
Squirrel, Tree (*Sciurus* spp.)
Squirrel, Western Gray (*Sciurus griseus*)
Starling (*Sturnus vulgaris*)
Starling, European (*Sturnus vulgaris*)
Swallow, Barn (*Hirundo rustica*)
Swan, Black (*Cygnus atratus*)
Swan, Mute (*Cygnus olor*)
Swan, Trumpeter (*Cygnus buccinator*)
Swan, Tundra (*Cygnus columbianus*)
Swan, Whooper (*Cygnus cygnus*)
Swift, Common (*Apus apus*)
Tasmanian Devil (*Sarcophilus harrisii*)
Thrush, Song (*Turdus philomelos*)
Thrush, Wood (*Hylocichla mustelina*)
Tick, American Dog (*Dermacentor variabilis*)
Tick, Blacklegged (*Ixodes scapularis*)
Tick, Brown Dog (*Rhipicephalus sanguineus*)
Tick, Cayenne (*Amblyomma cajennense*)
Tick, European Rabbit (*Ixodes ventalloi*)
Tick, European Sheep (*Ixodes ricinus*)
Tick, Gulf Coast (*Amblyomma maculatum*)
Tick, Lone Star (*Amblyomma americanum*)
Tick, Pacific Coast (*Dermacentor occidentalis*)
Tick, Rabbit (*Haemaphysalis leporispalustris*)
Tick, Rocky Mountain Wood (*Dermacentor andersoni*)
Tick, Squirrel (*Ixodes marxi*)
Tick, Western Blacklegged (*Ixodes pacificus*)
Tick, Woodchuck (*Ixodes cookei*)

Tiger (*Panthera tigris*)
Tit, Great (*Parus major*)
Titmouse, Tufted (*Baeolophus bicolor*)
Towhee, Eastern (*Pipilo erythrophthalmus*)
Towhee, Rufous-sided (*Pipilo erythrophthalmus*)
Turkey, Wild (*Meleagris gallopavo*)
Vole, Bank (*Myodes glareolus*)
Vole, Common (*Microtus arvalis*)
Vole, Eurasian Water (*Arvicola amphibius*)
Vole, Field (*Microtus agrestis*)
Vole, Long-tailed (*Microtus longicaudus*)
Vole, Meadow (*Microtus pennsylvanicus*)
Vole, Red-backed (*Myodes* spp.)
Vole, Woodland (*Microtus pinetorum*)
Vulture (*Gyps* spp.)
Vulture, Griffon (*Gyps fulvus*)
Wagtail, Pied (*Motacilla alba yarrellii*)
Walrus (*Odobenus rosmarus*)
Warbler, Myrtle (*Setophaga coronate*)
Warthog (*Phacochoerus aethiopicus*)
Water Buffalo (*Bubalus bubalis*)
Waterbuck (*Kobus ellipsiprymnus*)
Weasel (*Mustela* spp.)
Whale, Beluga (*Delphinapterus leucas*)
Whale, Fin (*Balaenoptera physalus*)
Whale, Minke (*Balaenoptera acutorostrata*)
Whale, Pilot (*Globicephala* spp.)
Wild Dog, African (*Lycaon pictus*)
Wildcat (*Felis silvestris*)
Wildebeest (*Connochaetes* spp.)
Wolf, Ethiopian (*Canis simensis*)
Wolf, Gray (*Canis lupus*)
Wolverine (*Gulo gulo*)
Woodrat, Allegheny (*Neotoma magister*)
Woodrat, Dusky-footed (*Neotoma fuscipes*)
Woodrat, Eastern (*Neotoma floridana*)
Woodchuck (*Marmota monax*)
Woodpecker, Downy (*Picoides pubescens*)
Wren, Eurasian (*Troglodytes troglodytes*)
Wren, House (*Troglodytes aedon*)
Yellowleg, Lesser (*Tringa flavipes*)
Zebra (*Equus* spp.)
Zebu (*Bos Taurus indicus*)

Appendix C: Photos of North American Ticks, Lice, and Fleas That Can Transmit Diseases from Animals to Humans

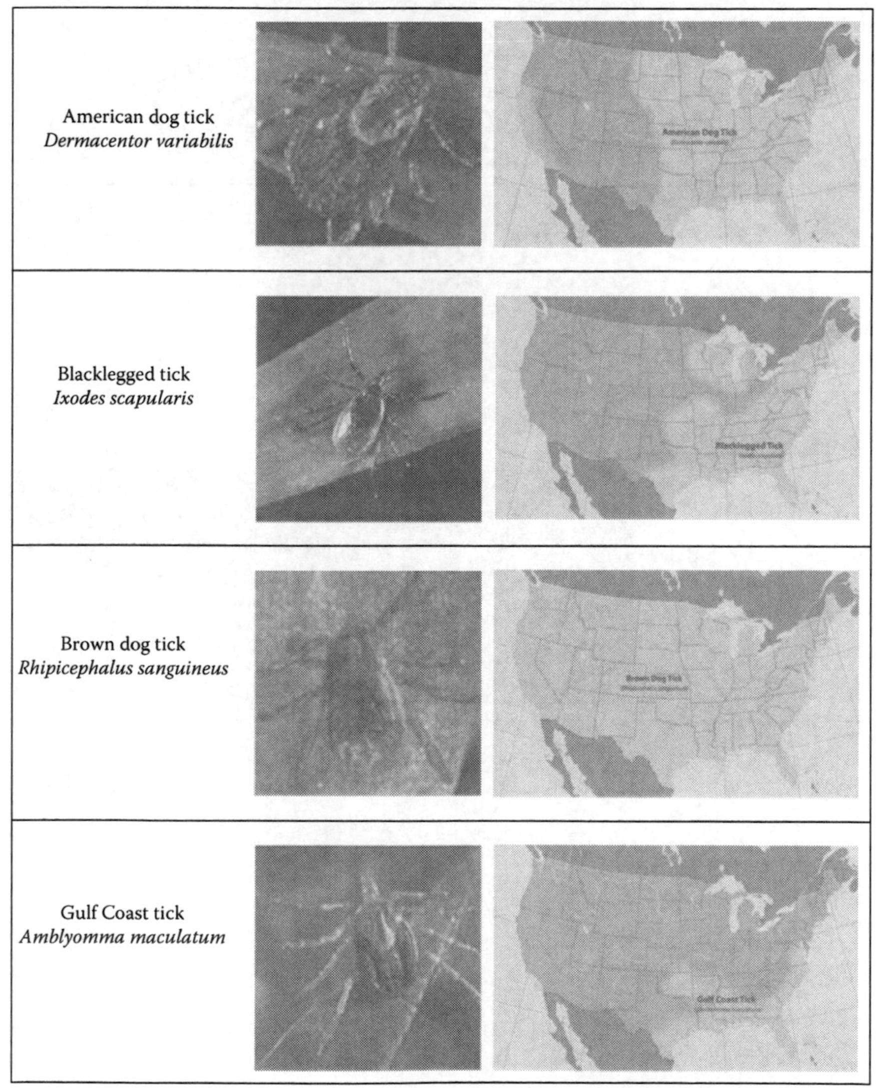

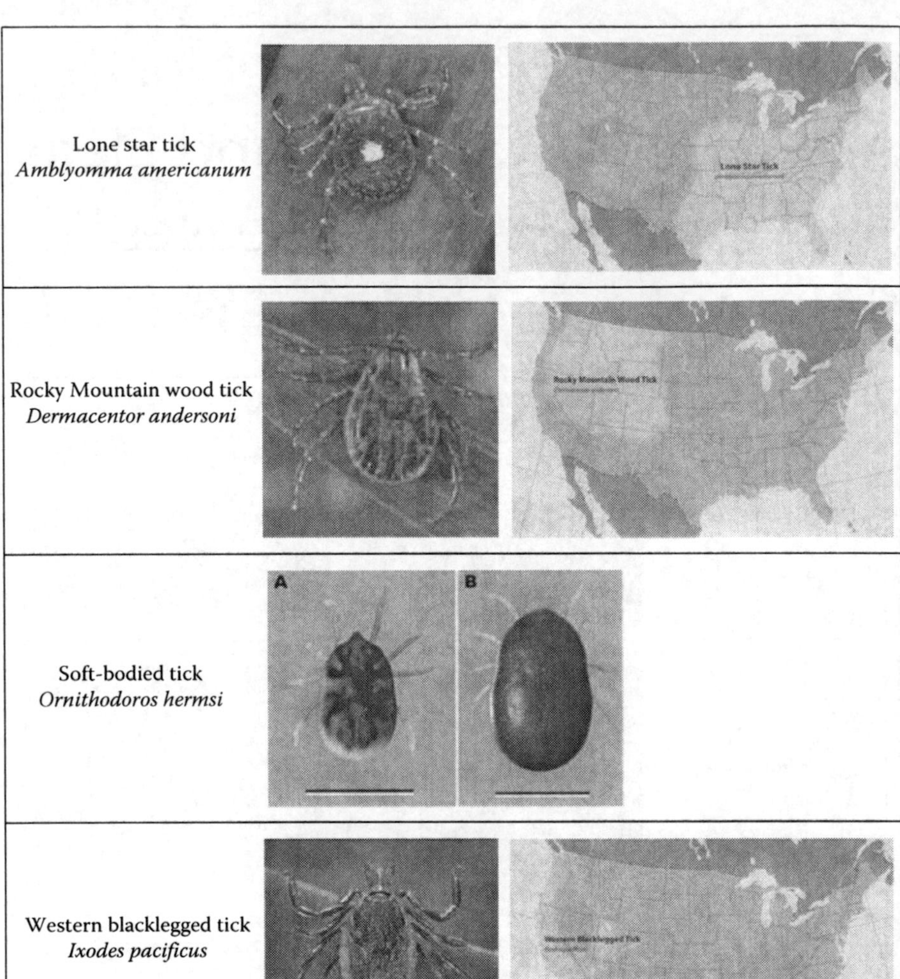
Lone star tick
Amblyomma americanum
Lone Star Tick
Rocky Mountain wood tick
Dermacentor andersoni
Rocky Mountain Wood Tick
Soft-bodied tick
Ornithodoros hermsi
A
B
Western blacklegged tick
Ixodes pacificus
Western Blacklegged Tick

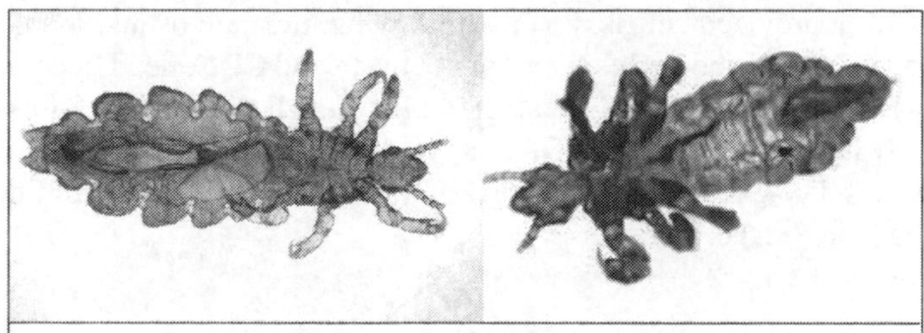

Head lice
Pediculus humanus capitis

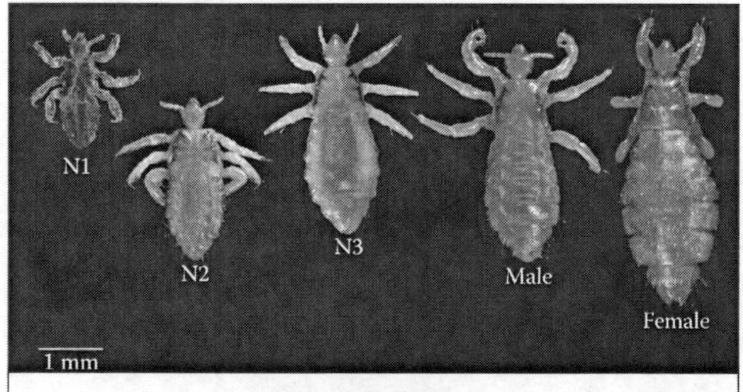

Human body lice
Pediculus humanus humanus

Pubic lice (crab lice)
Pthirus pubis

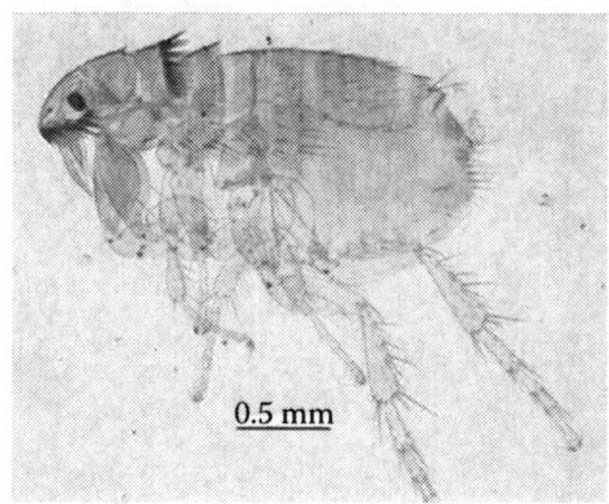

Cat flea
Ctenocephalides felis

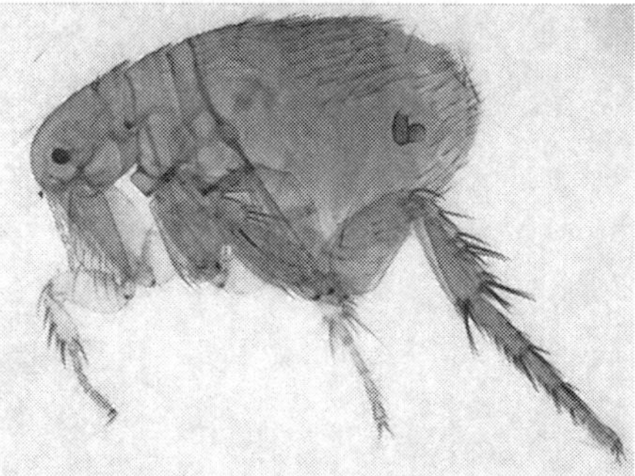

Oriental rat flea
Xenopsylla cheopis

The photos were provided courtesy of the following: tick photos and maps by James Gathany and CDC, soft-bodied tick by Tom Schwan and CDC, head lice photos (left) by Dennis D. Juranek and CDC, head lice photo (right) by DPDx Parasite Image Library, body lice photos by James Gathany and CDC, crab lice photo (left) by CDC and WHO, crab lice photo (right) by DPDx Parasite Image Library, and flea photos by DPDx Parasite Image Library.

Appendix D: Zoonotic Diseases Covered in This Book

Table D.1 on the following pages contains a comprehensive, alphabetized list of zoonotic diseases covered in this book. It includes the chapters of this book in which they appear as well as details on each disease's primary animals involved, geographic distribution, main mode of human infection, and number of reported cases in the United States during 2011.

TABLE D.1
Zoonotic Diseases Covered in this Book

Disease	Chapter	Pathogen	Primary Animals Involved	Geographic Distribution	Main Mode of Human Infection	Number of reported cases in U.S. during 2011
Anaplasmosis	18	*Anaplasma phagocytophilum*	White-footed mice, birds, mammals	North America, Europe	*Ixodes* ticks	2575
Anthrax	7	*Bacillus anthracis*	Mammalian herbivores	Worldwide	Skin contact, inhalation, ingestion	1
Baylisascariasis	28	Raccoon roundworms	Raccoons	North America, Europe, Japan	Ingestion	
Botulism	10	*Clostridium botulinum*	Waterbirds, horses, cattle	Worldwide	Ingestion	153
Brucellosis	3	*Brucella* spp.	Elk, bison, hogs, cattle	Worldwide	Ingestion, inhalation	79
Cache Valley virus (CVE)	22	CVE	Deer, elk, raccoons, livestock	North America	Mosquitoes	
California encephalitis (CE)	22	CE virus	Unknown	California	Mosquitoes	
Campylobacteriosis	10	*Campylobacter* spp.	Domestic and wild animals	Worldwide	Ingestion	
Chlamydia abortus	11	*Chlamydia abortus*	Livestock	Worldwide	Inhalation	
Chlamydia felis	11	*Chlamydia felis*	Cats	Worldwide	Physical contact	
Chlamydia pneumoniae	11	*Chlamydia pneumonia*	Mammals, reptiles, amphibians	Worldwide	Inhalation	
Chronic wasting disease (CWD)	27	CWD prions	Deer, elk	North America	Does not occur	
Clostridial diseases	10	*Clostridium perfringens*	Livestock, poultry, waterbirds, grouse	Worldwide	Ingestion	

Colorado tick fever (CTF)	22	CTF virus	Rodents, rabbits, hares, other mammals	Western United States, Canada above 4000′ elevation	Rocky Mountain wood tick	
Cryptococcosis	25	*Cryptococcus* spp.	Pigeons	Worldwide	Inhalation	9250
Deer tick virus (DTV)	21	DTV	White-footed mice	North America	*Ixodes* ticks	
Dengue	21	Dengue viruses	Monkeys	Hawaii, Florida, tropics	*Aedes* mosquitoes	254
Eastern equine encephalitis (EEE)	20	EEE virus	Birds, horses	New World	Mosquitoes	4
Ehrlichiosis	18	*Ehrlichia* spp.	White-tailed deer, dogs	North America, Brazil, Korea	Lone star tick	
Epidemic typhus	17	See typhus, epidemic				
Equine encephalitis	20	See eastern equine encephalitis, Venezuelan equine encephalitis, western equine encephalitis				
Everglades virus	20	Everglades virus	Cotton rats, rodents	Florida	*Culex* mosquitoes	
Escherichia coli	10	*Escherichia coli*	Mammalian herbivores	Worldwide	Ingestion	
Giardiasis	30	*Giardia intestinalis*	Mammals	Worldwide	Ingestion	16,747
Hantavirus	23	Hantaviruses	Deer mice, rodents	Western hemisphere	Inhalation	23
Haverhill fever	8	See rat-bite fever				
Hepatitis E	10	Hepatitis E virus	Pigs, rats	Worldwide	Ingestion, physical contact	
Histoplasmosis	26	*Histoplasma capsulatum*	Pigeons, blackbirds, bats	Worldwide	Inhalation	
Influenza	24	Influenza viruses	Waterbirds, poultry, pigs	Worldwide	Inhalation	
Jamestown Canyon virus (JCV)	22	JCV	White-tailed deer	North America	Mosquitoes	

(*Continued*)

TABLE D.1 (*Continued*)
Zoonotic Diseases Covered in this Book

Disease	Chapter	Pathogen	Primary Animals Involved	Geographic Distribution	Main Mode of Human Infection	Number of reported cases in U.S. during 2011
Japanese encephalitis (JE)	21	JE virus	Waterbirds, horses, pigs	Asia	*Culex* mosquitoes	
La Crosse encephalitis (LAC)	22	LAC virus	Squirrels, chipmunks	United States, Canada	*Aedes* mosquitoes	116
Leprosy	6	*Mycobacterium leprae*	Armadillos	Worldwide	Unknown	82
Listeriosis	10	*Listeria monocytogenes*	Poultry, wild birds, ungulates, feral hogs	Worldwide	Ingestion	870
Leptospirosis	12	*Leptospira* spp.	All mammals, especially rodents, livestock	Worldwide	Inhalation, physical contact	
Lyme disease	13	*Borrelia burgdorferi*	Deer, rodents	North America	*Ixodes* ticks	33,097
Mad cow disease	27	Mad cow prions	Cows	Great Britain, United States, Canada	Ingestion	
Maculatum infection	15	*Rickettsia parkeri*	Unknown	Southeast United States, Latin America	Gulf Coast tick, lone star tick	
Murine typhus	17	See typhus, murine				
Plague	2	*Yersinia pestis*	Rodents, ground squirrels, prairie dogs, predators	United States, South America, Asia, Africa	Fleas, inhalation, ingestion	3
Powassan encephalitis (POW)	21	POW virus	Woodchucks, skunks, squirrels	North America	*Ixodes* ticks	16

Psittacosis	11	*Chlamydia* spp.	Parrots, pigeons, waterbirds	Worldwide	Inhalation, physical contact	
Rabies	19	Rabies virus	Dogs, raccoons, skunks, bats, foxes	Worldwide	Bite, inhalation	6
Raccoon roundworm	28	See baylisascariasis				
Rat-bite fever, spirillary	8	See sodoku				
Rat-bite fever, streptobacillary	8	*Streptobacillus moniliformis*	Rats, rodents	Worldwide	Rodent bite, physical contact, ingestion	
Relapsing fever	14	See tick-borne relapsing fever				
Rickettsia felis	15	*Rickettsia felis*	Cats, dogs, rodents	California, Texas, worldwide	Cat flea	
Rickettsia amblyommii	15	*Rickettsia amblyommii*	Unknown	Eastern United States, Brazil	Lone star tick	
Rickettsial pox	15	*Rickettsia akari*	House mice, Norway rats, roof rats	U.S. cities, worldwide	Mites	
Rocky Mountain spotted fever	15	*Rickettsia rickettsii*	Small mammals, birds, dogs	New World	Dog tick, Rocky Mountain wood tick	2862
Q fever	16	*Coxiella burnetii*	Sheep, goats, mammals, birds	Worldwide	Inhalation, ingestion, arthropods	134
Salmonellosis	9	*Salmonella enterica*	Poultry, livestock, wild birds, wild mammals	Worldwide	Ingestion	51,887
Scrapie	27	Scrapie prions	Sheep, goats	Worldwide	Does not occur	
Sodoku	8	*Spirillum minus*	Rats, rodents	Asia	Bite, physical contact, ingestion	

(*Continued*)

TABLE D.1 (*Continued*)
Zoonotic Diseases Covered in this Book

Disease	Chapter	Pathogen	Primary Animals Involved	Geographic Distribution	Main Mode of Human Infection	Number of reported cases in U.S. during 2011
St. Louis encephalitis (SLE)	21	SLE virus	Birds	North America	*Culex* mosquitoes	6
Southern tick-associated rash	13	Unknown	Unknown	Eastern United States	Lone star tick	
Staphylococcosis	10	*Staphylococcus aureus*	Birds and mammals	Worldwide	Ingestion	
Swimmer's itch	30	Avian schistosomes	Waterbirds, muskrats, beaver, aquatic snails	Worldwide	Skin infection	
Tick-borne relapsing fever	14	*Borrelia* sp.	Tree squirrels, chipmunks	Western United States, Mexico, Africa, central Asia	Soft-bodied ticks (*Ornithodoros*)	
Typhus, epidemic	17	*Rickettsia prowazekii*	Flying squirrels	Eastern United States, worldwide	Human-to-human by body lice, squirrel-to-human unknown	
Typhus, murine	17	*Rickettsia typhi*	Rats, opossum, cats	Texas, southern United States, worldwide	Fleas	
Trichinellosis	29	*Trichinella* spp.	Pigs, bears, walruses	Worldwide	Meat consumption	15
Trichinosis	29	See trichinellosis				
Tuberculosis	4	*Mycobacterium* spp.	Deer, elk, cattle	Worldwide	Inhalation, ingestion	10,526

Tularemia	5	*Francisella tularensis*	Rabbits, hares, grouse, livestock	Northern hemisphere	Ingestion, inhalation, ticks, insects, physical contact	166
Venezuelan equine encephalitis (VEE)	20	VEE virus	Rodents, opossum, horses	Florida, Latin America	Mosquitoes	
West Nile virus (WNV)	21	WNV	Birds, horses	Worldwide	Mosquitoes	710
Western equine encephalitis (WEE)	20	WEE virus	Sparrows, finches, horses	New World	Mosquitoes	0
Yellow fever (YF)	21	YF virus	Primates	Africa, South America	Mosquitoes	0
Yersiniosis	10	*Yersinia pseudotuberculosis*, *Y. enterocolitica*	Birds and mammals	Worldwide	Ingestion	
364D rickettsiosis	15	*Rickettsia* sp. 364D	Unknown	California	Unknown, probably ticks	

Note: The actual number of cases of a disease is often much higher than the number of cases reported to the CDC; diseases missing data on the number of cases are not reportable diseases. Number of Rocky Mountain spotted fever (RMSP) includes all spotted fevers, most of which are RMSP.

Source for the number of reported cases is CDC 2013. California encephalitis includes all viruses in the California serogroup with the exception of La Crosse encephalitis, which is reported separately. Summary of Notifiable Diseases — United States, 2011. *Morbidity and Mortality Weekly Report* 60(53):1–117.

Index

C

D

E

G

H

I

J

K

L

M

N

O

P

Q

R

V

W

Y